Advances in Mechanics and Mathematics

Volume 39

More information about this series at http://www.springer.com/series/5613

Liqun Qi · Haibin Chen · Yannan Chen

Tensor Eigenvalues and Their Applications

Springer

Liqun Qi
Department of Applied Mathematics
The Hong Kong Polytechnic University
Hong Kong
Hong Kong

Yannan Chen
The Hong Kong Polytechnic University
Hong Kong
Hong Kong

Haibin Chen
School of Management Sciences
Qufu Normal University
Rizhao, Shandong
China

ISSN 1571-8689 ISSN 1876-9896 (electronic)
Advances in Mechanics and Mathematics
ISBN 978-981-10-8057-9 ISBN 978-981-10-8058-6 (eBook)
https://doi.org/10.1007/978-981-10-8058-6

Library of Congress Control Number: 2018934953

Printed on acid-free paper

This Springer imprint is published by the registered company Springer Nature Singapore Pte Ltd. part of Springer Nature
The registered company address is: 152 Beach Road, #21-01/04 Gateway East, Singapore 189721, Singapore

Preface

Tensors, as geometric objects that describe linear or multi-linear relations between geometric vectors, scalars, and other tensors, have provided a concise mathematical framework for formulating and solving practical physics problems in various areas such as relativity theory, fluid dynamics, solid mechanics, electromagnetism, etc. The concept of tensors can be traced back to the works by Carl Friedrich Gauss (1777–1855), Bernhard Riemann (1826–1866), Elwin Bruno Christoffel (1829–1900), etc., in the nineteenth century on differential geometry. It was further developed and analyzed by Gregorio Ricci-Curbastro (1853–1925), Tullio Levi-Civita (1873–1941), and others, in the very beginning of the twentieth century. A mathematical discipline on tensor analysis gradually emerged and was even applied in general relativity by the great scientist **Albert Einstein (1879–1955)** in 1916. This not only shows the great power of tensor analysis in theoretical physics but also starts the journey to widespread applications in continuum mechanics and many other areas in science and engineering.

When a coordinate basis or a frame of reference is given, a tensor can be represented as an organized multidimensional array of numerical values. In this case, tensors are treated as hypermatrices which are exactly the higher order generalization of matrices. Tensors that have been relatively systematically treated in the book *Tensor Analysis: Spectral Theory and Special Tensors* by Liqun Qi and Ziyan Luo [228] are actually hypermatrices. As for a great majority of references in tensor decomposition, tensor spectral theory, and spectral hypergraph theory, the word “tensor” is widely used for those multidimensional arrays. Following this habit, we will adopt the terminology of “tensor” both for multidimensional arrays and tensors as physical quantities. Different from the main concerns in the book [228] where great emphasis has been addressed on properties of tensor eigenvalues and special structured tensors in the setting of hypermatrices, more applications of tensors in both of the aforementioned two settings will be explored and discussed. These applications include multi-linear systems in numerical algebra, exponential data fitting in data science, tensor complementarity problems and tensor eigenvalue complementarity problems in optimization, higher order diffusion tensor imaging in medical imaging, liquid crystal study, piezoelectric effects, solid mechanics, quantum entanglement problems, etc. We hope that this book may provide a good base for further research on tensor eigenvalue applications in these and some more areas.

This book is divided into nine chapters. In Chap. 1, some preliminaries of tensor eigenvalues are given. In Chaps. 2–5, tensors are treated as hypermatrices just like in the book *Tensor Analysis: Spectral Theory and Special Tensors*, and more theoretical and practical applications are discussed. In the last four chapters, tensors take its original form of physical quantities and applications of tensors eigenvalues in physics and mechanics are elaborated, with a special and careful treatment on third-order tensors due to its important applications in physics and its nice properties in theoretical aspect, as seen in Sect. 7.1.

While tensors such as piezoelectric tensors and elasticity tensors have been used in physics and mechanics for more than one century, the study on spectral properties of these tensors is still very new. The fundamental principle of Galileo Galilei (1564–1642), who has played a pioneer role in the scientific revolution of the seventeenth century and is regarded as the father of science, is to study the rules and insights of the nature, while mathematics is the basic tool in this process. Inspired by this principle, the mathematical analysis on spectral properties will be stated for such tensors in physics and mechanics, aiming to get a better understanding in the involved applications. More spectral properties of tensors in physics and mechanics await being exploited even after this book.

We are grateful to David Gao and Ratiu Tudor for their support to include this book in their Springer book series “Advances in Mechanics and Mathematics”. We are thankful to Ramon Peng for his excellent editorial work. We are also grateful to Jingya Chang, Weiyang Ding, Zhenghai Huang, Ziyan Luo, Guofeng Zhang, Chen Ling, Yisheng Song, Shenglong Hu, Chen Ouyang, Jinjie Liu, Changqing Xu, Lejia Gu, and Zhongming Chen for their comments and proofreading of this book, and to Lieven De Lathauwer, Andrzej Cichocki, Kungching Chang, Avi Berman, Qingwen Wang, Donghui Li, Xiaoqing Jin, Lixing Han, Wen Li, Michael Ng, Jinyan Fan, Jiawang Nie, Yaotang Li, Chaoqian Li, and many others for their encouragements and supports. We are also thankful to our other research collaborators Yimin Wei, Maolin Che, Naihua Xiu, Seetharama Gowda, Hongjin He, Gaohang Yu, Yiju Wang, Deren Han, Ed Wu, Yuhong Dai, Hongyan Ni, Yi Xu, Epifanio Virga, Antal Jákli, Huihui Dai, Xinzhen Zhang, Hong Yan, Hua Xiang, Guyan Ni, Minru Bai, Daniel Braun, Fabian Bohnet-Waldraff, Olivier Giraud, etc. In particular, we have benefited from our discussion with Bernd Sturmfels on eigendiscriminants, and our discussion with Quanshui Zheng on mechanics.

Liqun Qi’s work was supported by the Hong Kong Research Grant Council (Grant No. PolyU 501913, 15302114, 15300715, and 15301716). Haibin Chen’s work was supported by the National Natural Science Foundation of China (Grant

No. 11601261) and Natural Science Foundation of Shandong Province (Grant No. ZR2016AQ12). Yannan Chen's work was supported by the National Natural Science Foundation of China (Grant No. 11401539, 11771405).

Hong Kong, Hong Kong — Liqun Qi
Qufu, China — Haibin Chen
Hong Kong, Hong Kong — Yannan Chen
September 2017

Contents

List of Figures

Chapter 1
Preliminaries

In this chapter, we review some basic knowledge about tensors.

1.1 Tensors (Hypermatrices) and Tensor Products

A tensor in Chapters Two-Five of this book will refer to a hypermatrix or a tentrix (cf. [100]), which is usually denoted as $\mathscr{A} = (a_{i_1 \ldots i_m})$ and represents a multi-array of entries $a_{i_1 \ldots i_m} \in \mathbf{F}$, where $i_j = 1, \ldots, n_j$ for $j = 1, \ldots, m$ and $\mathbf{F}$ is a field. In this book, we may involve either real tensors or complex tensors, i.e., $\mathbf{F} = \mathbf{R}$ or $\mathbf{C}$. In most cases, we only consider real tensors, which is the case when no specification to the field is made. Here, m is called the order of tensor $\mathscr{A}$ and $(n_1, \ldots, n_m)$ is the dimension of $\mathscr{A}$. When $n = n_1 = \cdots = n_m$, $\mathscr{A}$ is called an mth order n-dimensional tensor. The set of all mth order n-dimensional real tensors is denoted as $T_{m,n}$. Throughout this book, we assume that m and n are integers, and $m, n \geq 2$, unless otherwise stated. For any tensor $\mathscr{A} = (a_{i_1 \ldots i_m}) \in T_{m,n}$, if its entries $a_{i_1 \ldots i_m}$'s are invariant under any permutation of its indices, then $\mathscr{A}$ is called a **symmetric tensor**. The set of all mth order n-dimensional real symmetric tensors is denoted as $S_{m,n}$.

In Chapters Six-Nine, we will study tensors in physics, mechanics and engineering.

We will use small letters $x, y, a, b, \ldots$, for scalars, small bold letters $\mathbf{x}, \mathbf{y}, \ldots$, for vectors, capital letters $A, B, C, \ldots$, for matrices, calligraphic letters $\mathscr{A}, \mathscr{B}, \mathscr{C}, \ldots$, for tensors. In $\mathbf{R}^n$, we use $\mathbf{0}$ to denote the zero vector, $\mathbf{1}$ to denote the all 1 vector, and $\mathbf{1}^{(j)}$ to denote the jth unit vector. For simplicity, we denote $[n] := \{1, \ldots, n\}$. For a vector $\mathbf{x} \in \mathbf{R}^n$, we denote $\mathrm{supp}(\mathbf{x}) = \{j \in [n] : x_j \neq 0\}$, and call it the support of $\mathbf{x}$. We also denote $|\mathbf{x}|$ as a vector $\mathbf{y}$ in $\mathbf{R}^n$ such that $y_i = |x_i|$ for $i \in [n]$. For a finite set S, we use $|S|$ to denote its cardinality. We use $\mathscr{O}$ to denote the **zero tensor** in $T_{m,n}$,

L. Qi et al., *Tensor Eigenvalues and Their Applications*, Advances in Mechanics and Mathematics 39, https://doi.org/10.1007/978-981-10-8058-6_1

and $\mathcal{J}$ to denote the all 1 tensor in $T_{m,n}$, i.e., all entries of $\mathcal{J}$ are 1. We will omit the dependence on the dimension in the notation $\mathbf{0}$, $\mathbf{1}$, $\mathcal{O}$ and $\mathcal{J}$, as which will be clear from the context.

For a tensor $\mathscr{A} = (a_{i_1 \dots i_m}) \in T_{m,n}$, entries $a_{i \dots i}$ are called diagonal entries of $\mathscr{A}$, for $i \in [n]$. The other entries of $\mathscr{A}$ are called off-diagonal entries of $\mathscr{A}$. A tensor $\mathscr{A} \in T_{m,n}$ is called **diagonal** if all of its off-diagonal entries are zero. Clearly, a diagonal tensor is a symmetric tensor. A diagonal tensor with all of its diagonal entries as 1 is called the **identity tensor** in $T_{m,n}$, and denoted as $\mathscr{I}$. A tensor in $T_{m,n}$ is called a nonnegative tensor if all of its entries are nonnegative.

The most common tensor products include the tensor outer product and the inner product.

Tensor Outer Product: We use $\otimes$ to denote **tensor outer product**, that is, for any two tensors $\mathscr{A} = (a_{i_1 \dots i_m}) \in T_{m,n}$ and $\mathscr{B} = (b_{i_1 \dots i_p}) \in T_{p,n}$,

$$\mathscr{A} \otimes \mathscr{B} = \left(a_{i_1 \dots i_m} b_{i_{m+1} \dots i_{m+p}}\right) \in T_{m+p,n}. \tag{1.1}$$

Apparently, this tensor outer product is a binary operation and maps a tensor pair from $T_{m,n} \times T_{p,n}$ to an expanded order tensor in $T_{m+p,n}$. Invoking the definition of tensor outer product as described in (1.1), it is easy to check that

$$\mathbf{x}^{\otimes k} \equiv \underbrace{\mathbf{x} \otimes \cdots \otimes \mathbf{x}}_{k \text{ times}} = \left(x_{i_1} \cdots x_{i_k}\right) \in T_{k,n}. \tag{1.2}$$

Obviously, $\mathbf{x}^{\otimes k} \in S_{k,n}$ and it is called a **symmetric rank-one tensor** when $\mathbf{x} \neq \mathbf{0}$. We will abbreviate $\mathbf{x}^{\otimes k}$ as $\mathbf{x}^k$ for simplicity in the book. Analogous to the matrix case where k is specified to be 2, any tensor of the form $\alpha \mathbf{x}^{\otimes k}$ with any given $\alpha \in \mathbf{R} \backslash \{0\}$ and $\mathbf{x} \in \mathbf{R}^n \backslash \{\mathbf{0}\}$ is a symmetric rank-one tensor in $S_{k,n}$. More generally, let $\mathbf{x}^{(i)} = \left(x_1^{(i)}, \dots, x_n^{(i)}\right)^\top \in \mathbf{R}^n$ for $i \in [m]$ and $\alpha \in \mathbf{R}$. Then $\alpha \mathbf{x}^{(1)} \otimes \mathbf{x}^{(2)} \otimes \cdots \otimes \mathbf{x}^{(m)}$ is a tensor in $T_{m,n}$ with its $(i_1, \dots, i_m)$th entry as $\alpha x_{i_1}^{(1)} \cdots x_{i_m}^{(m)}$. Such a tensor (not necessarily symmetric) is called a rank-one tensor in $T_{m,n}$.

Inner Product: For any two tensors $\mathscr{A} = (a_{i_1 \dots i_m})$, $\mathscr{B} = (b_{i_1 \dots i_m}) \in T_{m,n}$, the **inner product** of $\mathscr{A}$ and $\mathscr{B}$, denoted as $\mathscr{A} \bullet \mathscr{B}$, is defined as

$$\mathscr{A} \bullet \mathscr{B} = \sum_{i_1, \dots, i_m = 1}^{n} a_{i_1 \dots i_m} \bar{b}_{i_1 \dots i_m}, \tag{1.3}$$

where $\bar{\alpha}$ is the complex conjugate of α. Analogous to the matrix case, the induced norm $\sqrt{\mathscr{A} \bullet \mathscr{A}}$ is called the **Frobenius norm** of $\mathscr{A}$, denoted as $\|\mathscr{A}\|_F$.

k-mode product: Let $\mathscr{A} = (a_{i_1\ldots i_m}) \in \mathbf{C}^{I_1\times\cdots\times I_m}$ be a tensor and $X^{(k)} = (x^{(k)}_{j_k i_k}) \in \mathbf{C}^{J_k I_k}$ be matrices for $k \in [n]$. We denote $\mathscr{A} \times_k X^{(k)} \in \mathbf{C}^{I_1\times\cdots\times I_{k-1}\times J_k\times I_{k+1}\times\cdots\times I_m}$ the k-mode product of the tensor $\mathscr{A}$ and a matrix $X^{(k)}$, whose elements are

$$(\mathscr{A} \times_k X^{(k)})_{i_1\ldots i_{k-1} j_k i_{k+1}\ldots i_m} = \sum_{i_k=1}^{I_k} a_{i_1\ldots i_k\ldots i_m} x^{(k)}_{j_k i_k}.$$

In the cases of vectors $\mathbf{x}^{(k)} = (x^{(k)}_{i_k}) \in \mathbf{C}^{I_k}$ for $k \in [m]$, we get a $(m-1)$th order tensor $\mathscr{A} \times_k \mathbf{x}^{(k)} \in \mathbf{C}^{I_1\times\cdots\times I_{k-1}\times I_{k+1}\times\cdots\times I_m}$ with elements

$$(\mathscr{A} \times_k \mathbf{x}^{(k)})_{i_1\ldots i_{k-1} i_{k+1}\ldots i_m} = \sum_{i_k=1}^{I_k} a_{i_1\ldots i_k\ldots i_m} x^{(k)}_{i_k}.$$

Positive Semi-Definiteness and Positive Definiteness: An n-dimensional homogeneous real polynomial form of degree m, $f(\mathbf{x})$, where $\mathbf{x} \in \mathbf{R}^n$, is equivalent to the tensor product of an n-dimensional tensor $\mathscr{A} = \left(a_{i_1\ldots i_m}\right)$ of order m, and the symmetric rank-one tensor $\mathbf{x}^m$:

$$f(\mathbf{x}) \equiv \mathscr{A}\mathbf{x}^m \equiv \mathscr{A} \bullet \mathbf{x}^m := \sum_{i_1,\ldots,i_m=1}^{n} a_{i_1\ldots i_m} x_{i_1} \cdots x_{i_m}. \tag{1.4}$$

The tensor $\mathscr{A}$ is called positive semi-definite (PSD) if $f(\mathbf{x}) \geq 0$ for all $\mathbf{x} \in \mathbf{R}^n$; and positive definite (PD) if $f(\mathbf{x}) > 0$ for all $\mathbf{x} \in \mathbf{R}^n$, $\mathbf{x} \neq \mathbf{0}$. Clearly, when m is odd, there is no nontrivial positive semi-definite tensor. It is easy to see that $\mathscr{A}\mathbf{x}^m$ defined in (1.4) is exactly $\mathscr{A} \bullet \mathbf{x}^m$.

Best Rank One Approximation: Given a tensor $\mathscr{A} \in T_{m,n}$, the *best rank one approximation* of $\mathscr{A}$ is a rank one tensor $\mathscr{B} \in T_{m,n}$ such that

$$\mathscr{B} \in \operatorname{argmin}\{\|\mathscr{A} - \mathscr{C}\|_F : \mathscr{C} \text{ is a rank one tensor}\}.$$

Or, in a more concrete form, $\mathscr{B} = \mathbf{x}^{(1)} \otimes \cdots \otimes \mathbf{x}^{(m)}$ such that

$$(\mathbf{x}^{(1)}, \ldots, \mathbf{x}^{(m)}) \in \operatorname{argmin}\{\|\mathscr{A} - \mathbf{y}^{(1)} \otimes \cdots \otimes \mathbf{y}^{(m)}\|_F : \mathbf{y}^{(i)} \in \mathbf{R}^n\}.$$

If we normalize each factor vector $\mathbf{y}^{(i)}$ in the above minimization problem, we get $\mathscr{B} = \lambda\mathbf{x}^{(1)} \otimes \cdots \otimes \mathbf{x}^{(m)}$ such that

$$(\lambda, \mathbf{x}^{(1)}, \ldots, \mathbf{x}^{(m)}) \in \operatorname{argmin}\{\|\mathscr{A} - \lambda\mathbf{y}^{(1)} \otimes \cdots \otimes \mathbf{y}^{(m)}\|_F : \mathbf{y}^{(i)} \in \mathbf{R}^n,\ \|\mathbf{y}^{(i)}\| = 1\},$$

which is further equivalent to

$$(\mathbf{x}^{(1)}, \ldots, \mathbf{x}^{(m)}) \in \operatorname{argmax}\{\mathscr{A} \bullet \mathbf{y}^{(1)} \otimes \cdots \otimes \mathbf{y}^{(m)} : \mathbf{y}^{(i)} \in \mathbf{R}^n,\ \|\mathbf{y}^{(i)}\| = 1\}$$
$$\text{and } \lambda = \mathscr{A} \bullet \mathbf{x}^{(1)} \otimes \cdots \otimes \mathbf{x}^{(m)}.$$

If $\mathscr{A}$ is a symmetric tensor, then it follows from Banach's theorem that in the best rank one approximation we can take $\mathbf{x} = \mathbf{x}^{(1)} = \cdots = \mathbf{x}^{(m)}$.

1.2 Eigenvalues of Tensors

In 2005, Qi [221] defined eigenvalues and eigenvectors of a real symmetric tensor, and explored their practical applications in determining positive definiteness of an even degree multivariate form.

By the tensor product, $\mathscr{A}\mathbf{x}^{m-1}$ for a vector $\mathbf{x} \in \mathbf{R}^n$ denotes a vector in $\mathbf{R}^n$, whose ith component is

$$\left(\mathscr{A}\mathbf{x}^{m-1}\right)_i \equiv \sum_{i_2,\ldots,i_m=1}^{n} a_{ii_2\ldots i_m} x_{i_2} \cdots x_{i_m}.$$

We call a number $\lambda \in \mathbf{C}$ an **eigenvalue** of $\mathscr{A}$ if it together with a nonzero vector $x \in \mathbf{C}^n$ forms a solution to the following system of homogeneous polynomial equations:

$$\left(\mathscr{A}\mathbf{x}^{m-1}\right)_i = \lambda x_i^{m-1}, \quad \forall\, i = 1, \ldots, n. \tag{1.5}$$

The vector $\mathbf{x}$ is called an **eigenvector** of $\mathscr{A}$ associated with the eigenvalue λ. We call an eigenvalue of $\mathscr{A}$ an **H-eigenvalue** of $\mathscr{A}$ if it has a real eigenvector $\mathbf{x}$. An eigenvalue which is not an H-eigenvalue is called an **N-eigenvalue**. A real eigenvector associated with an H-eigenvalue is called an **H-eigenvector**.

The concept of classical resultant can be found in textbooks such as [64, 65, 102]. Let $\mathbf{N}$ be the set of natural numbers and $\mathbf{Z}$ be the ring of integers. For $\boldsymbol{\alpha} \in \mathbf{N}^n$, define monomial $\mathbf{x}^{\boldsymbol{\alpha}} := \prod_{i=1}^n x_i^{\alpha_i}$ and $|\boldsymbol{\alpha}| = \sum_{i=1}^n \alpha_i$. For fixed positive degree d, let $\{u_{i,\boldsymbol{\alpha}} : |\boldsymbol{\alpha}| = d,\ i = 1, \ldots, n\}$ be the set of indeterminants, and $f_i := \sum_{|\boldsymbol{\alpha}|=d} c_{i,\boldsymbol{\alpha}} \mathbf{x}^{\boldsymbol{\alpha}}$ be a homogeneous polynomial of degree d in $\mathbf{C}[\mathbf{x}]$ for $i \in \{1, \ldots, n\}$. Then there exists a unique polynomial RES $\in \mathbf{Z}[\{u_{i,\boldsymbol{\alpha}}\}]$ called the resultant of degrees $(d, \ldots, d)$ satisfying the following properties:

(i) The system of polynomial equations $f_1 = \cdots = f_n = 0$ has a nontrivial solution in $\mathbf{C}^n$ if and only if $\mathrm{RES}(f_1, \ldots, f_n) := \mathrm{RES}|_{u_{i,\boldsymbol{\alpha}}=c_{i,\boldsymbol{\alpha}}} = 0$.
(ii) $\mathrm{RES}(x_1^{d_1}, \ldots, x_n^{d_n}) = 1$.
(iii) RES is an irreducible polynomial in $\mathbf{C}[\{u_{i,\boldsymbol{\alpha}}\}]$.

The resultant of (1.5) is a one-dimensional polynomial of λ. We call it the **characteristic polynomial** of $\mathscr{A}$.

Theorem 1.1 (Qi 2005)

We have the following conclusions on eigenvalues of an mth order n-dimensional symmetric tensor $\mathscr{A}$.

(a). A number $\lambda \in \mathbf{C}$ *is an eigenvalue of* $\mathscr{A}$ *if and only if it is a root of the characteristic polynomial.*

(b). The number of eigenvalues of $\mathscr{A}$ *is* $d = n(m-1)^{n-1}$. *Their product is equal to* $\det(\mathscr{A})$, *the resultant of* $\mathscr{A}\mathbf{x}^{m-1} = 0$.

(c). The sum of all the eigenvalues of $\mathscr{A}$ *is*

$$(m-1)^{n-1}\mathrm{tr}(\mathscr{A}),$$

where $\mathrm{tr}(\mathscr{A})$ *denotes the sum of all diagonal elements of* $\mathscr{A}$.

(d). If m is even, then $\mathscr{A}$ *always has H-eigenvalues.* $\mathscr{A}$ *is positive definite (positive semidefinite) if and only if all of its H-eigenvalues are positive (nonnegative).*

(e). The eigenvalues of $\mathscr{A}$ *lie in the following n disks:*

$$|\lambda - a_{ii\ldots i}| \leq \sum \left\{|a_{ii_2\ldots i_m}| : i_2, \ldots, i_m = 1, \ldots, n, \{i_2, \ldots, i_m\} \neq \{i, \ldots, i\}\right\},$$

for $i = 1, \ldots, n$.

In the same year, Qi [221] also defined another kind of eigenvalues for tensors. Their characteristic polynomial has a lower degree. More importantly, their structure is different from the structure described in Theorem 1.1.

Suppose that $\mathscr{A}$ is an mth order n-dimensional symmetric tensor. We say a complex number λ is an **E-eigenvalue** of $\mathscr{A}$ if there exists a complex vector $\mathbf{x}$ such that

$$\begin{cases} \mathscr{A}\mathbf{x}^{m-1} = \lambda \mathbf{x}, \\ \mathbf{x}^\top \mathbf{x} = 1. \end{cases} \tag{1.6}$$

In this case, we say that $\mathbf{x}$ is an **E-eigenvector** of the tensor $\mathscr{A}$ associated with the E-eigenvalue λ. If an E-eigenvalue has a real E-eigenvector, then we call it a **Z-eigenvalue** and call the real E-eigenvector a **Z-eigenvector**.

When m is even, the resultant of

$$\mathscr{A}\mathbf{x}^{m-1} - \lambda(\mathbf{x}^\top \mathbf{x})^{\frac{m-2}{2}}\mathbf{x} = 0$$

is a univariate polynomial of λ and is called the **E-characteristic polynomial** of $\mathscr{A}$. We say that $\mathscr{A}$ is **regular** if the following system has no nonzero complex solutions:

$$\begin{cases} \mathscr{A}\mathbf{x}^{m-1} = 0, \\ \mathbf{x}^\top \mathbf{x} = 0. \end{cases}$$

When m is odd, the E-characteristic polynomial is defined as the resultant of the system

$$\mathscr{A}\mathbf{x}^{m-1} - \lambda t^{m-2}\mathbf{x} = 0, \;\; \mathbf{x}^\top \mathbf{x} = t^2.$$

In this case, it can be shown that in the E-characteristic polynomial only powers of λ^2 appear.

Let $P = (p_{ij})$ be an $n \times n$ real matrix. Define $\mathscr{B} = P^m \mathscr{A}$ as another mth order n-dimensional tensor with entries

$$b_{i_1 i_2 \ldots i_m} = \sum_{j_1, j_2, \ldots, j_m = 1}^{n} p_{i_1 j_1} p_{i_2 j_2} \cdots p_{i_m j_m} a_{j_1 j_2 \ldots j_m}.$$

If P is an orthogonal matrix, then we say that $\mathscr{A}$ and $\mathscr{B}$ are **orthogonally similar**.

In the following, we summarize important properties on E/Z-eigenvalues of a real tensor.

Theorem 1.2 (Qi 2005)

We have the following conclusions on E-eigenvalues of an mth order n-dimensional symmetric tensor $\mathscr{A}$.

(a). When $\mathscr{A}$ is regular, a complex number is an E-eigenvalue of $\mathscr{A}$ if and only if it is a root of its E-characteristic polynomial.

(b). Z-eigenvalues always exist. An even order symmetric tensor is positive definite if and only if all of its Z-eigenvalues are positive.

(c). If $\mathscr{A}$ and $\mathscr{B}$ are orthogonally similar, then they have the same E-eigenvalues and Z-eigenvalues.

(d). If λ is the Z-eigenvalue of $\mathscr{A}$ with the largest absolute value and $\mathbf{x}$ is a Z-eigenvector associated with it, then $\lambda \mathbf{x}^m$ is the best rank-one approximation of $\mathscr{A}$, i.e.,

$$\|\mathscr{A} - \lambda \mathbf{x}^m\|_F = \sqrt{\|\mathscr{A}\|_F^2 - \lambda^2} = \min \left\{ \|\mathscr{A} - \alpha \mathbf{u}^m\|_F : \alpha \in \mathbf{R}, \mathbf{u} \in \mathbf{R}^n, \|\mathbf{u}\|_2 = 1 \right\},$$

where $\|\cdot\|_F$ is the Frobenius norm.

The tensors in theoretical physics and continuum mechanics are physical quantities which are invariant under co-ordinate system changes. A scalar associated with a tensor is an **invariant** of that tensor, if it keeps unchange under co-ordinate system changes. Theorem 1.2 (c) implies that E-eigenvalues and Z-eigenvalues are invariants of the tensor. Later research demonstrate that these eigenvalues, in particular Z-eigenvalues, have practical uses in physics and mechanics.

Independently, in 2005, Lek-Heng Lim also defined eigenvalues for tensors in his paper [177].

Lim [177] defined eigenvalues for general real tensors in the real field. The l^2-eigenvalues of tensors defined by Lim [177] are Z-eigenvalues of Qi [221]. The l^k-eigenvalues of tensors defined by Lim [177] are the same as H-eigenvalues in Qi [221] in the even order case, and different in the odd order case. Notably, Lim [177] proposed a multilinear generalization of the Perron-Frobenius theorem based upon the notion of l^k-eigenvalues (H-eigenvalues) of tensors.

1.3 Notes

The eigenvalues of tensors can be viewed from two independent perspectives. The first one is that we can regard the eigenvalues of tensors as formal generalizations of eigenvalues of matrices. As tensors are naturally corresponding to multilinear systems, the defining equations of eigenvalues thus generalize from linear equations to polynomial equations. The other one is that eigenvalues of a matrix are just the roots of the characteristic polynomial, which is the determinant of a certain parametrized matrix; whereas the eigenvalues of a tensor are the roots of the characteristic polynomial of a tensor, which is the determinant of a certain parametrized tensor (cf. [127, 221]). Certainly, the two aspects are equivalent somehow.

The original motivation for studying eigenvalues of tensors are scattered as follows. Qi [221] studies them for the positive definiteness of a polynomial form which has important applications in stability study of nonlinear autonomous system via Liapunov's direct method in automatic control. Lim [177] studies them from the generalizations of the variational characterizations of eigenvalues and singular values of matrices, and intends to study of spectral hypergraph theory.

Mathematical modelling and methodology based on higher order tensors have made great progress in various fields. A relatively systematic treatment of the basic theory of eigenvalues of tensors can be found in the book *Tensor Analysis: Spectral Theory and Special Tensors* [228]. That book also covers the discussion on four special types of tensors, namely nonnegative tensors [32, 34, 98, 129, 292, 296], positive semidefinite tensors [167, 188, 190], completely positive tensors [187, 230] and copositive tensors [44, 45, 157, 223], as well as spectral hypergraph theory via tensors. Plenty of references on these topics are collected in this book. Beyond that, recent developments and works are made in signal processing [293], automatica [11, 176], polynomial optimization [41, 150, 151, 181, 282], network analysis [16, 31, 67, 69, 111], the number of eigenvalues [26, 175], eigenvectors of tensors [160, 191, 316], spectra of tensors [36, 104, 149, 152, 153], singular value of tensors [242], tensor products and tensor norms [260, 274], eigenvalue inclusion sets [27, 154, 165, 172, 237, 273, 281, 311–314], Perron-Frobenius type theorems [39, 101, 130], numerical algorithms for eigenvalues of nonnegative tensors [46, 110, 294, 310], structured tensors [43, 54, 194, 252, 275], special tensors [68, 156, 173, 245, 272, 283, 297, 317], tensor approximations [35, 92, 213, 219, 295], polynomial optimization for tensor eigenvalues [47, 208], computing tensor eigenvalues using nonlinear programming [147, 298, 299, 315], tensor computation [48, 57, 88, 95, 113, 161, 174], tensor inverse [13, 258], spectral hypergraph theory [5, 7, 42, 55, 94, 109, 171, 210, 214, 241, 290, 301, 306, 307], etc.

Chapter 2
Multilinear Systems

A central problem in both pure and applied mathematics is solving various kinds of equations. Every progress in this discipline makes a big step in mathematics, especially in applied mathematics, such as Gaussian elimination method for linear equations, the simplex method for linear inequalities, and Gröbner bases for polynomial equations.

It is known that both linear equations and linear inequalities can be solved efficiently by well-developed methods together with sophisticated softwares. Systems of polynomial equations, on the other hand, are much more difficult to handle in general. Nonetheless, we can still confirm the existence of some particular solutions as well as efficient numerical methods for specific systems of polynomial equations.

In this chapter, we will consider such one scenario–*structured multilinear systems*. Let $\mathscr{A} \in T_{m,n}$ and $\mathbf{b} \in \mathbf{R}^n$. We call

$$\mathscr{A}\mathbf{x}^{m-1} = \mathbf{b}, \tag{2.1}$$

for solving $\mathbf{x}$, a **multilinear system**.

We will first study the system whose coefficient tensor $\mathscr{A}$ is an M-tensor, abbreviated as an M-equation. It can be proved that a nonsingular M-equation with a positive right-hand side always has a unique positive solution. With such a theoretical result, several iterative algorithms can be proposed for solving multilinear nonsingular M-equations, generalizing the classical iterative methods for solving linear systems. The results can be applied to some nonlinear differential equations and the inverse iteration for spectral radii of nonnegative tensors. Then, a homotopy method is presented for finding the unique positive solution to the multilinear system (2.1), where the related tensor is a nonsingular M-equation. Furthermore, the convergence of the method to the desired solution is proved. It should be noted that the corresponding M-equation here may not be symmetric.

For M-tensor equation system (2.1) with symmetric M-tensors, we will study a new tensor method which is based on the rank-1 approximation of the coefficient

L. Qi et al., *Tensor Eigenvalues and Their Applications*, Advances in Mechanics and Mathematics 39, https://doi.org/10.1007/978-981-10-8058-6_2

tensor. Furthermore, the local convergence property of the tensor method is also shown. Next, we study the multilinear system with general tensors. It is proved that the well known Jacobi, Gauss–Seidel and SOR methods for solving system of linear equations can be generalized to solve general multilinear equations. Under appropriate conditions, the proposed methods are shown to be globally and locally R-linearly convergent. Particularly, a Newton–Gauss–Seidel method will be given, and its convergence performance is better than that of the above methods.

2.1 Multilinear Systems Defined by M-Tensors

In this section, we study the multilinear systems (2.1), whose coefficient tensors are M-tensors. Recall that such multilinear systems are called M-equations. The M-equations were first studied by Ding and Wei [87]. An interesting result proved by Ding and Wei [87] is that a nonsingular M-equation with a positive right-hand side always has a unique positive solution. We will introduce this result here.

Now, we define Z-tensors, M-tensors, and H-tensors. They are extensions of Z-matrices, M-matrices, and H-matrices. We call a tensor $\mathscr{A} = (a_{i_1 i_2 \ldots i_m}) \in T_{m,n}$ a **Z-tensor** if all of its off-diagonal entries are nonpositive. Let the eigenvalues and eigenvectors of $\mathscr{A}$ be defined as in Chap. 1. The spectral radius of tensor $\mathscr{A}$ is defined by

$$\rho(\mathscr{A}) := \max\{|\lambda| : \lambda \text{ is an eigenvalue of tensor } \mathscr{A}\}.$$

A Z-tensor $\mathscr{A}$ is called an **M-tensor** if it can be written as $\mathscr{A} = s\mathscr{I} - \mathscr{B}$ with $s \geq \rho(\mathscr{B})$, where $\mathscr{B}$ is a nonnegative tensor. Furthermore, we call $\mathscr{A}$ a **nonsingular M-tensor** if $s > \rho(\mathscr{B})$ [82]. Nonsingular M-tensors are also called strong M-tensors [228, 305].

M-matrices are also called Minkowski matrices to memorize German mathematician Hermann Minkowski (1864–1909). Thus, M-tensors may also be called Minkowski tensors (Fig. 2.1).

If $\mathscr{A}$ is a Z-tensor, then the following statements are equivalent [82, 228, 285, 305]:

(a) $\mathscr{A}$ is a nonsingular M-tensor;
(b) The real part of each eigenvalue of $\mathscr{A}$ is positive;
(c) There exists positive vector $\mathbf{x}$ such that $\mathscr{A}\mathbf{x}^{m-1} > \mathbf{0}$;
(d) There exists nonnegative vector $\mathbf{x}$ with $\mathscr{A}\mathbf{x}^{m-1} > \mathbf{0}$,

where $\mathbf{x} > \mathbf{0}$ or $\mathbf{x} \geq \mathbf{0}$ means all its entries are positive or nonnegative, respectively. Thus, for any two vectors $\mathbf{x}, \mathbf{y} \in \mathbf{R}^n$, $\mathbf{x} \geq \mathbf{y}$ means that $\mathbf{x} - \mathbf{y} \geq \mathbf{0}$. A tensor is called an **H-tensor**, if it becomes an M-tensor when its diagonal entries are made nonnegative and its off-diagonal entries are made nonpositive by preserving their absolute values.

Fig. 2.1 Hermann Minkowski (1864–1909)

Following [87], we denote the set of all the solutions, the set of all nonnegative solutions, and the set of all positive solutions of multilinear systems $\mathscr{A}\mathbf{x}^{m-1} = \mathbf{b}$ by

$$\mathscr{A}^{-1}\mathbf{b} := \{\mathbf{x} \in \mathbf{R}^n \ : \ \mathscr{A}\mathbf{x}^{m-1} = \mathbf{b}\},$$

$$(\mathscr{A}^{-1}\mathbf{b})_+ := \{\mathbf{x} \in \mathbf{R}^n_+ \ : \ \mathscr{A}\mathbf{x}^{m-1} = \mathbf{b}\},$$

and

$$(\mathscr{A}^{-1}\mathbf{b})_{++} := \{\mathbf{x} \in \mathbf{R}^n_{++} \ : \ \mathscr{A}\mathbf{x}^{m-1} = \mathbf{b}\},$$

respectively, where $\mathbf{R}^n_+ (\mathbf{R}^n_{++})$ is a set containing all nonnegative real numbers (positive real numbers) of $\mathbf{R}^n$. By the famous Hilbert Nullstellensatz, the solution set $\mathscr{A}^{-1}\mathbf{b}$ is nonempty if and only if there is no contradictive equation. Note that $\mathscr{A}^{-1}\mathbf{b}$ is merely a notation.

We now consider properties of the positive solution set of an M-equation. As defined before, an M-equation is a multilinear system

$$\mathscr{A}\mathbf{x}^{m-1} = \mathbf{b}, \tag{2.2}$$

where $\mathscr{A} = s\mathscr{I} - \mathscr{B}$ is an M-tensor. In particular, we are interested in the existence of a nonnegative solution when the right-hand side $\mathbf{b}$ is nonnegative. It is easy to see that $\mathbf{x}$ is a fixed point of the following iteration

$$\mathbf{x}^{(k+1)} = T_{s,\mathscr{B},\mathbf{b}}(\mathbf{x}^{(k)}) := (s^{-1}\mathscr{B}(\mathbf{x}^{(k)})^{m-1} + s^{-1}\mathbf{b})^{[\frac{1}{m-1}]}, \quad k = 0, 1, 2, \ldots \tag{2.3}$$

if and only if it is a solution of the M-equation (2.2).

Alternatively, we may study the fixed point iteration (2.3). We now need some concepts about cones and increasing maps. Let $\mathbb{H}$ be a real Banach space. We call a nonempty closed convex set $\mathbb{D}$ in $\mathbb{H}$ a cone if for any $\mathbf{x} \in \mathbb{D}$ and $\lambda \geq 0$, it holds that $\lambda\mathbf{x} \in \mathbb{D}$ and $\mathbf{x} = \mathbf{0}$ if further $-\mathbf{x} \in \mathbb{D}$.

Note that a cone $\mathbb{D} \subset \mathbb{H}$ induces a semi-order in $\mathbb{H}$, i.e., $\mathbf{x} \leq \mathbf{y}$ if $\mathbf{y} - \mathbf{x} \in \mathbb{D}$. Suppose that $\{\mathbf{x}_n\}$ is an increasing series in $\mathbb{H}$ with an upper bound, i.e., there exists $\mathbf{y} \in \mathbb{H}$ such that $\mathbf{x}_n \leq \mathbf{x}_{n+1} \leq \mathbf{y}$ for $n = 1, 2, \ldots$. If there is a vector $\mathbf{x}^* \in \mathbb{H}$ such that $\|\mathbf{x}_n - \mathbf{x}^*\| \to 0$ $(n \to \infty)$, then we say that $\mathbb{D}$ is a regular cone. Let $T : \mathbb{D} \to \mathbb{H}$, where $\mathbb{D} \subset \mathbb{H}$. If $\mathbf{x} \leq \mathbf{y}$ for $\mathbf{x}, \mathbf{y} \in \mathbb{D}$ always implies $T(\mathbf{x}) \leq T(\mathbf{y})$, then we say that T is an increasing map on $\mathbb{D}$.

Hence, $T_{s,\mathscr{B},\mathbf{b}}$ is an increasing map on $\mathbf{R}^n_+$. For an increasing map on a regular cone, we may apply the following fixed-point theorem by Amann.

Theorem 2.1 (Amann 1976) *Let $\mathbb{D}$ be a regular cone in an ordered Banach space $\mathbb{H}$ and $[\mathbf{u}, \mathbf{v}] \subset \mathbb{H}$ be a bounded order interval. Suppose that $T : [\mathbf{u}, \mathbf{v}] \to \mathbb{H}$ is an increasing continuous map which satisfies*

$$\mathbf{u} \leq T(\mathbf{u}) \quad \text{and} \quad \mathbf{v} \geq T(\mathbf{v}).$$

Then, T has at least one fixed point in $[\mathbf{u}, \mathbf{v}]$. Moreover, there exists a minimal fixed point $\mathbf{x}_*$ and a maximal fixed point $\mathbf{x}^*$ in the sense that every fixed point $\bar{\mathbf{x}}$ satisfies $\mathbf{x}_* \leq \bar{\mathbf{x}} \leq \mathbf{x}^*$. Furthermore, we consider the following iterative method

$$\mathbf{x}^{(k+1)} = T(\mathbf{x}^{(k)}), \quad k = 0, 1, 2, \ldots.$$

The sequence $\{\mathbf{x}^{(k)}\}$ converges to $\mathbf{x}_*$ from below if the initial point $\mathbf{x}^{(0)} = \mathbf{u}$, i.e.,

$$\mathbf{u} = \mathbf{x}^{(0)} \leq \mathbf{x}^{(1)} \leq \mathbf{x}^{(2)} \leq \cdots \leq \mathbf{x}_*,$$

and converges to $\mathbf{x}^*$ from above if the initial point $\mathbf{x}^{(0)} = \mathbf{v}$, i.e.,

$$\mathbf{v} = \mathbf{x}^{(0)} \geq \mathbf{x}^{(1)} \geq \mathbf{x}^{(2)} \geq \cdots \geq \mathbf{x}^*.$$

With the above Amann fixed point theorem, Ding and Wei [87] proved the existence of positive solutions of the M-equations, and obtained the following theorem.

Theorem 2.2 *Suppose that $\mathscr{A}$ is a nonsingular M-tensor. Then for every positive vector $\mathbf{b}$, the M-equation $\mathscr{A}\mathbf{x}^{m-1} = \mathbf{b}$ has a unique positive solution.*

Proof As we discussed above, $\mathbf{x}$ is a nonnegative solution of the M-equation (2.2) if and only if it is a fixed point of

$$T_{s,\mathscr{B},\mathbf{b}} : \mathbf{R}^n_+ \to \mathbf{R}^n_+, \mathbf{x} \mapsto (s^{-1}\mathscr{B}\mathbf{x}^{m-1} + s^{-1}\mathbf{b})^{[\frac{1}{m-1}]}.$$

We note that $\mathbf{R}^n_+$ is a regular cone and $T_{s,\mathscr{B},\mathbf{b}}$ is an increasing continuous map. If $\mathscr{A}$ is a nonsingular M-tensor, i.e., $s > \rho(\mathscr{B})$, then there exists a positive vector $\mathbf{z} \in \mathbf{R}^n_{++}$ such that $\mathscr{A}\mathbf{z}^{m-1} > \mathbf{0}$. Let

$$\underline{\gamma} = \min_{i\in[n]} \frac{b_i}{(\mathscr{A}\mathbf{z}^{m-1})_i} \text{ and } \overline{\gamma} = \max_{i\in[n]} \frac{b_i}{(\mathscr{A}\mathbf{z}^{m-1})_i}.$$

Then we have $\underline{\gamma}\mathscr{A}\mathbf{z}^{m-1} \le \mathbf{b} \le \overline{\gamma}\mathscr{A}\mathbf{z}^{m-1}$. This implies that

$$\underline{\gamma}^{\frac{1}{m-1}}\mathbf{z} \le T_{s,\mathscr{B},\mathbf{b}}(\underline{\gamma}^{\frac{1}{m-1}}\mathbf{z}) \text{ and } \overline{\gamma}^{\frac{1}{m-1}}\mathbf{z} \le T_{s,\mathscr{B},\mathbf{b}}(\overline{\gamma}^{\frac{1}{m-1}}\mathbf{z}).$$

By the Amann fixed-point theorem, i.e., Theorem 2.1, there exists at least one fixed point $\bar{\mathbf{x}}$ of $T_{s,\mathscr{B},\mathbf{b}}$ with

$$\underline{\gamma}^{\frac{1}{m-1}}\mathbf{z} \le \bar{\mathbf{x}} \le \overline{\gamma}^{\frac{1}{m-1}}\mathbf{z},$$

which clearly is a positive vector if $\mathbf{b}$ is positive.

Furthermore, we may prove that the positive fixed point $\bar{\mathbf{x}}$ is unique if $\mathbf{b}$ is positive. Assume that there exist two positive fixed points $\mathbf{x}$ and $\mathbf{y}$, i.e.,

$$T_{s,\mathscr{B},\mathbf{b}}(\mathbf{x}) = \mathbf{x} > \mathbf{0} \text{ and } T_{s,\mathscr{B},\mathbf{b}}(\mathbf{y}) = \mathbf{y} > \mathbf{0}.$$

Let $\eta = \min_{i\in[n]} \frac{x_i}{y_i}$. Then $\mathbf{x} \ge \eta\mathbf{y}$ and $x_j = \eta y_j$ for some j. If $\eta < 1$, then $\mathscr{A}(\eta\mathbf{y})^{m-1} = \eta^{m-1}\mathbf{b} < \mathbf{b}$, which implies that

$$T_{s,\mathscr{B},\mathbf{b}}(\eta\mathbf{y}) = (s^{-1}\mathscr{B}(\eta\mathbf{y})^{m-1} + s^{-1}\mathbf{b})^{[\frac{1}{m-1}]} > \eta\mathbf{y}.$$

On the other hand, since $T_{s,\mathscr{B},\mathbf{b}}$ is nonnegative and increasing, we have

$$T_{s,\mathscr{B},\mathbf{b}}(\eta\mathbf{y})_j \le T_{s,\mathscr{B},\mathbf{b}}(\mathbf{x})_j = x_j = \eta y_j.$$

This forms a contradiction. Thus $\eta \ge 1$, which implies that $\mathbf{x} \ge \mathbf{y}$. Similarly, we may also show that $\mathbf{y} \ge \mathbf{x}$. Thus, we have $\mathbf{x} = \mathbf{y}$. This implies that the positive fixed point of $T_{s,\mathscr{B},\mathbf{b}}$ is unique, i.e., the positive solution to the M-equation $\mathscr{A}\mathbf{x}^{m-1} = \mathbf{b}$ is unique. The theorem is proved. □

By the equivalent condition (c) of nonsingular M-tensors, the above theorem indicates an equivalent condition for nonsingular M-tensors, which generalizes the "nonnegative inverse" property of M-matrix [17] to the tensor case.

Theorem 2.3 *A Z-tensor $\mathscr{A} \in T_{m,n}$ is a nonsingular M-tensor if and only if $(\mathscr{A}^{-1}\mathbf{b})_{++}$ has a unique element for every positive vector* $\mathbf{b}$.

Proof Suppose that $\mathscr{A} \in T_{m,n}$ is a nonsingular M-tensor. Then by Theorem 2.2 we have the existence and uniqueness of the element in the positive solution set $(\mathscr{A}^{-1}\mathbf{b})_{++}$.

On the other hand, let $\mathscr{A} \in T_{m,n}$ be a Z-tensor. If $(\mathscr{A}^{-1}\mathbf{b})_{++}$ has a unique element for every positive vector $\mathbf{b}$, then there is a positive vector $\mathbf{x}$ such that $\mathscr{A}\mathbf{x}^{m-1} > \mathbf{0}$. Then by the equivalent condition (c) of nonsingular M-tensors, $\mathscr{A}$ must also be a nonsingular M-tensor. □

Now, denote the unique positive solution of $\mathscr{A}\mathbf{x}^{m-1} = \mathbf{b}$ by $\mathscr{A}_{++}^{-1}\mathbf{b}$ for a nonsingular M-tensor $\mathscr{A}$ and a positive vector $\mathbf{b}$. Then $\mathscr{A}_{++}^{-1} : \mathbf{R}_{++}^n \to \mathbf{R}_{++}^n$ is an increasing map under partial order " $\geq$ " in the cone $\mathbf{R}_{++}^n$, that is, $\mathscr{A}_{++}^{-1}\widehat{\mathbf{b}} \geq \mathscr{A}_{++}^{-1}\widetilde{\mathbf{b}} > \mathbf{0}$ if $\widehat{\mathbf{b}} \geq \widetilde{\mathbf{b}} > \mathbf{0}$ [87].

By a similar argument, the following theorem holds for general M-equations.

Theorem 2.4 *Let $\mathscr{A} \in T_{m,n}$ be an M-tensor and $\mathbf{b} \in \mathbf{R}_+^n$. If there exists a nonnegative vector $\mathbf{v}$ such that $\mathscr{A}\mathbf{v}^{m-1} \geq \mathbf{b}$, then $(\mathscr{A}^{-1}\mathbf{b})_+$ is nonempty.*

Remark 2.1 In general, the nonnegative solution set $(\mathscr{A}^{-1}\mathbf{b})_+$ in the above theorem may not be a singleton, and these nonnegative solutions lay on a hypersurface in $\mathbf{R}^n$.

Now, we study the nonsingular M-equations with nonpositive right-hand sides. If the coefficient tensor is of even order, the case is simple. Let $\mathscr{A} \in T_{m,n}$ be an even-order (i.e., m is even) nonsingular M-tensor and $\mathbf{b}$ be a nonpositive vector. Then the M-equation $\mathscr{A}\mathbf{x}^{m-1} = \mathbf{b}$ is equivalent to the nonsingular M-equation $\mathscr{A}(-\mathbf{x})^{m-1} = -\mathbf{b}$ with nonnegative right-hand side. However, the case is totally different if the coefficient tensor $\mathscr{A}$ is of odd order. In that case, the following property of nonsingular M-tensors is needed.

Theorem 2.5 *A Z-tensor $\mathscr{A} \in T_{m,n}$ is a nonsingular M-tensor if and only if $\mathscr{A}$ does not reverse the sign of any vector; That is, if $\mathbf{x} \neq \mathbf{0}$ and $\mathbf{b} = \mathscr{A}\mathbf{x}^{m-1}$, then for some subscript i,*

$$x_i^{m-1} b_i > 0.$$

Proof Suppose that $\mathscr{A}$ is a nonsingular M-tensor. Then we will show that $\mathscr{A}$ does not reverse the sign of any vector. Assume that $\mathbf{x} \neq \mathbf{0}$ and $\mathbf{b} = \mathscr{A}\mathbf{x}^{m-1}$ with $x_i^{m-1} b_i \leq 0$ for all indices i. Let J be the largest index set such that $x_j \neq 0$ for all $j \in J$, and $\mathscr{A}^J$ is the corresponding leading sub-tensor of $\mathscr{A}$. Then

$$\mathbf{b}_J = \mathscr{A}^J \mathbf{x}_J^{m-1}.$$

Since $x_i^{m-1}b_i \le 0$ and $x_j \ne 0$ for all $j \in J$, there is a nonnegative diagonal tensor $\mathscr{D}^J$ such that

$$\mathbf{b}_J = -\mathscr{D}^J \mathbf{x}_J^{m-1}.$$

Thus we have $(\mathscr{A}^J + \mathscr{D}^J)\mathbf{x}_J^{m-1} = \mathbf{0}$, which forms a contradiction.

On the other hand, suppose that $\mathscr{A}$ is not a nonsingular M-tensor. Then it has an eigenvector $\mathbf{x}$ such that $\mathscr{A}\mathbf{x}^{m-1} = \lambda \mathbf{x}^{[m-1]}$ and $\lambda \le 0$. Then $\mathscr{A}$ reverses the sign of vector $\mathbf{x}$. This implies that if $\mathscr{A}$ does not reverse the sign of any vector, then $\mathscr{A}$ must be a nonsingular M-tensor. □

From the above theorem, it is easy to see that there is no real vector $\mathbf{x}$ such that $\mathbf{b} = \mathscr{A}\mathbf{x}^{m-1}$ is nonpositive when m is odd, since $\mathbf{x}^{[m-1]}$ is always nonnegative.

In (2.2), the left-hand side is a homogeneous form. Similar results can be established for some special equations with non-homogeneous left-hand sides. Consider the following equation

$$\mathscr{A}\mathbf{x}^{m-1} - \mathscr{B}_{m-1}\mathbf{x}^{m-2} - \cdots - \mathscr{B}_2\mathbf{x} = \mathbf{b} > \mathbf{0},$$

where $\mathscr{A} = s\mathscr{I} - \mathscr{B}_m$ is an mth-order nonsingular M-tensor and $\mathscr{B}_j \in T_{j,n}$ is a nonnegative tensor for $j = 2, 3, \ldots, m$. Assume that there exists a positive vector $\mathbf{v}$ such that

$$\mathscr{A}\mathbf{v}^{m-1} - \mathscr{B}_{m-1}\mathbf{v}^{m-2} - \cdots - \mathscr{B}_2\mathbf{v} > \mathbf{0}. \tag{2.4}$$

This condition is an extension of a parallel property of nonsingular M-tensors. Similar to the discussion in (2.3), we may discuss the fixed point iteration for analyzing the positive solution to the above equation

$$\mathbf{x}^{(k)} = F(\mathbf{x}^{(k-1)}), k = 1, 2, \ldots$$

where

$$F(\mathbf{x}) = [s^{-1}(\mathscr{B}_m\mathbf{x}^{m-1} + \mathscr{B}_{m-1}\mathbf{x}^{m-2} + \cdots + \mathscr{B}_2\mathbf{x} + \mathbf{b})]^{[\frac{1}{m-1}]}.$$

Then a similar argument can be applied to this fixed point iteration. Thus we may still conclude that this non-homogeneous equation has a unique positive solution for each positive right-hand side. Based upon this approach, we have the following theorem for such a kind of special equations with non-homogeneous left-hand sides.

Theorem 2.6 *Let $\mathscr{A} \in T_{m,n}$ be a Z-tensor and $\mathscr{B}_j \in T_{j,n}$ be a nonnegative tensor for $j = 2, 3, \ldots, m-1$. Then the equation*

$$\mathscr{A}\mathbf{x}^{m-1} - \mathscr{B}_{m-1}\mathbf{x}^{m-2} - \cdots - \mathscr{B}_2\mathbf{x} = \mathbf{b} \tag{2.5}$$

has a unique positive solution for every positive vector $\mathbf{b}$ if and only if $\mathscr{A}$ is a nonsingular M-tensor.

Proof Suppose that (2.5) has a unique positive solution $\mathbf{x}$ for every positive vector $\mathbf{b}$. Then we have

$$\mathscr{A}\mathbf{x}^{m-1} = \mathscr{B}_{m-1}\mathbf{x}^{m-2} + \cdots + \mathscr{B}_2\mathbf{x} + \mathbf{b}.$$

Since $\mathscr{B}_j$ is nonnegative for $j = 2, 3, \ldots, m-1$, we have

$$\mathscr{A}\mathbf{x}^{m-1} = \mathscr{B}_{m-1}\mathbf{x}^{m-2} + \cdots + \mathscr{B}_2\mathbf{x} + \mathbf{b} \geq \mathbf{b} > \mathbf{0}.$$

By the equivalent condition (c) for nonsingular M-tensors, we conclude that $\mathscr{A}$ is a nonsingular M-tensor.

On the other hand, suppose that $\mathscr{A}$ is a nonsingular M-tensor. By the equivalent condition (c) for nonsingular M-tensors, there is a positive vector $\mathbf{y}$ such that $\mathscr{A}\mathbf{y}^{m-1} > \mathbf{0}$. Since the order of $\mathscr{A}$ is higher than the order of $\mathscr{B}_j$ for $j = 2, 3, \ldots, m-1$, there is a positive number β such that $\mathscr{A}(\beta\mathbf{y})^{m-1} > \mathscr{B}_{m-1}(\beta\mathbf{y})^{m-2} + \cdots + \mathscr{B}_2(\beta\mathbf{y})$. This implies that the condition (2.4) holds. By the above discussion, this shows that the Eq. (2.5) has a unique positive solution for every positive vector $\mathbf{b}$. □

More extensions of above results can be found in [87].

2.2 Finding the Positive Solution of a Nonsingular M-Equation

Theorem 2.2 informs us that a nonsingular M-equation $\mathscr{A}\mathbf{x}^{m-1} = \mathbf{b}$ with a positive vector $\mathbf{b}$ always has a unique positive solution $\mathbf{x}$. In this section, we show how to find such a positive solution $\mathbf{x}$. Ding and Wei [87] proposed several classical iterative methods and Newton method for solving M-equations. Then Han [117] raised a homotopy method for solving multilinear systems with nonsymmetric M-tensors. Several numerical examples show the performance of the homotopy method.

Generally speaking, it is complicated to solve general multilinear equation systems. We follow [87] to take an approach for solving a system of polynomial equations by computing its triangular partition. As in [87], we now define the triangular parts of a tensor in $T_{m,n}$.

For a tensor $\mathscr{A} = (a_{i_1 i_2 \ldots i_m}) \in T_{m,n}$, we say that the entry $a_{i_1 i_2 \ldots i_m}$ is in the lower triangular part of $\mathscr{A}$ if $i_1 \in [n]$ and $i_2, \ldots, i_m \leq i_1$; Otherwise, we say that the entry is in the off-lower triangular part of $\mathscr{A}$. If $i_1 \in [n]$ and $i_2, \ldots, i_m < i_1$, then we say that the entries $a_{i_1 i_2 \ldots i_m}$ are in the strictly lower triangular part of $\mathscr{A}$. A tensor $\mathscr{A} \in T_{m,n}$ is lower triangular if all its entries in the off-lower triangular part are zero. We call the following multilinear system a lower triangular equation

$$\mathscr{L}\mathbf{x}^{m-1} = \mathbf{b},$$

if the coefficient tensor $\mathscr{L}$ is lower triangular.

Similarly, for a tensor $\mathscr{A} = (a_{i_1 i_2 \dots i_m}) \in T_{m,n}$, we say that the entries $a_{i_1 i_2 \dots i_m}$ are in the upper triangular part of $\mathscr{A}$ if $i_1 \in [n]$ and $i_2, \dots, i_m \geq i_1$, and other entries are in the off-upper triangular part of $\mathscr{A}$. If $i_1 \in [n]$ and $i_2, \dots, i_m > i_1$, then we say that the entries $a_{i_1 i_2 \dots i_m}$ are in the strictly upper triangular part of $\mathscr{A}$. A tensor $\mathscr{A} \in T_{m,n}$ is upper triangular if all its entries in the off-upper triangular part are zero. We call the following multilinear system an upper triangular equation

$$\mathscr{U}\mathbf{x}^{m-1} = \mathbf{b},$$

if the coefficient tensor $\mathscr{U}$ is upper triangular.

As in the matrix case, Ding and Wei considered the lower triangular tensor with all nonzero diagonal entries, where a nonsingular lower triangular equation can be solved by forward substitution [87].

Algorithm 2.1 (*Forward Substitution*) If $\mathscr{L} = (l_{i_1 i_2 \dots i_m}) \in T_{m,n}$ is lower triangular with entries in complex number field, and $\mathbf{b} \in \mathbf{C}^n$, then the algorithm overwrites $\mathbf{b}$ with one of the solutions to $\mathscr{L}\mathbf{x}^{m-1} = \mathbf{b}$.

b_1 = one of the $(m-1)$th roots of $b_1 / l_{11 \dots 1}$.
for $i = 2 : n$
  for $k = 1 : m$
    $p_k = \sum\{l_{ii_2 \dots i_m} \cdot \prod_{t=2,\dots,m; t \neq p_1,\dots,p_{k-1}} b_{i_t} \; : \; i_2, i_3, \dots, i_m \leq i,$

    $i_{p_1}, i_{p_2}, \dots, i_{p_{k-1}}$ are the only k-1 indices equal to $i\}$.
  end
  b_i = one of the roots of $p_1 + p_2 z + \cdots + p_m z^{m-1} = b_i$.
end

With this algorithm, the existence and uniqueness of a solution can be analyzed. By the fundamental theorem of algebra, the solution set of the lower triangular equation has $(m-1)^n$ elements in the complex number field (counted with multiplicity). When m is even, the polynomial equation of odd degree $m-1$ in the algorithm has at least one real solution, i.e., the real solution set $\mathscr{L}^{-1}\mathbf{b}$ is nonempty. If this polynomial equation has a unique real solution at each step, then the real solution set $\mathscr{L}^{-1}\mathbf{b}$ is a singleton. When m is odd, the existence of real solution is not guaranteed, since the degree of the polynomial equation is even. Even if the real solution exists, there are at least two elements in the solution set as x_1 has two choices.

One may solve the upper triangular equation for tensors with all nonzero diagonal entries by a similar back substitution algorithm.

Algorithm 2.2 (*Back Substitution*) If $\mathscr{U} = (u_{i_1 i_2 \dots i_m}) \in T_{m,n}$ is upper triangular with entries in complex number field, and $\mathbf{b} \in \mathbf{C}^n$, then the algorithm overwrites $\mathbf{b}$ with one of the solutions to $\mathscr{U}\mathbf{x}^{m-1} = \mathbf{b}$.

b_n = one of the $(m-1)$th roots of $b_n / u_{nn \dots n}$.
for $i = n-1 : -1 : 1$
  for $k = 1 : m$
    $p_k = \sum\{u_{ii_2 \dots i_m} \cdot \prod_{t=2,\dots,m; t \neq p_1,\dots,p_{k-1}} b_{i_t} \; : \; i_2, i_3, \dots, i_m \geq i,$

$i_{p_1}, i_{p_2}, \ldots, i_{p_{k-1}}$ are the only k-1 indices equal to $i\}$.
end
b_i = one of the roots of $p_1 + p_2 z + \cdots + p_m z^{m-1} = b_i$.
end

Even if there are some zero diagonal entries, a higher order triangular equation may still have solutions. This is different from the matrix case. In such a case, the polynomial equation of degree $m - 1$ reduces to a lower degree one but may still have solutions.

Next, we show how to find the unique positive solution for a special case such that the triangular equations with the coefficient tensor being a nonsingular M-tensor, i.e., its diagonal entries are positive and its off-diagonal entries are nonpositive. In the following proposition, we take the lower triangular M-equation $\mathscr{L}\mathbf{x}^{m-1} = \mathbf{b}$ as an example, and the upper triangular M-equation system can be easily proved similarly.

Proposition 2.1 *Suppose $\mathscr{L}$ is an mth order n dimensional lower triangular M-tensor. If* $\mathbf{b}$ *is a nonnegative vector, then* $\mathscr{L}\mathbf{x}^{m-1} = \mathbf{b}$ *has at least one nonnegative solution. Furthermore, if* $\mathbf{b}$ *is a positive vector, then* $\mathscr{L}\mathbf{x}^{m-1} = \mathbf{b}$ *has a unique positive solution.*

Proof By Algorithm 2.1, it is obvious that the coefficients $p_1, p_2, \ldots, p_{m-1}$ are nonpositive and p_m is positive in each step. Thus the companion matrix of $p_1 + p_2 t + \cdots + p_m t^{m-1} = b_i$ can be written as

$$C_i = \begin{pmatrix} 0 & 1 & 0 & \cdots & 0 \\ 0 & 0 & 1 & \cdots & 0 \\ \vdots & \vdots & \vdots & \ddots & \vdots \\ 0 & 0 & 0 & \cdots & 1 \\ \frac{b_i - p_1}{p_m} & \frac{-p_2}{p_m} & \frac{-p_3}{p_m} & \cdots & \frac{-p_{m-1}}{p_m} \end{pmatrix}.$$

It is not difficult to know that C_i is a nonnegative irreducible matrix when $b_i \geq 0$. Thus, in each step of Algorithm 2.1, the corresponding polynomial has at least one nonnegative solution $x_i = \rho(C_i)$ when $\mathbf{b} \geq \mathbf{0}$, where $\rho(C_i)$ denotes the spectral radius of matrix C_i. Similarly, when $\mathbf{b} > \mathbf{0}$, we know that the corresponding polynomial has at least one positive solution and the desired results hold. □

We now consider solution methods for solving the nonsingular M-equation $\mathscr{A}\mathbf{x}^{m-1} = \mathbf{b} > \mathbf{0}$, where $\mathscr{A} = s\mathscr{I} - \mathscr{B}$ is a nonsingular M-tensor, i.e., $\mathscr{B}$ is a nonnegative tensor and $s > \rho(\mathscr{B})$. By the discussion in Sect. 2.1, such an M-equation has a positive solution. The problem now is how to find such a positive solution, or more generally a nonnegative solution. Until now, there is no "LU factorization" results for general tensors. Hence, we study several iterative methods for solving such an M-equation.

The Jacobi method and the Gauss–Seidel(G-S) method for solving linear equations split the coefficient matrices of such linear equations in their iterations. Stimulated from such a splitting approach, we split the coefficient tensor $\mathscr{A}$ into

$\mathscr{A} = \mathscr{M} - \mathscr{N}$ such that the coefficient tensor $\mathscr{N}$ is nonnegative and the equation with coefficient tensor $\mathscr{M}$ is easy to solve. For the nonsingular M-equation $\mathscr{A}\mathbf{x}^{m-1} = \mathbf{b}$, we may take $\mathscr{M}$ as the diagonal part, the lower triangular part, or the upper triangular part of $\mathscr{A}$. Then $\mathscr{N}$ is nonnegative. It is clearly easy to solve the equation with a diagonal tensor as its coefficient tensor as long as all of the diagonal entries are nonzero. If $\mathscr{M}$ is the lower or upper triangular part of $\mathscr{A}$, the equation with $\mathscr{M}$ as its coefficient tensor is also easy to solve due to the early discussion on triangular M-equations. Then by applying the iteration

$$\mathbf{x}^{(k)} = \mathscr{M}_{++}^{-1}(\mathscr{N}(\mathbf{x}^{(k-1)})^{m-1} + \mathbf{b}), k = 1, 2, \ldots,$$

we obtain a nonnegative solution to the above M-equation if the iteration converges.

Let $\varphi : \mathbf{R}^n \to \mathbf{R}^n$. Suppose that $\mathbf{x}_*$ is a fixed point of $\varphi(\mathbf{x})$. We call x_* an attracting fixed point if there is $\delta > 0$ such that the sequence $\{\mathbf{x}^{(k)}\}$ defined by $\mathbf{x}^{(k+1)} = \varphi(\mathbf{x}^{(k)})$ converges to $\mathbf{x}_*$ for any $\mathbf{x}^{(0)}$ satisfying $\|\mathbf{x}^{(0)} - \mathbf{x}_*\| \leq \delta$. With such notation, we may state the following result of [234], which will be useful for our further discussion.

Theorem 2.7 (Rheinboldt 1998) *Suppose that $\mathbf{x}_*$ is a fixed point of the operator $\varphi : \mathbf{R}^n \to \mathbf{R}^n$, and $\nabla\varphi : \mathbf{R}^n \to \mathbf{R}^{n\times n}$ denotes the Jacobian of φ. Then $\mathbf{x}_*$ is an attracting fixed point if $\sigma := \rho(\nabla\varphi(\mathbf{x}_*)) < 1$. Furthermore, if $\sigma > 0$, then we have linear convergence of the iteration $\mathbf{x}^{(k+1)} = \varphi(\mathbf{x}^{(k)})$ to $\mathbf{x}_*$ with rate σ.*

We now derive the Jacobian $\nabla\varphi$ for the operator

$$\varphi(\mathbf{x}) = \mathscr{M}_{++}^{-1}(\mathscr{N}\mathbf{x}^{m-1} + \mathbf{b}).$$

Note that we may always modify $\mathscr{A}$ into a tensor $\widetilde{\mathscr{A}}$ such that $\widetilde{\mathscr{A}}\mathbf{x}^{m-1} = \mathscr{A}\mathbf{x}^{m-1}$ for all $\mathbf{x} \in \mathbf{R}^n$ and $\widetilde{\mathscr{A}}$ is symmetric on the last $(m-1)$ modes. Thus we may assume that $\mathscr{A}$ is symmetric on the last $(m-1)$ modes, such that the gradient $\nabla(\mathscr{A}\mathbf{x}^{m-1}) = (m-1)\mathscr{A}\mathbf{x}^{m-2}$, where $\mathscr{A}\mathbf{x}^{m-2}$ is a matrix with $(\mathscr{A}\mathbf{x}^{m-2})_{ij} = \sum_{i_3,\ldots,i_m} a_{iji_3\ldots i_m}x_{i_3}\cdots x_{i_m}$ (see [105]). Taking gradients on both sides of

$$\mathscr{M}\varphi(\mathbf{x})^{m-1} = \mathscr{N}\mathbf{x}^{m-1} + \mathbf{b},$$

we have

$$\mathscr{M}\varphi(\mathbf{x})^{m-2} \cdot \nabla\varphi(\mathbf{x}) = \mathscr{N}\mathbf{x}^{m-2}.$$

We now consider the positive fixed point $\mathbf{x}_*$ of φ. Note that the matrix $\mathscr{M}\varphi(\mathbf{x}_*)^{m-2} = \mathscr{M}\mathbf{x}_*^{m-2}$ is a nonsingular M-matrix, since $\mathbf{x}_* > \mathbf{0}$ and

$$\mathscr{M}\mathbf{x}_*^{m-2} \cdot \mathbf{x}_* = \mathscr{N}\mathbf{x}_*^{m-1} + \mathbf{b} \geq \mathbf{b} > \mathbf{0}.$$

Then the Jacobian of φ at $\mathbf{x}_*$,

$$\nabla\varphi(\mathbf{x}_*) = (\mathscr{M}\mathbf{x}_*^{m-2})^{-1}\mathscr{N}\mathbf{x}_*^{m-2},$$

is a nonnegative matrix. Since

$$\mathscr{N}\mathbf{x}_*^{m-2}\cdot\mathbf{x}_* = \mathscr{M}\mathbf{x}_*^{m-1} - \mathbf{b} \le \theta\mathscr{M}\mathbf{x}_*^{m-1}$$

with $0 \le \theta < 1$, we have $\nabla\varphi(\mathbf{x}_*)\cdot\mathbf{x}_* \le \theta^{\frac{1}{m-1}}\mathbf{x}_*$. Thus the spectral radius $\rho(\nabla\varphi(\mathbf{x}_*)) \le \theta^{\frac{1}{m-1}} < 1$, which implies that $\mathbf{x}_*$ is an attracting fixed point of φ.

By the above discussion, the solution is attainable. The next issue is to choose an initial vector which can ensure the convergence of the algorithm. Since $\mathscr{A}$ is a nonsingular M-tensor, we may take an initial vector $\mathbf{x}^{(0)}$ such that

$$\mathbf{0} < \mathscr{A}(\mathbf{x}^{(0)})^{m-1} \le \mathbf{b}.$$

Then we will prove that the iteration

$$\mathbf{x}^{(k)} = \mathscr{M}_{++}^{-1}(\mathscr{N}(\mathbf{x}^{(k-1)})^{m-1} + \mathbf{b}), k = 1, 2, \ldots,$$

converges to the positive solution of the nonsingular M-equation with a positive right-hand side **b**.

Since $\mathscr{A}(\mathbf{x}^{(0)})^{m-1} > \mathbf{0}$, we have $\mathscr{N}(\mathbf{x}^{(0)})^{m-1} < \mathscr{M}(\mathbf{x}^{(0)})^{m-1}$. Then there is $\alpha \in (0, 1)$ such that $\mathscr{N}(\mathbf{x}^{(0)})^{m-1} \le \alpha(\mathscr{M}(\mathbf{x}^{(0)})^{m-1})$. Now, pick a positive number β such that $\mathbf{b} \le \beta(\mathscr{M}(\mathbf{x}^{(0)})^{m-1})$. The first iteration step indicates that

$$\mathscr{M}(\mathbf{x}^{(0)})^{m-1} \le \mathscr{N}(\mathbf{x}^{(0)})^{m-1} + \mathbf{b} \le (\alpha+\beta)\mathscr{M}(\mathbf{x}^{(0)})^{m-1},$$

which further implies that

$$(\mathbf{x}^{(0)})^{[m-1]} \le (\mathbf{x}^{(1)})^{[m-1]} \le (\alpha+\beta)(\mathbf{x}^{(0)})^{[m-1]}$$

since $\mathscr{M}$ is a triangular M-tensor. We now assume that

$$(\mathbf{x}^{(k-1)})^{[m-1]} \le (\mathbf{x}^{(k)})^{[m-1]} \le (\alpha^k + \alpha^{k-1}\beta + \cdots + \beta)(\mathbf{x}^{(0)})^{[m-1]}.$$

Then the $(k+1)$th iteration step indicates that

$$\mathscr{N}(\mathbf{x}^{(k-1)})^{m-1} + \mathbf{b} \le \mathscr{N}(\mathbf{x}^{(k)})^{m-1} + \mathbf{b} \le [\alpha(\alpha^k + \alpha^{k-1}\beta + \cdots + \beta) + \beta]\mathscr{M}(\mathbf{x}^{(0)})^{m-1},$$

which further implies that

$$(\mathbf{x}^{(k)})^{[m-1]} \le (\mathbf{x}^{(k+1)})^{[m-1]} \le (\alpha^{k+1} + \alpha^k\beta + \cdots + \beta)(\mathbf{x}^{(0)})^{[m-1]}.$$

Then, the sequence $\{\mathbf{x}^{(k)}\}$ is increasing and has an upper bound $\left(\frac{\beta}{1+\alpha}\right)^{1/(m-1)}\mathbf{x}^{(0)}$. This implies that the sequence converges to a positive vector $\mathbf{x}_*$, which is the positive solution of the nonsingular M-equation $\mathscr{A}\mathbf{x}^{m-1} = \mathbf{b}$.

We may apply an SOR-like acceleration technique [105] to such a splitting method for solving a nonsingular M-equation. We may take a positive number ω such that the splitting method

$$\mathbf{x}^{(k)} = (\mathscr{M} - \omega\mathscr{I})_{++}^{-1}[(\mathscr{N} - \omega\mathscr{I})(\mathbf{x}^{(k-1)})^{m-1} + \mathbf{b}], k = 1, 2, \ldots,$$

converges faster. The acceleration takes effect because of a smaller positive ω in the splitting method. When selecting the parameter ω, we need to note the following three restrictions: 1. ω is positive; 2. $\mathscr{M} - \omega\mathscr{I}$ is still a nonsingular M-tensor; 3. $(\mathscr{N} - \omega\mathscr{I})(\mathbf{x}^{(k-1)})^{m-1} + \mathbf{b} > \mathbf{0}$ for all $k = 1, 2, \ldots$. Within these three restrictions, we may find a parameter ω which accelerates the iteration effectively.

For a symmetric nonsingular M-tensor $\mathscr{A}$, we may consider using the Newton method to compute the positive solution $\mathbf{x}$ of the nonsingular M-equation $\mathscr{A}\mathbf{x}^{m-1} = \mathbf{b} > \mathbf{0}$. Define

$$\varphi(\mathbf{x}) := \frac{1}{m}\mathscr{A}\mathbf{x}^m - \mathbf{x}^\top\mathbf{b}.$$

We see that $\varphi(\mathbf{x})$ is convex on $\Omega = \{\mathbf{x} > \mathbf{0} : \mathscr{A}\mathbf{x}^{m-1} > \mathbf{0}\}$ and its gradient is

$$\nabla\varphi(\mathbf{x}) = \mathscr{A}\mathbf{x}^{m-1} - \mathbf{b} =: -\mathbf{r}.$$

Therefore, computing the positive solution of the above symmetric nonsingular M-equation is equivalent to solving the optimization problem

$$\min_{\mathbf{x}\in\Omega} \varphi(\mathbf{x}).$$

Then we may apply the Newton method to solve the optimization problem. Note that the Hessian of φ is

$$\nabla^2\varphi(\mathbf{x}) = (m-1)\mathscr{A}\mathbf{x}^{m-2}.$$

When $\mathscr{A}\mathbf{x}^{m-1} > \mathbf{0}$, matrix $\mathscr{A}\mathbf{x}^{m-2}$ is a symmetric Z-matrix and

$$\mathscr{A}\mathbf{x}^{m-2} \cdot \mathbf{x} = \mathscr{A}\mathbf{x}^{m-1} > \mathbf{0}.$$

Hence, $\mathscr{A}\mathbf{x}^{m-2}$ is a symmetric nonsingular M-matrix, which is a positive definite matrix (see Chap. 2 of [17]). Thus the Newton direction

$$\mathbf{p}_k = -[\nabla^2\varphi(\mathbf{x}^{(k)})]^{-1}\nabla\varphi(\mathbf{x}^{(k)}) = \frac{1}{m-1}(\mathscr{A}(\mathbf{x}^{(k)})^{m-2})^{-1}\mathbf{r}_k$$

is a descending direction. We have the following iterative scheme:

$$\begin{cases} M_k = \mathscr{A}(\mathbf{x}^{(k)})^{m-2}, \\ \mathbf{r}_k = \mathbf{b} - M_k\mathbf{x}^{(k)}, \\ \mathbf{p}_k = \dfrac{1}{m-1}M_k^{-1}\mathbf{r}_k, \\ \mathbf{x}^{(k+1)} = \mathbf{x}^{(k)} + \lambda_k\mathbf{p}_k, \end{cases} \quad k = 0, 1, 2, \ldots.$$

In general, the computational cost for the Newton method is expensive. It is relatively cheaper in the case of higher-order tensors. First, we do not need to pay additional effort to compute the matrix $M_k = \mathscr{A}(\mathbf{x}^{(k)})^{m-2}$, as it is a byproduct of the computation of $\mathscr{A}(\mathbf{x}^{(k)})^{m-1}$. Second, the computational complexity of solving a linear system is $O(n^3)$, which is no larger than $O(n^m)$, the computational complexity of computing a tensor-vector product $\mathscr{A}\mathbf{x}^{m-1}$, when $m \geq 3$.

We may set the initial vector $\mathbf{x}^{(0)}$ such that

$$\mathbf{x}^{(0)} > \mathbf{0} \text{ and } \mathscr{A}(\mathbf{x}^{(0)})^{m-1} > \mathbf{0}.$$

Then the restriction $\mathbf{x} > \mathbf{0}$ in the optimization problem can be automatically satisfied in the procedure. This can be seen by rewriting

$$\mathbf{x}^{(k+1)} = M_k^{-1}\left(\frac{m-2}{m-1}\mathscr{A}(\mathbf{x}^{(k)})^{m-1} + \frac{1}{m-1}\mathbf{b}\right) > \mathbf{0},$$

as M_k is a nonsingular M-matrix, $\mathscr{A}\mathbf{x}_k^{m-1}$ and $\mathbf{b}$ are both positive vectors. For asymmetric M-equations, we still may apply such an iteration. Then the method may not work well, since the matrix $\mathscr{A}\mathbf{x}^{m-2}$ may not be positive definite.

Now, we present a homotopy method for solving multilinear systems with asymmetric M-tensors. As discussed above and some results in Sect. 2.1, the Jacobi and Gauss–Seidel methods are raised to find the unique positive solution of the multilinear system with M-tensors. And the Newton method is also presented and it is shown that the Newton method is much faster than the other iterative methods [87]. However, it is unclear that whether or not Newton method is implementable when the corresponding tensor $\mathscr{A}$ is not symmetric.

Recently, Han [117] proposed a homotopy method for the system (2.2) with a nonsymmetric M-tensor $\mathscr{A}$. Based on the Euler-Newton prediction-correction approach for path tracking, it is proved that the homotopy method has a better performance than the Newton method [117].

For the sake of simplicity, let $P(\mathbf{x})$ be defined as

$$P(\mathbf{x}) = \mathscr{A}\mathbf{x}^{m-1} - \mathbf{b} = \mathbf{0}, \tag{2.6}$$

where $\mathscr{A} \in T_{m,n}$ is a nonsingular M-tensor and $\mathbf{b} > \mathbf{0}$. If we choose $\mathscr{A} = \mathscr{I}$, then the system (2.6) reduces to

$$Q(\mathbf{x}) = \mathscr{I}\mathbf{x}^{m-1} - \mathbf{b} = \mathbf{0}, \tag{2.7}$$

which has a unique positive solution $\mathbf{x} = \mathbf{b}^{\frac{1}{m-1}}$. Hence, we can construct the following homotopy

$$H(\mathbf{x}, t) = (1-t)Q(\mathbf{x}) + tP(\mathbf{x}) = (t\mathscr{A} + (1-t)\mathscr{I})\mathbf{x}^{m-1} - \mathbf{b} = \mathbf{0}, \quad t \in [0, 1].$$

Suppose $\mathscr{A} = (a_{i_1 i_2 \ldots i_m})$. It should be noted that the partial derivatives matrix $\nabla_{\mathbf{x}} H(\mathbf{x}, t)$ plays an important role in the homotopy algorithm. Hence, to compute $\nabla_{\mathbf{x}} H(\mathbf{x}, t)$, we first present a partially symmetrized tensor $\hat{\mathscr{A}} = (\hat{a}_{i_1 i_2 \ldots i_m}) \in T_{m,n}$ which is defined by

$$\hat{a}_{i_1 i_2 \ldots i_m} = \frac{1}{(m-1)!} \sum_{\pi} a_{i_1 \pi(i_2 \ldots i_m)},$$

where the sum is over all the permutations of $\pi(i_2 \ldots i_m)$. Then, we have the following conclusion.

Lemma 2.1 *If $\mathscr{A} \in T_{m,n}$ is a nonsingular M-tensor, then $\hat{\mathscr{A}}$ is a nonsingular M-tensor.*

Proof Suppose that $\mathscr{A} = s\mathscr{I} - \mathscr{B}$, where $\mathscr{B}$ is a nonnegative tensor. By the definition of $\hat{\mathscr{A}}$, it is easy to know that $\hat{\mathscr{A}} = s\mathscr{I} - \hat{\mathscr{B}}$ with $\hat{\mathscr{B}}$ being nonnegative, which implies that $\hat{\mathscr{A}}$ is a Z-tensor. To prove that $\hat{\mathscr{A}}$ is a nonsingular M-tensor, as discussed in last section, $\hat{\mathscr{A}}$ is a nonsingular M-tensor if and only if there is a positive vector $\mathbf{y} \in \mathbf{R}^n$ such that $\mathscr{A}\mathbf{y}^{m-1}$ is positive. By a direct computation, it follows that

$$\hat{\mathscr{A}}\mathbf{x}^{m-1} = \mathscr{A}\mathbf{x}^{m-1}, \quad \text{for all } \mathbf{x} \in \mathbf{R}^n.$$

Thus, it is not difficult to know the desired result holds since $\mathscr{A}$ is a nonsingular M-tensor. □

By the proof of Lemma 2.1, we obtain that

$$\nabla_{\mathbf{x}} \mathscr{A}\mathbf{x}^{m-1} = \nabla_{\mathbf{x}} \hat{\mathscr{A}}\mathbf{x}^{m-1} = (m-1)\hat{\mathscr{A}}\mathbf{x}^{m-2}.$$

Combining this with the homotopy $H(\mathbf{x}, t)$, we have the following partial derivatives:

$$\nabla_{\mathbf{x}} H(\mathbf{x}, t) = (m-1)(t\hat{\mathscr{A}} + (1-t)\mathscr{I})\mathbf{x}^{m-2},$$

and

$$\nabla_t H(\mathbf{x}, t) = (\mathscr{A} - \mathscr{I})\mathbf{x}^{m-1}.$$

Then the following results hold.

Theorem 2.8 *Let $\mathscr{A} \in T_{m,n}$ be a nonsingular M-tensor and suppose* **b** *is a positive vector. Then there is a scalar $\tau_0 > 0$ such that the system $H(\mathbf{x}, t) = 0$ has a unique positive solution $\mathbf{x}(t)$ for any $t \in [0, 1 + \tau_0)$. Furthermore, the matrix*

$$\nabla_{\mathbf{x}} H(\mathbf{x}(t), t) = (m-1)(t\hat{\mathscr{A}} + (1-t)\mathscr{I})\mathbf{x}(t)^{m-2}$$

is nonsingular.

Proof Since $\mathscr{A}$ is a nonsingular M-tensor, it can be written as $\mathscr{A} = s\mathscr{I} - \mathscr{B}$, where $s > \rho(\mathscr{B})$ and $\mathscr{B}$ is nonnegative. For any $t \in [0, 1]$, it is obvious that the tensor $t\mathscr{A} + (1-t)\mathscr{I}$ is also a nonsingular M-tensor. Let

$$\tau_0 = \begin{cases} \frac{s-\rho(\mathscr{B})}{\rho(\mathscr{B})-s+2}, & \text{if } \rho(\mathscr{B}) - s + 2 > 0, \\ 1, & \text{if } \rho(\mathscr{B}) - s + 2 \le 0. \end{cases}$$

Then it holds that

$$\tau_0 > 0, \quad \frac{st+1-t}{t} > \frac{s+\rho(\mathscr{B})}{2} > \rho(\mathscr{B}) \text{ for any } t \in [1, 1+\tau_0),$$

which implies that

$$t\mathscr{A} + (1-t)\mathscr{I} = t\left(\frac{st+1-t}{t}\mathscr{I} - \mathscr{B}\right), \quad t \in [1, 1+\tau_0),$$

is a nonsingular M-tensor. By Theorem 2.2, we have that

$$H(\mathbf{x}, t) = (1-t)Q(\mathbf{x}) + tP(\mathbf{x}) = (t\mathscr{A} + (1-t)\mathscr{I})\mathbf{x}^{m-1} - \mathbf{b} = \mathbf{0}$$

has a unique positive solution $\mathbf{x}(t)$ for each $t \in [0, 1+\tau_0)$.

On the other hand, from Lemma 2.1, we know that

$$(t\hat{\mathscr{A}} + (1-t)\mathscr{I})\mathbf{x}(t)^{m-2}$$

is a Z-matrix, and

$$(t\hat{\mathscr{A}} + (1-t)\mathscr{I})\mathbf{x}(t)^{m-2}\mathbf{x}(t) = (t\hat{\mathscr{A}} + (1-t)\mathscr{I})\mathbf{x}(t)^{m-1} = \mathbf{b}$$

is a positive vector. Then by Theorem 2.3 of Chap. 6 in [17], it follows that

$$\nabla_{\mathbf{x}} H(\mathbf{x}(t), t) = (m-1)(t\hat{\mathscr{A}} + (1-t)\mathscr{I})\mathbf{x}(t)^{m-2}$$

is nonsingular and the desired results hold. □

By the Implicit Function Theorem and a continuation argument in [169], the following result holds automatically.

Corollary 2.1 *Assume that* $\mathscr{A} \in T_{m,n}$ *is a nonsingular M-tensor and* $\mathbf{b} \in \mathbf{R}^n$ *is a positive vector. Then the positive solution* $\mathbf{x}(t)$ *of* $H(\mathbf{x}, t) = 0$ *for any* $t \in [0, 1 + \tau_0)$ *forms a smooth curve in* $\mathbf{R}^n_{++}$.

Now, we present the main result for the homotopy method, which shows that the homotopy method is implementable for the system (2.2).

Theorem 2.9 *Let* $\mathscr{A} \in T_{m,n}$ *be a nonsingular M-tensor and suppose* $\mathbf{b} \in \mathbf{R}^n$ *is a positive vector. Assume* $\mathbf{x}(t)$ *is the solution curve obtained by solving the system* $H(\mathbf{x}, t) = 0$ *in* $\mathbf{R}^n_{++} \times [0, 1]$. *When we choose the initial point* $\mathbf{x}(0) = \mathbf{b}^{[\frac{1}{m-1}]}$, *then* $\mathbf{x}(1)$ *is the unique positive solution of the system (2.2).*

Proof From Corollary 2.1, we know that $\mathbf{x}(t)$ forms a smooth curve in $\mathbf{R}^n_{++}$ for any $t \in [0, 1 + \tau_0)$. By a straightforward computation, it follows that

$$\nabla_{\mathbf{x}} H(\mathbf{x}(t), t) \cdot \frac{d\mathbf{x}}{dt} + \nabla_t H(\mathbf{x}(t), t) = \mathbf{0}.$$

By Theorem 2.8, the above differential equation system is well defined for $t \in [0, 1 + \tau_0)$ since $\nabla_{\mathbf{x}} H(\mathbf{x}(t), t)$ is nonsingular. Hence, we can follow the curve by solving the system with the initial point $\mathbf{x}(0) = \mathbf{b}^{[\frac{1}{m-1}]}$, and it is easy to know that $\mathbf{x}(1)$ is the unique positive solution of the system (2.2). □

Based on Theorem 2.9, we now present the homotopy method below.

Algorithm 2.3 Finding the unique positive solution of (2.2) where $\mathscr{A}$ is a nonsingular M-tensor and $\mathbf{b}$ is positive.

Initialization: Take initial point $\mathbf{x}(0) = \left(b_1^{\frac{1}{m-1}}, b_2^{\frac{1}{m-1}}, \ldots, b_n^{\frac{1}{m-1}}\right)^\top$.

Path following: Solving the following system

$$\nabla_{\mathbf{x}} H(\mathbf{x}(t), t) \cdot \frac{d\mathbf{x}}{dt} + \nabla_t H(\mathbf{x}(t), t) = \mathbf{0},$$

with initial point $\mathbf{x}(0) \in \mathbf{R}^n_{++}$, and $\mathbf{x}(1)$ is the desired solution for system (2.2).

2.3 Tensor Methods for Solving Symmetric M-Tensor Systems

Now, we study a new tensor method for solving the M-tensor equation system (2.2). The tensor method was first introduced by Schnabel and Frank in [238], which was applied to solve nonlinear equation systems. Combining this idea with the results for the rank-1 approximation of tensors, Xie et al. proposed a new tensor method for solving the symmetric M-tensor equation system [291].

It is well known that the M-tensor equation system (2.2) is one of the special case for the following nonlinear equation problem:

$$\text{Given } f : \mathbf{R}^n \to \mathbf{R}^n, \quad \text{find } \mathbf{x}^* \in \mathbf{R}^n \text{ such that } f(\mathbf{x}^*) = \mathbf{0}, \tag{2.8}$$

where $f(\mathbf{x})$ is supposed to be at least once continuously differentiable. The classical Newton method for solving problem (2.8) is upon a linear model which is

$$L(\mathbf{x}^{(k)} + \mathbf{d}) = f(\mathbf{x}^{(k)}) + \nabla f(\mathbf{x}^{(k)})\mathbf{d},$$

where $\mathbf{d} \in \mathbf{R}^n$, $\mathbf{x}^{(k)}$ is the current iterate and $\nabla f(\mathbf{x}^{(k)})$ denotes the Jacobian matrix of f at $\mathbf{x}^{(k)}$. Under condition that $\nabla f(\mathbf{x}^{(k)})$ is nonsingular, we can get $\mathbf{d} = -\nabla f(\mathbf{x}^{(k)})^{-1} f(\mathbf{x}^{(k)})$ and

$$\mathbf{x}^{(k+1)} = \mathbf{x}^{(k)} + \mathbf{d} = \mathbf{x}^{(k)} - \nabla f(\mathbf{x}^{(k)})^{-1} f(\mathbf{x}^{(k)}). \tag{2.9}$$

Combining this with the fact that $\nabla f(\mathbf{x}^{(k)})$ is Lipschitz continuous in a neighborhood of $\mathbf{x}^*$, it follows that the sequence of iterates generated by (2.9) is quadratically convergent to $\mathbf{x}^*$ locally.

To present the tensor method for system (2.2), we first recall the tensor method for nonlinear equation system (2.8) introduced in [238]. The biggest difference between the classical Newton method and the tensor method in [238] is that a second order term is added to the linear model such that

$$L(\mathbf{x}^{(k)} + \mathbf{d}) = f(\mathbf{x}^{(k)}) + \nabla f(\mathbf{x}^{(k)})\mathbf{d} + \frac{1}{2}\mathscr{T}_k\mathbf{d}^2, \tag{2.10}$$

where $\mathscr{T}_k \in \mathbf{R}^{n\times n\times n}$ is intended to supply second-order information about $f(\mathbf{x})$ around $\mathbf{x}^{(k)}$. Thus, one can obtain that the simplest way to choose $\mathscr{T}_k$ is $\nabla^2 f(\mathbf{x}^{(k)})$, which implies that (2.10) constructs of the first three terms of the Taylor expansion of $f(\mathbf{x})$ around $\mathbf{x}^{(k)}$. However, if $\mathscr{T}_k = \nabla^2 f(\mathbf{x}^{(k)})$ in each iterates $\mathbf{x}^{(k)}$, several drawbacks show that it is not practical for computing (details see [238]), which means that at each iteration one has to solve a system with n quadratic equations in n unknowns.

To avoid the disadvantages above, Schnabel and Frank [238] proposed a new way to choose tensor $\mathscr{T}_k$ such that

$$\begin{cases} \min_{\mathscr{T}_k \in \mathbf{R}^{n\times n\times n}} & \|\mathscr{T}_k\|_F \\ s.t. & \mathscr{T}_k\mathbf{s}_i^2 = \mathbf{z}_i, \quad 1 \le i \le p, \end{cases} \tag{2.11}$$

where p is a very small number for past iterates $\mathbf{x}^{(k-1)}, \mathbf{x}^{(k-2)}, \ldots, \mathbf{x}^{(k-p)}$ and $\mathbf{s}_i = \mathbf{x}^{(k-i)} - \mathbf{x}^{(k)}$, $\mathbf{z}_i = 2(f(\mathbf{x}^{(k-i)}) - f(\mathbf{x}^{(k)}) - \nabla f(\mathbf{x}^{(k)})\mathbf{s}_i)$. Based on (2.11), we have the following result.

Theorem 2.10 (Schnabel, Frank 1984) *Let* $p \in [n]$, $\mathbf{s}_i, \mathbf{z}_i \in \mathbf{R}^n, i \in [p]$. *Suppose* $M = (M_{ij}) \in \mathbf{R}^{p \times p}$ *with* $M_{ij} = (\mathbf{s}_i^\top \mathbf{s}_j)^2, i, j \in [p]$. *Let* $Z = (\mathbf{z}_1, \ldots, \mathbf{z}_p) \in \mathbf{R}^{n \times p}$. *Assume* $\mathbf{s}_1, \mathbf{s}_2, \ldots, \mathbf{s}_p$ *are linearly independent, then* M *is positive definite and the solution of (2.11) is*

$$\mathscr{T}_k = \sum_{i=1}^{p} \mathbf{a}_i \otimes \mathbf{s}_i \otimes \mathbf{s}_i,$$

where $\mathbf{a}_i$ *is the* ith *column of* $A = ZM^{-1} \in \mathbf{R}^{n \times p}$, *and* $\otimes$ *is the outer product as discussed in Chap. 1.*

By Theorem 2.10, the Eq. (2.10) can be rewritten as

$$L(\mathbf{x}^{(k)} + \mathbf{d}) = f(\mathbf{x}^{(k)}) + \nabla f(\mathbf{x}^{(k)})\mathbf{d} + \frac{1}{2}\sum_{i=1}^{p}(\mathbf{s}_i^\top \mathbf{d})^2 \mathbf{a}_i. \tag{2.12}$$

In the following analysis, we denote $f(\mathbf{x}) = \mathscr{A}\mathbf{x}^{m-1} - \mathbf{b} \in \mathbf{R}^n$ in (2.8) with $\mathscr{A} = (a_{i_1 \ldots i_m}) \in T_{m,n}$ being a symmetric M-tensor. Then, one has that

$$\nabla f(\mathbf{x}) = (m-1)\mathscr{A}\mathbf{x}^{m-2} \in \mathbf{R}^{n \times n}, \quad \nabla^2 f(\mathbf{x}) = (m-1)(m-2)\mathscr{A}\mathbf{x}^{m-3} \in \mathbf{R}^{n \times n \times n},$$

where $\mathscr{A}\mathbf{x}^{m-k} \in T_{k,n}$ is also a symmetric tensor with entries

$$(\mathscr{A}\mathbf{x}^{m-k})_{i_1 i_2 \ldots i_k} = \sum_{i_{k+1}, \ldots, i_m \in [n]} a_{i_1 i_2 \ldots i_m} x_{i_{k+1}} \cdots x_{i_m}.$$

Now, we introduce the tensor method for finding $\mathbf{d} \in \mathbf{R}^n$ such that

$$L(\mathbf{x}^{(k)} + \mathbf{d}) = f(\mathbf{x}^{(k)}) + \nabla f(\mathbf{x}^{(k)})\mathbf{d} + \frac{1}{2}\sum_{i=1}^{p}(\mathbf{s}_i^\top \mathbf{d})^2 \mathbf{a}_i = \mathbf{0},$$

which can be written as

$$f(\mathbf{x}^{(k)}) + \nabla f(\mathbf{x}^{(k)})\mathbf{d} + \frac{1}{2}A(S^\top \mathbf{d})^{[2]} = \mathbf{0}, \tag{2.13}$$

where A and $\mathbf{s}_i$ are as given in Theorem 2.10, and $S = (\mathbf{s}_1, \mathbf{s}_2, \ldots, \mathbf{s}_p)$.

Algorithm 2.4 (*Schnabel, Frank 1984*)

Step (0): Given $\mathbf{x}^{(k)}, \mathbf{x}^{(k-1)}, \ldots \mathbf{x}^{(k-p)}$, $\mathbf{b}$, and $\mathscr{A} \in T_{m,n}$ is a symmetric M-tensor. Evaluate $S, A, J = \nabla f(\mathbf{x}^{(k)}) = \mathscr{A}(\mathbf{x}^{(k)})^{m-1} - \mathbf{b}$.

Step (1): Find an orthogonal matrix $Q \in \mathbf{R}^{n\times n}$ such that $\widehat{S} = Q^\top S$, where

$$\widehat{S} = \begin{pmatrix} \mathbf{0} \\ \widehat{S}_2 \end{pmatrix} \in \mathbf{R}^{n\times n}, \quad \widehat{S}_2 = \begin{pmatrix} 0 & \cdots & 0 & 0 & * \\ 0 & \cdots & 0 & * & * \\ 0 & \cdots & * & * & * \\ \vdots & \cdot^{\cdot^{\cdot}} & \vdots & \vdots & \vdots \\ * & \cdots & * & * & * \end{pmatrix},$$

and $\widehat{S}_2 \in \mathbf{R}^{p\times p}$ is an anti-triangular matrix.

Step (2): Calculate $\widehat{J} = JQ = (\widehat{J}_1, \widehat{J}_2) \in \mathbf{R}^{n\times n}$, where $\widehat{J}_1 \in \mathbf{R}^{n\times(n-p)}$ and $\widehat{J}_2 \in \mathbf{R}^{n\times p}$ are the first $n-p$ and last p columns of $\widehat{J}$, respectively. Denote

$$\widehat{\mathbf{d}} = Q^\top \mathbf{d} = \begin{pmatrix} \widehat{\mathbf{d}}_1 \\ \widehat{\mathbf{d}}_2 \end{pmatrix} \in \mathbf{R}^n,$$

where $\widehat{\mathbf{d}}_1 \in \mathbf{R}^{n-p}$ and $\widehat{\mathbf{d}}_2 \in \mathbf{R}^p$ are the first $n-p$ and last p components of $\widehat{J}$, respectively.

Step (3): Find an orthogonal $\widehat{Q} \in \mathbf{R}^{n\times n}$ and a permutation matrix $P \in \mathbf{R}^{(n-p)\times(n-p)}$ such that

$$\widehat{Q}\widehat{J}_1 P = \begin{pmatrix} * & * & * & \cdots & * & \cdots & * \\ 0 & * & * & \cdots & * & \cdots & * \\ 0 & 0 & * & \cdots & * & \cdots & * \\ \vdots & \vdots & \vdots & \ddots & \vdots & \cdots & \vdots \\ 0 & 0 & 0 & \cdots & * & \cdots & * \\ 0 & 0 & 0 & \cdots & 0 & \cdots & 0 \\ \vdots & \vdots & \vdots & \cdots & \vdots & \cdots & \vdots \\ 0 & 0 & 0 & \cdots & 0 & \cdots & 0 \end{pmatrix} = \begin{pmatrix} \widetilde{J}_1 \\ \mathbf{0} \end{pmatrix} \in \mathbf{R}^{n\times(n-p)},$$

where the number of zero rows is $q \geq p$, and $\widetilde{J}_1 \in \mathbf{R}^{(n-q)\times(n-p)}$ is in echelon form with a nonzero diagonal. Define $\widetilde{\mathbf{d}}_1 = P^\top \widehat{\mathbf{d}}_1 \in \mathbf{R}^{n-p}$.

Step (4): Calculate

$$\widehat{Q}\widehat{J}_2 = \begin{pmatrix} \widetilde{J}_2 \\ \widetilde{J}_3 \end{pmatrix}, \quad \widetilde{A} = \widetilde{Q}A = \begin{pmatrix} \widetilde{A}_1 \\ \widetilde{A}_2 \end{pmatrix}, \text{ and } \widetilde{f} = \widehat{Q}f = \begin{pmatrix} \widetilde{f}_1 \\ \widetilde{f}_2 \end{pmatrix},$$

where $\widetilde{J}_2, \widetilde{A}_1 \in \mathbf{R}^{(n-q)\times p}$, $\widetilde{f}_1 \in \mathbf{R}^{n-q}$ and $\widetilde{J}_3, \widetilde{A}_2 \in \mathbf{R}^{q\times p}$, $\widetilde{f}_2 \in \mathbf{R}^q$.

Step (5): Find a $\widehat{\mathbf{d}}_2$ such that

$$\widetilde{f}_2 + \widetilde{J}_3 \widehat{\mathbf{d}}_2 + \frac{1}{2}\widetilde{A}_2(\widehat{S}_2^\top \widehat{\mathbf{d}}_2)^{[2]} = 0. \tag{2.14}$$

Furthermore, we can solve it in the least squares sense which is

$$\min_{\widehat{\mathbf{d}}_2 \in \mathbf{R}^p} \|\widetilde{f}_2 + \widetilde{J}_3 \widehat{\mathbf{d}}_2 + \frac{1}{2}\widetilde{A}_2(\widehat{S}_2^\top \widehat{\mathbf{d}}_2)^{[2]}\|_2,$$

since the above quadratic equations (2.14) may have no solution.

Step (6): Find a $\widetilde{\mathbf{d}}_1$ such that

$$\widetilde{J}_1 \widetilde{\mathbf{d}}_1 = -\widetilde{f}_1 - \widetilde{J}_2 \widehat{\mathbf{d}}_2 - \frac{1}{2}\widetilde{A}_1(\widehat{S}_2^\top \widehat{\mathbf{d}}_2)^{[2]}.$$

Step (7): Calculate $\widehat{\mathbf{d}}_1 = P\widetilde{\mathbf{d}}_1$, $\mathbf{d} = Q\widehat{\mathbf{d}}$.

For Algorithm 2.4, it should be noted that Steps (1)-(2) aim to change the system

$$f + J\mathbf{d} + \frac{1}{2}A(S^\top \mathbf{d})^{[2]} = \mathbf{0}$$

to another system

$$f + \widehat{J}_1 \widehat{\mathbf{d}}_1 + \widehat{J}_2 \widehat{\mathbf{d}}_2 + \frac{1}{2}A(\widehat{S}_2^\top \widehat{\mathbf{d}}_2)^{[2]} = \mathbf{0},$$

which can be transformed to another system in Steps (3)-(4) such that

$$\begin{pmatrix} \widetilde{f}_1 \\ \widetilde{f}_2 \end{pmatrix} + \begin{pmatrix} \widetilde{J}_1 & \widetilde{J}_2 \\ \mathbf{0} & \widetilde{J}_3 \end{pmatrix} \begin{pmatrix} \widetilde{\mathbf{d}}_1 \\ \widehat{\mathbf{d}}_2 \end{pmatrix} + \frac{1}{2} \begin{pmatrix} \widetilde{A}_1 \\ \widetilde{A}_2 \end{pmatrix} (\widehat{S}_2^\top \widehat{\mathbf{d}}_2)^{[2]} = \mathbf{0}.$$

Then the system above can also be transformed to the system with $n - q$ equations in n unknowns and the system with q equations in p unknowns such that

$$\widetilde{f}_1 + \widetilde{J}_1 \widetilde{\mathbf{d}}_1 + \widetilde{J}_2 \widehat{\mathbf{d}}_2 + \frac{1}{2}\widetilde{A}_1(\widehat{S}_2^\top \widehat{\mathbf{d}}_2)^{[2]} = \mathbf{0},$$

$$\widetilde{f}_2 + \widetilde{J}_3 \widehat{\mathbf{d}}_2 + \frac{1}{2}\widetilde{A}_2(\widehat{S}_2^\top \widehat{\mathbf{d}}_2)^{[2]} = \mathbf{0}.$$

After obtaining a solution $\mathbf{d}$ to the system (2.13), we give a framework for a full tensor method.

Algorithm 2.5 (*Schnabel, Frank 1984*) Given $\mathbf{x}^{(k)}, \mathbf{x}^{(k-1)}, \ldots, \mathbf{x}^{(k-p)}$, $\mathscr{A}$, $\mathbf{b}$ and tol.

Step 1: Evaluate $f_k = f(\mathbf{x}^{(k)})$ and decide whether to stop, if not, go to Step 2.

Step 2: Evaluate S_k, A_k and $J_k = \nabla f(\mathbf{x}^{(k)})$.

Step 3: Find a solution $\mathbf{d}_k$ to the tensor model (2.13) by Algorithm 2.4.

Step 4: Update $\mathbf{x}^{(k+1)} = \mathbf{x}^{(k)} + \mathbf{d}_k$, and go to Step 1.

Next, based on the rank-1 approximation of the proposed tensor $\mathscr{A}$, we introduce a new tensor method for the system (2.2):

$$\mathscr{A}\mathbf{x}^{m-1} - \mathbf{b} = \mathbf{0}.$$

Recall the results from [12, 61], for a symmetric tensor $\mathscr{A} \in T_{m,n}$, we know that

$$\mathscr{A} \approx \sum_{i=1}^{p} \lambda_i \mathbf{t}^{(i)} \otimes \mathbf{t}^{(i)} \otimes \cdots \otimes \mathbf{t}^{(i)},$$

where $\mathbf{t}^{(i)} \in \mathbf{R}^n$ are unit vectors i.e. $\|\mathbf{t}^{(i)}\|_2 = 1$. Hence, the second derivative of $\mathscr{A}\mathbf{x}^{m-1}$ is that

$$\begin{aligned}\nabla^2 f(\mathbf{x}) =&(m-1)(m-2)\mathscr{A}\mathbf{x}^{m-3}\\ \approx&(m-1)(m-2)(\sum_{i=1}^{p} \lambda_i \mathbf{t}^{(i)} \otimes \mathbf{t}^{(i)} \otimes \cdots \otimes \mathbf{t}^{(i)})\mathbf{x}^{m-3}\\ =&(m-1)(m-2)\left[\sum_{i=1}^{p} \lambda_i ((\mathbf{t}^{(i)})^\top \mathbf{x})^{m-3} (\mathbf{t}^{(i)} \otimes \mathbf{t}^{(i)} \otimes \mathbf{t}^{(i)})\right].\end{aligned}$$

Hence, the tensor model (2.10) can be written as

$$\begin{aligned}L(\mathbf{x}^{(k)} + \mathbf{d}) =& f(\mathbf{x}^{(k)}) + \nabla f(\mathbf{x}^{(k)})\mathbf{d}\\ &+ \frac{1}{2}(m-1)(m-2)\left[\sum_{i=1}^{p} \lambda_i ((\mathbf{t}^{(i)})^\top \mathbf{x})^{m-3} ((\mathbf{t}^{(i)})^\top \mathbf{d})^2 \mathbf{t}^{(i)}\right].\end{aligned} \tag{2.15}$$

Then, for solving the tensor system $\mathscr{A}\mathbf{x}^{m-1} - \mathbf{b}$ with $\mathscr{A}$ being a symmetric M-tensor, we have the following algorithm.

Algorithm 2.6 Given $\mathbf{x}^{(k)}, \mathbf{x}^{(k-1)}, \ldots, \mathbf{x}^{(k-p)}, \mathscr{A}, \mathbf{b}$ and tol. Compute λ_i and $\mathbf{t}^{(i)}$ such that

$$\mathscr{A} \approx \sum_{i=1}^{p} \lambda_i \mathbf{t}^{(i)} \otimes \mathbf{t}^{(i)} \otimes \cdots \otimes \mathbf{t}^{(i)}.$$

Step 1: Evaluate $f_k = f(\mathbf{x}^{(k)})$ and decide whether to stop, if not, go to Step 2.

Step 2: Compute $J_k = \nabla f(\mathbf{x}^{(k)})$.

Step 3: Find a solution $\mathbf{d}_k$ to the tensor model (2.15) according to Algorithm 2.4 by setting $S_k = (\mathbf{t}^{(1)}, \mathbf{t}^{(2)}, \ldots, \mathbf{t}^{(p)})$ and $A_k = (\mathbf{a}_1^{(k)}, \mathbf{a}_2^{(k)}, \ldots, \mathbf{a}_p^{(k)})$, where $\mathbf{a}_i^{(k)} = (m-1)(m-2)\lambda_i ((\mathbf{t}^{(i)})^\top \mathbf{x})^{m-3} \mathbf{t}^{(i)}$.

Step 4: Update $\mathbf{x}^{(k+1)} = \mathbf{x}^{(k)} + \mathbf{d}_k$, and go to Step 1.

To prove the convergence of the new tensor method, we first present several lemmas, which will be used in the following analysis.

Lemma 2.2 *For any* $\mathscr{A} = (a_{i_1 i_2 \ldots i_m}) \in T_{m,n}$ *and* $\mathbf{x} \in \mathbf{R}^n$, *it holds that*
(1) $\|\mathscr{A}\mathbf{x}\|_F \leq \|\mathscr{A}\|_F \|\mathbf{x}\|_2$;
(2) $\|\mathscr{A}\mathbf{x}^k\|_F \leq \|\mathscr{A}\|_F \|\mathbf{x}\|_2^k$, *where* $1 \leq k \leq m$.

Proof By the notion of Frobenius norm and Cauchy–Schwarz inequality, it follows that

$$\begin{aligned}\|\mathscr{A}\mathbf{x}\|_F^2 &= \sum_{i_1,\ldots,i_{m-1}=1}^{n} |\sum_{i_m=1}^{n} a_{i_1 i_2 \ldots i_m} x_{i_m}|^2 \\ &\leq \sum_{i_1,\ldots,i_{m-1}=1}^{n} \left(\sum_{i_m=1}^{n} |a_{i_1 i_2 \ldots i_m}|^2\right)\left(\sum_{i_m=1}^{n} |x_{i_m}|^2\right) \\ &= \|\mathscr{A}\|_F^2 \|\mathbf{x}\|_2^2.\end{aligned}$$

Hence, we have $\|\mathscr{A}\mathbf{x}\|_F \leq \|\mathscr{A}\|_F \|\mathbf{x}\|_2$.

On the other hand, since $\mathscr{A}\mathbf{x}^k = (\mathscr{A}\mathbf{x}^{k-1})\mathbf{x}$, it follows from (1) that

$$\begin{aligned}\|\mathscr{A}\mathbf{x}^k\|_F &= \|(\mathscr{A}\mathbf{x}^{k-1})\mathbf{x}\|_F \leq \|\mathscr{A}\mathbf{x}^{k-1}\|_F \|\mathbf{x}\|_2 \\ &= \|(\mathscr{A}\mathbf{x}^{k-2})\mathbf{x}\|_F \|\mathbf{x}\|_2 \leq \|\mathscr{A}\mathbf{x}^{k-2}\|_F \|\mathbf{x}\|_2^2 \\ &= \|(\mathscr{A}\mathbf{x}^{k-3})\mathbf{x}\|_F \|\mathbf{x}\|_2^2 \leq \cdots \leq \|\mathscr{A}\|_F \|\mathbf{x}\|_2^k,\end{aligned}$$

which implies the desired result holds. □

From Lemma 2.2, we have the following results.

Lemma 2.3 *Suppose* $\mathscr{A} \in T_{3,n}$ *and* $\mathbf{x}, \mathbf{y} \in \mathbf{R}^n$. *Then*

$$\|\mathscr{A}\mathbf{x}^2 - \mathscr{A}\mathbf{y}^2\|_2 \leq \|\mathscr{A}\|_F (\|\mathbf{x}\|_2 + \|\mathbf{y}\|_2) \|\mathbf{x} - \mathbf{y}\|_2.$$

Proof By Lemma 2.2, we have

$$\|\mathscr{A}\mathbf{x} - \mathscr{A}\mathbf{y}\|_F = \|\mathscr{A}(\mathbf{x} - \mathbf{y})\|_F \leq \|\mathscr{A}\|_F \|\mathbf{x} - \mathbf{y}\|_2.$$

Then we obtain that

$$\begin{aligned}\|\mathscr{A}\mathbf{x}^2 - \mathscr{A}\mathbf{y}^2\|_2 &= \|(\mathscr{A}\mathbf{x})\mathbf{x} - (\mathscr{A}\mathbf{y})\mathbf{x} + (\mathscr{A}\mathbf{y})\mathbf{x} - (\mathscr{A}\mathbf{y})\mathbf{y}\|_2 \\ &\leq \|(\mathscr{A}\mathbf{x} - \mathscr{A}\mathbf{y})\mathbf{x}\|_2 + \|\mathscr{A}\mathbf{y}(\mathbf{x} - \mathbf{y})\|_2 \\ &\leq \|\mathscr{A}\mathbf{x} - \mathscr{A}\mathbf{y}\|_F \|\mathbf{x}\|_2 + \|\mathscr{A}\mathbf{y}\|_F \|\mathbf{x} - \mathbf{y}\|_2 \\ &\leq \|\mathscr{A}\|_F \|\mathbf{x} - \mathbf{y}\|_2 \|\mathbf{x}\|_2 + \|\mathscr{A}\|_F \|\mathbf{x} - \mathbf{y}\|_2 \|\mathbf{y}\|_2 \\ &= \|\mathscr{A}\|_F (\|\mathbf{x}\|_2 + \|\mathbf{y}\|_2) \|\mathbf{x} - \mathbf{y}\|_2,\end{aligned}$$

and the desired result holds. □

Combining Lemma 2.2 with the fact that $f(\mathbf{x}) = \mathscr{A}\mathbf{x}^{m-1} - \mathbf{b}$ is continuous differentiable, we have the following results.

Lemma 2.4 (Xie, Jin, Wei 2017) *For the system $f(\mathbf{x}) = \mathscr{A}\mathbf{x}^{m-1} - \mathbf{b} = \mathbf{0}$, assume it has a solution $\mathbf{x}^*$ and $\nabla f(\mathbf{x}) \in \mathbf{R}^{n\times n}$ is nonsingular in a neighborhood Ω of $\mathbf{x}^*$. Then there exist scalars $\varepsilon > 0$, $K > 0$ and $M_k > 0$, $k \in [m-1]$ such that*

$$\|(\nabla f(\mathbf{x}))^{-1}\|_F \le K, \quad \|\nabla^k f(\mathbf{x})\|_F \le M_k$$

for all $\|\mathbf{x} - \mathbf{x}^*\|_2 \le \varepsilon$, where $\nabla^k f(\mathbf{x})$ is the kth derivative of $f(\mathbf{x})$.

Now we show the local convergence of the tensor method under reasonable conditions.

Theorem 2.11 *Assume $\mathbf{d}_k = \mathbf{x}^{(k+1)} - \mathbf{x}^{(k)}$ obtained from Algorithm 2.6 satisfies*

$$f(\mathbf{x}^{(k)}) + \nabla f(\mathbf{x}^{(k)})\mathbf{d}_k + \frac{1}{2}\nabla^2 f(\mathbf{x}^{(k)})(\mathbf{d}_k)^2 = \mathscr{E}_k(\mathbf{d}_k)^3, \tag{2.16}$$

where $\mathscr{E}_k$ is a fourth order tensor with $\|\mathscr{E}_k\|_F$ being bounded. Suppose the sequence $\{\mathbf{x}^{(k)}\}$ generated from Algorithm 2.6 converges to $\mathbf{x}^*$, which is the roof of the equation $f(\mathbf{x}) = \mathscr{A}\mathbf{x}^{m-1} - \mathbf{b} = \mathbf{0}$. Suppose $\nabla f(\mathbf{x}) \in \mathbf{R}^{n\times n}$ is nonsingular in a neighborhood Ω of $\mathbf{x}^*$. Then there is a constant C such that

$$\|\mathbf{x}^{(k+1)} - \mathbf{x}^*\|_2 \lesssim C\|\mathbf{x}^{(k)} - \mathbf{x}^*\|_2^3.$$

Proof By the Taylor expansion, we know that

$$\begin{aligned} f(\mathbf{x}^*) = \ & f(\mathbf{x}^{(k)}) + \nabla f(\mathbf{x}^{(k)})(\mathbf{x}^* - \mathbf{x}^{(k)}) + \frac{1}{2}\nabla^2 f(\mathbf{x}^{(k)})(\mathbf{x}^* - \mathbf{x}^{(k)})^2 \\ & + \frac{1}{6}\nabla^3 f(\xi^{(k)})(\mathbf{x}^* - \mathbf{x}^{(k)})^3. \end{aligned} \tag{2.17}$$

Since $f(\mathbf{x}^*) = \mathbf{0}$ and $\nabla f(\mathbf{x})$ is nonsingular, (2.17) becomes

$$\begin{aligned} -(\nabla f(\mathbf{x}^{(k)}))^{-1} f(\mathbf{x}^{(k)}) =& (\mathbf{x}^* - \mathbf{x}^{(k)}) + \frac{1}{2}(\nabla f(\mathbf{x}^{(k)}))^{-1}[\nabla^2 f(\mathbf{x}^{(k)})(\mathbf{x}^* - \mathbf{x}^{(k)})^2] \\ & + \frac{1}{6}(\nabla f(\mathbf{x}^{(k)}))^{-1}[\nabla^3 f(\xi^{(k)})(\mathbf{x}^* - \mathbf{x}^{(k)})^3]. \end{aligned}$$

By (2.16) it holds that

$$\begin{aligned} -(\nabla f(\mathbf{x}^{(k)}))^{-1} f(\mathbf{x}^{(k)}) =& (\mathbf{x}^{(k+1)} - \mathbf{x}^{(k)}) + \frac{1}{2}(\nabla f(\mathbf{x}^{(k)}))^{-1}[\nabla^2 f(\mathbf{x}^{(k)})(\mathbf{x}^{(k+1)} - \mathbf{x}^{(k)})^2] \\ & - (\nabla f(\mathbf{x}^{(k)}))^{-1}[\mathscr{E}_k(\mathbf{x}^{(k+1)} - \mathbf{x}^{(k)})^2]. \end{aligned}$$

Based on the above equation, we further have that

$$\begin{aligned}&(\mathbf{x}^* - \mathbf{x}^{(k)}) + \frac{1}{2}(\nabla f(\mathbf{x}^{(k)}))^{-1}[\nabla^2 f(\mathbf{x}^{(k)})(\mathbf{x}^* - \mathbf{x}^{(k)})^2]\\&+ \frac{1}{6}(\nabla f(\mathbf{x}^{(k)}))^{-1}[\nabla^3 f(\xi^{(k)})(\mathbf{x}^* - \mathbf{x}^{(k)})^3]\\&= x^{k+1} - \mathbf{x}^{(k)} + \frac{1}{2}(\nabla f(\mathbf{x}^{(k)}))^{-1}[\nabla^2 f(\mathbf{x}^{(k)})(\mathbf{x}^{(k+1)} - \mathbf{x}^{(k)})^2]\\&- (\nabla f(\mathbf{x}^{(k)}))^{-1}[\mathscr{E}_k(\mathbf{x}^{(k+1)} - \mathbf{x}^{(k)})^3],\end{aligned}$$

which implies that

$$\begin{aligned}\mathbf{x}^{(k+1)} - \mathbf{x}^* =& \frac{1}{2}(\nabla f(\mathbf{x}^{(k)}))^{-1}[\nabla^2 f(\mathbf{x}^{(k)})(\mathbf{x}^* - \mathbf{x}^{(k)})^2]\\&- \frac{1}{2}(\nabla f(\mathbf{x}^{(k)}))^{-1}[\nabla^2 f(\mathbf{x}^{(k)})(\mathbf{x}^{(k+1)} - \mathbf{x}^{(k)})^2]\\&+ \frac{1}{6}(\nabla f(\mathbf{x}^{(k)}))^{-1}[\nabla^3 f(\xi^{(k)})(\mathbf{x}^* - \mathbf{x}^{(k)})^3]\\&+ (\nabla f(\mathbf{x}^{(k)}))^{-1}[\mathscr{E}_k(\mathbf{x}^{(k+1)} - \mathbf{x}^{(k)})^3].\end{aligned}$$

Taking the Euclidean norm on both sides, we get

$$\begin{aligned}\|\mathbf{x}^{(k+1)} - \mathbf{x}^*\|_2 \le& \|\frac{1}{2}(\nabla f(\mathbf{x}^{(k)}))^{-1}[\nabla^2 f(\mathbf{x}^{(k)})(\mathbf{x}^* - \mathbf{x}^{(k)})^2]\\&- \frac{1}{2}(\nabla f(\mathbf{x}^{(k)}))^{-1}[\nabla^2 f(\mathbf{x}^{(k)})(\mathbf{x}^{(k+1)} - \mathbf{x}^{(k)})^2]\|_2\\&+ \|\frac{1}{6}(\nabla f(\mathbf{x}^{(k)}))^{-1}[\nabla^3 f(\xi^{(k)})(\mathbf{x}^* - \mathbf{x}^{(k)})^3]\|_2\\&+ \|(\nabla f(\mathbf{x}^{(k)}))^{-1}[\mathscr{E}_k(\mathbf{x}^{(k+1)} - \mathbf{x}^{(k)})^3]\|_2.\end{aligned} \tag{2.18}$$

On the other hand, recall the notion of mode-1 tensor-matrix product in Chap. 1, then we have

$$(\nabla f(\mathbf{x}^{(k)}))^{-1}[\nabla^2 f(\mathbf{x}^{(k)})(\mathbf{x}^* - \mathbf{x}^{(k)})^2] = [\nabla^2 f(\mathbf{x}^{(k)}) \times_1 (\nabla f(\mathbf{x}^{(k)}))^{-1}](\mathbf{x}^* - \mathbf{x}^{(k)})^2.$$

By Lemma 2.3, one obtains that

$$\begin{aligned}&\|(\nabla f(\mathbf{x}^{(k)}))^{-1}[\nabla^2 f(\mathbf{x}^{(k)})(\mathbf{x}^* - \mathbf{x}^{(k)})^2] - \frac{1}{2}(\nabla f(\mathbf{x}^{(k)}))^{-1}[\nabla^2 f(\mathbf{x}^{(k)})(\mathbf{x}^{(k+1)} - \mathbf{x}^{(k)})^2]\|_2\\&= \frac{1}{2}\|[\nabla^2 f(\mathbf{x}^{(k)}) \times_1 (\nabla f(\mathbf{x}^{(k)}))^{-1}](\mathbf{x}^* - \mathbf{x}^{(k)})^2 - [\nabla^2 f(\mathbf{x}^{(k)}) \times_1 (\nabla f(\mathbf{x}^{(k)}))^{-1}](\mathbf{x}^{(k+1)} - \mathbf{x}^{(k)})^2\|_2\\&\le \frac{1}{2}\|\nabla^2 f(\mathbf{x}^{(k)}) \times_1 (\nabla f(\mathbf{x}^{(k)}))^{-1}\|_F(\|\mathbf{x}^* - \mathbf{x}^{(k)}\|_2 + \|\mathbf{x}^{(k+1)} - \mathbf{x}^{(k)}\|_2)\|\mathbf{x}^{(k+1)} - \mathbf{x}^*\|_2.\end{aligned}$$

Therefore, from (2.18), it holds that

$$\begin{aligned}
\|\mathbf{x}^{(k+1)}-\mathbf{x}^*\|_2 \leq & \frac{1}{2}\|\nabla^2 f(\mathbf{x}^{(k)})\times_1(\nabla f(\mathbf{x}^{(k)}))^{-1}\|_F(\|\mathbf{x}^*-\mathbf{x}^{(k)}\|_2+\|\mathbf{x}^{(k+1)}-\mathbf{x}^{(k)}\|_2)\|\mathbf{x}^{(k+1)}-\mathbf{x}^*\|_2 \\
&+\frac{1}{6}\|\nabla^3 f(\xi^{(k)})\times_1(\nabla f(\mathbf{x}^{(k)}))^{-1}\|_F\|\mathbf{x}^{(k)}-\mathbf{x}^*\|_2^3 \\
&+\|\mathscr{E}_{k\times 1}(\nabla f(\mathbf{x}^{(k)}))^{-1}\|_F\|\mathbf{x}^{(k+1)}-\mathbf{x}^{(k)}\|_2^3 \\
\leq & \frac{1}{2}\|\nabla^2 f(\mathbf{x}^{(k)})\|_F\|(\nabla f(\mathbf{x}^{(k)}))^{-1}\|_F(\|\mathbf{x}^*-\mathbf{x}^{(k)}\|_2+\|\mathbf{x}^{(k+1)}-\mathbf{x}^{(k)}\|_2)\|\mathbf{x}^{(k+1)}-\mathbf{x}^*\|_2 \\
&+\frac{1}{6}\|\nabla^3 f(\xi^{(k)})\|_F\|(\nabla f(\mathbf{x}^{(k)}))^{-1}\|_F\|\mathbf{x}^{(k)}-\mathbf{x}^*\|_2^3 \\
&+\|\mathscr{E}_k\|_F\|(\nabla f(\mathbf{x}^{(k)}))^{-1}\|_F\|\mathbf{x}^{(k+1)}-\mathbf{x}^{(k)}\|_2^3.
\end{aligned} \tag{2.19}$$

It is obvious that $\|\mathbf{x}^{(k)}-\mathbf{x}^*\|_2\to 0$, $\|\mathbf{x}^{(k+1)}-\mathbf{x}^{(k)}\|_2\to 0$ as $k\to\infty$ since $\mathbf{x}^{(k)}\to\mathbf{x}^*$. And we further have

$$\frac{1}{2}\|\nabla^2 f(\mathbf{x}^{(k)})\|_F\|(\nabla f(\mathbf{x}^{(k)}))^{-1}\|_F(\|\mathbf{x}^*-\mathbf{x}^{(k)}\|_2+\|\mathbf{x}^{(k+1)}-\mathbf{x}^{(k)}\|_2)\leq\frac{1}{2}.$$

Then we get that

$$\begin{aligned}
\|\mathbf{x}^{(k+1)}-\mathbf{x}^{(k)}\|_2^3 =& \|\mathbf{x}^{(k+1)}-\mathbf{x}^*+\mathbf{x}^*-\mathbf{x}^{(k)}\|_2^2 \\
\leq & (\|\mathbf{x}^{(k+1)}-\mathbf{x}^*\|_2+\|\mathbf{x}^{(k)}-\mathbf{x}^*\|_2)^3 \\
\approx & 8\|\mathbf{x}^{(k)}-\mathbf{x}^*\|_2^3.
\end{aligned}$$

By (2.19), one has that

$$\|\mathbf{x}^{(k+1)}-\mathbf{x}^*\|_2 \lesssim \left(\frac{1}{3}\|\nabla^3 f(\xi^{(k)})\|_F+16\|\mathscr{E}_k\|_F\right)\|(\nabla f(\mathbf{x}^{(k)}))^{-1}\|_F\|\mathbf{x}^{(k)}-\mathbf{x}^*\|_2^3.$$

Let $k\to\infty$ in the right side of the equation above, it follows that

$$\left(\frac{1}{3}\|\nabla^3 f(\mathbf{x}^*)\|_F+16\|\mathscr{E}_k\|_F\right)\|(\nabla f(\mathbf{x}^*))^{-1}\|_F. \tag{2.20}$$

By Lemma 2.4 and the fact that $\|\mathscr{E}_k\|_F$ is bounded, we can always take a constant C which is larger than the value in (2.20). Then we know that $\|\mathbf{x}^{(k+1)}-\mathbf{x}^*\|_2\lesssim C\|\mathbf{x}^{(k)}-\mathbf{x}^*\|_2^3$ and the desired results hold. □

It was noted by Xie et al. in [291] that the tensor method has a faster local convergence rate than the classical Newton method.

2.4 Solution Methods for General Multilinear Systems

In previous sections, we studied M-tensor equation systems with symmetric M-tensors and nonsymmetric M-tensors. In this section, we study solution methods for a general multilinear system

$$\mathscr{A}\mathbf{x}^{m-1} = \mathbf{b}. \tag{2.21}$$

The methods studied in this section include Jacobi, the Gauss–Seidel and the successive over-relaxation (SOR) methods. The content of this section is based upon Li et al. in [168].

To move on, we need the following concepts. The diagonal tensors and triangular tensors defined here have a different version with previous sections.

Definition 2.1 A tensor $\mathscr{D} = (d_{i_1 i_2 \dots i_m}) \in T_{m,n}$ is called a **diagonal face tensor** if its entries satisfy

$$d_{i_1 i_2 \dots i_m} = 0, \text{ if } i_2 \neq i_1.$$

A tensor $\mathscr{L} = (l_{i_1 i_2 \dots i_m}) \in T_{m,n}$ is called a **lower half tensor** if its entries satisfy

$$l_{i_1 i_2 \dots i_m} = 0, \text{ if } i_2 > i_1,$$

It is called a **strictly lower half tensor** if

$$l_{i_1 i_2 \dots i_m} = 0, \text{ if } i_2 \geq i_1.$$

We may define **upper half tensors** and **strictly upper half tensors** similarly.

Suppose that $\mathscr{D}$, $\mathscr{L}$ and $\mathscr{L}'$ are the diagonal face, lower half and strictly lower half parts of a tensor $\mathscr{A}$, respectively. Then we have the following observations:
(1) If $\mathscr{A}$ is symmetric, then its diagonal face part $\mathscr{D}$ is also symmetric;
(2) The numbers of entries of a (strictly) lower half tensor and an (strictly) upper half tensor are equal;
(3) For any $\mathbf{x} \in \mathbf{R}^n$, $\mathscr{D}\mathbf{x}^{m-2}$, $\mathscr{L}\mathbf{x}^{m-2}$ and $\mathscr{L}'\mathbf{x}^{m-2}$ are the diagonal, lower and strictly lower triangle parts of $\mathscr{A}\mathbf{x}^{m-2}$, respectively. In particular, for any $\mathbf{x} \in \mathbf{R}^n$, we have

$$\mathscr{D}\mathbf{x}^{m-2} = \mathrm{diag}(\mathscr{A}\mathbf{x}^{m-2}),$$

$$\mathscr{L}\mathbf{x}^{m-2} = \begin{pmatrix} \mathscr{A}_{11}\mathbf{x}^{m-2} & 0 & \cdots & 0 \\ \mathscr{A}_{21}\mathbf{x}^{m-2} & \mathscr{A}_{22}\mathbf{x}^{m-2} & \cdots & 0 \\ \vdots & \vdots & \ddots & 0 \\ \mathscr{A}_{n1}\mathbf{x}^{m-2} & \mathscr{A}_{n2}\mathbf{x}^{m-2} & \cdots & \mathscr{A}_{nn}\mathbf{x}^{m-2} \end{pmatrix}$$

and

$$\mathscr{L}'\mathbf{x}^{m-2} = \begin{pmatrix} 0 & 0 & \cdots & 0 \\ \mathscr{A}_{21}\mathbf{x}^{m-2} & 0 & \cdots & 0 \\ \vdots & \vdots & \ddots & 0 \\ \mathscr{A}_{n1}\mathbf{x}^{m-2} & \mathscr{A}_{n2}\mathbf{x}^{m-2} & \cdots & 0 \end{pmatrix},$$

where $\mathscr{A}_{ij} = (a_{iji_3i_4\ldots i_m})_{i_3,\ldots,i_m=1}^n$, $i, j = 1, 2, \ldots, n$.

For any nonsingular matrix $E \in \mathbf{R}^{n\times n}$ and any nonzero scalar α, we see that $\mathbf{x}$ is a solution of the multilinear system $\mathscr{A}\mathbf{x}^{m-1} - \mathbf{b} = 0$ if and only if it solves the following fixed point problem

$$\mathbf{x} = \left(I + \alpha E^{-1}\mathscr{A}\mathbf{x}^{m-2}\right)\mathbf{x} - \alpha E^{-1}\mathbf{b}.$$

According to this, we set the following iterative scheme:

$$\mathbf{x}^{(k+1)} = \left(I + \alpha_k E_k^{-1}\mathscr{A}\left(\mathbf{x}^{(k)}\right)^{m-2}\right)\mathbf{x}^{(k)} - \alpha_k E_k^{-1}\mathbf{b}. \tag{2.22}$$

When $m = 2$, the iterative scheme (2.22) is an iterative method for solving the system of linear equations, with a fixed iterative matrix. If $m \geq 3$, the iterative matrix $I + \alpha_k E_k^{-1}\mathscr{A}(\mathbf{x}^{(k)})^{m-2}$ is nonlinear with respect to $\mathbf{x}^{(k)}$. However, we may still prove the following convergence theorem in a way similar to the case $m = 2$, such as in [269].

Theorem 2.12 *Suppose that the sequence $\{\mathbf{x}^{(k)}\}$ is generated by the iterative scheme (2.22). If there are constants $\rho \in (0, 1)$ and $M > 0$ such that*

$$\sigma\left(I + \alpha_k E_k^{-1}\mathscr{A}\left(\mathbf{x}^{(k)}\right)^{m-2}\right) \leq \rho, \quad |\alpha_k|\|E_k^{-1}\mathbf{b}\| \leq M,$$

then the iterative sequence $\{\mathbf{x}^{(k)}\}$ converges to a solution of the multilinear system (2.21), where $\sigma(A)$ denotes the spectral radius of matrix A.

The choice of matrix E_k in the iterative scheme (2.22) plays a central role in this iterative method for solving multilinear system (2.21). For solving a system of linear equations, typical choices of matrix E_k include the diagonal part, the lower triangle part of the coefficient matrix, or some appropriate lower triangle matrix, corresponding to the Jacobi, the Gauss–Seidel or the successive over-relaxation (SOR) method. This idea was recently extended to solve multilinear system (2.21) [168].

Let $\mathscr{A} \in S_{m,n}$. Then the multilinear system (2.21) is the first order necessary condition of the minimization problem

$$\min f(\mathbf{x}) = \frac{1}{m}\mathscr{A}\mathbf{x}^m - \mathbf{b}^\top\mathbf{x}. \tag{2.23}$$

Split $\mathcal{A}$ to $\mathcal{A} = \mathcal{M} - \mathcal{U}$. We have

$$f(\mathbf{x}) = \frac{1}{m}\mathcal{M}\mathbf{x}^m - \frac{1}{m}\mathcal{U}\mathbf{x}^m - \mathbf{b}^\top\mathbf{x}.$$

Li, Xie and Xu [168] approximate the first term of $f(\mathbf{x})$ by a quadratic function and the second term by a linear function. Then there is the following approximate problem

$$\begin{aligned}
\min f\left(\mathbf{x}^{(k)}\right) =& \frac{1}{2}(m-1)\left(\mathbf{x}-\mathbf{x}^{(k)}\right)^\top \mathcal{M}\left(\mathbf{x}^{(k)}\right)^{m-2}\left(\mathbf{x}-\mathbf{x}^{(k)}\right) \\
&+ \left(\mathbf{x}-\mathbf{x}^{(k)}\right)^\top\left(\mathcal{M}\left(\mathbf{x}^{(k)}\right)^{m-1} - \mathcal{U}(\mathbf{x}^{(k)})^{m-1} - \mathbf{b}\right) \\
=& \frac{1}{2}(m-1)\left(\mathbf{x}-\mathbf{x}^{(k)}\right)^\top \mathcal{M}\left(\mathbf{x}^{(k)}\right)^{m-2}\left(\mathbf{x}-\mathbf{x}^{(k)}\right) \\
&+ \left(\mathbf{x}-\mathbf{x}^{(k)}\right)^\top\left(\mathcal{A}\left(\mathbf{x}^{(k)}\right)^{m-1} - \mathbf{b}\right).
\end{aligned}$$

Denote the solution of this approximate problem by $\mathbf{x}^{(k+1)}$. We have

$$\begin{aligned}
\mathbf{x}^{(k+1)} =& \mathbf{x}^{(k)} - \frac{1}{m-1}\left(\mathcal{M}\left(\mathbf{x}^{(k)}\right)^{m-2}\right)^{-1}\left(\mathcal{A}\left(\mathbf{x}^{(k)}\right)^{m-1} - \mathbf{b}\right) \\
=& \left(I - \frac{1}{m-1}\left(\mathcal{M}\left(\mathbf{x}^{(k)}\right)^{m-2}\right)^{-1}\mathcal{A}\left(\mathbf{x}^{(k)}\right)^{m-1}\right)\mathbf{x}^{(k)} + \frac{1}{m-1}\left(\mathcal{M}\left(\mathbf{x}^{(k)}\right)^{m-2}\right)^{-1}\mathbf{b} \\
=& H_k\mathbf{x}^{(k)} + \mathbf{b}_k,
\end{aligned} \tag{2.24}$$

where

$$H_k = I - \frac{1}{m-1}\left(\mathcal{M}\left(\mathbf{x}^{(k)}\right)^{m-2}\right)^{-1}\mathcal{A}\left(\mathbf{x}^{(k)}\right)^{m-1},\quad \mathbf{b}_k = \frac{1}{m-1}\left(\mathcal{M}\left(\mathbf{x}^{(k)}\right)^{m-2}\right)^{-1}\mathbf{b}.$$

By Theorem 2.12, a sufficient condition for the convergence of the iterative scheme is that there are constants $\rho \in (0, 1)$ and $C > 0$ such that

$$\sigma(H_k) \le \rho, \text{ and } \|\mathbf{b}_k\| \le C.$$

In (2.24), setting $\mathcal{M} = \mathcal{D}$, the diagonal face part of $\mathcal{A}$, we have the **Jacobi method**, which is an extension of the Jacobi method for solving the system of linear equations. Similarly, setting $\mathcal{M} = \mathcal{L}$, the lower half part of $\mathcal{A}$, we have the **Gauss–Seidel method**. And setting $\mathcal{M} = \frac{1}{\omega}(\mathcal{D} + \omega\mathcal{L}')$, we have the **SOR method**, where ω is a relaxation parameter. We may write the Jacobi, Gauss–Seidel, and SOR methods as follows:

- The Jacobi method

$$\mathbf{x}^{(k+1)} = \mathbf{x}^{(k)} - \frac{1}{m-1}\left(\mathscr{D}\left(\mathbf{x}^{(k)}\right)^{m-2}\right)^{-1}\left(\mathscr{A}\left(\mathbf{x}^{(k)}\right)^{m-1} - \mathbf{b}\right),\quad k = 0, 1, \ldots.$$

- The Gauss–Seidel method

$$\mathbf{x}^{(k+1)} = \mathbf{x}^{(k)} - \frac{1}{m-1}\left(\mathscr{L}\left(\mathbf{x}^{(k)}\right)^{m-2}\right)^{-1}\left(\mathscr{A}\left(\mathbf{x}^{(k)}\right)^{m-1} - \mathbf{b}\right),\quad k = 0, 1, \ldots.$$

- The SOR method

$$\mathbf{x}^{(k+1)} = \mathbf{x}^{(k)} - \frac{\omega}{m-1}\left(\left(\mathscr{D} + \omega\mathscr{L}'\right)\left(\mathbf{x}^{(k)}\right)^{m-2}\right)^{-1}\left(\mathscr{A}\left(\mathbf{x}^{(k)}\right)^{m-1} - \mathbf{b}\right),\quad k = 0, 1, \ldots.$$

We may also write the above iterative methods in the form (2.22) corresponding to $\alpha_k = \frac{-1}{m-1}$,

$$E_k = \mathscr{D}\left(\mathbf{x}^{(k)}\right)^{m-2},\ \mathscr{L}\left(\mathbf{x}^{(k)}\right)^{m-2} \text{ and } \frac{1}{\omega}(\mathscr{D} + \omega\mathscr{L}')\left(\mathbf{x}^{(k)}\right)^{m-2},$$

respectively. Then the Gauss–Seidel method is an SOR method that corresponds to $\omega = 1$.

Theorem 2.13 *Let $\mathscr{A} \in S_{m,n}$ and $\mathbf{x}^*$ be a solution of the tensor equation (2.21) such that $\mathscr{A}(\mathbf{x}^*)^{m-2}$ is positive definite. Then the following statements hold.*
(1) There is a neighborhood $U(\mathbf{x}^*)$ of $\mathbf{x}^*$ such that, as long as $\mathbf{x}_0 \in U(\mathbf{x}^*)$, the sequence $\{\mathbf{x}^{(k)}\}$ generated by the Gauss–Seidel method converges to $\mathbf{x}^*$.
(2) If the matrix $2\mathscr{D}(\mathbf{x}^*)^{m-2} - \mathscr{A}(\mathbf{x}^*)^{m-2}$ is positive definite, then there is a neighborhood $U(\mathbf{x}^*)$ of $\mathbf{x}^*$ such that, as long as $\mathbf{x}_0 \in U(\mathbf{x}^*)$, the sequence $\{\mathbf{x}^{(k)}\}$ generated by the Jacobi method converges to $\mathbf{x}^*$.
(3) If $0 < \omega < 2$, then there is a neighborhood $U(\mathbf{x}^*)$ of $\mathbf{x}^*$ such that, as long as $\mathbf{x}_0 \in U(\mathbf{x}^*)$, the sequence $\{\mathbf{x}^{(k)}\}$ generated by the SOR method converges to $\mathbf{x}^*$.
In either case, the convergence rate of $\{\mathbf{x}^{(k)}\}$ is at least r-linear.

Proof We only prove the convergence of the SOR method. As discussed above, the Gauss–Seidel method is a special case of the SOR method when $\omega = 1$. The convergence of the Jacobi method can be proved similarly. Let

$$L_k(\omega) = (\mathscr{D} + \omega\mathscr{L}')\left(\mathbf{x}^{(k)}\right)^{m-2} = (1-\omega)\mathscr{D}\left(\mathbf{x}^{(k)}\right)^{m-2} + \omega\mathscr{L}\left(\mathbf{x}^{(k)}\right)^{m-2}$$

and

$$\mathbf{d}_k = -\frac{\omega}{m-1}L_k(\omega)^{-1}\left(\mathscr{A}\left(\mathbf{x}^{(k)}\right)^{m-1} - \mathbf{b}\right) = -\frac{\omega}{m-1}L_k(\omega)^{-1}\nabla f\left(\mathbf{x}^{(k)}\right).$$

We may write the SOR method as $\mathbf{x}^{(k+1)} = \mathbf{x}^{(k)} + \mathbf{d}_k$. By the definition of $\mathscr{D}$ and $\mathscr{L}$, we have the following equality,

$$\mathbf{p}^\top L_k(\omega)\mathbf{p} = \frac{1}{2}\omega\mathbf{p}^\top\left(\mathscr{A}\left(\mathbf{x}^{(k)}\right)^{m-2}\right)\mathbf{p} + \left(1-\frac{1}{2}\omega\right)\mathbf{p}^\top\left(\mathscr{D}\left(\mathbf{x}^{(k)}\right)^{m-2}\right)\mathbf{p} > 0,\ \forall\,\mathbf{p}\in\mathbf{R}^n. \tag{2.25}$$

Since $f(\mathbf{x}) = \frac{1}{m}\mathscr{A}\mathbf{x}^m - \mathbf{b}^\top\mathbf{x}$, by Taylor's expansion, we have

$$\begin{aligned}
f(\mathbf{x}^{(k+1)}) =& f(\mathbf{x}^{(k)}) + \mathbf{d}_k^\top\nabla f(\mathbf{x}^{(k)}) + \frac{1}{2}(m-1)\mathbf{d}_k^\top\nabla^2 f(\widetilde{\mathbf{x}}^{(k)})\mathbf{d}_k\\
=& f(\mathbf{x}^{(k)}) - \frac{1}{\omega}(m-1)\mathbf{d}_k^\top L_k(\omega)\mathbf{d}_k + \frac{1}{2}(m-1)\mathbf{d}_k^\top(\mathscr{A}(\widetilde{\mathbf{x}}^{(k)})^{m-2})\mathbf{d}_k\\
=& f(\mathbf{x}^{(k)}) - (m-1)\left(\frac{1}{\omega}\mathbf{d}_k^\top L_k(\omega)\mathbf{d}_k - \frac{1}{2}\mathbf{d}_k^\top(\mathscr{A}(\mathbf{x}^{(k)})^{m-2})\mathbf{d}_k\right) + \frac{m-1}{2}\mathbf{d}_k^\top\varepsilon_k\mathbf{d}_k\\
=& f(\mathbf{x}^{(k)}) - \frac{1}{\omega}\left(1-\frac{1}{2}\omega\right)(m-1)\mathbf{d}_k^\top(\mathscr{D}(\mathbf{x}^{(k)})^{m-2})\mathbf{d}_k + \frac{m-1}{2}\mathbf{d}_k^\top\varepsilon_k\mathbf{d}_k,
\end{aligned}$$

where

$$\widetilde{\mathbf{x}}^{(k)} = \mathbf{x}^{(k)} + \theta_k(\mathbf{x}^{(k+1)} - \mathbf{x}^{(k)})$$

with $\theta_k \in (0,1)$ and $\varepsilon_k = \mathscr{A}(\widetilde{\mathbf{x}}^{(k)})^{m-2} - \mathscr{A}(\mathbf{x}^{(k)})^{m-2}$ tends to zero if $\mathbf{x}^{(k)}$ converges. We may show by induction that

$$\{\mathbf{d}_k^\top\mathscr{D}(\mathbf{x}^{(k)})^{m-2}\mathbf{d}_k\} \to 0, \quad \text{and } \{\mathbf{d}_k\} \to 0,$$

or equivalently

$$\lim_{k\to\infty}(\mathscr{L}(\mathbf{x}^{(k)})^{m-2})^{-1}(\mathscr{A}(\mathbf{x}^{(k)})^{m-1} - \mathbf{b}) = 0.$$

This, combined with the fact that $\mathscr{L}(\mathbf{x}^*)^{m-2}$ is positive definite, implies $\mathscr{A}(\mathbf{x}^{(k)})^{m-1} - \mathbf{b} \to 0$.

Since $\mathscr{D}(\mathbf{x}^*)^{m-2}$ is positive definite, the convergence rate of $\{\mathbf{x}^{(k)}\}$ is at least r-linear. The theorem is proved. □

In the remainder of this section, we will globalize the method. We only globalize the Gauss–Seidel method. The Jacobi and SOR methods can be globalized similarly. We assume that for any $\mathbf{x} \neq \mathbf{0}$ the matrix $\mathscr{A}\mathbf{x}^{m-2}$ is positive semidefinite with positive diagonals.

We rewrite the Gauss–Seidel method as

$$\mathbf{x}^{(k+1)} = \mathbf{x}^{(k)} + \mathbf{d}_k,$$

where $\mathbf{d}_k$ is the solution of the system of linear equations

$$(m-1)L_k\mathbf{d} + \nabla f(\mathbf{x}^{(k)}) = \mathbf{0}, \tag{2.26}$$

f is defined by (2.23) and $L_k = \mathscr{L}(\mathbf{x}^{(k)})^{m-2}$. Since $\mathscr{A}(\mathbf{x}^{(k)})^{m-2}$ is positive semidefinite with positive diagonals, by (2.25) with $\omega = 1$, we have

$$\mathbf{d}_k^\top L_k \mathbf{d}_k \geq \frac{1}{2}\mathbf{d}_k^\top (\mathscr{D}(\mathbf{x}^{(k)})^{m-2}\mathbf{d}_k > 0, \tag{2.27}$$

which further implies

$$\nabla f(\mathbf{x}^{(k)})^\top \mathbf{d}_k = -(m-1)\mathbf{d}_k^\top L_k \mathbf{d}_k < 0.$$

This indicates that the Gauss–Seidel direction is a descent direction of the objective function. Then we may globalize the method by using a line search process. This results in the following damped Gauss–Seidel method,

$$\mathbf{x}^{(k+1)} = \mathbf{x}^{(k)} + \alpha_k \mathbf{d}_k,$$

where $\mathbf{d}_k$ is the Gauss–Seidel direction and the steplength α_k is obtained by some line search process such as the Armijo or the Wolfe line search process.

The Armijo line search finds a steplength $\alpha_k = \rho^i$ with $\rho^i \in (0, 1)$, where i is the smallest positive integer such that

$$f(\mathbf{x}^{(k)} + \rho^i \mathbf{d}_k) \leq f(\mathbf{x}^{(k)}) + \sigma_1 \rho^i \nabla f(\mathbf{x}^{(k)})^\top \mathbf{d}_k, \tag{2.28}$$

where $\sigma_1 \in (0, 1)$. The Wolfe line search finds a steplength α_k such that

$$\begin{cases} f(\mathbf{x}^{(k)} + \alpha_k \mathbf{d}_k) \leq f(\mathbf{x}^{(k)}) + \sigma_1 \alpha_k \nabla f(\mathbf{x}^{(k)})^\top \mathbf{d}^k, \\ \nabla f(\mathbf{x}^{(k)} + \alpha_k \mathbf{d}_k)^\top \mathbf{d}_k \geq \sigma_2 \nabla f(\mathbf{x}^{(k)})^\top \mathbf{d}^k, \end{cases} \tag{2.29}$$

where $0 < \sigma_1 < \sigma_2 < 1$.

Theorem 2.14 *Suppose that for any $\mathbf{x} \neq \mathbf{0}$, the matrix $\mathscr{A}\mathbf{x}^{m-2}$ is positive semidefinite. Then every accumulation point of the sequence $\{\mathbf{x}^{(k)}\}$ generated by the damped Gauss–Seidel method with Armijo or Wolfe line search is a solution of the minimization problem (2.23).*

Proof Since the matrix $\mathscr{A}\mathbf{x}^{m-2}$ is positive semidefinite for any $\mathbf{x} \neq \mathbf{0}$, f is convex. For a continuously differentiable convex function, every stationary point is a global minimizer. Hence, it suffices to prove that every accumulation point of $\{\mathbf{x}^{(k)}\}$ is a stationary point of the problem.

Suppose that $\mathbf{x}^*$ is an arbitrary accumulation point of $\{\mathbf{x}^{(k)}\}$. That is to say, there exists a subsequence $\{\mathbf{x}^{(k)} : k \in K\}$ that converges to $\mathbf{x}^*$. By (2.26) and (2.27), there are constants $C_2 \geq C_1 > 0$ such that the inequalities

$$C_1\|\nabla f(\mathbf{x}^{(k)})\| \leq \|\mathbf{d}_k\| \leq C_2\|\nabla f(\mathbf{x}^{(k)})\| \tag{2.30}$$

hold for all sufficiently large $k \in K$. In particular, the sequence $\{\|\mathbf{d}_k\|\}$ is bounded. Without loss of generality, assume that $\{\mathbf{d}_k : k \in K\}_K$ converges to some $\mathbf{d}^*$. Letting $k \to \infty$ in (2.26) with $k \in K$, we have

$$(m-1)\mathscr{L}(\mathbf{x}^*)^{m-2}\mathbf{d}^* + \nabla f(\mathbf{x}^*) = \mathbf{0}. \tag{2.31}$$

By (2.28) or the first condition in (2.29), we also have

$$\lim_{k\to\infty} \alpha_k \nabla f(\mathbf{x}^{(k)})^\top \mathbf{d}_k \to 0.$$

In the case where Armijo line search is used, if $\limsup_{k\in K} \alpha_k > 0$, then we have

$$\lim_{k\to\infty} \nabla f(\mathbf{x}^{(k)})^\top \mathbf{d}_k \to 0.$$

This combined with (2.31) indicates $\nabla f(\mathbf{x}^*) = \mathbf{0}$. If $\liminf_{k\in K} \alpha_k = 0$, then when k is sufficiently large, we have

$$f(\mathbf{x}^{(k)} + \alpha_k \rho^{-1}\mathbf{d}_k) - f(\mathbf{x}^{(k)}) > \sigma_1 \alpha_k \rho^{-1} \nabla f(\mathbf{x}^{(k)})^\top \mathbf{d}_k.$$

Multiplying both sides of the last inequality by $\alpha_k^{-1}\rho$ and letting $k \to \infty$ with $k \in K$, we have

$$\nabla f(\mathbf{x}^*)^\top \mathbf{d}^* \geq 0.$$

This combined with (2.31) again indicates $\nabla f(\mathbf{x}^*) = 0$, i.e., $\mathbf{x}^*$ is a stationary point of the problem.

In the case that Wolfe line search is used, let θ_k be the angle between $-\nabla f(\mathbf{x}^{(k)})$ and $\mathbf{d}_k$. By the Zoutendijk condition

$$\sum_{k=0}^{\infty} \|\nabla f(\mathbf{x}^{(k)})\|^2 \cos^2\theta_k < \infty,$$

we only need to prove that $\{\cos\theta_k\}_K$ is bounded away from zero. In fact, we have

$$\cos\theta_k = \frac{-\nabla f(\mathbf{x}^{(k)})^\top \mathbf{d}_k}{\|\nabla f(\mathbf{x}^{(k)})\|\,\|\mathbf{d}_k\|} = \frac{(m-1)\mathbf{d}_k^\top L_k \mathbf{d}_k}{\|\nabla f(\mathbf{x}^{(k)})\|\,\|\mathbf{d}_k\|} \geq \frac{\frac{1}{2}(m-1)\mathbf{d}_k^\top \mathscr{D}_k \mathbf{d}_k}{C_1^{-1}\|\mathbf{d}_k\|^2},$$

where the inequality follows from (2.27) and (2.30). Since $\{\mathscr{D}_k\}_{k\in K} \to \mathscr{D}^*$ is positive definite, the last inequality indicates that there exists a constant $\delta > 0$ such that $\cos\theta_k \geq \delta$ for all sufficiently large $k \in K$. Hence, we have $\lim_{k\to\infty, k\in K} \|\nabla f(\mathbf{x}^{(k)})\| = 0$, i.e., $\mathbf{x}^*$ is a stationary point of the problem. The theorem is proved. □

The Newton method for solving the multilinear system (2.21) generates a sequence $\{\mathbf{x}^{(k)}\}$ with

$$\mathbf{x}^{(k+1)} = \mathbf{x}^{(k)} + \mathbf{d}_k,$$

where

$$(m-1)\mathscr{A}(\mathbf{x}^{(k)})^{m-2}\mathbf{d}_k + \mathscr{A}(\mathbf{x}^{(k)})^{m-1} - \mathbf{b} = \mathbf{0}.$$

This indicates

$$(m-1)\mathscr{A}(\mathbf{x}^{(k)})^{m-2}(\mathbf{x}^{(k)} + \mathbf{d}_k) - (m-2)\mathscr{A}(\mathbf{x}^{(k)})^{m-1} - \mathbf{b} = \mathbf{0}.$$

It implies that

$$\begin{aligned}\mathbf{x}^{(k+1)} &= \frac{m-2}{m-1}(\mathscr{A}(\mathbf{x}^{(k)})^{m-2})^{-1}\mathscr{A}(\mathbf{x}^{(k)})^{m-1} + \frac{1}{m-1}(\mathscr{A}(\mathbf{x}^{(k)})^{m-2})^{-1}\mathbf{b}\\ &= \frac{m-2}{m-1}\mathbf{x}^{(k)} + \frac{1}{m-1}(\mathscr{A}(\mathbf{x}^{(k)})^{m-2})^{-1}\mathbf{b}.\end{aligned}$$

Therefore, by letting $\mathbf{z}^{(k+1)} = (\mathscr{A}(\mathbf{x}^{(k)})^{m-2})^{-1}\mathbf{b}$, we may also write the Newton method as

$$\begin{cases}\mathbf{x}^{(k+1)} = \dfrac{m-2}{m-1}\mathbf{x}^{(k)} + \dfrac{1}{m-1}\mathbf{z}^{(k+1)},\\ \mathscr{A}(\mathbf{x}^{(k)})^{m-2}\mathbf{z}^{(k+1)} = \mathbf{b}.\end{cases}$$

Thus, we may develop the Newton–Gauss–Seidel method by getting an estimate of $\mathbf{z}^{(k+1)}$ in one step or multi-step Gauss–Seidel iterations. The one step Newton–Gauss–Seidel method generates $\mathbf{x}^{(k+1)}$ by

$$\begin{cases}\mathbf{x}^{(k+1)} = \dfrac{m-2}{m-1}\mathbf{x}^{(k)} + \dfrac{1}{m-1}\mathbf{y}^{(k+1)},\\ \mathbf{y}^{(k+1)} = \mathbf{y}^{(k)} - (\mathscr{L}(\mathbf{x}^{(k)})^{m-2})^{-1}(\mathscr{A}(\mathbf{x}^{(k)})^{m-2}\mathbf{y}^{(k)} - \mathbf{b}),\end{cases} \tag{2.32}$$

where $\mathbf{y}^{(k)}$ is an arbitrary estimation to $\mathbf{z}^{(k+1)}$. If we make $\mathbf{y}^{(k)} = \mathbf{x}^{(k)}$, then the above one step Newton–Gauss–Seidel iteration reduces to the Gauss–Seidel iteration. In the case $m = 2$, we have $\mathbf{x}^{(k+1)} = \mathbf{y}^{(k+1)}$, and hence the iterative method is just the Gauss–Seidel method for solving a system of linear equation.

We write (2.32) in a compact form. Then we have

$$\begin{pmatrix} I & -\frac{1}{m-1}I \\ 0 & \mathscr{L}(\mathbf{x}^{(k)})^{m-2}\end{pmatrix}\begin{pmatrix}\mathbf{x}^{(k+1)}\\ \mathbf{y}^{(k+1)}\end{pmatrix} = \begin{pmatrix}\frac{m-2}{m-1}I & 0\\ 0 & \mathscr{U}(\mathbf{x}^{(k)})^{m-2}\end{pmatrix}\begin{pmatrix}\mathbf{x}^{(k)}\\ \mathbf{y}^{(k)}\end{pmatrix} + \begin{pmatrix}\mathbf{0}\\ \mathbf{b}\end{pmatrix},$$

where $\mathscr{U} = \mathscr{L} - \mathscr{A}$. This implies

$$\begin{aligned}
\begin{pmatrix} \mathbf{x}^{(k+1)} \\ \mathbf{y}^{(k+1)} \end{pmatrix} &= \begin{pmatrix} I & -\frac{1}{m-1} I \\ 0 & \mathscr{L}(\mathbf{x}^{(k)})^{m-2} \end{pmatrix}^{-1} \begin{pmatrix} \frac{m-2}{m-1} I & 0 \\ 0 & \mathscr{U}(\mathbf{x}^{(k)})^{m-2} \end{pmatrix} \begin{pmatrix} \mathbf{x}^{(k)} \\ \mathbf{y}^{(k)} \end{pmatrix} \\
&+ \begin{pmatrix} I & -\frac{1}{m-1} I \\ 0 & \mathscr{L}(\mathbf{x}^{(k)})^{m-2} \end{pmatrix}^{-1} \begin{pmatrix} \mathbf{0} \\ \mathbf{b} \end{pmatrix} \\
&= \begin{pmatrix} \frac{m-2}{m-1} I & \frac{1}{m-1} (\mathscr{L}(\mathbf{x}^{(k)})^{m-2})^{-1} \mathscr{U}(\mathbf{x}^{(k)})^{m-2} \\ 0 & (\mathscr{L}(\mathbf{x}^{(k)})^{m-2})^{-1} \mathscr{U}(\mathbf{x}^{(k)})^{m-2} \end{pmatrix} \begin{pmatrix} \mathbf{x}^{(k)} \\ \mathbf{y}^{(k)} \end{pmatrix} \\
&+ \begin{pmatrix} \frac{1}{m-1} (\mathscr{L}(\mathbf{x}^{(k)})^{m-2})^{-1} \mathbf{b} \\ (\mathscr{L}(\mathbf{x}^{(k)})^{m-2})^{-1} \mathbf{b} \end{pmatrix} \\
&= B_k \begin{pmatrix} \mathbf{x}^{(k)} \\ \mathbf{y}^{(k)} \end{pmatrix} + \mathbf{b}_k,
\end{aligned}$$

where

$$\begin{aligned}
B_k &= \begin{pmatrix} \frac{m-2}{m-1} I & \frac{1}{m-1} (\mathscr{L}(\mathbf{x}^{(k)})^{m-2})^{-1} \mathscr{U}(\mathbf{x}^{(k)})^{m-2} \\ 0 & (\mathscr{L}(\mathbf{x}^{(k)})^{m-2})^{-1} \mathscr{U}(\mathbf{x}^{(k)})^{m-2} \end{pmatrix} \\
&= I - \begin{pmatrix} \frac{1}{m-1} I & \frac{1}{m-1} (\mathscr{L}(\mathbf{x}^{(k)})^{m-2})^{-1} \mathscr{U}(\mathbf{x}^{(k)})^{m-2} \\ 0 & (\mathscr{L}(\mathbf{x}^{(k)})^{m-2})^{-1} \mathscr{A}(\mathbf{x}^{(k)})^{m-2} \end{pmatrix},
\end{aligned}$$

and

$$\mathbf{b}_k = \begin{pmatrix} \frac{1}{m-1} (\mathscr{L}(\mathbf{x}^{(k)})^{m-2})^{-1} \mathbf{b} \\ (\mathscr{L}(\mathbf{x}^{(k)})^{m-2})^{-1} \mathbf{b} \end{pmatrix}.$$

Since the matrix B_k is block upper triangular, its eigenvalues are $\frac{m-2}{m-1}$ together with eigenvalues of the Gauss–Seidel matrix $(\mathscr{L}(\mathbf{x}^{(k)})^{m-2})^{-1} \mathscr{U}(\mathbf{x}^{(k)})^{m-2}$. Thus, the convergence of the iterative scheme is the same as the one of the Gauss–Seidel method.

Note that we may also use multi-step Gauss–Seidel steps to obtain an approximate Newton direction. Then we have the following iterative scheme:

$$\begin{cases} \mathbf{x}^{(k+1)} = \dfrac{m-2}{m-1} \mathbf{x}^{(k)} + \dfrac{1}{m-1} \mathbf{y}^{(k+1)} \\ \mathbf{y}^{(k+j)} = \mathbf{y}^{(k+j-1)} - (\mathscr{L}(\mathbf{x}^{(k)})^{m-2})^{-1} (\mathscr{A}(\mathbf{x}^{(k)})^{m-2} \mathbf{y}^{(k+j-1)} - \mathbf{b}), \quad j = 1, 2, \ldots, i. \end{cases}$$

Li et al. [168] presented several numerical experiments to show that the multistep Newton–Gauss–Seidel method works quite well.

2.5 Notes

The multilinear nonsingular M-equations and some extensions have been investigated carefully in this chapter. Some well-known iterative methods for solving system of linear equations are extended to the tensor equations. Under reasonable conditions, the convergence of the proposed methods are established. Furthermore, a Newton–Gauss–Seidel method, a homotopy method and a new tensor method are also proposed for solving the M-tensor equation system.

Section 2.1 Conclusions in this section were first presented by Ding and Wei in [87]. M-tensors are Z-tensors, which was defined by Zhang, Qi, and Zhou in [305]. Then, Z-tensors and M-tensors were further studied by Ding, Qi, and Wei in [82]. H-matrices were extended to H-tensors in [82]. Section 5.4 of [228] was devoted to M-tensors and H-tensors, and the bibliography of [228] includes up-to-date references on applications of Z-tensors, M-tensors and H-tensors.

Section 2.2 Several numerical methods are studied in this section. The forward substitution method and the back substitution method for multilinear systems with M-tensors were first proposed by Ding and Wei in [87]. Theorem 2.7 was cited from [234] by Rheinboldt. Another method for mulitilinear system is the homotopy method, which was proposed by Han in [117].

Section 2.3 A tensor method for multilinear system is discussed in this section. To solve nonlinear equation systems, the tensor method was first introduced by Schnabel and Frank in [238]. Combining this idea with the results for the rank-1 approximation of tensors, Xie, Jin, and Wei proposed a new tensor method for solving the symmetric M-tensor equation system [291]. Theorem 2.10 and Algorithms 2.4–2.5 were proved by Schnabel, Frank in 1984 [238].

Section 2.4 In this part, several methods are studied for solving multilinear system with tensors with general structures. The methods include Jacobi, the Gauss–Seidel and the successive over-relaxation (SOR) methods, which are well known basic iterative methods for solving system of linear equations. Here, they were extended to solve the multilinear system by Li, Xie, and Xu in [168].

Besides above, there are some other results on tensor equation systems or having close relationships with tensor equation systems [25], which are different from the M-tensor equation system in this chapter. Now, we will describe the main idea of the related papers in three parts.

1. Li and Ng studied the sparse nonnegative tensor equation system in [170], where the related tensors $\mathscr{A}_k = (a^{(k)}_{i_1 i_2 \dots i_m})$, $k \in [m]$ are nonnegative mth order $n_1 \times n_2 \times \cdots \times n_m$ dimensional stochastic tensors with respect to the kth dimension:

$$\sum_{i_k=1}^{n_k} a^{(k)}_{i_1 i_2 \dots i_m} = 1 \quad \text{for all } i_l \in [n_l],\ l \in [m],\ l \neq k.$$

Here the tensor $\mathscr{A}_k$ may not be symmetric. Hence, the tensor equation model studied in [170] is that

$$\begin{cases} \mathbf{y}^{(1)} = (1-\alpha_1)\mathscr{A}_1 \times_1 \mathbf{y}^{(2)}\mathbf{y}^{(3)}\cdots\mathbf{y}^{(m)} + \alpha_1\mathbf{z}^{(1)} \\ \mathbf{y}^{(2)} = (1-\alpha_2)\mathscr{A}_2 \times_2 \mathbf{y}^{(1)}\mathbf{y}^{(3)}\cdots\mathbf{y}^{(m)} + \alpha_2\mathbf{z}^{(2)} \\ \cdots\cdots\cdots \\ \mathbf{y}^{(m)} = (1-\alpha_m)\mathscr{A}_m \times_m \mathbf{y}^{(1)}\cdots\mathbf{y}^{(m-1)} + \alpha_m\mathbf{z}^{(m)} \end{cases} \tag{2.33}$$

where $\mathbf{z}^{(k)} \in \mathbf{R}^{n_k}$ are given nonnegative vectors such that $\sum_{i=1}^{n_k} z_i^{(k)} = 1$ for $k \in [m]$ and $\alpha_k \in [0, 1]$ for $k \in [m]$ are given scalars. It should be noted that the system (2.33) has a strong practical background such as community discovery for online social media, and information retrieval that can assist in many data mining tasks.

First of all, Li and Ng proved that there always exist solutions $\mathbf{y}^{(1)}, \mathbf{y}^{(2)}, \ldots, \mathbf{y}^{(m)}$ satisfying (2.33) with

$$\sum_{i=1}^{n_k} y_i^{(k)} = 1, \quad \forall k \in [m],$$

and furthermore, the solution vectors $\mathbf{y}^{(k)}$, $k \in [m]$ are positive vectors if the related tensors $\mathscr{A}_k$, $k \in [m]$ are irreducible. If suitable conditions are added to scalars α_k such that

$$\frac{m-2}{m-1} < \alpha_k < 1, \ \forall\, k \in [m], \tag{2.34}$$

then it is proved that the solutions $\mathbf{y}^{(k)}$, $k \in [m]$ of (2.33) are unique, and the results above construct the main theoretical results in [170].

According to the main results listed above, Jacobi and Gauss–Seidel methods are developed for solving tensor equation system (2.33). In the iterative process, the multiplication of tensors with vectors are required and the cost depends on the number of non-zeros in the tensors. Hence, the proposed Jacobi and Gauss–Seidel methods are not expensive especially when the tensors are very sparse. Furthermore, the linear convergence of the Jacobi and Gauss–Seidel methods are established, and experimental results on information retrieval by query search and community discovery in networks are also reported to demonstrate the effectiveness of the proposed methods. Interested readers may check the details in [170].

2. Since the polynomials corresponding to high order tensors have more complex structures than polynomials corresponding to matrices, it is not easy to define the inverse like the matrix case. But it is possible to define generalized inverse of tensors in some sense. With the help of the notion of tensor equation systems, Sun et al. [259] studied the $\{i\}$-inverse ($i = 1; 2; 5$) and group inverse of tensor $\mathscr{A}$ defined in complex number field. Then, by the definition of $\{1\}$-inverse, some solutions for the tensor equation system

$$\mathscr{A}\mathbf{x}^{m-1} = \mathbf{b}$$

are given under reasonable conditions. It should be noted that all tensors and vectors studied in [259] are defined in complex number field $\mathbf{C}$, and the definition of $\{i\}$-inverse for tensors is defined based on the tensor multiplication given by Shao in [240], which means that the multiplication $\mathscr{A}\mathscr{B}$ of an order k dimension n tensor $\mathscr{A}$ and an order t dimension n tensor $\mathscr{B}$ is a tensor with order $k+t-1$ dimension n.

Now we give a short description for the main results of [259]. Suppose $\mathscr{A}$ is a tensor with order t dimension $m \times n \times \cdots \times n$. If there is a tensor $\mathscr{X}$ with order k dimension $n \times m \times \cdots \times m$ satisfying

$$\mathscr{A}\mathscr{X}(\mathscr{A}\mathbf{y}^{m-1})^{[\frac{1}{s}]} = \mathscr{A}\mathbf{y}^{t-1}, \quad s = (t-1)(k-1),$$

for all vectors $\mathbf{y} \in \mathbf{C}^n$, then the tensor $\mathscr{X}$ is called a $\{1\}$-inverse of $\mathscr{A}$ with order k. It is not difficult to check that the $\{1\}$-inverse of tensors reduces to the $\{1\}$-inverse of matrices when $\mathscr{A}$ and $\mathscr{X}$ are matrices.

For tensor $\mathscr{A}$ given as above, if it has at least one $\{1\}$-inverse tensor $\mathscr{X}$ with order k and the equation system $\mathscr{A}\mathbf{x}^{t-1} = \mathbf{b}$ is solvable, then it is proved by Sun et al. in [259] that the proposed equation system has a solution $\mathbf{x} = \mathscr{X}(\mathbf{b}^{[\frac{1}{s}]})^{k-1}$, where $s = (t-1)(k-1)$. By this, we know that one can get the solution of the system directly if the $\{1\}$-inverse of the related coefficient tensor can be computed. Now, the question is how to compute the $\{1\}$-inverse tensor for a given tensor? The authors of [259] did not present any methods to compute the $\{1\}$-inverse for a general given tensor. This may be an interesting work in the future.

However, for some tensors with special structure, we can obtain the $\{1\}$-inverse by a direct computation. For example, in [259], Sun et al. showed that, a diagonal tensor $\mathscr{A} \in T_{m,n}$ with diagonal entries $a_1, a_2, \ldots, a_n \in \mathbf{C}$ has an k-order $\{1\}$-inverse diagonal tensor with diagonal entries being $(a_1^{\dagger})^{\frac{1}{t-1}}, (a_2^{\dagger})^{\frac{1}{t-1}}, \ldots, (a_n^{\dagger})^{\frac{1}{t-1}}$, where

$$a^{\dagger} = \begin{cases} a^{-1}, & \text{if } a \neq 0 \\ 0, & \text{if } a = 0. \end{cases}$$

Furthermore, it shows how to characterize the representations for the $\{1\}$-inverse and group inverse of some block tensors [259], which extends the results on generalized inverses of matrices. Since the results about block tensor are uncorrelated with the contents of this chapter, we do not want to go far for the detail. Interested readers may recall the related paper [259].

3. Very recently, Liu et al. studied another kind of sparse nonnegative tensor systems for data sciences in [183], which can be seen as a further research for the paper [170]. For the tensor equation system (2.33), the uniqueness of solutions is proved under conditions (2.34). However, in some practical problems, these parameters $\alpha_k > 0, k \in [m]$ would tend to 1. Thus a question can be raised naturally: Can we exploit any better upper bounds for scalars α_k? Motivated by this, Liu et al. devoted to study the uniqueness of solution for system (2.33), where explicit upper bounds for parameters α_k are given [183]. In that paper, two different techniques are introduced to get better upper bound: the parameter method and the optimal set method.

On the other hand, with the two techniques, Liu et al. also considered the system (2.33) with a perturbation, which means the related stochastic tensors $\mathscr{A}_k, k \in [m]$ are replaced by $\mathscr{A}_k + \Delta\mathscr{A}_k, k \in [m]$ respectively. Here $\mathscr{A}_k + \Delta\mathscr{A}_k, k \in [m]$ are still stochastic tensors. Then the authors presented a new perturbation tensor equation system, and studied the unique property of the solution of the new system, which has important applications in sensitivity analysis [183].

2.6 Exercise

1 Let $\mathscr{A} = \mathscr{I} - \mathscr{B}$ be a $3 \times 3 \times 3$ singular M-tensor, where $\mathscr{B}$ is a nonnegative tensor with $\mathscr{B}\mathbf{1}^2 = \mathbf{1}$, and $b_{ijk} = 0$ if $i \in \{2, 3\}$ and either $j = 1$ or $k = 1$. Apparently, the subtensor $\mathscr{A}_{2:3,2:3,2:3}$ is also a singular M-tensor, which satisfies that $\mathscr{A}_{2:3,2:3,2:3}\mathbf{1}^2 = \mathbf{0}$. For each right-hand side $\mathbf{b} = (b_1, 0, 0)$, what is the form of the nonnegative solutions to $\mathscr{A}\mathbf{x}^2 = \mathbf{b}$?
2 Suppose $\mathscr{A} \in T_{3,n}$ is a third order symmetric M-tensor. Define $\mathscr{A} = s\mathscr{I} - \mathscr{B}$, where $\mathscr{B} = (b_{ijk})$ with entries such that

$$b_{ijk} = |\sin(i + j + k)|.$$

Let $s = n^2$. Prove $\mathscr{A}$ is a symmetric nonsingular M-tensor.

Chapter 3
Hankel Tensor Computation and Exponential Data Fitting

Hankel structures are widely used in real-world problems arising from signal processing, automatic control, and geophysics. For example, a Hankel matrix was formulated to analyze the time-domain signals in nuclear magnetic resonance spectroscopy, which is crucial for brain tumour detection [268]. Papy et al. [217, 218] improved this method by using a high order Hankel tensor to replace the Hankel matrix. Ding et al. [83] found a fast computational technique for computing products of a Hankel tensor and vectors.

Exploring the special structure, we first define two classes of tensors: Hankel tensors and anti-circulant tensors. They share the same generating vector. Second, we show that an anti-circulant tensor has a diagonal decomposition using fast Fourier transforms. Then, we introduce the fast Hankel tensor–vector product.

For real-valued symmetric Hankel tensors, we propose an optimization approach for computing extreme eigenvalues and associated eigenvectors. First, we consider a spherical optimization problem, whose objective is a homogeneous polynomial involved in the Hankel tensor. The first-order stationary point of the optimization problem is an eigenvector of the tensor. Second, we propose a gradient-descent method for solving the spherical optimization, where the Cayley transform is used to preserve the spherical constraint. Third, using the Łojasiewicz property of the homogeneous polynomial involved in a tensor, we prove that the total sequence of iterates generated by the gradient-descent method converges to an eigenvector and the corresponding objective converges to the associated eigenvalue. When we start the algorithm from plenty of randomly initial points sampled from the unit sphere uniformly, we could touch the extreme eigenvalue of the Hankel tensor with a higher probability.

Finally, we study the application of Hankel tensors in exponential data fitting. For signals modeled as a sum of exponentially damped sinusoids, we arrange them as a Hankel tensor which has an inherent Vandermonde decomposition. The subspace spanned by the Vandermonde factor is approximated by the higher-order orthogonal iterations (HOOI). Here, the fast Hankel tensor–vector product accelerates HOOI

L. Qi et al., *Tensor Eigenvalues and Their Applications*, Advances in Mechanics and Mathematics 39, https://doi.org/10.1007/978-981-10-8058-6_3

significantly. Using the shift-invariance property of the Vandermonde factor, we compute multiple poles of signals. Then, amplitudes are obtained by solving a linear Vandermonde system.

3.1 Fast Hankel Tensor–Vector Product

As a kind of structured tensors, the Hankel tensor is completely determined by a compact generating vector. We start with the definitions of Hankel tensors and anti-circulant tensors.

Definition 3.1 A tensor $\mathscr{H} = (h_{i_1 i_2 \ldots i_m}) \in T_{m,n}$ is called a **Hankel tensor** if its elements satisfy

$$h_{i_1 i_2 \ldots i_m} = v_{i_1+i_2+\cdots+i_m-m} \qquad \text{for } i_j = 1, \ldots, n,\ j = 1, \ldots, m.$$

The vector $\mathbf{v} \equiv (v_0, v_1, \ldots, v_{m(n-1)})^\top$ with length $\ell \equiv m(n-1)+1$ is called the **generating vector** of $\mathscr{H}$. Using the same generating vector $\mathbf{v}$, we define an mth order ℓ-dimensional **anti-circulant tensor** $\mathscr{C} = (c_{i_1 i_2 \ldots i_m})$ whose elements are

$$c_{i_1 i_2 \ldots i_m} = v_{(i_1+i_2+\cdots+i_m-m \bmod \ell)} \qquad \text{for } i_j = 1, \ldots, \ell,\ j = 1, \ldots, m.$$

Clearly, Hankel tensors are symmetric tensors. We remark here that the Hankel tensor $\mathscr{H}$ could be viewed as a sub-tensor of the anti-circulant tensor $\mathscr{C}$. Owing to the same generating vector $\mathbf{v}$, we find

$$c_{i_1 i_2 \ldots i_m} = h_{i_1 i_2 \ldots i_m} \qquad \text{for } i_j = 1, \ldots, n,\ j = 1, \ldots, m.$$

For example, a third order two dimensional Hankel tensor with a generating vector $\mathbf{v} = (0, 1, 2, 3)^\top$ is

$$\mathscr{H} = \left(\begin{array}{cc|cc} 0 & 1 & 1 & 2 \\ 1 & 2 & 2 & 3 \end{array}\right).$$

Obviously, it is a sub-tensor of the following anti-circulant tensor with the same order and a larger dimension

$$\mathscr{C} = \left(\begin{array}{cccc|cccc|cccc|cccc} 0&1&2&3&1&2&3&0&2&3&0&1&3&0&1&2 \\ 1&2&3&0&2&3&0&1&3&0&1&2&0&1&2&3 \\ 2&3&0&1&3&0&1&2&0&1&2&3&1&2&3&0 \\ 3&0&1&2&0&1&2&3&1&2&3&0&2&3&0&1 \end{array}\right).$$

The generating vector $\mathbf{v}$ is exactly the first column of the anti-circulant tensor $\mathscr{C}$, i.e., $\mathbf{v} = \mathscr{C}(\mathbf{1}^{(1)})^{m-1}$.

The anti-circulant tensor enjoys a diagonalization via the **Fourier matrix**. Let $\omega \equiv \exp(-\frac{2\pi \mathrm{i}}{\ell}) \in \mathbf{C}$ be a primitive ℓth root of unity, where i is the imaginary unit. An ℓ-by-ℓ Fourier matrix is defined as

$$F_\ell = (\omega^{jk})_{j,k=0,1,\ldots,\ell-1}.$$

Note that $F_\ell = F_\ell^\top$. By some calculation, we have

$$F_\ell^H F_\ell = \ell I,$$

where $(\cdot)^H$ means the conjugate transpose.

For example, in the case of $\ell = 4$, we have $\omega = \exp(-\frac{2\pi \mathrm{i}}{4}) = -\mathrm{i}$ and

$$F_4 = \begin{pmatrix} 1 & 1 & 1 & 1 \\ 1 & -\mathrm{i} & -1 & \mathrm{i} \\ 1 & -1 & 1 & -1 \\ 1 & \mathrm{i} & -1 & -\mathrm{i} \end{pmatrix}.$$

Next, we give the following theorem on anti-circulant tensors [83].

Theorem 3.1 *A tensor $\mathscr{C} \in T_{m,n}$ is an anti-circulant tensor if and only if it can be diagonalized by the Fourier matrix, i.e.,*

$$\mathscr{C} = \mathscr{D} F_\ell^m \equiv \mathscr{D} \times_1 F_\ell \times_2 F_\ell \cdots \times_m F_\ell,$$

where $\mathscr{D}$ is a diagonal tensor with diagonal elements $\mathrm{diag}(\mathscr{D}) = F_\ell^{-1}\mathbf{v}$ and $\mathbf{v} = \mathscr{C}(\mathbf{1}^{(1)})^{m-1}$ is the generating vector of the anti-circulant tensor $\mathscr{C}$.

Proof First, we prove that $\mathscr{D} F_\ell^m$ is an anti-circulant tensor. The $(i_1, \ldots, i_m)$th element of $\mathscr{D} F_\ell^m$ is

$$(\mathscr{D} F_\ell^m)_{i_1 \ldots i_m} = (\mathscr{D} F_\ell^m) \times_1 \mathbf{1}^{(i_1)} \cdots \times_m \mathbf{1}^{(i_m)} = \mathscr{D} \times_1 (F_\ell \mathbf{1}^{(i_1)}) \cdots \times_m (F_\ell \mathbf{1}^{(i_m)})$$

for indices $i_1, \ldots, i_m = 1, 2, \ldots, \ell$. Let $\mathrm{diag}(\mathscr{D}) = (d_0, d_1, \ldots, d_{\ell-1})^\top$. We get

$$\begin{aligned}(\mathscr{D} F_\ell^m)_{i_1 \ldots i_m} &= \sum_{j=0}^{\ell-1} d_j \omega^{j(i_1-1)} \cdots \omega^{j(i_m-1)} \\ &= \sum_{j=0}^{\ell-1} d_j \omega^{j(i_1+\cdots+i_m-m)} \\ &= \sum_{j=0}^{\ell-1} d_j \omega^{j(i_1+\cdots+i_m-m \bmod \ell)}.\end{aligned}$$

Hence, $\mathscr{D}F_\ell^m$ is an anti-circulant tensor with a generating vector $(\sum_{j=0}^{\ell-1} d_j\omega^{jk})_{k=0,\ldots,\ell-1}$.

On the other hand, if $\operatorname{diag}(\mathscr{D}) = F_\ell^{-1}\mathbf{v}$, by some calculations, we have

$$(\mathscr{D}F_\ell^m)_{i_1\ldots i_m} = (F_\ell^{-1}\mathbf{v})^\top(F_\ell \mathbf{1}^{(1+(i_1+\cdots+i_m-m \bmod \ell))}) = \mathbf{v}_{(i_1+\cdots+i_m-m \bmod \ell)}$$

for $i_1,\ldots,i_m = 1,2,\ldots,\ell$. Since $\mathbf{v}$ is the generating vector of the anti-circulant tensor $\mathscr{C}$, the theorem is valid. □

It is well-known that the fast Fourier transform `fft` and its inverse transform `ifft` could be employed to compute products involving a Fourier matrix and its inverse, respectively. The computational cost of an (inverse) Fourier matrix–vector product is about $\mathscr{O}(\ell \log \ell)$ multiplications, which is cheap when the dimension ℓ is large.

Now, we are going to show how to compute the Hankel tensor–vector product using Theorem 3.1. The basic idea is to enlarge the Hankel tensor to an anti-circulant tensor. For a given vector $\mathbf{x} \in \mathbf{R}^n$, we define another vector $\mathbf{y} \in \mathbf{R}^\ell$ such that

$$\mathbf{y} \equiv \begin{pmatrix} \mathbf{x} \\ \mathbf{0}_{\ell-n} \end{pmatrix},$$

where $\ell = m(n-1)+1$ and $\mathbf{0}_{\ell-n}$ is a zero vector with length $\ell - n$. Then, we have

$$\mathscr{H}\mathbf{x}^m = \mathscr{C}\mathbf{y}^m = \mathscr{D}(F_\ell\mathbf{y})^m = \mathtt{ifft}(\mathbf{v})^\top\left(\mathtt{fft}(\mathbf{y})^{\circ m}\right).$$

To obtain $\mathscr{H}\mathbf{x}^{m-1}$, we first compute

$$\mathscr{C}\mathbf{y}^{m-1} = F_\ell\left(\mathscr{D}(F_\ell\mathbf{y})^{m-1}\right) = \mathtt{fft}\left(\mathtt{ifft}(\mathbf{v}) \circ (\mathtt{fft}(\mathbf{y})^{\circ(m-1)})\right).$$

Then, elements of the vector $\mathscr{H}\mathbf{x}^{m-1}$ is the leading n elements of $\mathscr{C}\mathbf{y}^{m-1}$. Here, $\circ$ denotes the Hadamard product such that $(A \circ B)_{ij} = A_{ij}B_{ij}$ if A and B have the same size. Furthermore, we denote

$$A^{\circ k} = \underbrace{A \circ \cdots \circ A}_{k \text{ times}}$$

as the Hadamard product of k copies of A.

Since the computations of $\mathscr{H}\mathbf{x}^m$ and $\mathscr{H}\mathbf{x}^{m-1}$ require 2 and 3 `fft`/`iffts`, respectively, the cost is about $\mathscr{O}(mn\log(mn))$ multiplications and hence cheap. Another advantage of this approach is that we only need to store and work with the compact generating vector of the Hankel tensor.

3.2 Computing Eigenvalues of a Hankel Tensor

We consider the **generalized eigenvalue** [33, 86] of a real-valued Hankel tensor $\mathscr{H} \in S_{m,n}$ defined with the following system:

$$\mathscr{H}\mathbf{x}^{m-1} = \lambda \mathscr{B}\mathbf{x}^{m-1}, \tag{3.1}$$

where m is even, $\mathscr{B} \in S_{m,n}$ is a positive definite tensor. If there is a scalar $\lambda \in \mathbf{R}$ and a vector $\mathbf{x} \in \mathbf{R}^n \backslash \{\mathbf{0}\}$ satisfying the above system, we call λ a generalized eigenvalue of $\mathscr{H}$ and $\mathbf{x}$ its associated generalized eigenvector. Particularly, λ is an H-eigenvalue of $\mathscr{H}$ and $\mathbf{x}$ is its associated H-eigenvector if $\mathscr{B} = \mathscr{I}$ where $\mathscr{I}\mathbf{x}^{m-1} = \mathbf{x}^{[m-1]}$ [221]. If $\mathscr{B}$ is a tensor $\mathscr{E} \in S_{m,n}$ such that $\mathscr{E}\mathbf{x}^{m-1} = \|\mathbf{x}\|^{m-2}\mathbf{x}$, then λ is called a Z-eigenvalue of $\mathscr{H}$ and $\frac{\mathbf{x}}{\|\mathbf{x}\|}$ is its associated Z-eigenvector. Obviously, computations on these tensors such like $\mathscr{B}$ are straightforward.

To compute a generalized eigenvalue and its associated eigenvector, we consider an optimization model with a spherical constraint

$$\min \quad f(\mathbf{x}) \equiv \frac{\mathscr{H}\mathbf{x}^m}{\mathscr{B}\mathbf{x}^m} \qquad \text{s.t.} \quad \|\mathbf{x}\| = 1. \tag{3.2}$$

To keep variables $\mathbf{x}$ away from zero, we introduce the spherical constraint $\|\mathbf{x}\| = 1$. Since the objective $f(\mathbf{x})$ is zero-order homogeneous, we do not need to consider the multiplier of the spherical constraint (See Theorem 3.2). By some calculations, we get its gradient

$$\mathbf{g}(\mathbf{x}) = \frac{m}{\mathscr{B}\mathbf{x}^m}\left(\mathscr{H}\mathbf{x}^{m-1} - \frac{\mathscr{H}\mathbf{x}^m}{\mathscr{B}\mathbf{x}^m}\mathscr{B}\mathbf{x}^{m-1}\right) \tag{3.3}$$

and Hessian

$$\begin{aligned} H(\mathbf{x}) = {} & \frac{m(m-1)\mathscr{H}\mathbf{x}^{m-2}}{\mathscr{B}\mathbf{x}^m} - \frac{m(m-1)\mathscr{H}\mathbf{x}^m\mathscr{B}\mathbf{x}^{m-2} + m^2(\mathscr{H}\mathbf{x}^{m-1} \circledcirc \mathscr{B}\mathbf{x}^{m-1})}{(\mathscr{B}\mathbf{x}^m)^2} \\ & + \frac{m^2\mathscr{H}\mathbf{x}^m(\mathscr{B}\mathbf{x}^{m-1} \circledcirc \mathscr{B}\mathbf{x}^{m-1})}{(\mathscr{B}\mathbf{x}^m)^3}, \end{aligned} \tag{3.4}$$

where $\mathbf{x} \circledcirc \mathbf{y} \equiv \mathbf{x}\mathbf{y}^\top + \mathbf{y}\mathbf{x}^\top$.

The relationship between eigenvalues of a Hankel tensor and the spherical optimization problem (3.2) is established via the following theorem.

Theorem 3.2 *Let m be even and $\mathbf{x}_* \in \mathbf{S}_{n-1} \equiv \{\mathbf{x} \in \mathbf{R}^n \mid \mathbf{x}^\top\mathbf{x} = 1\}$. Then, $\mathbf{x}_*$ is a first-order stationary point of $f(\mathbf{x})$, i.e., $\mathbf{g}(\mathbf{x}_*) = \mathbf{0}$, if and only if there exists a $\lambda_* \in \mathbf{R}$ such that $(\lambda_*, \mathbf{x}_*)$ satisfies (3.1). Indeed, $\lambda_* = f(\mathbf{x}_*)$ is the H-eigenvalue (resp., Z-eigenvalue) and $\mathbf{x}_*$ is the associated H-eigenvector (resp., Z-eigenvector) if $\mathscr{B} = \mathscr{I}$ (resp., $\mathscr{B} = \mathscr{E}$).*

Proof Since $\mathscr{B}$ is positive definite and $\mathbf{x}_*$ belongs to a compact set $\mathbf{S}_{n-1}$, $\mathscr{B}\mathbf{x}_*^m$ has a finite upper bound. Thus, $\lambda_* = \frac{\mathscr{H}\mathbf{x}_*^m}{\mathscr{B}\mathbf{x}_*^m} = f(\mathbf{x}_*)$ and the theorem is valid. □

Next, we construct a curvilinear search path on the sphere $\mathbf{S}_{n-1}$ via the **Cayley transform** [37, 105]. Suppose that the current iterate is $\mathbf{x}_k \in \mathbf{S}_{n-1}$ and the next iterate is $\mathbf{x}_{k+1}$. To preserve the spherical constraint $\mathbf{x}_{k+1}^\top \mathbf{x}_{k+1} = \mathbf{x}_k^\top \mathbf{x}_k = 1$, we try to choose an orthogonal matrix $Q \in \mathbf{R}^{n\times n}$, whose eigenvalues do not contain -1, such that

$$\mathbf{x}_{k+1} = Q\mathbf{x}_k. \tag{3.5}$$

Using the Cayley transform, the matrix

$$Q = (I + W)^{-1}(I - W) \tag{3.6}$$

is orthogonal if and only if $W \in \mathbf{R}^{n\times n}$ is skew-symmetric. Now, our task is to select a skew-symmetric matrix W such that $\mathbf{g}(\mathbf{x}_k)^\top(\mathbf{x}_{k+1} - \mathbf{x}_k) < 0$. For simplicity, we take

$$W = \mathbf{a}\mathbf{b}^\top - \mathbf{b}\mathbf{a}^\top, \tag{3.7}$$

where $\mathbf{a}, \mathbf{b} \in \mathbf{R}^n$ are two undetermined vectors. From (3.5) and (3.6), we have

$$\mathbf{x}_{k+1} - \mathbf{x}_k = -W(\mathbf{x}_k + \mathbf{x}_{k+1}).$$

Then, by (3.7), it yields that

$$\mathbf{g}(\mathbf{x}_k)^\top(\mathbf{x}_{k+1} - \mathbf{x}_k) = -[(\mathbf{g}(\mathbf{x}_k)^\top\mathbf{a})\mathbf{b}^\top - (\mathbf{g}(\mathbf{x}_k)^\top\mathbf{b})\mathbf{a}^\top](\mathbf{x}_k + \mathbf{x}_{k+1}).$$

Since the objective $f(\cdot)$ in (3.2) is zero-order homogeneous, we have

$$\mathbf{x}_k^\top\mathbf{g}(\mathbf{x}_k) = 0, \tag{3.8}$$

which means that the gradient $\mathbf{g}(\mathbf{x}_k)$ of $\mathbf{x}_k \in \mathbf{S}_{n-1}$ is located in the tangent plane of $\mathbf{S}_{n-1}$ at $\mathbf{x}_k$. For convenience, we choose

$$\mathbf{a} = \mathbf{x}_k \quad \text{and} \quad \mathbf{b} = -\alpha\mathbf{g}(\mathbf{x}_k). \tag{3.9}$$

Here, α is a positive parameter, which serves as a step size. According to this selection and (3.8), we obtain

$$\begin{aligned}\mathbf{g}(\mathbf{x}_k)^\top(\mathbf{x}_{k+1} - \mathbf{x}_k) &= -\alpha\|\mathbf{g}(\mathbf{x}_k)\|^2\mathbf{x}_k^\top(\mathbf{x}_k + \mathbf{x}_{k+1}) \\ &= -\alpha\|\mathbf{g}(\mathbf{x}_k)\|^2(1 + \mathbf{x}_k^\top Q\mathbf{x}_k).\end{aligned}$$

Since -1 is not an eigenvalue of the orthogonal matrix Q, we have $1 + \mathbf{x}_k^\top Q\mathbf{x}_k > 0$ for $\mathbf{x}_k^\top\mathbf{x}_k = 1$. Therefore, the conclusion $\mathbf{g}(\mathbf{x}_k)^\top(\mathbf{x}_{k+1} - \mathbf{x}_k) < 0$ holds for any positive step size α. We summarize the iterative process in the following Theorem.

Theorem 3.3 *Suppose that the new iterate* $\mathbf{x}_{k+1}$ *is generated by (3.5), (3.6), (3.7), and (3.9). Then, the iterative scheme is*

$$\mathbf{x}_{k+1}(\alpha) = \frac{1-\alpha^2\|\mathbf{g}(\mathbf{x}_k)\|^2}{1+\alpha^2\|\mathbf{g}(\mathbf{x}_k)\|^2}\mathbf{x}_k - \frac{2\alpha}{1+\alpha^2\|\mathbf{g}(\mathbf{x}_k)\|^2}\mathbf{g}(\mathbf{x}_k). \tag{3.10}$$

This theorem means that we do not need to generate the orthogonal matrix Q explicitly. It is cheap to preserve iterates on the unit sphere using (3.10) by the Cayley transform. Whereafter, we devote to choosing a suitable step size α by an inexact curvilinear search. At the beginning, we give a useful theorem.

Theorem 3.4 *Suppose that the new iterate* $\mathbf{x}_{k+1}(\alpha)$ *is generated by (3.10). Then, we have*

$$\left.\frac{\mathrm{d}f(\mathbf{x}_{k+1}(\alpha))}{\mathrm{d}\alpha}\right|_{\alpha=0} = -2\|\mathbf{g}(\mathbf{x}_k)\|^2.$$

Proof By some calculations, we get $\mathbf{x}_{k+1}(0) = \mathbf{x}_k$ and $\mathbf{x}'_{k+1}(0) = -2\mathbf{g}(\mathbf{x}_k)$. Therefore,

$$\left.\frac{\mathrm{d}f(\mathbf{x}_{k+1}(\alpha))}{\mathrm{d}\alpha}\right|_{\alpha=0} = \mathbf{g}(\mathbf{x}_{k+1}(0))^\top\mathbf{x}'_{k+1}(0) = -2\|\mathbf{g}(\mathbf{x}_k)\|^2.$$

□

According to Theorem 3.4, for any constant $\eta \in (0, 2)$, there exists a positive scalar $\tilde{\alpha}$ such that for all $\alpha \in (0, \tilde{\alpha}]$,

$$f(\mathbf{x}_{k+1}(\alpha)) - f(\mathbf{x}_k) \le -\eta\alpha\|\mathbf{g}(\mathbf{x}_k)\|^2.$$

Hence, the curvilinear search process is well-defined.

Algorithm 1 An optimization algorithm

1: Give the generating vector $\mathbf{v}$ of a Hankel tensor $\mathscr{H}$, the tensor $\mathscr{B}$, an initial unit iterate $\mathbf{x}_1$, parameters $\eta \in (0, \frac{1}{2}]$, $\beta \in (0, 1)$, $\bar{\alpha}_1 = 1 \le \alpha_{\max}$, and $k \leftarrow 1$.
2: **while** the sequence of iterates does not converge **do**
3: Compute $\mathscr{H}\mathbf{x}_k^m$ and $\mathscr{H}\mathbf{x}_k^{m-1}$ by the fast Hankel tensor–vector product introduced in Sect. 1.
4: Calculate $\mathscr{B}\mathbf{x}_k^m$, $\mathscr{B}\mathbf{x}_k^{m-1}$, $\lambda_k = f(\mathbf{x}_k) = \frac{\mathscr{H}\mathbf{x}_k^m}{\mathscr{B}\mathbf{x}_k^m}$ and $\mathbf{g}(\mathbf{x}_k)$ by (3.3).
5: Choose the smallest nonnegative integer ℓ and determine $\alpha_k = \beta^\ell\bar{\alpha}_k$ such that

$$f(\mathbf{x}_{k+1}(\alpha_k)) \le f(\mathbf{x}_k) - \eta\alpha_k\|\mathbf{g}(\mathbf{x}_k)\|^2, \tag{3.11}$$

where $\mathbf{x}_{k+1}(\alpha)$ is calculated by (3.10).
6: Update the iterate $\mathbf{x}_{k+1} = \mathbf{x}_{k+1}(\alpha_k)$.
7: Choose an initial step size $\bar{\alpha}_{k+1} \in (0, \alpha_{\max}]$ for the next iteration.
8: $k \leftarrow k+1$.
9: **end while**

Now, we present the optimization algorithm formally in Algorithm 1 for the smallest generalized eigenvalue of a Hankel tensor and its associated eigenvector. If we try to compute the largest generalized eigenvalue of a Hankel tensor, we only need to change the objective in (3.2) to

$$\hat{f}(\mathbf{x}) = -\frac{\mathscr{H}\mathbf{x}^m}{\mathscr{B}\mathbf{x}^m}.$$

3.3 Convergence Analysis

If Algorithm 1 terminates finitely, there exists a positive integer k such that $\mathbf{g}(\mathbf{x}_k) = 0$. According to Theorem 3.2, $f(\mathbf{x}_k)$ is a generalized eigenvalue and $\mathbf{x}_k$ is its associated generalized eigenvector.

Next, we assume that Algorithm 1 generates an infinite sequence of iterates. Since $\mathscr{B}$ is positive definite, $f(\mathbf{x})$ is twice continuously differentiable. Owing to the compactness of $\mathbf{S}_{n-1}$, we have the following lemma.

Lemma 3.1 *There exists a positive constant $M > 1$ such that for all $\mathbf{x} \in \mathbf{S}_{n-1}$,*

$$|f(\mathbf{x})| \leq M, \quad \|\mathbf{g}(\mathbf{x})\| \leq M, \quad \text{and} \quad \|H(\mathbf{x})\| \leq M.$$

Theorem 3.5 *Suppose that the infinite sequence $\{\lambda_k\}$ is generated by Algorithm 1. Then, the sequence $\{\lambda_k\}$ is monotonously decreasing. And there exists a λ_* such that*

$$\lim_{k\to\infty} \lambda_k = \lambda_*.$$

Proof As $\lambda_k = f(\mathbf{x}_k)$ is monotonously decreasing and has a lower bound, the infinite sequence $\{\lambda_k\}$ must converge to a unique λ_*. □

This theorem means that the sequence of generalized eigenvalues converges. To show the convergence of iterates, we first prove that step sizes bound away from zero.

Lemma 3.2 *Suppose that the step size α_k is generated by Algorithm 1. Then, for all k, we get*

$$\alpha_k \geq \frac{(2-\eta)\beta}{5M} \equiv \alpha_{\min} > 0. \tag{3.12}$$

Proof Let $\underline{\alpha} \equiv \frac{2-\eta}{5M}$. According to Algorithm 1, it is sufficient to prove that the inequality (3.11) holds if $\alpha_k \in (0, \underline{\alpha}]$.

From (3.10) and (3.8), we get

$$\|\mathbf{x}_{k+1}(\alpha) - \mathbf{x}_k\| = \frac{2\alpha\|\mathbf{g}(\mathbf{x}_k)\|}{\sqrt{1+\alpha^2\|\mathbf{g}(\mathbf{x}_k)\|^2}}. \tag{3.13}$$

From the mean value theorem, (3.10), (3.8), and (3.13), we have

$$\begin{aligned}
& f(\mathbf{x}_{k+1}(\alpha)) - f(\mathbf{x}_k) \le \mathbf{g}(\mathbf{x}_k)^\top(\mathbf{x}_{k+1}(\alpha) - \mathbf{x}_k) + \frac{1}{2}M\|\mathbf{x}_{k+1}(\alpha) - \mathbf{x}_k\|^2 \\
&= \frac{1}{1+\alpha^2\|\mathbf{g}(\mathbf{x}_k)\|^2}\left(-2\alpha^2\|\mathbf{g}(\mathbf{x}_k)\|^2\mathbf{g}(\mathbf{x}_k)^\top\mathbf{x}_k - 2\alpha\|\mathbf{g}(\mathbf{x}_k)\|^2 + \frac{M}{2}4\alpha^2\|\mathbf{g}(\mathbf{x}_k)\|^2\right) \\
&\le \frac{\alpha\|\mathbf{g}(\mathbf{x}_k)\|^2}{1+\alpha^2\|\mathbf{g}(\mathbf{x}_k)\|^2}(4\alpha M - 2).
\end{aligned}$$

It can be verified that $4\alpha M - 2 \le -\eta(1+\alpha^2 M^2)$ for all $\alpha \in (0, \underline{\alpha}]$. Therefore, we have

$$f(\mathbf{x}_{k+1}(\alpha)) - f(\mathbf{x}_k) \le \frac{-\eta(1+\alpha^2M^2)}{1+\alpha^2\|\mathbf{g}(\mathbf{x}_k)\|^2}\alpha\|\mathbf{g}(\mathbf{x}_k)\|^2 \le -\eta\alpha\|\mathbf{g}(\mathbf{x}_k)\|^2.$$

□

Theorem 3.6 *Suppose that the infinite sequence $\{\mathbf{x}_k\}$ is generated by Algorithm 1. Then, the sequence $\{\mathbf{x}_k\}$ has at least an accumulation point. And we have*

$$\lim_{k\to\infty}\|\mathbf{g}(\mathbf{x}_k)\| = 0. \tag{3.14}$$

That is to say, every accumulation point of $\{\mathbf{x}_k\}$ is a generalized eigenvector whose associated generalized eigenvalue is λ_*.

Proof From Lemma 3.1, (3.11) and (3.12), we have

$$2M \ge f(\mathbf{x}_1) - \lambda_* = \sum_{k=1}^{\infty} f(\mathbf{x}_k) - f(\mathbf{x}_{k+1}) \ge \sum_{k=1}^{\infty}\eta\alpha_k\|\mathbf{g}(\mathbf{x}_k)\|^2 \ge \eta\alpha_{\min}\sum_{k=1}^{\infty}\|\mathbf{g}(\mathbf{x}_k)\|^2.$$

It yields that

$$\sum_k\|\mathbf{g}(\mathbf{x}_k)\|^2 \le \frac{2M}{\eta\alpha_{\min}} < +\infty. \tag{3.15}$$

Thus, the limit (3.14) holds.

Let $\mathbf{x}_\infty$ be an accumulation point of $\{\mathbf{x}_k\}$. Then $\mathbf{x}_\infty \in \mathbf{S}_{n-1}$ and $\|\mathbf{g}(\mathbf{x}_\infty)\| = 0$. According to Theorem 3.2, $\mathbf{x}_\infty$ is a generalized eigenvector whose associated eigenvalue is $f(\mathbf{x}_\infty) = \lambda_*$. □

We remark that the objective $f(\mathbf{x})$ in (3.2) is a semi-algebraic function since its graph

$$\text{Graph} f = \{(\mathbf{x}, \lambda) : \mathscr{H}\mathbf{x}^k - \lambda \mathscr{B}\mathbf{x}^k = 0\}$$

is a semi-algebraic set. Therefore, $f(\mathbf{x})$ satisfies the following Łojasiewicz property.

Theorem 3.7 (Łojasiewicz property) *Suppose that* $\mathbf{x}_*$ *is a stationary point of* $f(\mathbf{x})$. *Then, there is a neighborhood* $\mathbf{U}$ *of* $\mathbf{x}_*$, *an exponent* $\theta \in [0, 1)$, *and a positive constant* C_L *such that for all* $\mathbf{x} \in \mathbf{U}$, *the following inequality holds*

$$|f(\mathbf{x}) - f(\mathbf{x}_*)|^{\theta} \le C_L \|\mathbf{g}(\mathbf{x})\|. \tag{3.16}$$

Here, we define $0^0 \equiv 0$.

Using Łojasiewicz property [22], we will prove that the infinite sequence of iterates $\{\mathbf{x}_k\}$ converges to a unique accumulation point.

Lemma 3.3 *Suppose that* $\mathbf{x}_*$ *is a stationary point of* $f(\mathbf{x})$, *and* $\mathbf{B}(\mathbf{x}_*, \rho) = \{\mathbf{x} \in \mathbf{R}^n : \|\mathbf{x} - \mathbf{x}_*\| \le \rho\} \subseteq \mathbf{U}$ *is a neighborhood of* $\mathbf{x}_*$. *Let* $\mathbf{x}_1$ *be an initial point satisfying*

$$\rho > \rho(\mathbf{x}_1) \equiv \frac{2C_L}{\eta(1-\theta)} |f(\mathbf{x}_1) - f(\mathbf{x}_*)|^{1-\theta} + \|\mathbf{x}_1 - \mathbf{x}_*\|. \tag{3.17}$$

Then, the following assertions hold:

$$\mathbf{x}_k \in \mathbf{B}(\mathbf{x}_*, \rho), \qquad k = 1, 2, \ldots, \tag{3.18}$$

and

$$\sum_{k=1}^{\infty} \|\mathbf{x}_{k+1} - \mathbf{x}_k\| \le \frac{2C_L}{\eta(1-\theta)} |f(\mathbf{x}_1) - f(\mathbf{x}_*)|^{1-\theta}. \tag{3.19}$$

Proof We proceed by induction. Obviously, $\mathbf{x}_1 \in \mathbf{B}(\mathbf{x}_*, \rho)$.

Now, we assume that $\mathbf{x}_i \in \mathbf{B}(\mathbf{x}_*, \rho)$ for all $i = 1, \ldots, k$. Hence, Łojasiewicz property holds in these points. Let

$$\phi(t) \equiv \frac{C_L}{1-\theta} |t - f(\mathbf{x}_*)|^{1-\theta}.$$

It is clear that $\phi(t)$ is a concave function for $t > f(\mathbf{x}_*)$. Therefore, for $i = 1, \ldots, k$, we have

$$\begin{aligned}
\phi(f(\mathbf{x}_i)) - \phi(f(\mathbf{x}_{i+1})) &\ge \phi'(f(\mathbf{x}_i))(f(\mathbf{x}_i) - f(\mathbf{x}_{i+1})) \\
&= C_L |f(\mathbf{x}_i) - f(\mathbf{x}_*)|^{-\theta} (f(\mathbf{x}_i) - f(\mathbf{x}_{i+1})) \\
[\text{Łojasiewicz property}] &\ge \|\mathbf{g}(\mathbf{x}_i)\|^{-1} (f(\mathbf{x}_i) - f(\mathbf{x}_{i+1})) \\
[\text{by (3.11)}] &\ge \|\mathbf{g}(\mathbf{x}_i)\|^{-1} \eta \alpha_i \|\mathbf{g}(\mathbf{x}_i)\|^2
\end{aligned}$$

$$
\begin{aligned}
&\geq \frac{\eta\alpha_i \|\mathbf{g}(\mathbf{x}_i)\|}{\sqrt{1+\alpha_i^2\|\mathbf{g}(\mathbf{x}_i)\|^2}} \\
[\text{by } (3.13)] &= \frac{\eta}{2}\|\mathbf{x}_{i+1}-\mathbf{x}_i\|.
\end{aligned}
$$

Then,

$$
\begin{aligned}
\|\mathbf{x}_{k+1}-\mathbf{x}_*\| &\leq \sum_{i=1}^{k}\|\mathbf{x}_{i+1}-\mathbf{x}_i\| + \|\mathbf{x}_1-\mathbf{x}_*\| \\
&\leq \frac{2}{\eta}\sum_{i=1}^{k}[\phi(f(\mathbf{x}_i))-\phi(f(\mathbf{x}_{i+1}))] + \|\mathbf{x}_1-\mathbf{x}_*\| \\
&\leq \frac{2}{\eta}\phi(f(\mathbf{x}_1)) + \|\mathbf{x}_1-\mathbf{x}_*\| \\
&< \rho.
\end{aligned}
$$

Hence, we get $\mathbf{x}_{k+1} \in \mathbf{B}(\mathbf{x}_*, \rho)$ and (3.18) holds. Moreover,

$$
\sum_{k=1}^{\infty}\|\mathbf{x}_{k+1}-\mathbf{x}_k\| \leq \frac{2}{\eta}\sum_{k=1}^{\infty}[\phi(f(\mathbf{x}_k))-\phi(f(\mathbf{x}_{k+1}))] \leq \frac{2}{\eta}\phi(f(\mathbf{x}_1)).
$$

The inequality (3.19) is valid. □

Theorem 3.8 *Suppose that Algorithm 1 generates an infinite sequence of iterates* $\{\mathbf{x}_k\}$. *Then,*

$$
\sum_{k=1}^{\infty}\|\mathbf{x}_{k+1}-\mathbf{x}_k\| < +\infty.
$$

Hence, the total sequence $\{\mathbf{x}_k\}$ *has a finite length. Moreover, there exists a stationary point* $\mathbf{x}_* \in \mathbf{S}_{n-1}$ *such that*

$$
\lim_{k\to\infty}\mathbf{x}_k = \mathbf{x}_*.
$$

Proof Owing to the compactness of $\mathbf{S}_{n-1}$, there exists an accumulation point $\mathbf{x}_*$ of iterates $\{\mathbf{x}_k\}$. By Theorem 3.6, $\mathbf{x}_*$ is also a stationary point. Then, there exists an iteration K such that $\rho(\mathbf{x}_K) < \rho$. Hence, by Lemma 3.3, we have $\sum_{k=K}^{\infty}\|\mathbf{x}_{k+1} - \mathbf{x}_k\| < \infty$. This inequality implies that $\{\mathbf{x}_k\}$ is a Cauchy sequence. □

For the target of computing the smallest generalized eigenvalue of a Hankel tensor, we start Algorithm 1 from plenty of randomly initial points. Then, we regard the resulting smallest objective value as the smallest generalized eigenvalue of this tensor. The following theorem reveals the successful probability of this strategy.

Theorem 3.9 *Suppose that we start Algorithm 1 from N initial points which are sampled from $\mathbf{S}_{n-1}$ uniformly and regard the resulting smallest objective value as* *the smallest generalized eigenvalue. Then, there exists a constant* $\varsigma \in (0, 1]$ *such that the probability of obtaining the smallest generalized eigenvalue is at least*

$$1 - (1 - \varsigma)^N. \tag{3.20}$$

Therefore, if the number of samples N is large enough, we obtain the smallest generalized eigenvalue with a high probability.

Proof Suppose that $\mathbf{x}^*$ is an generalized eigenvector corresponding to the smallest generalized eigenvalue and $\mathbf{B}(\mathbf{x}^*, \rho) \subseteq \mathbf{U}$ is a neighborhood as defined in Lemma 3.3. Since the function $\rho(\cdot)$ in (3.17) is continuous and satisfies $\rho(\mathbf{x}^*) = 0 < \rho$, there exists a neighborhood $\mathbf{V}(\mathbf{x}^*) \equiv \{\mathbf{x} \in \mathbf{S}_{n-1} : \rho(\mathbf{x}) < \rho\} \subseteq \mathbf{U}$. That means, if an initial point $\mathbf{x}_1$ happens to be sampled from $\mathbf{V}(\mathbf{x}^*)$, the total sequence of iterates $\{\mathbf{x}_c\}$ converges to $\mathbf{x}^*$ by Lemma 3.3 and Theorem 3.8. Next, we consider the probability of this random event.

Let S and A be hypervolumes of $(n-1)$ dimensional solids $\mathbf{S}_{n-1}$ and $\mathbf{V}(\mathbf{x}^*)$ respectively. (Specifically, the "area" of the surface of $\mathbf{S}_{n-1}$ in $\mathbf{R}^n$ is S and the "area" of the surface of $\mathbf{V}(\mathbf{x}^*) \subseteq \mathbf{S}_{n-1}$ in $\mathbf{R}^n$ is A. Hence, $A \leq S$.) Then, S and A are positive. By the geometric probability model, the probability of one randomly initial point $\mathbf{x}_1 \in \mathbf{V}(\mathbf{x}^*)$ is

$$\varsigma \equiv \frac{A}{S} > 0.$$

In fact, once $\{\mathbf{x}_k\} \cap \mathbf{V}(\mathbf{x}^*) \neq \emptyset$, we could obtain the smallest generalized eigenvalue.

By the binomial distribution with parameters N and ς, we obtain the probabilistic estimation (3.20) straightforwardly. □

3.4 Exponential Data Fitting

Exponential data fitting is a subspace-based approach for signals modeled as a sum of exponentially damped sinusoids [23, 217, 218]. For example, we consider a one-dimensional noiseless complex signal with length N:

$$x_n = \sum_{k=1}^{K} a_k \exp(\mathrm{i}\varphi_k) \exp((-\alpha_k + \mathrm{i}\omega_k) n \Delta t) \qquad \text{for } n = 0, 1, \ldots, N-1,$$

where i is the imaginary unit, Δt is the sampling time interval. The amplitudes a_k, the phases φ_k, the damping factors α_k, and the pulsations ω_k are under-determined parameters. Signals $\{x_n\}_{n=0,1,\ldots,N-1}$ could be written in a compact form

$$x_n = \sum_{k=1}^{K} c_k z_k^n,$$

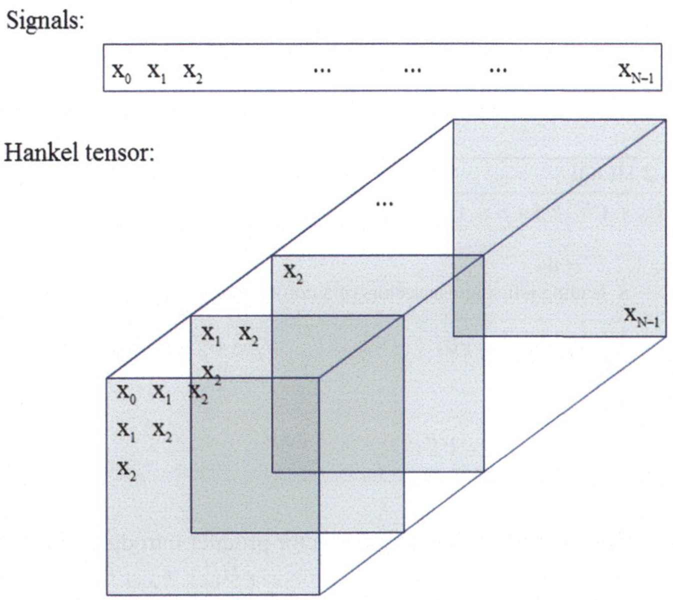

Fig. 3.1 A Hankel tensor of the signals

where $c_k = a_k \exp(\mathrm{i}\varphi_k)$ is the complex amplitude and $z_k = \exp((-\alpha_k + \mathrm{i}\omega_k)\Delta t)$ is the pole of the signal. We first introduce the method for estimating poles z_k for $k = 1, \ldots, K$. Then, complex amplitudes c_k are obtained by solving a Vandermonde system straightforwardly.

Using the generating vector $\mathbf{x} = (x_0, x_1, \ldots, x_{N-1})^\top$, We establish the Hankel tensor $\mathscr{H} \in \mathbf{C}^{I_1 \times I_2 \times \cdots \times I_m}$, where $I_1 + I_2 + \cdots + I_m = N + m - 1$; See Fig. 3.1 for an example. This Hankel tensor has a Vandermonde decomposition [217, 224]

$$\mathscr{H} = \mathscr{D} \times_1 Z_1 \times_2 Z_2 \cdots \times_m Z_m,$$

where $\mathscr{D}$ is a diagonal matrix and

$$Z_p = \begin{pmatrix} 1 & 1 & \cdots & 1 \\ z_1 & z_2 & \cdots & z_K \\ \vdots & \vdots & & \vdots \\ z_1^{I_p-1} & z_2^{I_p-1} & \cdots & z_K^{I_p-1} \end{pmatrix} \quad \text{for } p = 1, \ldots, m \tag{3.21}$$

are Vandermonde matrices. However, it is not straightforward to find the Vandermonde decomposition for a given Hankel tensor. Since K is usually a small integer, we turn to the best rank-$(K, K, \ldots, K)$ approximation of the Hankel tensor

$$\min \ \|\mathscr{H} - \mathscr{A} \times_1 U_1 \times_2 U_2 \cdots \times_m U_m\|_F^2,$$

where $\mathscr{A} \in \mathbf{C}^{K\times K\times\cdots\times K}$ is a core tensor and matrices $U_p \in \mathbf{C}^{I_p\times K}$ for $p = 1, \ldots, m$ have orthonormal columns. The best rank-$(K, K, \ldots, K)$ approximation of $\mathscr{H}$ could be computed by the following higher-order orthogonal iterations (HOOI) [75].

Algorithm 2 HOOI

1: Initialize $U_p \in \mathbf{C}^{I_p\times K}$ for $p = 1, \ldots, m$.
2: **repeat**
3:    **for** $p = 1, \ldots, m$ **do**
4:       $U_p \leftarrow K$ leading left singular vectors of a unfold matrix

$$\text{Unfold}_p(\mathscr{H} \times_1 U_1^H \cdots \times_{p-1} U_{p-1}^H \times_{p+1} U_{p+1}^H \cdots \times_m U_m^H).$$

5:    **end for**
6: **until** convergence
7: $\mathscr{A} = \mathscr{H} \times_1 U_1^H \times_2 U_2^H \cdots \times_m U_m^H$.

Now, we apply the fast Hankel tensor–vector product introduced in Sect. 3.1 for computing the tensor product $\mathscr{H} \times_1 U_1^H \cdots \times_{p-1} U_{p-1}^H \times_{p+1} U_{p+1}^H \cdots \times_m U_m^H \in \mathbf{C}^{K\times\cdots\times K I_p K\times\cdots\times K}$. For example, we address the computation of the $(k_1, \ldots, k_{p-1}, :, k_{p+1}, \ldots, k_m)$ column of the tensor $\mathscr{H} \times_1 U_1^H \cdots \times_{p-1} U_{p-1}^H \times_{p+1} U_{p+1}^H \cdots \times_m U_m^H$. Recalling that $\mathbf{x} \in \mathbf{C}^N$ is the generating vector of the Hankel tensor $\mathscr{H}$, we consider the anti-circulant tensor $\mathscr{C} \in \mathbf{C}^{N\times\cdots\times N}$ and matrices

$$V_p = \begin{pmatrix} \bar{U}_p \\ 0_{(N-I_p)\times K} \end{pmatrix} = (\mathbf{v}_1^{(p)}, \ldots, \mathbf{v}_K^{(p)}) \in \mathbf{C}^{N\times K}$$

for $p \in [m]$, where $\mathbf{v}_k^{(p)}$ is the kth column of a matrix V_p. We note that the $(k_1, \ldots, k_{p-1}, :, k_{p+1}, \ldots, k_m)$ column of the tensor $\mathscr{H} \times_1 U_1^H \cdots \times_{p-1} U_{p-1}^H \times_{p+1} U_{p+1}^H \cdots \times_m U_m^H$ is the first I_p elements of the following vector

$$\begin{aligned}
&\mathscr{C} \times_1 \mathbf{v}_{k_1}^{(1)} \cdots \times_{p-1} \mathbf{v}_{k_{p-1}}^{(p-1)} \times_{p+1} \mathbf{v}_{k_{p+1}}^{(p+1)} \cdots \times_m \mathbf{v}_{k_m}^{(m)} \\
&= \text{diag}(F_N^{-1}\mathbf{x}) \times_1 (F_N\mathbf{v}_{k_1}^{(1)}) \cdots \times_{p-1} (F_N\mathbf{v}_{k_{p-1}}^{(p-1)}) \times_p F_N \times_{p+1} (F_N\mathbf{v}_{k_{p+1}}^{(p+1)}) \cdots \times_m (F_N\mathbf{v}_{k_m}^{(m)}) \\
&= F_N \left((F_N^{-1}\mathbf{x}) \circ (F_N\mathbf{v}_{k_1}^{(1)}) \circ \cdots \circ (F_N\mathbf{v}_{k_{p-1}}^{(p-1)}) \circ (F_N\mathbf{v}_{k_{p+1}}^{(p+1)}) \circ \cdots \circ (F_N\mathbf{v}_{k_m}^{(m)})\right) \\
&= \texttt{fft}\left(\texttt{ifft}(\mathbf{x}) \circ \texttt{fft}(\mathbf{v}_{k_1}^{(1)}) \circ \cdots \circ \texttt{fft}(\mathbf{v}_{k_{p-1}}^{(p-1)}) \circ \texttt{fft}(\mathbf{v}_{k_{p+1}}^{(p+1)}) \circ \cdots \circ \texttt{fft}(\mathbf{v}_{k_m}^{(m)})\right).
\end{aligned}$$

The first equality holds owing to Theorem 3.1. When the Hankel structure in $\mathscr{H}$ exploited, the computational cost of the tensor product is reduced significantly.

We argue that the subspaces spanned by columns of U_p and columns of Z_p are usually the same in the absence of noise. Specifically, there is a non-singular matrix $T \in \mathbf{C}^{K\times K}$ such that

$$U_p = Z_p T.$$

Since Z_p in (3.21) is a Vandermonde matrix and enjoys the shift-invariance property

$$Z_p{}^{\uparrow} = Z_{p_{\downarrow}} S,$$

where the up (down) arrow placed behind a matrix stands for deleting the top (bottom) row of the matrix and $S = \text{diag}(z_1, z_2, \ldots, z_K)$ is a diagonal matrix. Hence, we have

$$U_p{}^{\uparrow} = U_{p_{\downarrow}} T^{-1} S T.$$

We solve the square matrix $T^{-1}ST$ from the above system and the poles $z_1, z_2, \ldots, z_K$ of signals are eigenvalues of $T^{-1}ST$.

3.5 Notes

Generally speaking, Henkel tensors are neither sparse nor with lower-rank. However, Hankel tensors enjoy its particular properties. A theory of positive semi-definite Hankel tensors has been developed. In particular, the theory of strong Hankel tensors and inheritance properties of Hankel tensors were developed in [84, 224]. These contents can be found in Sects. 5.7 and 6.3 of [228].

Section 3.1 The main theoretical result in this chapter is Theorem 3.1, which was proposed by Ding, Qi and Wei in [83]. Using the same compact generating vector, we could handle the Hankel tensor in a large anti-circulant tensor, which contains the former as a sub-tensor. Theorem 3.1 means that the anti-circulant tensor could be diagonalized by the fast Fourier transform, which is rather cheap in computation. In this way, we obtain the fast Hankel tensor–vector product. Here, we gave a new proof for Theorem 3.1, which was established directly from the structure of the Fourier matrix.

Section 3.2 The computation of eigenvalues of large-scale tensors is a challenging problem. For special tensors and structured tensors, there exist several algorithms. For the largest H-eigenvalue of nonnegative tensors, Ng, Qi and Zhou [197] proposed the NQZ algorithm. Chen, Qi and Wang [37] established the ACSA algorithm for large-scale Hankel tensors. Chang, Chen and Qi [30] proposed the CEST algorithm for large-scale sparse tensors arising from a hypergraph. Indeed, ACSA and CEST work for H- and Z-eigenvalues of general even-order symmetric tensors, if the product of the involved tensor and arbitrary vectors could be computed efficiently.

We addressed the ASCA algorithm in Sect. 3.2, owing to the fast Hankel tensor–vector product introduced in Sect. 3.2. The Cayley transform used there could be replaced by a simple projection onto the unit sphere in implementation.

Section 3.3 Chen, Qi and Wang [37] pointed out that the objective function (3.2) of an optimization model for tensor eigenvalues satisfied the Łojasiewicz inequality; See also [22]. Hence, the convergence analysis of optimization algorithms presented here does not need additional assumptions. Chang, Chen and Qi [30] gave Theorem 3.9

based on a simple global strategy. Chen, Qi and Wang [37] analyzed the linear convergence rate of ACSA algorithm, which is dependent on the exponent in the Łojasiewicz inequality.

Section 3.4 Exponential data fitting using Hankel tensors was proposed by Papy, De Lathauwer and Van Huffel [217, 218]. Ding, Qi and Wei [83] gave the fast algorithms, which improve the efficiency of exponential data fitting notably.

3.6 Exercises

1 Prove that the generating vector $\mathbf{v}$ of an anti-circulant tensor $\mathscr{C}$ is its first column, i.e., $\mathbf{v} = \mathscr{C}(\mathbf{1}^{(1)})^{m-1}$.

2 Derive formulas of the gradient (3.3) and the Hessian (3.4) of the objective $f(\mathbf{x})$ in (3.2).

3 Try to prove Theorem 3.3 using the Sherman–Morrison–Woodbury formula.

Chapter 4
Tensor Complementarity Problems

Complementarity problems encompass several important classes of mathematical optimization problems, e.g., linear programming, quadratic programming, linear conic optimization problems, etc. Actually, we always solve an optimization problem via its optimality condition, which usually turns out to be a complementarity problem, e.g., KKT system.

The tensor complementarity problem, as a generalization of the linear complementarity problem and a further bridge towards the more general nonlinear complementarity problem, was proposed very recently. A tensor complementarity problem (TCP) can be formulated as follows: finding $\mathbf{x} \in \mathbf{R}^n$ such that

$$\mathbf{x} \geq \mathbf{0}, \ \mathscr{A}\mathbf{x}^{m-1} + \mathbf{q} \geq \mathbf{0} \ \text{and} \ \mathbf{x}^\top(\mathscr{A}\mathbf{x}^{m-1} + \mathbf{q}) = 0, \tag{4.1}$$

where $\mathscr{A} \in T_{m,n}$ and $\mathbf{q} \in \mathbf{R}^n$. It is denoted as $TCP(\mathbf{q}, \mathscr{A})$. In this chapter, we will give a comprehensive study on this topic.

As an impetus, we will consider a practical application for tensor complementarity problem, i.e., a class of m person noncooperative games, where the utility function of every player is given by a homogeneous polynomial defined by the payoff tensor of that player. Recall the bimatrix game, where the utility function of every player is given by a quadratic form defined by the payoff matrix of that player. In this sense, we say that the m person noncooperative game is a natural extension of the bimatrix game. We will call such a problem the multilinear game in this chapter. After reformulating the multilinear game to a tensor complementarity problem, it will be shown that finding a Nash equilibrium point of the multilinear game is equivalent to finding a solution of the resulted tensor complementarity problem. Especially, we will present an explicit relationship between the solutions of a multilinear game and its corresponding tensor complementarity problem, which builds a bridge connecting these two classes of problems.

Starting from the Sect. 4.2, the existence of a solution to the tensor complementarity problem will be studied. For structured tensors, such as symmetric positive-

L. Qi et al., *Tensor Eigenvalues and Their Applications*, Advances in Mechanics and Mathematics 39, https://doi.org/10.1007/978-981-10-8058-6_4

definite tensors and copositive tensors, we will derive the existence theorems on solutions of these TCPs. We will prove that a unique solution of the TCP exists under the condition of diagonalizable tensors. Moreover, we introduce several classes of tensors, such as Q-tensors, Z-tensors, semi-positive tensors, P-tensors, R-tensors, P_0-tensors, R_0-tensors, etc. It can be shown that the followings are sub-classes of Q-tensors: P-tensors, R-tensors, strictly semi-positive tensors and semi-positive R_0-tensors. We will show that a nonnegative tensor is a Q-tensor if and only if all of its principal diagonal entries are positive, and a symmetric nonnegative tensor is a Q-tensor if and only if it is strictly copositive. It is verified that the zero vector is the unique feasible solution of the tensor complementarity problem $TCP(\mathbf{q}, \mathscr{A})$ for $\mathbf{q} \geq \mathbf{0}$ if $\mathscr{A}$ is a nonnegative Q-tensor. It will be shown that within the scope of strong P_0-tensors or nonnegative tensors, four classes of tensors, i.e., R_0-tensors, R-tensors, ER-tensors and Q-tensors, are all equivalent. We will construct several examples to show that several famous results related to Q-matrices cannot be extended to the tensor space.

We will also study in details the (strictly) semi-positive tensors. It will be proved that a real tensor is strictly semi-positive if and only if the corresponding tensor complementarity problem has a unique solution for any nonnegative vector and that a real tensor is semi-positive if and only if the corresponding tensor complementarity problem has a unique solution for any positive vector. It is shown that a real symmetric tensor is a (strictly) semi-positive tensor if and only if it is (strictly) copositive. Global error bound analysis for tensor complementarity problem will be defined by a strictly semi-positive tensor. For strictly semi-positive tensor, we will prove that all $H^+(Z^+)$-eigenvalues of each principal sub-tensor are positive. We will introduce two new constants associated with $H^+(Z^+)$-eigenvalues of a strictly semi-positive tensor. With the help of these two constants, we will establish upper bounds of an important quantity whose positivity is a necessary and sufficient condition for a general tensor to be a strictly semi-positive tensor. The monotonicity and boundedness of such a quantity are established as well.

The next subclass to be examined is the Z-tensor. Z-tensors are tensors with nonpositive off-diagonal entries. We will describe various equivalent conditions for a Z-tensor to have the Q-property. These conditions/properties include the strong M-tensor property, the S-property, positive stable property, strict semi-monotonicity property, etc. Based on degree-theoretic ideas, we will prove some refined results for even order tensors. A sufficient and easily checkable condition for a strong M-tensor to have unique complementarity solutions is also established. In particular, a special type of tensor complementarity problems with Z-tensors will be considered. Under some mild conditions, we will show that to pursuit the sparsest solutions is equivalent to solving a polynomial optimization problem with a linear objective function. The involved conditions guarantee the desired exact relaxation and enable us to achieve a global optimal solution to the relaxed non-convex polynomial programming problem.

The final subclass of tensors to be studied is the exceptionally regular tensor (ER-tensor). We will show that an exceptionally regular tensor can be an R-tensor under some conditions. We will also show that within the scope of the semi-positive tensors, the class of exceptionally regular tensors coincides with the class of R-tensors. In addition, we will consider the tensor complementarity problem with an exceptionally regular tensor or an R-tensor or a P_0+R_0-tensor, and show that the solution sets of these classes of tensor complementarity problems are nonempty and compact.

This chapter will be closed with the discussion on the property of global uniqueness and solvability (GUS-property) for a class of TCPs. We will present a class of TCPs whose solution set is nonempty and compact. In particular, we will introduce a class of related structured tensors, and verify that the corresponding TCP has the GUS-property.

4.1 Preliminaries for Tensor Complementarity Problems

Our aim in this brief section is to recall some basic definitions and essential conclusions in nonlinear complementarity problems, which are useful in the study of tensor complementarity problems.

As shown in the beginning of this chapter, the TCP problem (4.1) is a natural extension of the linear complementarity problem (LCP), which consists in finding a vector $\mathbf{x} \in \mathbf{R}^n$ that satisfies a certain system of inequalities:

$$\mathbf{x} \geq \mathbf{0}, \quad A\mathbf{x} + \mathbf{q} \geq \mathbf{0}, \quad \mathbf{x}^\top (A\mathbf{x} + \mathbf{q}) = 0.$$

The LCP above has been a subject with a rich mathematical theory, a variety of algorithms, and a wide range of applications in applied science and technology. More of this early history about LCPs can be found in [63]. Recently, with an emerging interest in the assets of multilinear algebra concentrated on the higher order tensors, more and more researchers pay their attention to the TCP (4.1). One of the typical applications for TCP is for the m person game problem, which can be reformulated and solved by a corresponding TCP. We will show the detail of the application in Sect. 4.2.

On the other hand, since the polynomial function defined by a tensor is a nonlinear function, the TCP is a special instance of the nonlinear complementarity problem, which has also been studied extensively in dealing with existence, uniqueness, computation, and applications [91]. Therefore, the entire theory of nonlinear complementarity problems is applicable to tensor complementarity problems. However, by the homogeneity of function corresponded by tensors, some special results for TCP can be obtained.

In the following, we will list some useful notions and conclusions from the nonlinear complementarity problem.

Let F be a mapping from $\mathbf{R}^n$ to itself. Denote the nonlinear complementarity problem by $NCP(F)$, which aims to find a vector $\mathbf{x}^* \in \mathbf{R}^n_+$ such that

$$F(\mathbf{x}^*) \in \mathbf{R}^n_+, \quad (\mathbf{x}^*)^\top F(\mathbf{x}^*) = 0.$$

Let $\nabla F(\mathbf{x})$ denote the Jacobian matrix for $F(\mathbf{x})$. The following lemma is about the existence and uniqueness of solutions for the nonlinear complementarity problem.

Lemma 4.1 (Cottle 1966) *Let $F : \mathbf{R}^n_+ \to \mathbf{R}^n$ be continuously differentiable. Suppose that there exists $\delta \in (0, 1)$ such that all principal minors of the Jacobian matrix $\nabla F(\mathbf{x})$ are bounded between δ and δ^{-1}, for all $\mathbf{x} \in \mathbf{R}^n_+$. Then, the $NCP(F)$ has a unique solution.*

The notion below for copositive mapping is motivated by the definition of copositive matrix.

Definition 4.1 Let $X \subseteq \mathbf{R}^n$. Define the mapping $F : X \to \mathbf{R}^n$. Then,
(1) F is **copositive** with respect to X, if and only if

$$\mathbf{x}^\top (F(\mathbf{x}) - F(\mathbf{0})) \geq 0, \quad \forall\, \mathbf{x} \in X.$$

(2) F is **strictly copositive** with respect to X, if and only if

$$\mathbf{x}^\top (F(\mathbf{x}) - F(\mathbf{0})) > 0, \quad \forall\, \mathbf{x} \in X, \mathbf{x} \neq \mathbf{0}.$$

(3) F is **strongly copositive** with respect to X, if and only if there is a scalar $\alpha > 0$ such that

$$\mathbf{x}^\top (F(\mathbf{x}) - F(\mathbf{0})) \geq \alpha \|\mathbf{x}\|_2^2, \quad \forall\, \mathbf{x} \in X.$$

Suppose the conditions in Lemma 4.1 hold. If further the mapping F is strictly copositive, then the following result holds.

Lemma 4.2 (Moré 1974) *Suppose $F : \mathbf{R}^n_+ \to \mathbf{R}^n$ is continuous and strictly copositive with respect to $\mathbf{R}^n_+$. If there exists a mapping $c : \mathbf{R}_+ \to \mathbf{R}$ such that $c(\lambda) \to \infty$ as $\lambda \to \infty$, and for all $\lambda \geq 1, \mathbf{x} \geq \mathbf{0}$,*

$$\mathbf{x}^\top (F(\lambda \mathbf{x}) - F(\mathbf{0})) \geq c(\lambda) \mathbf{x}^\top (F(\mathbf{x}) - F(\mathbf{0})), \tag{4.2}$$

then the problem $NCP(F)$ has a nonempty, compact solution set.

Another useful notion is **d**-regular mapping.

Definition 4.2 (*Harker, Pang 1990*) For any vector $\mathbf{x} \in \mathbf{R}^n_+$, we define the index sets

$$I_+(\mathbf{x}) = \{i \,:\, x_i > 0\} \text{ and } I_0(\mathbf{x}) = \{i \,:\, x_i = 0\}.$$

Let $\mathbf{d} \in \mathbf{R}^n$ be an arbitrary vector. The mapping $G : \mathbf{R}^n \to \mathbf{R}^n$ is said to be **d**-regular if the following system has no solution in $(\mathbf{x}, t) \in \mathbf{R}^n_+ \times \mathbf{R}_+$ with $\mathbf{x} \neq \mathbf{0}$,

$$\begin{aligned} G_i(\mathbf{x}) + t d_i &= 0, \quad & i \in I_+(\mathbf{x}), \\ G_i(\mathbf{x}) + t d_i &\geq 0, \quad & i \in I_0(\mathbf{x}). \end{aligned} \tag{4.3}$$

If F is strictly copositive with respect to $\mathbf{R}^n_+$, then it is not difficult to know that the mapping $G(\cdot) = F(\cdot) - F(\mathbf{0})$ is $\mathbf{d}$-regular for any $\mathbf{d} > \mathbf{0}$. The following lemma presents a solution existence result for the nonlinear complementarity problem with $\mathbf{d}$-regular mapping.

Lemma 4.3 (Karamardian 1972) *Let F be a continuous mapping from $\mathbf{R}^n$ into itself. Let $G(\mathbf{x}) = F(\mathbf{x}) - F(\mathbf{0})$. Suppose that G is positively homogeneous of degree $\alpha > 0$ and that G is $\mathbf{d}$-regular for some $\mathbf{d} > \mathbf{0}$. Then, the problem $NCP(F)$ has a nonempty, compact solution set.*

If further $F : S \to \mathbf{R}^{n+1}$ is continuous with S being a compact set, we have the following results.

Lemma 4.4 (Berman, Plemmons 1994) *Let $S = \{\mathbf{x} \in \mathbf{R}^{n+1}_+ \mid \sum_{i=1}^{n+1} x_i = 1\}$. Assume that $F : S \to \mathbf{R}^{n+1}$ is continuous on S. Then there exists $\bar{\mathbf{x}} \in S$ such that*

$$\begin{aligned}
&\mathbf{x}^\top F(\bar{\mathbf{x}}) \geq \bar{\mathbf{x}}^\top F(\bar{\mathbf{x}}), \quad \forall\, \mathbf{x} \in S, \\
&(F(\bar{\mathbf{x}}))_k = \min_{i \in [n+1]} (F(\bar{\mathbf{x}}))_i = \omega \ \ if \ \bar{x}_k > 0, \\
&(F(\bar{\mathbf{x}}))_k \geq \omega \ \ if \ \bar{x}_k = 0.
\end{aligned}$$

Now, we recall the concept of Z-function, which is introduced below.

Definition 4.3 A mapping $F : \mathbf{R}^n \to \mathbf{R}^n$ is said to be a **Z-function** if for every $\mathbf{x}$, $\mathbf{y}, \mathbf{z} \in \mathbf{R}^n_+$ with $\min\{\mathbf{x}, \mathbf{y} - \mathbf{z}\} = 0$, we have

$$\mathbf{x}^\top (F(\mathbf{y}) - F(\mathbf{z})) \leq 0.$$

Here $\min\{\mathbf{x}, \mathbf{y} - \mathbf{z}\} \in \mathbf{R}^n$ and its ith element equals $\min\{x_i, y_i - z_i\}$.

Then, we have the following results for the Jacobian matrix of a Z-function.

Proposition 4.1 (Isac 1992) *A Gâteaux continuous differentiable function F : $\mathbf{R}^n \to \mathbf{R}^n$ is a Z-function if and only if $\nabla F(\mathbf{x})$ is a Z-matrix for any $\mathbf{x} \in \mathbf{R}^n_+$.*

Lemma 4.5 (Isac 1992) *If $F : \mathbf{R}^n \to \mathbf{R}^n$ is a Z-function, then the following implication holds:*

$$\mathbf{x} \in \mathbf{R}^n_+, \mathbf{y} \in \mathbf{R}^n_+, \mathbf{x}^\top \mathbf{y} = 0 \Rightarrow \mathbf{x}^\top (F(\mathbf{y}) - F(\mathbf{0})) \leq 0. \tag{4.4}$$

Moreover, if $F(\mathbf{x}) = A\mathbf{x}$ is a linear function, then A is a Z-matrix, which is equivalent to the following implication:

$$\mathbf{x} \in \mathbf{R}^n_+, \ \mathbf{y} \in \mathbf{R}^n_+, \ \mathbf{x}^\top \mathbf{y} = 0 \Rightarrow \mathbf{x}^\top A\mathbf{y} \leq 0. \tag{4.5}$$

At last, we recall the concept of P-function.

Definition 4.4 Let F be a mapping from $K \subseteq \mathbf{R}^n$ to $\mathbf{R}^n$. We say that F is

(1) a **P-function** if and only if for each pair of distinct vectors $\mathbf{x}$ and $\mathbf{y}$ in K,

$$\max_{i\in[n]} (x_i - y_i)(F_i(\mathbf{x}) - F_i(\mathbf{y})) > 0;$$

(2) a **uniform P-function** if and only if there exists a constant $\mu > 0$ such that for each pair of vectors $\mathbf{x}$ and $\mathbf{y}$ in K,

$$\max_{i\in[n]} (x_i - y_i)(F_i(\mathbf{x}) - F_i(\mathbf{y})) \geq \mu\|\mathbf{x} - \mathbf{y}\|^2.$$

The solution set of NCP for P-functions [196] is described below.

Lemma 4.6 (Moré 1974) *Let $F : \mathbf{R}^n_+ \to \mathbf{R}^n$ be a P-function, then the corresponding complementarity problem*

$$\mathbf{x} \geq \mathbf{0},\ F(\mathbf{x}) \geq \mathbf{0},\ \mathbf{x}^\top F(\mathbf{x}) = 0,$$

has no more than one solution.

4.2 An m Person Noncooperative Game

In this section, a class of m person noncooperative games is considered, where the utility function of every player is a homogeneous polynomial of degree n defined by the payoff tensor of that player. The model considered here is a natural extension of the bimatrix game, which is called the multilinear game in this section. We will study the multilinear game by reformulating it as a tensor complementarity problem. It is shown that finding a Nash equilibrium point of the multilinear game is equivalent to finding a solution of the resulted tensor complementarity problem. In particular, an explicit corresponding relation between the solutions of these two classes of problems is exhibited. Furthermore, the resulted tensor complementarity problem is reformulated as a parameterized smooth equation, which can be solved by a smoothing-type algorithm.

Suppose $n_1, n_2, \ldots, n_m$ and m are positive integers, and $m > 2$ unless specifically stated. Recall the definition of the k-mode (vector) product for tensors. For any tensor $\mathscr{B} := (b_{i_1 i_2 \ldots i_m}) \in \mathbf{R}^{n_1\times n_2\times\cdots\times n_m}$ and vectors $\mathbf{u}^{(k)} \in \mathbf{R}^{n_k}$ for $k \in [m]$, we use $\mathscr{B}\mathbf{u}^{(1)}\mathbf{u}^{(2)}\cdots\mathbf{u}^{(m)}$ and $\mathscr{B}\mathbf{u}^{(2)}\mathbf{u}^{(3)}\cdots\mathbf{u}^{(m)}$ to denote $\mathscr{B}\times_1\mathbf{u}^{(1)}\times_2\mathbf{u}^{(2)}\cdots\times_m\mathbf{u}^{(m)}$ and $\mathscr{B}\times_2\mathbf{u}^{(2)}\times_3\mathbf{u}^{(3)}\cdots\times_m\mathbf{u}^{(m)}$ respectively. In component sense, they are

$$\mathscr{B}\mathbf{u}^{(1)}\mathbf{u}^{(2)}\cdots\mathbf{u}^{(m)} = \sum_{i_1=1}^{n_1}\sum_{i_2=1}^{n_2}\cdots\sum_{i_m=1}^{n_m} b_{i_1 i_2\ldots i_m} u^{(1)}_{i_1} u^{(2)}_{i_2}\cdots u^{(m)}_{i_m}$$

and

$$\mathscr{B}\mathbf{u}^{(2)}\mathbf{u}^{(3)}\cdots\mathbf{u}^{(m)} = \begin{pmatrix} \sum_{i_2=1}^{n_2}\cdots\sum_{i_m=1}^{n_m} b_{1i_2\ldots i_m}u_{i_2}^{(2)}\cdots u_{i_m}^{(m)} \\ \sum_{i_2=1}^{n_2}\cdots\sum_{i_m=1}^{n_m} b_{2i_2\ldots i_m}u_{i_2}^{(2)}\cdots u_{i_m}^{(m)} \\ \vdots \\ \sum_{i_2=1}^{n_2}\cdots\sum_{i_m=1}^{n_m} b_{n_1i_2\ldots i_m}u_{i_2}^{(2)}\cdots u_{i_m}^{(m)} \end{pmatrix}.$$

Another useful equation is that

$$\begin{aligned} &\frac{\partial}{\partial \mathbf{u}^k}\mathscr{B}\mathbf{u}^{(1)}\mathbf{u}^{(2)}\cdots\mathbf{u}^{(m)} \\ &= \begin{pmatrix} \sum_{i_1=1}^{n_1}\cdots\sum_{i_{k-1}=1}^{n_{k-1}}\sum_{i_{k+1}=1}^{n_{k+1}}\cdots\sum_{i_m=1}^{n_m} b_{i_1\ldots i_{k-1}1i_{k+1}\ldots i_m}u_{i_1}^{(1)}\cdots u_{i_{k-1}}^{(k-1)}u_{i_{k+1}}^{(k+1)}\cdots u_{i_m}^{(m)} \\ \vdots \\ \sum_{i_1=1}^{n_1}\cdots\sum_{i_{k-1}=1}^{n_{k-1}}\sum_{i_{k+1}=1}^{n_{k+1}}\cdots\sum_{i_m=1}^{n_m} b_{i_1\ldots i_{k-1}n_ki_{k+1}\ldots i_m}u_{i_1}^{(1)}\cdots u_{i_{k-1}}^{(k-1)}u_{i_{k+1}}^{(k+1)}\cdots u_{i_m}^{(m)} \end{pmatrix}. \end{aligned}$$

For any tensor $\mathscr{B} := (b_{i_1i_2\ldots i_m}) \in \mathbf{R}^{n_1\times n_2\times\cdots\times n_m}$ and for any $k \in [m]$, we define $\bar{\mathscr{B}}^k := (\bar{b}_{i_ki_1i_2\ldots i_m}) \in \mathbf{R}^{n_k\times n_1\times\cdots\times n_{k-1}\times n_{k+1}\times\cdots\times n_m}$ as

$$\bar{b}_{i_ki_1i_2\ldots i_{k-1}i_{k+1}\ldots i_m} = b_{i_1i_2\ldots i_m},\quad \forall\, i_j \in [n_j] \text{ and } j \in [m]. \tag{4.6}$$

Then, the following results hold automatically.

Proposition 4.2 *Let $\mathscr{B} := (b_{i_1i_2\ldots i_m}) \in \mathbf{R}^{n_1\times n_2\times\cdots\times n_m}$ and $\bar{\mathscr{B}}^k$ be defined by (4.6) for $k \in [m]$. Then, $\bar{\mathscr{B}}^1 = \mathscr{B}$. For any $k \in [m]$, it holds that*

$$\frac{\partial}{\partial \mathbf{u}^{(k)}}\mathscr{B}\mathbf{u}^{(1)}\mathbf{u}^{(2)}\cdots\mathbf{u}^{(m)} = \bar{\mathscr{B}}^k\mathbf{u}^{(1)}\cdots\mathbf{u}^{(k-1)}\mathbf{u}^{(k+1)}\cdots\mathbf{u}^{(m)}$$

and

$$(\mathbf{u}^{(k)})^\top(\bar{\mathscr{B}}^k\mathbf{u}^{(1)}\cdots\mathbf{u}^{(k-1)}\mathbf{u}^{(k+1)}\cdots\mathbf{u}^{(m)}) = \mathscr{B}\mathbf{u}^{(1)}\mathbf{u}^{(2)}\cdots\mathbf{u}^{(m)}.$$

For a square tensor $\mathscr{B} \in T_{m,n}$ and any vector $\mathbf{u} \in \mathbf{R}^n$, $\mathscr{B}\mathbf{u}^{m-1}$ is exactly the vector $\mathscr{B} \times_2 \mathbf{u} \times_3 \cdots \times_m \mathbf{u} \in \mathbf{R}^n$, which is obtained by replacing each $\mathbf{u}^{(k)}$ with $\mathbf{u}$ in $\mathscr{B}\mathbf{u}^{(1)}\mathbf{u}^{(2)}\cdots\mathbf{u}^{(m)}$.

We now introduce the basic notion of m person noncooperative game, which is called multilinear game in this section. In the following analysis, denote $n := \sum_{j=1}^m n_j$ and let $\mathbf{1}_n \in \mathbf{R}^n$ be the all-one vector. Suppose $\mathbf{x} = (\mathbf{x}^{(k)})_{k\in[m]} \in \mathbf{R}^n$ and

$\mathbf{x}^* = (\mathbf{x}^{(k^*)})_{k\in[m]} \in \mathbf{R}^n$ are vectors such that

$$\mathbf{x} = \begin{pmatrix} \mathbf{x}^{(1)} \\ \mathbf{x}^{(2)} \\ \vdots \\ \mathbf{x}^{(m)} \end{pmatrix}, \quad \mathbf{x}^* = \begin{pmatrix} \mathbf{x}^{(1^*)} \\ \mathbf{x}^{(2^*)} \\ \vdots \\ \mathbf{x}^{(m^*)} \end{pmatrix} \in \mathbf{R}^{n_1} \times \mathbf{R}^{n_2} \times \cdots \times \mathbf{R}^{n_m} = \mathbf{R}^n.$$

The **multilinear game** is a noncooperative game with a finite number of players, where each player has finite number of pure strategies. The basic assumptions for this kind of game are listed below.
(1) There are m players denoted by player 1, player 2, ..., player m.
(2) For any $k \in [m]$, player k has n_k pure strategies. Suppose the pure strategy set of player k is $[n_k]$.
(3) Let $\mathscr{A}^k = (a^{(k)}_{i_1 i_2 \ldots i_m})$ be the payoff tensor of player k, $k \in [m]$, which means that for any $i_j \in [n_j]$ with any $j \in [m]$, if player 1 plays his i_1-th pure strategy, player 2 plays his i_2-th strategy, ..., player m plays his i_m-th strategy, then the payoffs of player 1, player 2, ..., player m are $a^{(1)}_{i_1 i_2 \ldots i_m}, a^{(2)}_{i_1 i_2 \ldots i_m}, \ldots, a^{(m)}_{i_1 i_2 \ldots i_m}$, respectively.
(4) Suppose that $\mathbf{x}^{(k)} = (x_{i_j}) \in \mathbf{R}^{n_k}$ represents a mixed strategy of player $k \in [m]$, where $x_{i_j} \geq 0$ is the relative probability that player k plays his i_j-th pure strategy, i.e., $\mathbf{x}^{(k)} \in \Omega_k := \{\mathbf{x} \in \mathbf{R}^{n_k} : \mathbf{x} \geq \mathbf{0} \text{ and } \mathbf{1}_{n_k}^\top \mathbf{x} = 1\}$.

For any $k \in [m]$, the utility function of player k is

$$\mathscr{A}^k \mathbf{x}^{(1)} \mathbf{x}^{(2)} \cdots \mathbf{x}^{(m)} = \sum_{i_1=1}^{n_1} \sum_{i_2=1}^{n_2} \cdots \sum_{i_m=1}^{n_m} a^{(k)}_{i_1 i_2 \ldots i_m} x^{(1)}_{i_1} x^{(2)}_{i_2} \cdots x^{(m)}_{i_m}. \tag{4.7}$$

For any $k \in [m]$, if $\mathbf{x}^{(k)}$ is a mixed strategy of player k satisfying (4) in the above, then we call $\mathbf{x} = (\mathbf{x}^{(k)})_{k\in[m]} \in \mathbf{R}^n$ a joint mixed strategy. Furthermore, we call $\mathbf{x}^* = (\mathbf{x}^{(k^*)})_{k\in[m]} \in \mathbf{R}^n$ a Nash equilibrium point of the multilinear game if it satisfies

$$\mathscr{A}^k \mathbf{x}^{(1^*)} \mathbf{x}^{(2^*)} \cdots \mathbf{x}^{(m^*)} \geq \mathscr{A}^k \mathbf{x}^{(1^*)} \mathbf{x}^{(2^*)} \cdots \mathbf{x}^{(k-1^*)} \mathbf{x}^{(k)} \mathbf{x}^{(k+1^*)} \cdots \mathbf{x}^{(m^*)},$$

for any joint mixed strategy $\mathbf{x} = (\mathbf{x}^{(k)})_{k\in[m]} \in \mathbf{R}^n$. Clearly, for any $k \in [m]$, $\mathbf{x}^* \in \mathbf{R}^n$ is a **Nash equilibrium point** of the multilinear game if and only if $\mathbf{x}^*$ is an optimal solution of the following optimization problem:

$$\begin{cases} \max\limits_{\mathbf{x}^{(k)} \in \mathbf{R}^{n_k}} & \mathscr{A}^k \mathbf{x}^{(1^*)} \mathbf{x}^{(2^*)} \cdots \mathbf{x}^{(k-1^*)} \mathbf{x}^{(k)} \mathbf{x}^{(k+1^*)} \cdots \mathbf{x}^{(m^*)} \\ \text{s. t.} & \mathbf{x}^{(k)} \geq \mathbf{0}, \ \mathbf{1}_{n_k}^\top \mathbf{x}^{(k)} = 1. \end{cases} \tag{4.8}$$

It should be noted that the entries of the payoff tensor can always be assumed negative, i.e., $a^{(k)}_{i_1 i_2 \ldots i_m} < 0$ for any $k \in [m]$ and any $i_j \in [n_j]$ with all $j \in [m]$. The reason is that we can choose a sufficiently large $c > 0$ such that $a^{(k)}_{i_1 i_2 \ldots i_m} - c < 0$, and for any joint mixed strategy $\mathbf{x} = (\mathbf{x}^{(k)})_{k \in [m]} \in \mathbf{R}^n$, it holds that for any $k \in [m]$,

$$\sum_{i_1=1}^{n_1} \sum_{i_2=1}^{n_2} \cdots \sum_{i_m=1}^{n_m} (a^{(k)}_{i_1 i_2 \ldots i_m} - c) x^{(1)}_{i_1} x^{(2)}_{i_2} \cdots x^{(m)}_{i_m} = \mathscr{A}^k \mathbf{x}^{(1)} \mathbf{x}^{(2)} \cdots \mathbf{x}^{(m)} - c.$$

Therefore, $\mathbf{x}^*$ is a Nash equilibrium point of the multilinear game with payoff tensors $\mathscr{A}^k$ if and only if $\mathbf{x}^*$ is a Nash equilibrium point of the multilinear game with payoff tensors $\mathscr{A}^k - c\mathscr{J}$, where $\mathscr{J} \in \mathbf{R}^{n_1 \times n_2 \times \cdots \times n_m}$ is a tensor whose all entries are 1.

To reformulate the multilinear game as a specific complementarity problem (4.1), a new tensor

$$\mathscr{A} = (a_{i_1 i_2 \ldots i_m}) \in T_{m,n}, \tag{4.9}$$

is constructed by all the payoff tensors $\mathscr{A}^k$, where

$$a_{i_1 i_2 \ldots i_m} = \begin{cases} a^{(1)}_{i_1 (i_2 - n_1) \ldots (i_m - \Sigma_{j=1}^{m-1} n_j)} \\ \quad \text{if } i_1 \in [n_1], i_2 \in [n_1 + n_2] \backslash [n_1], \ldots, i_m \in [\Sigma_{j=1}^{m} n_j] \backslash [\Sigma_{j=1}^{m-1} n_j], \\ a^{(2)}_{(i_1 - n_1) i_2 (i_3 - n_1 - n_2) \ldots (i_m - \Sigma_{j=1}^{m-1} n_j)} \\ \quad \text{if } i_1 \in [n_1 + n_2] \backslash [n_1], i_2 \in [n_1], \ldots, i_m \in [\Sigma_{j=1}^{m} n_j] \backslash [\Sigma_{j=1}^{m-1} n_j], \\ a^{(k)}_{(i_1 - \Sigma_{j=1}^{k-1} n_j)(i_2 - n_1) \ldots (i_{k-1} - \Sigma_{j=1}^{k-2} n_j) i_k (i_{k+1} - \Sigma_{j=1}^{k} n_j) \ldots (i_m - \Sigma_{j=1}^{m-1} n_j)} \\ \quad \text{if } k \neq 2, i_1 \in [\Sigma_{j=1}^{k} n_j] \backslash [\Sigma_{j=1}^{k-1} n_j], i_2 \in [n_1], \ldots, \\ \quad i_k \in [\Sigma_{j=1}^{k-1} n_j] \backslash [\Sigma_{j=1}^{k-2} n_j], i_{k+1} \in [\Sigma_{j=1}^{k+1} n_j] \backslash [\Sigma_{j=1}^{k} n_j], \ldots, \\ \quad i_m \in [\Sigma_{j=1}^{m} n_j] \backslash [\Sigma_{j=1}^{m-1} n_j], \\ 0 \quad \text{otherwise.} \end{cases}$$

Let $\bar{\mathscr{A}}^k := (\bar{a}^{(k)}_{i_1 i_2 \ldots i_m}) \in \mathbf{R}^{n_k \times n_1 \times n_2 \times \cdots \times n_{k-1} \times n_{k+1} \times \cdots \times n_m}$, $k \in [m]$, be defined as

$$\bar{a}^{(k)}_{i_k i_1 \ldots i_{k-1} i_{k+1} \ldots i_m} = a^{(k)}_{i_1 i_2 \ldots i_m}, \ \forall\, i_j \in [n_j], \ j \in [m].$$

For any $\mathbf{x} = (\mathbf{x}^{(k)})_{k \in [m]} \in \mathbf{R}^n$, it is not difficult to see that

$$\mathscr{A} \mathbf{x}^{m-1} = \begin{pmatrix} \bar{\mathscr{A}}^1 \mathbf{x}^{(2)} \cdots \mathbf{x}^{(m)} \\ \vdots \\ \bar{\mathscr{A}}^k \mathbf{x}^{(1)} \cdots \mathbf{x}^{(k-1)} \mathbf{x}^{(k+1)} \cdots \mathbf{x}^{(m)} \\ \vdots \\ \bar{\mathscr{A}}^m \mathbf{x}^{(1)} \mathbf{x}^{(2)} \cdots \mathbf{x}^{(m-1)} \end{pmatrix}. \tag{4.10}$$

Combining (4.9) with (4.10), we consider the following tensor complementarity problem TCP($\mathbf{q}$, $\mathscr{A}$):

$$\text{Find } \mathbf{y} \in \mathbf{R}^n \text{ such that } \mathbf{y} \geq \mathbf{0},\ \mathscr{A}\mathbf{y}^{m-1} + \mathbf{q} \geq \mathbf{0},\ \mathbf{y}^\top(\mathscr{A}\mathbf{y}^{m-1} + \mathbf{q}) = 0, \tag{4.11}$$

where $\mathscr{A} \in T_{m,n}$ is defined as (4.9) and $\mathbf{q} = -\mathbf{1}_n \in \mathbf{R}^n$, i.e., all entries of $\mathbf{q}$ are -1.

The following theorem shows the relationship between the multilinear game and the corresponding tensor complementarity problem.

Theorem 4.1 *Suppose $\mathbf{x}^* = (\mathbf{x}^{(k^*)})_{k\in[m]} \in \mathbf{R}^n$ is a Nash equilibrium point of a multilinear game, i.e., $\mathbf{x}^*$ is a solution of the optimization problem (4.8). Then the following assertions hold.*
(1) $\mathbf{y}^* = (\mathbf{y}^{(k^*)})_{k\in[m]} \in \mathbf{R}^n$ is a solution of the TCP($\mathbf{q}$, $\mathscr{A}$) in (4.11), where

$$\mathbf{y}^{(k^*)} := \sqrt[m-1]{\frac{(-\mathscr{A}^k\mathbf{x}^{(1^*)}\mathbf{x}^{(2^*)}\cdots\mathbf{x}^{(m^*)})^{m-2}}{\prod_{i\in[m]\setminus\{k\}}(-\mathscr{A}^i\mathbf{x}^{(1^*)}\mathbf{x}^{(2^*)}\cdots\mathbf{x}^{(m^*)})}}\mathbf{x}^{(k^*)},\ \ \textit{for } k \in [m]. \tag{4.12}$$

(2) If $\mathbf{y}^*$ is a solution of TCP($\mathbf{q}$, $\mathscr{A}$) in (4.11), then $\mathbf{y}^{(k^*)} \neq \mathbf{0}$ for all $k \in [m]$; and the vector $\mathbf{x}^* = (\mathbf{x}^{(k^*)})_{k\in[m]}$ with

$$\mathbf{x}^{(k^*)} := \frac{\mathbf{y}^{(k^*)}}{\mathbf{1}_{n_k}^\top\mathbf{y}^{(k^*)}} \tag{4.13}$$

is a Nash equilibrium point of the multilinear game.

Proof (1) Assume $\mathbf{x}^*$ is a Nash equilibrium point of the multilinear game. For any $k \in [m]$, by the KKT conditions of (4.8), there exist $\lambda_k^* \in \mathbf{R}$ and a nonnegative vector $\mu^{k^*} \in \mathbf{R}^{n_k}$ such that

$$-\mathscr{A}^k\mathbf{x}^{(1^*)}\cdots\mathbf{x}^{(k-1^*)}\mathbf{x}^{(k+1^*)}\cdots\mathbf{x}^{(m^*)} - \lambda_k^*\mathbf{1}_{n_k} - \mu^{k^*} = 0 \tag{4.14}$$

and

$$\mathbf{1}_{n_k}^\top\mathbf{x}^{(k^*)} = 1,\ \ \mathbf{x}^{(k^*)} \geq \mathbf{0},\ \ \mu^{k^*} \geq \mathbf{0},\ \ (\mu^{k^*})^\top\mathbf{x}^{(k^*)} = 0. \tag{4.15}$$

By (4.14), we obtain that for any $k \in [m]$,

$$-\mathscr{A}^k\mathbf{x}^{(1^*)}\cdots\mathbf{x}^{(k^*)}\mathbf{x}^{(k+1^*)}\cdots\mathbf{x}^{(m^*)} - \lambda_k^*\mathbf{1}_{n_k}^\top\mathbf{x}^{(k^*)} - (\mu^{k^*})^\top\mathbf{x}^{(k^*)} = 0,$$

which together with (4.15) implies that

$$-\mathscr{A}^k\mathbf{x}^{(1^*)}\cdots\mathbf{x}^{(k^*)}\mathbf{x}^{(k+1^*)}\cdots\mathbf{x}^{(m^*)} = \lambda_k^*\mathbf{1}_{n_k}^\top\mathbf{x}^{(k^*)} + (\mu^{k^*})^\top\mathbf{x}^{(k^*)} = \lambda_k^*. \tag{4.16}$$

By the fact that $\mathbf{x}^{(k^*)} \geq \mathbf{0}$, $\mathbf{x}^{(k^*)} \neq \mathbf{0}$ and $-a^{(k)}_{i_1i_2\dots i_m} > 0$ for any $k \in [m], i_j \in [n_j], j \in [m]$, we have

$$\lambda_k^* = -\mathscr{A}^k\mathbf{x}^{(1^*)}\cdots\mathbf{x}^{(k^*)}\mathbf{x}^{(k+1^*)}\cdots\mathbf{x}^{(m^*)} > 0,\ \ \forall\, k \in [m].$$

Thus,

$$\mathbf{y}^{(k^*)} = \sqrt[m-1]{\frac{(\lambda_k^*)^{(m-2)}}{\prod_{i\in[m]\setminus\{k\}} \lambda_i^*}}\mathbf{x}^{(k^*)} \geq \mathbf{0}. \tag{4.17}$$

Moreover, by (4.10), (4.14), (4.15), and (4.17), it follows that

$$\mathscr{A}(\mathbf{y}^*)^{m-1} + \mathbf{q} = \begin{pmatrix} -\bar{\mathscr{A}}^1\mathbf{y}^{(2^*)}\cdots\mathbf{y}^{(m^*)} - \mathbf{1}_{n_1} \\ \vdots \\ -\bar{\mathscr{A}}^k\mathbf{y}^{(1^*)}\cdots\mathbf{y}^{(k-1^*)}\mathbf{y}^{(k+1^*)}\cdots\mathbf{y}^{(m^*)} - \mathbf{1}_{n_k} \\ \vdots \\ -\bar{\mathscr{A}}^n\mathbf{y}^{(1^*)}\mathbf{y}^{(2^*)}\cdots\mathbf{y}^{(m-1^*)} - \mathbf{1}_{n_m} \end{pmatrix}$$

$$= \begin{pmatrix} \frac{-1}{\lambda_1^*}\bar{\mathscr{A}}^1\mathbf{x}^{(2^*)}\cdots\mathbf{x}^{(m^*)} - \mathbf{1}_{n_1} \\ \vdots \\ \frac{-1}{\lambda_k^*}\bar{\mathscr{A}}^k\mathbf{x}^{(1^*)}\cdots\mathbf{x}^{(k-1^*)}\mathbf{x}^{(k+1^*)}\cdots\mathbf{x}^{(k^*)} - \mathbf{1}_{n_k} \\ \vdots \\ \frac{-1}{\lambda_m^*}\bar{\mathscr{A}}^m\mathbf{x}^{(1^*)}\mathbf{x}^{(2^*)}\cdots\mathbf{x}^{(m-1^*)} - \mathbf{1}_{n_m} \end{pmatrix}$$

$$= \begin{pmatrix} \frac{1}{\lambda_1^*}(\lambda_1^*\mathbf{1}_{n_1} + \mu^{1^*}) - \mathbf{1}_{n_1} \\ \vdots \\ \frac{1}{\lambda_k^*}(\lambda_k^*\mathbf{1}_{n_k} + \mu^{k^*}) - \mathbf{1}_{n_k} \\ \vdots \\ \frac{1}{\lambda_m^*}(\lambda_m^*\mathbf{1}_{n_m} + \mu^{m^*}) - \mathbf{1}_{n_m} \end{pmatrix} = \begin{pmatrix} \frac{\mu^{1^*}}{\lambda_1^*} \\ \vdots \\ \frac{\mu^{k^*}}{\lambda_k^*} \\ \vdots \\ \frac{\mu^{m^*}}{\lambda_m^*} \end{pmatrix} \geq \mathbf{0}.$$

Furthermore, by (4.17), (4.16) and (4.15), we obtain that

$$(\mathbf{y}^*)^\top(\mathscr{A}(\mathbf{y}^*)^{m-1}+\mathbf{q}) = \begin{pmatrix} \mathbf{y}^{(1^*)} \\ \vdots \\ \mathbf{y}^{(k^*)} \\ \vdots \\ \mathbf{y}^{(m^*)} \end{pmatrix}^\top \begin{pmatrix} -\bar{\mathscr{A}}^1\mathbf{y}^{(2^*)}\cdots\mathbf{y}^{(k^*)} - \mathbf{1}_{n_1} \\ \vdots \\ -\bar{\mathscr{A}}^k\mathbf{y}^{(1^*)}\cdots\mathbf{y}^{(k-1^*)}\mathbf{y}^{(k+1^*)}\cdots\mathbf{y}^{(m^*)} - \mathbf{1}_{n_k} \\ \vdots \\ -\bar{\mathscr{A}}^m\mathbf{y}^{(1^*)}\mathbf{y}^{(2^*)}\cdots\mathbf{y}^{(m-1^*)} - \mathbf{1}_{n_m} \end{pmatrix}$$

$$\begin{aligned}
&= \sum_{k=1}^m (\mathbf{y}^{(k^*)})^\top(-\bar{\mathscr{A}}^k\mathbf{y}^{(1^*)}\cdots\mathbf{y}^{(k-1^*)}\mathbf{y}^{(k+1^*)}\cdots\mathbf{y}^{(m^*)} - \mathbf{1}_{n_k}) \\
&= \sum_{k=1}^m \left\{ -\sqrt[m-1]{\frac{1}{\prod_{i\in[m]}\lambda_i^*}}\mathscr{A}^k\mathbf{x}^{(1^*)}\cdots\mathbf{x}^{(m^*)} - \sqrt[m-1]{\frac{(\lambda_k^*)^{m-2}}{\prod_{i\in[m]\setminus\{k\}}\lambda_i^*}}\mathbf{1}_{n_k}^\top\mathbf{x}^{(k^*)} \right\} \\
&= \sum_{k=1}^m \sqrt[m-1]{\frac{(\lambda_k^*)^{m-2}}{\prod_{i\in[m]\setminus\{k\}}\lambda_i^*}}(1-\mathbf{1}_{n_k}^\top\mathbf{x}^{(k^*)}) \\
&= 0.
\end{aligned}$$

Thus $\mathbf{y}^* = (\mathbf{y}^{(k^*)})_{k\in[m]} \in \mathbf{R}^n$ defined by (4.12) is a solution of TCP($\mathbf{q}$, $\mathscr{A}$) (4.11), and the conclusion (1) holds.

(2) If $\mathbf{y}^* = (\mathbf{y}^{(k^*)})_{k\in[m]} \in \mathbf{R}^n$ is a solution of TCP($\mathbf{q}$, $\mathscr{A}$), then

$$\begin{cases} \begin{pmatrix} \mathbf{y}^{(1^*)} \\ \vdots \\ \mathbf{y}^{(k^*)} \\ \vdots \\ \mathbf{y}^{(m^*)} \end{pmatrix} \geq \mathbf{0}, \begin{pmatrix} -\bar{\mathscr{A}}^1\mathbf{y}^{(2^*)}\cdots\mathbf{y}^{(m^*)} - \mathbf{1}_{n_1} \\ \vdots \\ -\bar{\mathscr{A}}^k\mathbf{y}^{(1^*)}\cdots\mathbf{y}^{(k-1^*)}\mathbf{y}^{(k+1^*)}\cdots\mathbf{y}^{(m^*)} - \mathbf{1}_{n_k} \\ \vdots \\ -\bar{\mathscr{A}}^m\mathbf{y}^{(1^*)}\mathbf{y}^{(2^*)}\cdots\mathbf{y}^{(m-1^*)} - \mathbf{1}_{n_m} \end{pmatrix} \geq \mathbf{0}, \\ \begin{pmatrix} \mathbf{y}^{(1^*)} \\ \vdots \\ \mathbf{y}^{(k^*)} \\ \vdots \\ \mathbf{y}^{(m^*)} \end{pmatrix}^\top \begin{pmatrix} -\bar{\mathscr{A}}^1\mathbf{y}^{(2^*)}\cdots\mathbf{y}^{(m^*)} - \mathbf{1}_{n_1} \\ \vdots \\ -\bar{\mathscr{A}}^k\mathbf{y}^{(1^*)}\cdots\mathbf{y}^{(k-1^*)}\mathbf{y}^{(k+1^*)}\cdots\mathbf{y}^{(m^*)} - \mathbf{1}_{n_k} \\ \vdots \\ -\bar{\mathscr{A}}^m\mathbf{y}^{(1^*)}\mathbf{y}^{(2^*)}\cdots\mathbf{y}^{(m-1^*)} - \mathbf{1}_{n_m} \end{pmatrix} = 0. \end{cases} \tag{4.18}$$

By the second inequality of (4.18), it is obvious that $\mathbf{y}^{(k^*)} \neq \mathbf{0}$ for all $k \in [m]$.

Suppose $\mathbf{x}^* \in \mathbf{R}^n$ is defined by (4.13). It is sufficient to prove that, for each $k \in [m]$, there exists $\lambda_k^* \in \mathbf{R}$ and a nonnegative vector $\mu^{k^*} \in \mathbf{R}^{n_k}$ such that

$$-\bar{\mathscr{A}}^k\mathbf{x}^{(1^*)}\cdots\mathbf{x}^{(k-1^*)}\mathbf{x}^{(k+1^*)}\cdots\mathbf{x}^{(m^*)} - \lambda_k^*\mathbf{1}_{n_k} - \mu^{k^*} = 0 \tag{4.19}$$

and

$$\mathbf{1}_{n_k}^\top \mathbf{x}^{(k^*)} = 1, \ \ \mathbf{x}^{(k^*)} \geq \mathbf{0}, \ \ \mu^{k*} \geq \mathbf{0}, \ \ (\mu^{k*})^\top \mathbf{x}^{(k^*)} = 0. \tag{4.20}$$

Since $\mathbf{y}^{(k^*)} \geq \mathbf{0}$ and $\mathbf{y}^{(k^*)} \neq \mathbf{0}$ for all $k \in [m]$, by (4.13) and (4.18), we know that $\mathbf{1}_{n_k}^\top \mathbf{x}^{(k^*)} > 0$ and

$$-\mathscr{A}^k \frac{\mathbf{y}^{(1^*)}}{\mathbf{1}_{n_1}^\top \mathbf{y}^{(1^*)}} \frac{\mathbf{y}^{(2^*)}}{\mathbf{1}_{n_2}^\top \mathbf{y}^{(2^*)}} \cdots \frac{\mathbf{y}^{(m^*)}}{\mathbf{1}_{n_m}^\top \mathbf{y}^{(m^*)}} - \frac{1}{\prod_{i \in [m] \setminus \{k\}} \mathbf{1}_{n_i}^\top \mathbf{y}^{(i^*)}} = 0, \ \ \forall\, k \in [m],$$

which is equivalent with

$$-\mathscr{A}^k \mathbf{x}^{(1^*)} \mathbf{x}^{(2^*)} \ldots \mathbf{x}^{(m^*)} - \frac{1}{\prod_{i \in [m] \setminus \{k\}} \mathbf{1}_{n_i}^\top \mathbf{y}^{(i^*)}} = 0, \ \ \forall\, k \in [m]. \tag{4.21}$$

Since $-\bar{\mathscr{A}}^k \mathbf{y}^{(1^*)} \ldots \mathbf{y}^{(k-1^*)} \mathbf{y}^{(k+1^*)} \ldots \mathbf{y}^{(m^*)} - \mathbf{1}_{n_k} \geq \mathbf{0}$ for all $k \in [m]$,

$$-\bar{\mathscr{A}}^k \mathbf{x}^{(1^*)} \ldots \mathbf{x}^{(k-1^*)} \mathbf{x}^{(k+1^*)} \ldots \mathbf{x}^{(m^*)} - \frac{\mathbf{1}_{n_k}}{\prod_{i \in [m] \setminus \{k\}} \mathbf{1}_{n_i}^\top \mathbf{y}^{(i^*)}} \geq \mathbf{0}, \ \ \forall\, k \in [m].$$

Thus, there exists a vector $\mu^{k*} \geq \mathbf{0}$ satisfying

$$-\bar{\mathscr{A}}^k \mathbf{x}^{(1^*)} \ldots \mathbf{x}^{(k-1^*)} \mathbf{x}^{(k+1^*)} \ldots \mathbf{x}^{(m^*)} - \frac{\mathbf{1}_{n_k}}{\prod_{i \in [m] \setminus \{k\}} \mathbf{1}_{n_i}^\top \mathbf{y}^{(i^*)}} - \mu^{k*} = \mathbf{0}, \ \ \forall\, k \in [m].$$

Combining this with (4.21), for any $k \in [m]$, it follows that

$$\begin{aligned}
(\mu^{k*})^\top \mathbf{x}^{(k^*)} &= (\mathbf{x}^{k^*})^\top \left(\bar{\mathscr{A}}^k \mathbf{x}^{(1^*)} \ldots \mathbf{x}^{(k-1^*)} \mathbf{x}^{(k+1^*)} \ldots \mathbf{x}^{(m^*)} - \frac{\mathbf{1}_{n_k}}{\prod_{i \in [m] \setminus \{k\}} \mathbf{1}_{n_i}^\top \mathbf{y}^{(i^*)}} \right) \\
&= \mathscr{A}^k \mathbf{x}^{(1^*)} \mathbf{x}^{(2^*)} \ldots \mathbf{x}^{(m^*)} - \frac{1}{\prod_{i \in [m] \setminus \{k\}} \mathbf{1}_{n_i}^\top \mathbf{y}^{(i^*)}} \\
&= 0,
\end{aligned}$$

which implies that (4.19) and (4.20) are true for

$$\lambda_k^* = \frac{1}{\prod_{i \in [m] \setminus \{k\}} \mathbf{1}_{n_i}^\top \mathbf{y}^{(i^*)}},$$

and the desired results hold. □

It is commonly acknowledged that the smoothing-type algorithm is an effective method for solving linear and nonlinear complementarity problems. In remainder of this section, we apply a smoothing-type algorithm to solve the TCP($\mathbf{q}$, $\mathscr{A}$) (4.11). Let $\mathscr{A}^k$, $k \in [m]$ be the payoff tensors of the multilinear game. Suppose $\mathscr{A} \in T_{m,n}$

is defined by (4.9), and $\bar{\mathscr{A}}^k$, $k \in [m]$ are defined as (4.10). Let $H : \mathbf{R}^{1+2n} \to \mathbf{R}^{1+2n}$ be a function with

$$H(\mu, \mathbf{y}, \mathbf{s}) := \begin{pmatrix} \mu \\ \mathbf{s} - F(\mathbf{y}) \\ \Phi(\mu, \mathbf{y}, \mathbf{s}) + \mu \mathbf{y} \end{pmatrix},$$

where $\mu \in \mathbf{R}$, $\mathbf{s} \in \mathbf{R}^n$, $F(\mathbf{y}) = \mathscr{A}\mathbf{y}^{m-1} + \mathbf{q}$ satisfy (4.11), and the mapping $\Phi(\mu, \mathbf{y}, \mathbf{s}) = (\phi(\mu, y_1, s_1), \ldots, \phi(\mu, y_n, s_n))^\top$ such that

$$\phi(\mu, y_i, s_i) = y_i + s_i - \sqrt{(y_i - s_i)^2 + 4\mu}, \ \forall\, i \in [n].$$

It is obvious that the TCP($\mathbf{q}$, $\mathscr{A}$) is equivalent with the following problem:

$$\mathbf{y} \geq \mathbf{0}, \ \ \mathbf{s} = F(\mathbf{y}) \geq \mathbf{0}, \ \ \mathbf{y}^\top \mathbf{s} = 0. \tag{4.22}$$

Thus, it can be verified that $(\mathbf{y}, \mathbf{s})$ is a solution of (4.22) if and only if $H(0, \mathbf{y}, \mathbf{s}) = 0$. Since the function H is continuously differentiable for any $(\mu, \mathbf{y}, \mathbf{s}) \in \mathbf{R}^{1+2n}$ with $\mu > 0$, the solution of (4.22) can be found by solving the system $H(\mu, \mathbf{y}, \mathbf{s}) = 0$ with $\mu \to 0$. The Newton-type method could be applied for $H(\mu, \mathbf{y}, \mathbf{s}) = 0$ at each iteration. Particularly, we use the following algorithm to compute a solution of TCP($\mathbf{q}$, $\mathscr{A}$).

Algorithm 4.1 (A Smoothing-type Algorithm)

Step 0 Choose $\delta, \sigma \in (0, 1)$, and $\mu_0 > 0$. Let $(\mathbf{y}^{(0)}, \mathbf{s}^{(0)}) \in \mathbf{R}^{2n}$ be an arbitrary vector. Set $\mathbf{z}^{(0)} := (\mu_0, \mathbf{y}^{(0)}, \mathbf{s}^{(0)})$. Choose $\beta > 1$ such that $\|H(\mathbf{z}^{(0)})\| \leq \beta\mu_0$. Set $\mathbf{e}^{(0)} := (1, 0, \ldots, 0) \in \mathbf{R}^{1+2n}$ and $k := 0$.

Step 1 If $\|H(\mathbf{z}^k)\| = 0$, stop.

Step 2 Compute $\Delta\mathbf{z}^{(k)} = (\Delta\mu_k, \Delta\mathbf{x}^{(k)}, \Delta\mathbf{s}^{(k)}) \in \mathbf{R} \times \mathbf{R}^n \times \mathbf{R}^n$ by

$$H(\mathbf{z}^{(k)}) + H'(\mathbf{z}^{(k)})\Delta\mathbf{z}^{(k)} = (1/\beta)\|H(\mathbf{z}^{(k)})\|\mathbf{e}^{(0)}.$$

Step 3 Let λ_k be the maximum of the values $1, \delta, \delta^2, \ldots,$ such that

$$\|H(\mathbf{z}^{(k)} + \lambda_k \Delta\mathbf{z}^{(k)})\| \leq [1 - \sigma(1 - 1/\beta)\lambda_k]\|H(\mathbf{z}^{(k)})\|.$$

Step 4 Set $\mathbf{z}^{(k+1)} := \mathbf{z}^{(k)} + \lambda_k \Delta\mathbf{z}^{(k)}$ and $k := k + 1$. Go to step **0**.

The global convergence for Algorithm 4.1 can be obtained under suitable assumptions, and one may refer to [135] for details.

4.3 Positive Definite Tensors for Tensor Complementarity Problems

In this section, we study the existence and uniqueness theorem for solutions of tensor complementarity problems with respect to structured tensors in the even order case. The corresponding tensors are positive definite or copositive. The main objective models that we study are TCP($\mathbf{q}$, $\mathscr{A}$) (4.1), and TCP($\mathbf{q}$, $\{\mathscr{A}_k\}$), which is to find $\mathbf{x} \in \mathbf{R}^n_+$ such that

$$F(\mathbf{x}) = \sum_{k=1}^{m/2} \mathscr{A}_k \mathbf{x}^{m-(2k-1)} + \mathbf{q} \in \mathbf{R}^n_+, \quad \mathbf{x}^\top \left(\sum_{k=1}^{m/2} \mathscr{A}_k \mathbf{x}^{m-(2k-1)} + \mathbf{q} \right) = 0, \tag{4.23}$$

where m is an even integer, $\mathscr{A}_k \in T_{m-(2k-2),n}$, $k \in [m/2]$ and $\mathbf{q} \in \mathbf{R}^n_+$. It is obvious that problem (4.23) reduces to (4.1) when $k = 1$. For simplicity, we consider the problem (4.1) in detail, and similar results can be obtained for problem (4.23). We always assume that m is even.

For tensor complementarity problem (4.1), the cornerstone for necessary conditions of the existence of its solution is the following optimization problem:

$$\begin{cases} \min & \mathscr{A}\mathbf{x}^m + \langle \mathbf{q}, \mathbf{x} \rangle \\ \text{s. t.} & \mathscr{A}\mathbf{x}^{m-1} + \mathbf{q} \in \mathbf{R}^n_+, \ \mathbf{x} \in \mathbf{R}^n_+. \end{cases} \tag{4.24}$$

Let $\Omega(\mathbf{q}, \mathscr{A}) = \{\mathbf{x} \in \mathbf{R}^n_+ \ : \ \mathscr{A}\mathbf{x}^{m-1} + \mathbf{q} \in \mathbf{R}^n_+\}$. If $\Omega(\mathbf{q}, \mathscr{A}) \neq \emptyset$, we see that (4.24) is feasible, which is equivalent to the feasibility of (4.1). Apparently, if $\mathbf{x}^*$ minimizes the problem (4.24) and $\mathscr{A}(\mathbf{x}^*)^m + \mathbf{q}^\top \mathbf{x}^* = 0$, then $\mathbf{x}^*$ is a solution of (4.1). By the first order necessary optimal conditions for nonlinear programming problems and the famous M-F constraint qualification [193], we have the following necessary optimal conditions for (4.1).

Theorem 4.2 *Suppose $\mathscr{A} \in T_{m,n}$ and m is even. Assume $\Omega(\mathbf{q}, \mathscr{A}) \neq \emptyset$, and the M-F constraint qualification holds in $\mathbf{x}^*$, which is a local solution to (4.24). Then there exists a vector $\mathbf{u}^*$ of multipliers satisfying the following conditions*

$$\begin{cases} \mathbf{q} + m\mathscr{A}(\mathbf{x}^*)^{m-1} - (m-1)\mathscr{A}(\mathbf{x}^*)^{m-2}\mathbf{u}^* \geq \mathbf{0}, \\ (\mathbf{x}^*)^\top(\mathbf{q} + m\mathscr{A}(\mathbf{x}^*)^{m-1} - (m-1)\mathscr{A}(\mathbf{x}^*)^{m-2}\mathbf{u}^*) = 0, \\ \mathbf{u}^* \geq \mathbf{0}, \\ (\mathbf{u}^*)^\top(\mathbf{q} + \mathscr{A}(\mathbf{x}^*)^{m-1}) = 0. \end{cases} \tag{4.25}$$

Finally, the vectors $\mathbf{x}^*$ and $\mathbf{u}^*$ satisfy

$$(m-1)(\mathbf{x}^* - \mathbf{u}^*)_i(\mathscr{A}(\mathbf{x}^*)^{m-2}(\mathbf{x}^* - \mathbf{u}^*))_i \leq 0, \ i \in [n]. \tag{4.26}$$

Proof Since $\Omega(\mathbf{q}, \mathscr{A}) \neq \emptyset$, the optimal solution $\mathbf{x}^*$ and a suitable vector $\mathbf{u}^*$ of multipliers will satisfy the KKT conditions (4.25). To prove (4.26), we consider the inner product

$$(\mathbf{x}^*)^\top \left(\mathbf{q} + m\mathscr{A}(\mathbf{x}^*)^{m-1} - (m-1)\mathscr{A}(\mathbf{x}^*)^{m-2}\mathbf{u}^*\right) = 0.$$

Combining this with the fact that $\mathbf{x}^* \geq \mathbf{0}$, $\mathbf{q} + \mathscr{A}(\mathbf{x}^*)^{m-1} \geq \mathbf{0}$, we know that

$$(m-1)(\mathbf{x}^*)_i(\mathscr{A}(\mathbf{x}^*)^{m-2}(\mathbf{x}^*-\mathbf{u}^*))_i \le 0,\ \forall\, i\in[n]. \tag{4.27}$$

Similarly, by the feasibility of $\mathbf{x}^*$ and by the complementarity conditions $(\mathbf{u}^*)_i(\mathbf{q}+\mathscr{A}(\mathbf{x}^*)^{m-1})_i = 0$ for all $i\in[n]$, it holds that

$$-(m-1)(\mathbf{u}^*)_i(\mathscr{A}(\mathbf{x}^*)^{m-2}(\mathbf{x}^*-\mathbf{u}^*))_i \le 0,\ \forall\, i\in[n]. \tag{4.28}$$

By (4.27) and (4.28), we obtain (4.26) and the desired results hold. □

By employing Theorem 4.2, a sufficient condition for solving the TCP($\mathbf{q}$, $\mathscr{A}$) (4.1) follows immediately.

Theorem 4.3 *Let $\mathscr{A}\in T_{m,n}$ be an even order tensor. Suppose the M-F constraint qualification holds in a local solution $\mathbf{x}^*\neq\mathbf{0}$ of (4.24). If $\mathscr{A}(\mathbf{x}^*)^{m-2}$ is positive definite, then $\mathbf{x}^*$ is an optimal solution of TCP($\mathbf{q}$, $\mathscr{A}$) (4.1).*

Proof By Theorem 4.2, there exists a vector $\mathbf{u}^*\ge\mathbf{0}$ such that (4.25) holds. It then follows from (4.26) that

$$(\mathbf{x}^*-\mathbf{u}^*)^\top(\mathscr{A}(\mathbf{x}^*)^{m-2}(\mathbf{x}^*-\mathbf{u}^*)) \le 0,$$

which further implies that $\mathbf{x}^*=\mathbf{u}^*$ since $\mathscr{A}(\mathbf{x}^*)^{m-2}$ is a positive definite matrix. Together with (4.25), the desired result follows. □

Since the main content of this chapter is about tensor complementarity problem, we will study some properties of the mapping $F(\mathbf{x})=\mathscr{A}\mathbf{x}^{m-1}+\mathbf{q}$, where tensor $\mathscr{A}$ is selected from sets of some structured tensors. Before that, we recall the definition of copositive tensors, which is first defined in [223].

Definition 4.5 Let $\mathscr{A}\in T_{m,n}$. Then $\mathscr{A}$ is called a **copositive (strictly copositive)** tensor if $\mathscr{A}\mathbf{x}^m\ge 0$ ($\mathscr{A}\mathbf{x}^m>0$) for any $\mathbf{x}\in\mathbf{R}^n_+$ ($\mathbf{x}\in\mathbf{R}^n_+, \mathbf{x}\neq\mathbf{0}$).

Combining Definitions 4.1–4.5 with the notion of H(Z)-eigenvalues for symmetric tensors, we give the following results.

Theorem 4.4 *Let m be even. Suppose $\mathscr{A}\in S_{m,n}$ is a given symmetric tensor. Then the following results hold for $\mathbf{x}\in\mathbf{R}^n_+$.*

(1) If $\mathscr{A}$ is (strictly) copositive, then the mapping $F(\mathbf{x})$ is (strictly) copositive with respect to $\mathbf{R}^n_+$.
(2) If $\mathscr{A}$ is positive definite, then the mapping $F(\mathbf{x})$ is strongly copositive with respect to $\mathbf{R}^n_+$ when $\alpha\le\lambda_{\min}\|\mathbf{x}\|_2^{m-2}(\le\lambda_{\min}\frac{\|\mathbf{x}\|_m^m}{\|\mathbf{x}\|_2^2})$, where $\lambda_{\min}$ is the smallest Z-eigenvalue (H-eigenvalue) of $\mathscr{A}$.

Proof By Definitions 4.2–4.5, we know that

$$\mathbf{x}^\top(F(\mathbf{x})-F(\mathbf{0}))=\mathscr{A}\mathbf{x}^m$$

since $F(\mathbf{x}) = \mathscr{A}\mathbf{x}^{m-1} + \mathbf{q}$, the conclusion (1) holds.

To prove (2), by conditions that $\mathscr{A}$ is positive definite and by Theorem 1.2, the smallest Z-eigenvalue (H-eigenvalue) of $\mathscr{A}$ is positive. Therefore, when $\lambda_{\min}$ is the smallest Z-eigenvalue (H-eigenvalue), it follows that

$$\mathbf{x}^\top(F(\mathbf{x}) - F(\mathbf{0})) = \mathscr{A}\mathbf{x}^m \geq \lambda_{\min}\|\mathbf{x}\|_2^{m-2}\|\mathbf{x}\|_2^2 \left(\lambda_{\min}\frac{\|\mathbf{x}\|_m^m}{\|\mathbf{x}\|_2^2}\|\mathbf{x}\|_2^2\right),$$

and the desired results hold. □

Now, we present the notion of diagonalizable tensor below.

Definition 4.6 Let $\mathscr{A} = (a_{i_1 i_2 \ldots i_m}) \in S_{m,n}$. $\mathscr{A}$ is called **diagonalizable** if and only if $\mathscr{A}$ can be represented as

$$\mathscr{A} = \mathscr{D}\times_1 B\times_2 B\cdots\times_m B,$$

where $\mathscr{D} = (d_{i_1 i_2 \ldots i_m})$ is a diagonal tensor and $B = (b_{ij}) \in \mathbf{R}^{n\times n}$ is invertible.

Note that a diagonalizable tensor is not necessarily a diagonal tensor. Denote $D_{m,n}$ as the set of all diagonalizable tensors.

For $\mathbf{x} \neq \mathbf{0}$, we present a property for the Jacobian matrix $\nabla F(\mathbf{x})$ of $F(\mathbf{x}) = \mathscr{A}\mathbf{x}^{m-1} + \mathbf{q}$ under the condition that $\mathscr{A} \in D_{m,n}$ is positive semi-definite.

Theorem 4.5 *Let m be even. For any $\mathbf{x} \in \mathbf{R}^n\backslash\{\mathbf{0}\}$, the Jacobian matrix $\nabla F(\mathbf{x})$ is positive semi-definite if $\mathscr{A}$ is a positive semi-definite diagonalizable tensor.*

Proof Since $\mathscr{A}$ is diagonalizable, for any $\mathbf{x} \in \mathbf{R}^n$, it holds that

$$\begin{aligned}\mathscr{A}\mathbf{x}^m &= (\mathscr{D}\times_1 B\times_2 B\cdots\times_m B)\mathbf{x}^m = \mathscr{D}(B^\top\mathbf{x})^m \\ &= \mathscr{D}\mathbf{y}^m = \sum_{i=1}^n d_i y_i^m,\end{aligned}$$

where $\mathbf{y} = B^\top\mathbf{x}$ and all diagonal entries of $\mathscr{D}$ are $d_1, d_2, \ldots, d_n$. By condition that $\mathscr{A}$ is positive semi-definite, we know that $d_i \geq 0$ for all $i \in [n]$ since B is invertible and m is even. Since $F(\mathbf{x}) = \mathscr{A}\mathbf{x}^{m-1} + \mathbf{q}$, we have $\nabla F(\mathbf{x}) = (m-1)\mathscr{A}\mathbf{x}^{m-2}$ and

$$\mathscr{A}\mathbf{x}^{m-2} = \mathscr{D}(B^\top\mathbf{x})^{m-2}\times_1 B\times_2 B,$$

which implies that

$$\mathbf{z}^\top\nabla F(\mathbf{x})\mathbf{z} = \frac{1}{m}\sum_{i=1}^n d_i y_i^{m-2}\tilde{z}_i^2 \geq 0, \quad \forall\, \mathbf{z} \in \mathbf{R}^n,$$

where $\tilde{\mathbf{z}} = B^\top\mathbf{z} = (\tilde{z}_1, \tilde{z}_2, \ldots, \tilde{z}_n)^\top$. Thus, the Jacobian $\nabla F(\mathbf{x})$ is positive semi-definite for any $\mathbf{x} \in \mathbf{R}^n$. □

Theorem 4.6 *Let m be even. Suppose $\mathscr{A} \in S_{m,n}$ is a symmetric tenor. Then, we have the following results.*

(1) If $\mathscr{A}$ is positive definite, then (4.1) has a nonempty, compact solution set.
(2) If $\mathscr{A}$ is strictly copositive with respect to $\mathbf{R}^n_+$, then the TCP($\mathbf{q}$, $\mathscr{A}$) (4.1) has a nonempty compact solution set.

Proof Since all positive definite tensors are strictly copositive tensors, it is enough to prove (2). From Theorem 4.4, we have $F(\mathbf{x}) = \mathscr{A}\mathbf{x}^{m-1} + \mathbf{q}$ is strictly copositive. Then $G(\mathbf{x}) = F(\mathbf{x}) - F(\mathbf{0})$ is $\mathbf{d}$-regular for any $\mathbf{d} > \mathbf{0}$ from Definition 4.2.

Moreover, it holds that $G(\lambda\mathbf{x}) = \lambda^{m-1}G(\mathbf{x})$ for $\lambda > 0$. Hence, by Lemma 4.3, if $\mathscr{A}$ is strictly copositive with respect to $\mathbf{R}^n_+$, the problem (4.1) has a nonempty, compact solution set. □

In the above analysis, the solvability of (4.1) is considered in the even order case. Analogously, the following theorems verify the solvability of TCP($\mathbf{q}$, $\{\mathscr{A}_k\}$) (4.23).

Theorem 4.7 *Let m be even. Suppose that $\mathscr{A}_k \in S_{m-(2k-2),n}$ with $k \in [m/2]$. For TCP($\mathbf{q}$, $\{\mathscr{A}_k\}$) (4.23), the following results hold.*

(1) If $\mathscr{A}_k$ ($k \in [m/2-1]$) are diagonalizable and positive semi-definite and $\mathscr{A}_{m/2}$ is positive definite, then the problem TCP($\mathbf{q}$, $\{\mathscr{A}_k\}$) (4.23) has a unique solution;
(2) If each $\mathscr{A}_k$ is positive semi-definite and there exists $k_0 \in [m/2]$ such that $\mathscr{A}_{k_0}$ is positive definite, then the problem TCP($\mathbf{q}$, $\{\mathscr{A}_k\}$) (4.23) has a nonempty, compact solution set;
(3) If each $\mathscr{A}_k$ is strictly copositive with respect to $\mathbf{R}^n_+$, then the problem TCP($\mathbf{q}$, $\{\mathscr{A}_k\}$) (4.23) has a nonempty, compact solution set.

Proof (1) By conditions, for any $k \in [m/2-1]$ we can derive that $\mathscr{A}_k\mathbf{x}^{m-2k}$ is symmetric and positive semi-definite with respect to $\mathbf{x} \in \mathbf{R}^n$ due to Theorem 4.5. As $\mathscr{A}_{m/2}$ is symmetric and positive definite, we obtain that

$$\nabla F(\mathbf{x}) = (m-1)\mathscr{A}_1\mathbf{x}^{m-2} + (m-3)\mathscr{A}_2\mathbf{x}^{m-4} + \cdots + \mathscr{A}_{\frac{m}{2}}$$

is symmetric and positive definite for $\mathbf{x} \in \mathbf{R}^n$, where $F(\mathbf{x})$ is defined in (4.23). Then, by Lemma 4.1, the problem TCP($\mathbf{q}$, $\{\mathscr{A}_k\}$) (4.23) has a unique solution, and the desired result (1) holds.

(2) By conditions, if $k_0 = m/2$, then we have

$$\mathbf{x}^\top(F(\mathbf{x}) - F(\mathbf{0})) = \sum_{k=1}^{m/2} \mathscr{A}_k\mathbf{x}^{m-2k+2} \geq \lambda_1\|\mathbf{x}\|_2^2 > 0,$$

where λ_1 is the smallest eigenvalue of $\mathscr{A}_{m/2}$ for all $\mathbf{x} \in \mathbf{R}^n_+$, $\mathbf{x} \neq \mathbf{0}$. If $k_0 \in [m/2-1]$, then it holds that

$$\mathbf{x}^\top(F(\mathbf{x}) - F(\mathbf{0})) = \sum_{k=1}^{m/2} \mathscr{A}_k\mathbf{x}^{m-2k+2} \geq \lambda_2\|\mathbf{x}\|_2^2 > 0,$$

where λ_2 is the smallest Z-eigenvalue of $\mathscr{A}_{k_0}$ for all $\mathbf{x} \in \mathbf{R}^n_+, \mathbf{x} \neq \mathbf{0}$ (Meanwhile, we can also consider the case when λ_2 is the smallest H-eigenvalue of $\mathscr{A}_{k_0}$). Thus, $F(\mathbf{x})$ is strictly copositive. Let $c(\lambda) = \lambda$ with $\alpha = 1$ and $\lambda \geq 1$, we obtain that $c(\lambda) \to \infty$ as $\lambda \to \infty$ and

$$\mathbf{x}^\top (F(\lambda \mathbf{x}) - F(\mathbf{0})) \geq c(\lambda)\mathbf{x}^\top (F(\mathbf{x}) - F(\mathbf{0})).$$

By Lemma 4.2, we obtain that the problem (4.23) has a nonempty, compact solution set.

(3) From Theorem 4.4, it follows that

$$F(\mathbf{x}) = \sum_{k=1}^{m/2} \mathscr{A}_k \mathbf{x}^{m-(2k-1)} + \mathbf{q}$$

is strictly copositive. Let $c(\lambda)$ be defined as in the proof (2) with $\lambda \geq 1$. Then it holds that $c(\lambda) \to \infty$ as $\lambda \to \infty$ and

$$\mathbf{x}^\top (F(\lambda \mathbf{x}) - F(\mathbf{0})) \geq c(\lambda)\mathbf{x}^\top (F(\mathbf{x}) - F(\mathbf{0})).$$

By Lemma 4.2, the problem (4.23) has a nonempty, compact solution set and the desired results hold. □

In fact, the constraints of tensors $\mathscr{A}_k$ given in (1) of Theorem 4.7 can be weakened. Therefore, a more general result is given as follows.

Theorem 4.8 *Let m be even. Suppose that $\mathscr{A}_k \in S_{m-(2k-2),n}$ with $k \in [m/2 - 1]$ and $\mathscr{A}_{m/2}$ is a square matrix. For Problem (4.23), if all $\mathscr{A}_k$ are diagonalizable and positive semi-definite and there exists $\delta \in (0, 1)$ such that all principal minors of $\mathscr{A}_{m/2}$ are bounded between δ and δ^{-1}, then the problem (4.23) has a unique solution.*

Proof Since there exists $\delta \in (0, 1)$ such that all principal minors of $\mathscr{A}_{m/2}$ are bounded between δ and δ^{-1}, the real part of every eigenvalue of $\mathscr{A}_{m/2}$ is positive. Hence, for all nonzero vector $\mathbf{x}$, we can derive

$$\mathbf{x}^\top \mathscr{A}_{m/2}\mathbf{x} > 0.$$

By the assumptions, we can obtain that the Jacobian matrix $\nabla F(\mathbf{x})$ of $F(\mathbf{x})$ given in (4.23) is positive definite. Thus, the TCP($\mathbf{q}, \{\mathscr{A}_k\}$) has a unique solution. □

4.4 P and P_0-Tensors

P and P_0 matrices have a long history and wide applications in mathematical sciences, such as linear complementarity problems, variational inequalities and nonlinear complementarity problems [63, 91]. Fiedler and Pták first studied P matrices systematically in [97]. It is an important class of special matrices whose determinants of all submatrices are positive. An important criteria for checking P (P_0) matrix is that a symmetric matrix is a P (P_0) matrix if and only if it is positive (semi-)definite. With an emerging interest in the assets of multilinear algebra concentrated on the higher-order tensors, the concepts of P and P_0 matrices are extended to P-tensors and P_0-tensors. In this section, some interesting properties of P-tensors and P_0-tensors are studied.

Definition 4.7 Let $\mathscr{A} = (a_{i_1 \ldots i_m}) \in T_{m,n}$. We say that A is

(1) a **P_0-tensor** iff for any vector $\mathbf{x}$ in $\mathbf{R}^n \backslash \{\mathbf{0}\}$, there exists $i \in [n]$ such that $x_i \neq 0$ and

$$x_i \left(\mathscr{A}\mathbf{x}^{m-1}\right)_i \geq 0;$$

(2) a **P-tensor** iff for any vector $\mathbf{x}$ in $\mathbf{R}^n \backslash \{\mathbf{0}\}$,

$$\max_{i \in [n]} x_i \left(\mathscr{A}\mathbf{x}^{m-1}\right)_i > 0.$$

It is clear that, the P and P_0-tensors in Definition 4.7 reduce to P and P_0 matrices when $m = 2$.

Now we present several basic properties for P(P_0)-tensors, especially the spectral properties.

Theorem 4.9 *(1) All H-eigenvalues and Z-eigenvalues of an even order P(P_0)-tensor $\mathscr{A}$ are positive (nonnegative).*
(2) A symmetric tensor with even order is a P(P_0)-tensor if and only if it is positive (semi-)definite.
(3) There does not exist an odd order symmetric P-tensor. If an odd order nonsymmetric P-tensor exists, then it has no Z-eigenvalues.
(4) An odd order P_0-tensor has no nonzero Z-eigenvalues.
(5) All principal sub-tensors of a P(P_0)-tensor are P(P_0)-tensors. Furthermore, all diagonal entries of a P(P_0)-tensor are positive (nonnegative).

Proof (1) Let m be even and $\mathscr{A} \in T_{m,n}$. Suppose that λ is an H-eigenvalue of $\mathscr{A}$. If $\mathscr{A}$ is a P-tensor, then by the definition of H-eigenvalues, there exists a nonzero vector $\mathbf{x} \in \mathbf{R}^n$ satisfying

$$\mathscr{A}\mathbf{x}^{m-1} = \lambda \mathbf{x}^{[m-1]}.$$

By Definition 4.7, there exists $i \in [n]$ such that

$$0 < x_i(\mathscr{A}\mathbf{x}^{m-1})_i = \lambda x_i^m,$$

which implies that $\lambda > 0$ since m is even. If $\mathscr{A}$ is a P_0-tensor, then $\lambda \geq 0$. Similarly, we can prove the result for the Z-eigenvalue case, and the desired results (1) hold.
(2) By Theorems 1.1 and 1.2, a symmetric tensor is positive (semi-)definite if and only if all its H(Z)-eigenvalues are positive (nonnegative). Thus, from (1), all even order symmetric P(P_0)-tensors are positive (semi-)definite tensor.

On the other hand, it is obvious that positive (semi-)definite tensors are $P(P_0)$-tensors, and the desired results hold.
(3) Suppose $\mathscr{A} \in S_{m,n}$ is a symmetric tensor. Then it always has Z-eigenvalues according to Theorem 1.2. Suppose λ is an Z-eigenvalue of $\mathscr{A}$. Then, there is a vector $\mathbf{x} \in \mathbf{R}^n$, $\mathbf{x}^\top \mathbf{x} = 1$ such that

$$\mathscr{A}\mathbf{x}^{m-1} = \lambda \mathbf{x}.$$

By Definition 4.7, there exists an $i \in [n]$ such that

$$0 < x_i(\mathscr{A}\mathbf{x}^{m-1})_i = \lambda x_i^2, \tag{4.29}$$

which implies that $\lambda > 0$. When m is odd, it is not difficult to know that $-\lambda$ is also a Z-eigenvalue of $\mathscr{A}$ with Z-eigenvector $-\mathbf{x}$. Thus there is an index $j \in [n]$ such that

$$0 < -x_j(\mathscr{A}\mathbf{x}^{m-1})_j = (-\lambda)(-x_j)^2,$$

which contradicts (4.29) since $\lambda > 0$. Thus, odd order symmetric P-tensors do not exist.

If there is a nonsymmetric P-tensor, by the definition of Z-eigenvalue, it is clear that the tensor does not have any Z-eigenvalue and the desired result follows.
(4) The result can be proved similarly with the proof of (3).
(5) First of all, the second conclusion is obvious from Definition 4.7 by setting $\mathbf{x}$ an arbitrary unit vector. On the other hand, let $\mathscr{A}^J$ be a principal sub-tensor of $\mathscr{A}$, where $J \subseteq [n]$, $|J| = r$. For each nonzero vector $\mathbf{x} = (x_{j_1}, \ldots, x_{j_r})^\top \in \mathbf{R}^r$ with $j_1, j_2, \ldots, j_r \in J$, we may choose $\mathbf{x}^* = (x_1^*, x_2^*, \ldots, x_n^*)^\top \in \mathbf{R}^n$ with $x_i^* = x_i$ for $i \in J$ and $x_i^* = 0$ otherwise. Assume that $\mathscr{A}$ is a P-tensor, then there exists $j \in J$ such that

$$0 < x_j^*(\mathscr{A}(\mathbf{x}^*)^{m-1})_j = x_j(\mathscr{A}^J\mathbf{x}^{m-1})_j,$$

which implies that $\mathscr{A}^J$ is a P-tensor. The case for P_0-tensors can be proved similarly, and the desired results hold. □

Now we introduce a sufficient and necessary condition for a given tensor to be a P-tensor.

Theorem 4.10 *Let $\mathscr{A} \in T_{m,n}$. Then $\mathscr{A}$ is a P-tensor if and only if for any $\mathbf{x} \in \mathbf{R}^n, \mathbf{x} \neq \mathbf{0}$, there exists a matrix $D_\mathbf{x} \in \mathbf{R}^{n\times n}$ such that $\mathbf{x}^\top D_\mathbf{x}(\mathscr{A}\mathbf{x}^{m-1})$ is positive, where $D_\mathbf{x}$ is a diagonal matrix with positive diagonal entries.*

Proof For necessity, we take any $\mathbf{x} \in \mathbf{R}^n, \mathbf{x} \neq \mathbf{0}$. By Definition 4.7, there is at least a $k \in [n]$ such that $x_k(\mathscr{A}\mathbf{x}^{m-1})_k > 0$. Thus, for any enough small $\varepsilon > 0$, it follows that

$$x_k(\mathscr{A}\mathbf{x}^{m-1})_k + \varepsilon\left(\sum_{j\in[n], j\neq k} x_j\left(\mathscr{A}\mathbf{x}^{m-1}\right)_j\right) > 0.$$

Denote $D_{\mathbf{x}} = \mathrm{diag}(d_1, d_2, \ldots, d_n)$ with $d_k = 1$ and $d_j = \varepsilon$ for $j \neq k$. Then the inequality above implies that

$$\mathbf{x}^\top D_{\mathbf{x}}(\mathscr{A}\mathbf{x}^{m-1}) > 0,$$

and the necessary condition holds.

Now we prove the sufficiency. For each nonzero $\mathbf{x} \in \mathbf{R}^n$, there exists a matrix $D_{\mathbf{x}} = \mathrm{diag}(d_1, d_2, \ldots, d_n)$ with $d_i > 0$ for all $i \in [n]$ such that

$$0 < \mathbf{x}^\top D_{\mathbf{x}}(\mathscr{A}\mathbf{x}^{m-1}) = \sum_{i=1}^{n} d_i(x_i(\mathscr{A}\mathbf{x}^{m-1})_i).$$

Since $d_i > 0$, $i \in [n]$, there exists at least one $i \in [n]$ such that $x_i(\mathscr{A}\mathbf{x}^{m-1})_i > 0$. Otherwise, $x_i(\mathscr{A}\mathbf{x}^{m-1})_i \leq 0$ for all $i \in [n]$. Then $\sum_{i=1}^{n} d_i(x_i(\mathscr{A}\mathbf{x}^{m-1})_i) \leq 0$, which is a contradiction. Hence $\mathscr{A}$ is a P-tensor, and the desired conclusion follows. □

Let $\mathscr{A} \in T_{m,n}$. For any $\mathbf{x} \in \mathbf{R}^n$, define an operator $T_{\mathscr{A}} : \mathbf{R}^n \to \mathbf{R}^n$ such that

$$T_{\mathscr{A}}(\mathbf{x}) := \begin{cases} \|\mathbf{x}\|_2^{2-m}\mathscr{A}\mathbf{x}^{m-1}, & \mathbf{x} \neq \mathbf{0} \\ \mathbf{0}, & \mathbf{x} = \mathbf{0}. \end{cases} \tag{4.30}$$

When m is even, define another operator $F_{\mathscr{A}} : \mathbf{R}^n \to \mathbf{R}^n$ such that

$$F_{\mathscr{A}}(\mathbf{x}) := (\mathscr{A}\mathbf{x}^{m-1})^{[\frac{1}{m-1}]}, \quad \forall\, \mathbf{x} \in \mathbf{R}^n. \tag{4.31}$$

It should be noted that both $T_{\mathscr{A}}$ and $F_{\mathscr{A}}$ are continuous and positively homogeneous, i.e., $T_{\mathscr{A}}(t\mathbf{x}) = tT_{\mathscr{A}}(\mathbf{x})$ and $F_{\mathscr{A}}(t\mathbf{x}) = tF_{\mathscr{A}}(\mathbf{x})$ for $t > 0$.

Based on operators (4.30) and (4.31), we define another two quantities by

$$\alpha(T_{\mathscr{A}}) := \min_{\|\mathbf{x}\|_\infty = 1} \max_{i \in [n]} x_i(T_{\mathscr{A}}(\mathbf{x}))_i, \tag{4.32}$$

and

$$\alpha(F_{\mathscr{A}}) := \min_{\|\mathbf{x}\|_\infty = 1} \max_{i \in [n]} x_i(F_{\mathscr{A}}(\mathbf{x}))_i. \tag{4.33}$$

It is clear that $\alpha(T_{\mathscr{A}})$ is well defined for any positive integer m, and $\alpha(F_{\mathscr{A}})$ is well defined for even integer m. In the following analysis, the monotonicity and boundedness of such two quantities will be established when $\mathscr{A}$ is a P (P_0)-tensor. Furthermore, it is shown that $\mathscr{A}$ is a P (P_0)-tensor if and only if $\alpha(T_{\mathscr{A}})$ is positive (nonnegative), and when m is even, $\mathscr{A}$ is a P(P_0)-tensor if and only if $\alpha(F_{\mathscr{A}})$ is positive (nonnegative).

To move on, we introduce a new operator norm $\|\cdot\|_\infty$ for any continuous, positive homogeneous operator $T : \mathbf{R}^n \to \mathbf{R}^n$ such that

$$\|T\|_\infty := \max_{\|\mathbf{x}\|_\infty=1} \|T(\mathbf{x})\|_\infty.$$

Apparently, we have $\|T(\mathbf{x})\|_\infty \le \|T\|_\infty \|\mathbf{x}\|_\infty$ for any $\mathbf{x} \in \mathbf{R}^n$.

Lemma 4.7 *Let $\mathscr{A} = (a_{i_1\ldots i_m}) \in T_{m,n}$. Then*

(1) $\|T_{\mathscr{A}}\|_\infty \le \max_{i\in[n]} \left(\sum_{i_2,\ldots,i_m=1}^{n} |a_{ii_2\ldots i_m}| \right)$;

(2) $\|F_{\mathscr{A}}\|_\infty \le \max_{i\in[n]} \left(\sum_{i_2,\ldots,i_m=1}^{n} |a_{ii_2\ldots i_m}| \right)^{\frac{1}{m-1}}$ *when m is even.*

Proof (1) By the fact that $\|\mathbf{x}\|_2 \ge \|\mathbf{x}\|_\infty$, we obtain that

$$\begin{aligned}
\|T_{\mathscr{A}}\|_\infty &= \max_{\|\mathbf{x}\|_\infty=1} \|T_{\mathscr{A}}(\mathbf{x})\|_\infty \\
&= \max_{\|\mathbf{x}\|_\infty=1} \max_{1\le i\le n} \left| \|\mathbf{x}\|_2^{-(m-2)} \sum_{i_2,\ldots,i_m=1}^{n} a_{ii_2\ldots i_m} x_{i_2} x_{i_3} \cdots x_{i_m} \right| \\
&\le \max_{\|\mathbf{x}\|_\infty=1} \|\mathbf{x}\|_\infty^{-(m-2)} \max_{1\le i\le n} \left(\sum_{i_2,\ldots,i_m=1}^{n} |a_{ii_2\ldots i_m}||x_{i_2}||x_{i_3}| \cdots |x_{i_m}| \right) \\
&\le \max_{\|\mathbf{x}\|_\infty=1} \|\mathbf{x}\|_\infty^{-(m-2)} \|\mathbf{x}\|_\infty^{m-1} \max_{1\le i\le n} \left(\sum_{i_2,\ldots,i_m=1}^{n} |a_{ii_2\ldots i_m}| \right) \\
&= \max_{1\le i\le n} \left(\sum_{i_2,\ldots,i_m=1}^{n} |a_{ii_2\ldots i_m}| \right).
\end{aligned}$$

And the desired result (1) holds.
(2) When m is even. By (4.33), it follows that

$$\begin{aligned}
\|F_{\mathscr{A}}\|_\infty &= \max_{\|\mathbf{x}\|_\infty=1} \|F_{\mathscr{A}}(\mathbf{x})\|_\infty \\
&= \max_{\|\mathbf{x}\|_\infty=1} \max_{1\le i\le n} \left| \left(\sum_{i_2,\ldots,i_m=1}^{n} a_{ii_2\ldots i_m} x_{i_2} x_{i_3} \cdots x_{i_m} \right)^{\frac{1}{m-1}} \right| \\
&\le \max_{\|\mathbf{x}\|_\infty=1} \max_{1\le i\le n} \left(\sum_{i_2,\ldots,i_m=1}^{n} |a_{ii_2\ldots i_m}||x_{i_2}||x_{i_3}| \cdots |x_{i_m}| \right)^{\frac{1}{m-1}} \\
&\le \max_{\|\mathbf{x}\|_\infty=1} \max_{1\le i\le n} \left(\sum_{i_2,\ldots,i_m=1}^{n} |a_{ii_2\ldots i_m}| \|\mathbf{x}\|_\infty^{m-1} \right)^{\frac{1}{m-1}}
\end{aligned}$$

$$= \max_{1\le i\le n}\left(\sum_{i_2,\ldots,i_m=1}^{n} |a_{ii_2\ldots i_m}|\right)^{\frac{1}{m-1}}.$$

So, the desired results hold. □

Combining (4.32) and (4.33) with Definition 4.7, we have the following results.

Theorem 4.11 *Suppose $\mathscr{D} \in T_{m,n}$ is a nonnegative diagonal tensor with diagonal entries $d_1, d_2, \ldots, d_n$. Let $\mathscr{A} = (a_{i_1\ldots i_m}) \in T_{m,n}$ be a P_0-tensor. Then the following results hold.*
(1) If m is even, $\alpha(T_{\mathscr{A}}) \le \alpha(T_{\mathscr{A}+\mathscr{D}})$;
(2) Suppose $\mathscr{A}^J$ is an arbitrary sub-tensor of $\mathscr{A}$ with $J \subseteq [n]$, $|J| = r$, then

$$\alpha(T_{\mathscr{A}}) \le \alpha(T_{\mathscr{A}^J});$$

(3) Assume m is even and $\mathscr{A}^J$ is an arbitrary sub-tensor of $\mathscr{A}$, then

$$\alpha(F_{\mathscr{A}}) \le \alpha(F_{\mathscr{A}^J});$$

(4) $\alpha(T_{\mathscr{A}}) \le \max_{i\in[n]}\left(\sum_{i_2,\ldots,i_m=1}^{n} |a_{ii_2\ldots i_m}|\right)$;

(5) If m is even, then $\alpha(F_{\mathscr{A}}) \le \max_{i\in[n]}\left(\sum_{i_2,\ldots,i_m=1}^{n} |a_{ii_2\ldots i_m}|\right)^{\frac{1}{m-1}}$.

Proof (1) From Definition 4.7, $\mathscr{A}$ is a P_0-tensor implies that $\mathscr{A}+\mathscr{D}$ is a P_0-tensor. Since m is even, then $x_i^m > 0$ for $x_i \ne 0$, and

$$\begin{aligned}
\alpha(T_{\mathscr{A}}) &= \min_{\|\mathbf{x}\|_\infty=1}\left\{\max_{i\in[n]} x_i(T_{\mathscr{A}}(\mathbf{x}))_i\right\}\\
&= \min_{\|\mathbf{x}\|_\infty=1}\left\{\|\mathbf{x}\|_2^{2-m}\max_{i\in[n]} x_i(\mathscr{A}\mathbf{x}^{m-1})_i\right\}\\
&\le \min_{\|\mathbf{x}\|_\infty=1}\left\{\|\mathbf{x}\|_2^{2-m}\max_{i\in[n]}\left\{x_i(\mathscr{A}\mathbf{x}^{m-1})_i + d_i x_i^m\right\}\right\}\\
&= \min_{\|\mathbf{x}\|_\infty=1}\left\{\max_{i\in[n]} x_i\left(\|\mathbf{x}\|_2^{2-m}(\mathscr{A}+\mathscr{D})\mathbf{x}^{m-1}\right)_i\right\}\\
&= \min_{\|\mathbf{x}\|_\infty=1}\left\{\max_{i\in[n]} x_i\left(T_{\mathscr{A}+\mathscr{D}}(\mathbf{x})\right)_i\right\}\\
&= \alpha(T_{\mathscr{A}+\mathscr{D}}).
\end{aligned}$$

(2) Let $\mathscr{A}^J$ be a principal sub-tensor of $\mathscr{A}$. For any $\mathbf{z} = (z_1, \ldots, z_r)^\top \in \mathbf{R}^r$, $\mathbf{z} \ne \mathbf{0}$, define $\mathbf{y}(\mathbf{z}) = (y_1(\mathbf{z}), y_2(\mathbf{z}), \ldots, y_n(\mathbf{z}))^\top \in \mathbf{R}^n$ with $y_i(\mathbf{z}) = z_i$ for $i \in J$ and $y_i(\mathbf{z}) = 0$ otherwise. So it holds that $\|\mathbf{z}\|_\infty = \|\mathbf{y}(\mathbf{z})\|_\infty$ and $\|\mathbf{z}\|_2 = \|\mathbf{y}(\mathbf{z})\|_2$. Hence,

$$
\begin{aligned}
\alpha(T_{\mathscr{A}}) &= \min_{\|\mathbf{x}\|_\infty=1}\left\{\max_{i\in[n]} x_i(T_{\mathscr{A}}(\mathbf{x}))_i\right\}\\
&= \min_{\|\mathbf{x}\|_\infty=1}\left\{\|\mathbf{x}\|_2^{2-m}\max_{i\in[n]} x_i(\mathscr{A}\mathbf{x}^{m-1})_i\right\}\\
&\le \min_{\|\mathbf{y}(\mathbf{z})\|_\infty=1}\left\{\|\mathbf{y}(\mathbf{z})\|_2^{2-m}\max_{i\in[n]}\{\mathbf{y}(\mathbf{z})_i(\mathscr{A}\mathbf{y}(\mathbf{z})^{m-1})_i\right\}\\
&= \min_{\|\mathbf{z}\|_\infty=1}\left\{\max_{i\in[n]} z_i(\|\mathbf{z}\|_2^{2-m}\mathscr{A}^J\mathbf{z}^{m-1})_i\right\}\\
&= \min_{\|\mathbf{z}\|_\infty=1}\left\{\max_{i\in[n]} z_i(T_{\mathscr{A}^J}(\mathbf{z}))_i\right\}\\
&= \alpha(T_{\mathscr{A}^J}).
\end{aligned}
$$

(3) It can be proved similarly as the proof of (2).

(4) For any $\mathbf{x}=(x_1,\ldots,x_n)^\top\in\mathbf{R}^n$, it holds that

$$x_i(T_{\mathscr{A}}(\mathbf{x}))_i\le\|\mathbf{x}\|_\infty\|T_{\mathscr{A}}(\mathbf{x})\|_\infty\le\|T_{\mathscr{A}}\|_\infty\|\mathbf{x}\|_\infty^2,\quad\text{for } i\in[n].$$

Then

$$\max_{i\in[n]} x_i(T_{\mathscr{A}}(\mathbf{x}))_i\le\|T_{\mathscr{A}}\|_\infty\|\mathbf{x}\|_\infty^2.$$

Therefore, we have

$$\alpha(T_{\mathscr{A}})=\min_{\|\mathbf{x}\|_\infty=1}\{\max_{i\in[n]} x_i(T_{\mathscr{A}}(\mathbf{x}))_i\}\le\|T_{\mathscr{A}}\|_\infty.$$

By Lemma 4.7, the desired conclusion follows.

(5) Similar to (4), the result can be proved. □

Now, based upon $\alpha(F_{\mathscr{A}})$ and $\alpha(T_{\mathscr{A}})$, we derive a necessary and sufficient condition for a given tensor to be a P(P$_0$)-tensor.

Theorem 4.12 *Suppose $\mathscr{A}\in T_{m,n}$ is a given tensor. Then the following results hold.*

(1) $\mathscr{A}$ is a P(P$_0$)-tensor if and only if $\alpha(T_{\mathscr{A}})$ is positive (nonnegative);

(2) If m is even, $\mathscr{A}$ is a P (P$_0$)-tensor if and only if $\alpha(F_{\mathscr{A}})$ is positive (nonnegative).

Proof We only present the prove for P-tensors, as the P$_0$-tensor case can be proved similarly.

(1) Assume $\mathscr{A}$ is a P-tensor. Then for each nonzero vector $\mathbf{x}\in\mathbf{R}^n$, there exists $i\in[n]$ such that

$$\max_{i\in[n]} x_i(\mathscr{A}\mathbf{x}^{m-1})_i>0,$$

which implies that

$$\max_{i\in[n]} x_i(\|\mathbf{x}\|_2^{2-m}\mathscr{A}\mathbf{x}^{m-1})_i=\|\mathbf{x}\|_2^{2-m}\max_{i\in[n]} x_i(\mathscr{A}\mathbf{x}^{m-1})_i>0.$$

Therefore it holds that

$$\alpha(T_{\mathscr{A}}) = \min_{\|\mathbf{x}\|_\infty=1} \left\{ \max_{i\in[n]} x_i(T_{\mathscr{A}}(\mathbf{x}))_i \right\} > 0.$$

On the other hand, if $\alpha(T_{\mathscr{A}}) > 0$, then it is obvious that for each nonzero $\mathbf{y} \in \mathbf{R}^n$,

$$\max_{i\in[n]} \left(\frac{\mathbf{y}}{\|\mathbf{y}\|_\infty}\right)_i \left(T_{\mathscr{A}}\left(\frac{\mathbf{y}}{\|\mathbf{y}\|_\infty}\right)\right)_i \geq \alpha(T_{\mathscr{A}}) > 0.$$

Hence,

$$\max_{i\in[n]} y_i(T_{\mathscr{A}}(\mathbf{y}))_i = \max_{i\in[n]} y_i \left(\|\mathbf{y}\|_2^{2-m} \mathscr{A}\mathbf{y}^{m-1}\right)_i > 0.$$

Thus $y_j(\mathscr{A}\mathbf{y}^{m-1})_j > 0$ for some $j \in [n]$, which implies that $\mathscr{A}$ is a P-tensor.
(2) Let m be even. Suppose $\mathscr{A}$ is a P-tensor. Then for each nonzero $\mathbf{x} \in \mathbf{R}^n$, there exists an $i \in [n]$ such that $x_i(\mathscr{A}\mathbf{x}^{m-1})_i > 0$, which implies that

$$0 < x_i^{\frac{1}{m-1}} (\mathscr{A}\mathbf{x}^{m-1})_i^{\frac{1}{m-1}} = x_i^{\frac{2-m}{m-1}} \left(x_i(\mathscr{A}\mathbf{x}^{m-1})_i^{\frac{1}{m-1}}\right).$$

Note that $x_i^{\frac{2-m}{m-1}} > 0$ for $x_i \neq 0$ since m is even, and thus

$$0 < x_i(\mathscr{A}\mathbf{x}^{m-1})_i^{\frac{1}{m-1}} = x_i(F_{\mathscr{A}}(\mathbf{x}))_i,$$

which means that

$$\max_{i\in[n]} x_i(F_{\mathscr{A}}(\mathbf{x}))_i > 0, \ \forall\, \mathbf{x} \in \mathbf{R}^n, \ \mathbf{x} \neq \mathbf{0}.$$

Thus it follows that

$$\alpha(F_{\mathscr{A}}) = \min_{\|\mathbf{x}\|_\infty=1} \{\max_{i\in[n]} x_i(F_{\mathscr{A}}(\mathbf{x}))_i\} > 0.$$

On the other hand, if $\alpha(F_{\mathscr{A}}) > 0$, then it is obvious that

$$\max_{i\in[n]} \left(\frac{\mathbf{y}}{\|\mathbf{y}\|_\infty}\right)_i \left(F_{\mathscr{A}}\left(\frac{\mathbf{y}}{\|\mathbf{y}\|_\infty}\right)\right)_i \geq \alpha(F_{\mathscr{A}}) > 0, \ \forall\, \mathbf{y} \in \mathbf{R}^n, \ \mathbf{y} \neq \mathbf{0}.$$

Thus, there exists a $j \in [n]$ such that

$$y_j(F_{\mathscr{A}}(\mathbf{y}))_j = y_j(\mathscr{A}\mathbf{y}^{m-1})_j^{\frac{1}{m-1}} > 0,$$

and hence

$$y_j^{m-2}(y_j(\mathscr{A}\mathbf{y}^{m-1})_j) = y_j^{m-1}(\mathscr{A}\mathbf{y}^{m-1})_j > 0.$$

Since m is even, we have $y_j^{m-2} > 0$, and it holds that $y_j(\mathscr{A}\mathbf{y}^{m-1})_j > 0$, i.e., $\mathscr{A}$ is a P-tensor. □

As stated in Theorem 4.9 that there exists no odd order symmetric P-tensor. To make up this deficiency, another extension of $P(P_0)$ matrix was defined in [81]. To avoid confusion, we call it $P'(P'_0)$-tensor in the following analysis.

Definition 4.8 An mth order n dimensional tensor $\mathscr{A}$ is called a **P′-tensor** if for each nonzero $\mathbf{x} \in \mathbf{R}^n$ there exist some indices $i \in [n]$ such that

$$x_i^{m-1}(\mathscr{A}\mathbf{x}^{m-1})_i > 0. \tag{4.34}$$

An mth order n dimensional tensor $\mathscr{A}$ is called a **$\mathbf{P}'_0$-tensor**, if for each nonzero $\mathbf{x} \in \mathbf{R}^n$ there exists some index $i \in [n]$ such that

$$x_i \neq 0 \text{ and } x_i^{m-1}(\mathscr{A}\mathbf{x}^{m-1})_i \geq 0. \tag{4.35}$$

It is worth pointing out that the concepts of P′-tensors and P'_0-tensors are well defined for both even and odd order tensors. Particularly, when m is even, P′- and P'_0-tensors are exactly P- and P_0-tensors respectively as defined in Definition 4.7. That is because x_i^{m-1} and x_i have the same sign for any $x_i \in \mathbf{R}$.

Properties on P′- and P'_0-tensors will be presented below without proof.

Theorem 4.13 *Let $\mathscr{A} \in T_{m,n}$.*
(1) If $\mathscr{A}$ is a P′-tensor (P'_0-tensor), then it has no nonpositive (negative) H-eigenvalues.
(2) If $\mathscr{A}$ is P′-tensor (P'_0-tensor), then each of its principal sub-tensor is also a P′-tensor (P'_0-tensor respectively).

Similar to Theorem 4.10, we have the following sufficient and necessary condition for a given tensor to be P′-tensor.

Theorem 4.14 *Let $\mathscr{A} \in T_{m,n}$. Then $\mathscr{A}$ is a P′-tensor if and only if for each nonzero vector $\mathbf{x} \in \mathbf{R}^n$, there exists a positive diagonal matrix $D_\mathbf{x}$ such that*

$$(D_\mathbf{x}\mathbf{x}^{[m-1]})^\top \mathscr{A}\mathbf{x}^{m-1} > 0.$$

Now we characterize the solution set of TCP $(\mathbf{q}, \mathscr{A})$ (4.1) with $\mathscr{A}$ being a P′-tensor. Before that, an important lemma is needed.

Lemma 4.8 (Moré 1974) *Let $f : \mathbf{R}^n \to \mathbf{R}^n$ be a continuous mapping on the rectangular cone $K_\mathbf{R}$, and assume that for each $\mathbf{x} \neq \mathbf{0}$ in $K_\mathbf{R}$,*

$$\max_{i=1,2,\ldots,n} x_i(f_i(\mathbf{x}) - f_i(\mathbf{0})) > 0.$$

If the mapping $g : \mathbf{R}^n \to \mathbf{R}^n$ defined by $g(\mathbf{x}) = f(\mathbf{x}) - f(\mathbf{0})$ is positively homogeneous, then for each $\mathbf{z} \in \mathbf{R}^n$, there is an $\mathbf{x}^* \geq_{K_\mathbf{R}} \mathbf{0}$ with

$$f(\mathbf{x}^*) \geq_{K_{\mathbf{R}}^*} \mathbf{z} \text{ and } \langle \mathbf{x}^*, f(\mathbf{x}^*) - \mathbf{z} \rangle = 0. \tag{4.36}$$

Moreover, the set of $\mathbf{x}^* \geq_{K_{\mathbf{R}}} \mathbf{0}$ *which satisfies (4.36) is compact.*

Set $K_{\mathbf{R}} = \mathbf{R}^n_+$ and $f(\mathbf{x}) = \mathscr{A}\mathbf{x}^{m-1} + \mathbf{q}$ in the above lemma, where $\mathscr{A}$ is a P′-tensor. Since $f(\mathbf{0}) = \mathbf{q}$, the definition of P′ tensors and $f(\mathbf{x}) - f(\mathbf{0}) = \mathscr{A}\mathbf{x}^{m-1}$ being homogeneous imply that $f(\mathbf{x})$ satisfies the conditions of Lemma 4.8. Hence, the following theorem about the tensor complementarity problems with P′-tensors follows readily.

Theorem 4.15 *Suppose $\mathscr{A}$ is a P′-tensor. Then the tensor complementarity problem (4.1) has a nonempty compact solution set.*

4.5 Tensor Complementarity Problems and Semi-positive Tensors

In this section, we study the relationships between the uniqueness of the solution of TCP and (strictly) semi-positive tensors. Several necessary conditions or sufficient conditions are given. Moreover, sufficient and necessary conditions for tensor being (strictly) semi-positive are presented. We will show that the tensor complementarity problem with a positive vector **q** has unique solution if and only if the corresponding tensor is semi-positive. It is proved that a symmetric tensor is (strictly) semi-positive if and only if it is (strictly) copositive.

First of all, we introduce (strictly) semi-positive tensors as below.

Definition 4.9 Suppose $\mathscr{A} = (a_{i_1 \ldots i_m}) \in T_{m,n}$. $\mathscr{A}$ is said to be

(1) **semi-positive** iff for each $\mathbf{x} \geq 0$ and $\mathbf{x} \neq \mathbf{0}$, there exists an index $k \in [n]$ such that

$$x_k > 0 \text{ and } \left(\mathscr{A}\mathbf{x}^{m-1}\right)_k \geq 0;$$

(2) **strictly semi-positive** iff for each $\mathbf{x} \geq \mathbf{0}$ and $\mathbf{x} \neq \mathbf{0}$, there exists an index $k \in [n]$ such that

$$x_k > 0 \text{ and } \left(\mathscr{A}\mathbf{x}^{m-1}\right)_k > 0;$$

The following results can be obtained directly, and the proof is omitted.

Proposition 4.3 *Suppose $\mathscr{A} = (a_{i_1 i_2 \ldots i_m}) \in T_{m,n}$ is a given tensor. Then the following results hold.*

(1) $a_{ii\ldots i} \geq 0$ for all $i \in [n]$ if $\mathscr{A}$ is semi-positive;

(2) $a_{ii\ldots i} > 0$ for all $i \in [n]$ if $\mathscr{A}$ is strictly semi-positive;

(3) There exists $k \in [n]$ such that $\Sigma^n_{i_2,\ldots,i_m=1} a_{ki_2\ldots i_m} \geq 0$ if $\mathscr{A}$ is semi-positive;

(4) There exists $k \in [n]$ such that $\Sigma^n_{i_2,\ldots,i_m=1} a_{ki_2\ldots i_m} > 0$ if $\mathscr{A}$ is strictly semi-positive;

(5) Each principal sub-tensor of a semi-positive tensor is semi-positive;

(6) Each principal sub-tensor of a strictly semi-positive tensor is strictly semi-positive.

The following two theorems present main results of this section, i.e., a real tensor $\mathscr{A}$ is a (strictly) semi-positive tensor if and only if the tensor complementarity problem TCP($\mathbf{q}, \mathscr{A}$) has a unique solution for each $\mathbf{q} > \mathbf{0}$ ($\mathbf{q} \geq \mathbf{0}$).

Theorem 4.16 *Suppose $\mathscr{A} = (a_{i_1\ldots i_m}) \in T_{m,n}$ is a given tensor. The following statements are equivalent:*
(1) $\mathscr{A}$ is semi-positive;
(2) The TCP $(\mathbf{q}, \mathscr{A})$ has a unique solution for every $\mathbf{q} > \mathbf{0}$;
(3) For any index set $N \subset [n]$, the system

$$\begin{cases} \mathscr{A}^N(\mathbf{x}^{(N)})^{m-1} < \mathbf{0} \\ \mathbf{x}^{(N)} \geq \mathbf{0} \end{cases} \tag{4.37}$$

has no solution, where $\mathbf{x}^{(N)} \in \mathbf{R}^{|N|}$ and $\mathscr{A}^N$ is a principal sub-tensor of $\mathscr{A}$.

Proof (1) $\Rightarrow$ (2). Since $\mathbf{q} > \mathbf{0}$, it is obvious that $\mathbf{0}$ is a solution of TCP $(\mathbf{q}, \mathscr{A})$. Suppose that there exists a vector $\mathbf{q}' > \mathbf{0}$ such that TCP $(\mathbf{q}', \mathscr{A})$ has a solution $\mathbf{x} \neq \mathbf{0}$. Since $\mathscr{A}$ is semi-positive, there is an index $k \in [n]$ such that

$$x_k > 0 \text{ and } \left(\mathscr{A}\mathbf{x}^{m-1}\right)_k \geq 0.$$

Then $q'_k + \left(\mathscr{A}\mathbf{x}^{m-1}\right)_k > 0$, and thus

$$\mathbf{x}^\top(\mathbf{q}' + \mathscr{A}\mathbf{x}^{m-1}) > 0,$$

which contradicts the assumption that $\mathbf{x}$ solves TCP $(\mathbf{q}', \mathscr{A})$, and the desired result holds.
(2) $\Rightarrow$ (3). By contradiction, assume there is a subset $N \subset [n]$ such that the system (4.37) has a solution $\bar{\mathbf{x}}^{(N)} \neq \mathbf{0}$. Let $\bar{\mathbf{x}} = (\bar{x}_1, \bar{x}_2, \ldots, \bar{x}_n)^\top \in \mathbf{R}^n$ such that

$$\bar{x}_i = \begin{cases} \bar{x}_i^{(N)}, i \in N \\ 0, \quad i \in [n] \setminus N. \end{cases}$$

Let $\mathbf{q} = (q_1, q_2, \ldots, q_n)^\top \in \mathbf{R}^n$ with entries

$$\begin{cases} q_i = -\left(\mathscr{A}^N(\bar{\mathbf{x}}^{(N)})^{m-1}\right)_i = -\left(\mathscr{A}\bar{\mathbf{x}}^{m-1}\right)_i, & i \in N \\ q_i > \max\{0, -\left(\mathscr{A}\bar{\mathbf{x}}^{m-1}\right)_i\}, & i \in [n] \setminus N. \end{cases}$$

Hence, $\mathbf{q} > \mathbf{0}$ and $\bar{\mathbf{x}}$ solves the TCP $(\mathbf{q}, \mathscr{A})$, which contradicts (2) and the desired result holds.

(3)$\Rightarrow$ (1). For any $\mathbf{x} \in \mathbf{R}^n_+, \mathbf{x} \neq \mathbf{0}$, assume that $\mathbf{x} = (x_1, x_2, \ldots, x_n)^\top \in \mathbf{R}^n$ with $N \subset [n]$ such that

$$\begin{cases} x_i > 0, & i \in N \\ x_i = 0, & i \in [n] \setminus N. \end{cases}$$

Since the system (4.37) has no solution, there exists an index $k \in N \subset [n]$ such that

$$x_k > 0 \text{ and } \left(\mathscr{A}\mathbf{x}^{m-1}\right)_k \geq 0,$$

which implies that $\mathscr{A}$ is semi-positive from Definition 4.9. □

By a similar analysis as that of Theorem 4.16 with proper changes in the inequalities, the following conclusions about strictly semi-positive tensors can be established.

Theorem 4.17 *Let $\mathscr{A} = (a_{i_1 \ldots i_m}) \in T_{m,n}$. The following statements are equivalent:*
(1) $\mathscr{A}$ is strictly semi-positive;
(2) The TCP $(\mathbf{q}, \mathscr{A})$ has a unique solution for every $\mathbf{q} \geq \mathbf{0}$;
(3) For any set $N \subset [n]$, the system

$$\begin{cases} \mathscr{A}^N (\mathbf{x}^{(N)})^{m-1} \leq \mathbf{0} \\ \mathbf{x}^{(N)} \geq \mathbf{0}, \ \mathbf{x}^{(N)} \neq \mathbf{0} \end{cases} \tag{4.38}$$

has no solution.

Now, we show the relationship between the semi-positive tensor and copositive tensor.

Theorem 4.18 *Suppose $\mathscr{A} = (a_{i_1 i_2 \ldots i_m}) \in S_{m,n}$ is a given symmetric tensor. Then $\mathscr{A}$ is a semi-positive tensor if and only if it is copositive; $\mathscr{A}$ is strictly semi-positive if and only if it is strictly copositive.*

Proof Suppose $\mathscr{A}$ is copositive, by Definition 4.5, it holds that

$$\mathscr{A}\mathbf{x}^m = \mathbf{x}^\top \mathscr{A}\mathbf{x}^{m-1} \geq 0, \ \ \forall\, \mathbf{x} \in \mathbf{R}^n_+. \tag{4.39}$$

Hence, $\mathscr{A}$ is semi-positive. Otherwise there is a vector $\mathbf{x} \in \mathbf{R}^n_+$ such that for all $k \in [n]$

$$\left(\mathscr{A}\mathbf{x}^{m-1}\right)_k < 0 \text{ when } x_k > 0.$$

This implies that

$$\mathscr{A}\mathbf{x}^m = \mathbf{x}^\top \mathscr{A}\mathbf{x}^{m-1} = \sum_{k=1}^{n} x_k \left(\mathscr{A}\mathbf{x}^{m-1}\right)_k < 0,$$

which contradicts (4.39).

If $\mathscr{A}$ is semi-positive, by Definition 4.5 and the homogeneity of $\mathscr{A}\mathbf{x}^m$, it is enough to prove that

$$f(\mathbf{x}) = \mathscr{A}\mathbf{x}^m \geq 0, \quad \text{for } \mathbf{x} \in S,$$

where $S = \{\mathbf{x} \in \mathbf{R}^n_+ \mid \Sigma^n_{i=1} x_i = 1\}$. It is obvious that $f : S \to \mathbf{R}$ is continuous, and the set S is compact. So there exists $\tilde{\mathbf{y}} \in S$ such that

$$f(\tilde{\mathbf{y}}) = \mathscr{A}\tilde{\mathbf{y}}^m = \min_{\mathbf{x}\in S} \mathscr{A}\mathbf{x}^m. \tag{4.40}$$

Since $\tilde{\mathbf{y}} \geq \mathbf{0}$ and $\tilde{\mathbf{y}} \neq \mathbf{0}$, without loss of generality, let

$$\tilde{\mathbf{y}} = (\tilde{y}_1, \tilde{y}_2, \ldots, \tilde{y}_l, 0, \ldots, 0)^\top, \quad \tilde{y}_i > 0, \ i \in [l], \ 1 \leq l \leq n.$$

Let $\tilde{\mathbf{w}} = (\tilde{y}_1, \tilde{y}_2, \ldots, \tilde{y}_l)^\top$. Suppose $\mathscr{B} = \mathscr{A}^{[l]}$ is a principal sub-tensor of $\mathscr{A}$. Then $\tilde{\mathbf{w}} > \mathbf{0}$, $\sum_{i=1}^{l} \tilde{y} = 1$, and

$$f(\tilde{\mathbf{y}}) = \mathscr{A}\tilde{\mathbf{y}}^m = \mathscr{B}\tilde{\mathbf{w}}^m = \min_{\mathbf{x}\in S} \mathscr{A}\mathbf{x}^m. \tag{4.41}$$

Since $\tilde{\mathbf{w}} \in \mathbf{R}^l$ is a relative interior point of $\mathbf{R}^l_+$, $\tilde{\mathbf{w}}$ is a local minimizer of the following optimization problem

$$\min_{\mathbf{z}\in\mathbf{R}^l} \mathscr{B}\mathbf{z}^m \quad \text{s.t.} \quad \Sigma^l_{i=1} z_i = 1.$$

Recall the KKT conditions of the optimization problem, it follows that there exists $\mu \in \mathbf{R}$ satisfying

$$\nabla(\mathscr{B}\tilde{\mathbf{w}}^m - \mu(\Sigma^l_{i=1}\tilde{y}_i - 1)) = m\mathscr{B}\tilde{\mathbf{w}}^{m-1} - \mu\mathbf{1} = 0,$$

where $\mathbf{1} = (1, 1, \ldots, 1)^\top \in \mathbf{R}^l$. Hence, $\mathscr{B}\tilde{\mathbf{w}}^{m-1} = \frac{\mu}{m}\mathbf{1}$. Let $\lambda = \frac{\mu}{m}$. Then we obtain

$$\mathscr{B}\tilde{\mathbf{w}}^m = \tilde{\mathbf{w}}^\top \mathscr{B}\tilde{\mathbf{w}}^{m-1} = \lambda.$$

By (4.41), it follows that $\mathscr{A}\tilde{\mathbf{y}}^m = \lambda$, which implies that for all $\tilde{y}_k > 0$,

$$(\mathscr{A}\tilde{\mathbf{y}}^{m-1})_k = (\mathscr{B}\tilde{\mathbf{w}}^{m-1})_k = \lambda.$$

The fact that $\mathscr{A}$ is semi-positive and $\tilde{\mathbf{y}}_i > 0, i \in [l]$ implies

$$(\mathscr{A}\tilde{\mathbf{y}}^{m-1})_k \geq \mathbf{0}$$

for all $k \in [l]$, which means that $\lambda \geq 0$. Therefore, it holds that

$$\min_{\mathbf{x}\in S} \mathscr{A}\mathbf{x}^m = \mathscr{A}\tilde{\mathbf{y}}^m = \lambda \geq 0,$$

which leads to $\mathscr{A}$ is copositive and the desired result holds.

Similar to the proof above, the second result follows. □

The solution analysis for symmetric strictly copositive tensors, such as Theorem 4.6, can be applied to strictly semi-positive tensors as well.

Next, we present the properties of a quantity $\beta(\mathscr{A})$, which is defined based on a strictly semi-positive tensor as

$$\beta(\mathscr{A}) := \min_{\|\mathbf{x}\|_\infty=1, \mathbf{x}\geq \mathbf{0}} \max_{i\in[n]} x_i(\mathscr{A}\mathbf{x}^{m-1})_i, \tag{4.42}$$

The quantity also plays an important role in studying the error boundedness of TCP, which will be further stated in Sect. 4.7.

Theorem 4.19 *Suppose $\mathscr{D} \in S_{m,n}$ is a nonnegative diagonal tensor with diagonal entries $d_i \geq 0, i \in [n]$. Let $\mathscr{A} = (a_{i_1\ldots i_m}) \in T_{m,n}$ be a semi-positive tensor. Then the following conditions hold.*
(1) $\beta(\mathscr{A}) \leq \beta(\mathscr{A}+\mathscr{D})$;
(2) $\beta(\mathscr{A}) \leq \beta(\mathscr{A}^J)$ for any $J \subseteq [n]$;
(3) $\beta(\mathscr{A}) \leq n^{\frac{m-2}{2}} \|T_\mathscr{A}\|_\infty$;
(4) $\beta(\mathscr{A}) \leq \|F_\mathscr{A}\|_\infty^{m-1}$ when m is even.

Proof (1) Since $\mathscr{A}$ is semi-positive, by Definition 4.9, it follows that $\mathscr{A}+\mathscr{D}$ is a semi-positive tensor. Hence $\beta(\mathscr{A}+\mathscr{D})$ is well defined and

$$\begin{aligned}
\beta(\mathscr{A}) &= \min_{\substack{\|\mathbf{x}\|_\infty=1\\ \mathbf{x}\geq\mathbf{0}}} \max_{i\in[n]} x_i(\mathscr{A}\mathbf{x}^{m-1})_i\\
&\leq \min_{\substack{\|\mathbf{x}\|_\infty=1\\ \mathbf{x}\geq\mathbf{0}}} \max_{i\in[n]} \left(x_i(\mathscr{A}\mathbf{x}^{m-1})_i + d_i x_i^m\right)\\
&= \min_{\substack{\|\mathbf{x}\|_\infty=1\\ \mathbf{x}\geq\mathbf{0}}} \max_{i\in[n]} x_i\left((\mathscr{A}+\mathscr{D})\mathbf{x}^{m-1}\right)_i\\
&= \beta(\mathscr{A}+\mathscr{D}).
\end{aligned}$$

(2) Let $\mathscr{A}^J$ be a sub-tensor of $\mathscr{A}$ with $|J| = r$ and

$$J = \{j_t \mid j_1 < j_2 < \cdots < j_r\}.$$

For any $\mathbf{z} = (z_1, \ldots, z_r)^\top \in \mathbf{R}^r_+, \mathbf{z} \neq \mathbf{0}$, define $\mathbf{y}(\mathbf{z}) = (y_1(\mathbf{z}), y_2(\mathbf{z}), \ldots, y_n(\mathbf{z}))^\top \in \mathbf{R}^n_+$ with $y_{j_i}(\mathbf{z}) = z_i$, $j_i \in J$ and $y_i(\mathbf{z}) = 0$ for $i \notin J$. So $\|\mathbf{z}\|_\infty = \|\mathbf{y}(\mathbf{z})\|_\infty$, and we have

$$\begin{aligned}
\beta(\mathscr{A}) &= \min_{\substack{\|\mathbf{x}\|_\infty=1\\ \mathbf{x}\geq\mathbf{0}}} \max_{i\in[n]} x_i(\mathscr{A}\mathbf{x}^{m-1})_i\\
&\leq \min_{\substack{\|\mathbf{y}(\mathbf{z})\|_\infty=1\\ \mathbf{y}(\mathbf{z})\geq\mathbf{0}}} \max_{i\in[n]} (\mathbf{y}(\mathbf{z}))_i(\mathscr{A}(\mathbf{y}(\mathbf{z}))^{m-1})_i\\
&= \min_{\substack{\|\mathbf{z}\|_\infty=1\\ \mathbf{z}\geq\mathbf{0}}} \max_{i\in[r]} z_i(\mathscr{A}^J\mathbf{z}^{m-1})_i\\
&= \beta(\mathscr{A}^J).
\end{aligned}$$

(3) By (4.30), for any vector $\mathbf{x} = (x_1, \ldots, x_n)^\top \in \mathbf{R}^n \setminus \{\mathbf{0}\}$ and any $i \in [n]$,

$$\begin{aligned} x_i(\mathscr{A}\mathbf{x}^{m-1})_i &= x_i\left(\|\mathbf{x}\|_2^{m-2} T_{\mathscr{A}}(\mathbf{x})\right)_i \\ &\le \|\mathbf{x}\|_2^{m-2} \|\mathbf{x}\|_\infty \|T_{\mathscr{A}}(\mathbf{x})\|_\infty \\ &\le \|\mathbf{x}\|_2^{m-2} \|T_{\mathscr{A}}\|_\infty \|\mathbf{x}\|_\infty^2. \end{aligned}$$

From $\|\mathbf{x}\|_2 \le \sqrt{n}\|\mathbf{x}\|_\infty$, we have

$$\max_{i\in[n]} x_i(\mathscr{A}\mathbf{x}^{m-1})_i \le \|\mathbf{x}\|_2^{m-2} \|T_{\mathscr{A}}\|_\infty \|\mathbf{x}\|_\infty^2 \le n^{\frac{m-2}{2}} \|T_{\mathscr{A}}\|_\infty \|\mathbf{x}\|_\infty^m.$$

Therefore

$$\beta(\mathscr{A}) = \min_{\substack{\|\mathbf{x}\|_\infty=1 \\ \mathbf{x}\ge\mathbf{0}}} \max_{i\in[n]} x_i(\mathscr{A}\mathbf{x}^{m-1})_i \le n^{\frac{m-2}{2}} \|T_{\mathscr{A}}\|_\infty.$$

(4) From (4.31), it holds that for each nonzero vector $\mathbf{x} = (x_1, \ldots, x_n)^\top \in \mathbf{R}^n_+$ and each $i \in [n]$,

$$x_i(\mathscr{A}\mathbf{x}^{m-1})_i = x_i(F_{\mathscr{A}}(\mathbf{x}))_i^{m-1} \le \|\mathbf{x}\|_\infty \|F_{\mathscr{A}}(\mathbf{x})\|_\infty^{m-1} \le \|F_{\mathscr{A}}\|_\infty^{m-1} \|\mathbf{x}\|_\infty^m.$$

Then

$$\max_{i\in[n]} x_i(\mathscr{A}\mathbf{x}^{m-1})_i \le \|F_{\mathscr{A}}\|_\infty^{m-1} \|\mathbf{x}\|_\infty^m,$$

and thus

$$\beta(\mathscr{A}) = \min_{\substack{\|\mathbf{x}\|_\infty=1 \\ \mathbf{x}\ge\mathbf{0}}} \max_{i\in[n]} x_i(\mathscr{A}\mathbf{x}^{m-1})_i \le \|F_{\mathscr{A}}\|_\infty^{m-1},$$

which ends the proof. □

Now we give another characterization for (strictly) semi-positive tensors with respect to the quantity $\beta(\mathscr{A})$.

Theorem 4.20 *Let $\mathscr{A} \in T_{m,n}$. Then*

(1) $\mathscr{A}$ is a strictly semi-positive tensor if and only if $\beta(\mathscr{A}) > 0$;
(2) $\beta(\mathscr{A}) \ge 0$ if $\mathscr{A}$ is a semi-positive tensor.

Proof (1) If $\mathscr{A}$ is strictly semi-positive, by Definition 4.9, it holds that for each $\mathbf{x} = (x_1, x_2, \ldots, x_n)^\top \in \mathbf{R}^n_+ \setminus \{\mathbf{0}\}$, there exists $k \in [n]$ such that

$$x_k > 0 \text{ and } (\mathscr{A}\mathbf{x}^{m-1})_k > 0. \tag{4.43}$$

Then, we have

$$\max_{i\in[n]} x_i(\mathscr{A}\mathbf{x}^{m-1})_i > 0.$$

Therefore

$$\beta(\mathscr{A}) = \min_{\substack{\|\mathbf{x}\|_\infty=1 \\ \mathbf{x}\geq \mathbf{0}}} \max_{i\in[n]} x_i(\mathscr{A}\mathbf{x}^{m-1})_i > 0.$$

On the other hand, if $\beta(\mathscr{A}) > 0$, for each $\mathbf{y} \in \mathbf{R}^n_+$ and $\mathbf{y} \neq \mathbf{0}$, if follows that

$$\max_{i\in[n]} \left(\frac{\mathbf{y}}{\|\mathbf{y}\|_\infty}\right)_i \left(\mathscr{A}\left(\frac{\mathbf{y}}{\|\mathbf{y}\|_\infty}\right)^{m-1}\right)_i \geq \beta(\mathscr{A}) > 0.$$

By $\|\mathbf{y}\|_\infty > 0$, we have

$$\max_{i\in[n]} y_i(\mathscr{A}\mathbf{y}^{m-1})_i > 0.$$

Hence, $y_k(\mathscr{A}\mathbf{y}^{m-1})_k > 0$ for some $k \in [n]$, i.e., $\mathscr{A}$ is a strictly semi-positive tensor.

(2) We can prove the conclusion similarly as (1). □

4.6 Tensor Complementarity Problems and Q-Tensors

Q-matrix is a type of long lasting and widely applied matrices in mathematical world, which has been generalized to higher order case as Q-tensor recently. In this section, we will study Q-tensors and their TCP solutions. Two famous conclusions related to Q-matrices are extended to the tensor space. Within the class of strong P_0-tensors or nonnegative tensors, it is proved that several of those subclasses are equivalent with Q-tensors. Moreover, the solvability of the tensor complementarity problem corresponding to Q-tensors are discussed.

We first give the definitions of $R(R_0)$-tensor and Q-tensor.

Definition 4.10 Let $\mathscr{A} = (a_{i_1\ldots i_m}) \in T_{m,n}$. Then

(1) $\mathscr{A}$ is an **R-tensor** if the following system is inconsistent

$$\begin{cases} 0 \neq \mathbf{x} \geq 0, \ t \geq 0, \\ \left(\mathscr{A}\mathbf{x}^{m-1}\right)_i + t = 0 \text{ if } x_i > 0, \\ \left(\mathscr{A}\mathbf{x}^{m-1}\right)_j + t \geq 0 \text{ if } x_j = 0; \end{cases} \tag{4.44}$$

(2) $\mathscr{A}$ is an **$\mathbf{R_0}$-tensor** if the system (4.44) is inconsistent for $t = 0$.

(3) $\mathscr{A}$ is a **Q-tensor** if the TCP $(\mathbf{q}, \mathscr{A})$ has a solution for all $\mathbf{q} \in \mathbf{R}^n$.

It is obvious that R-tensors are R_0-tensors. On the other hand, the following lemma shows that each strictly semi-positive tensor is a Q-tensor.

Lemma 4.9 *Suppose $\mathscr{A} \in S_{m,n}$ is a strictly semi-positive tensor, then $\mathscr{A}$ is a Q-tensor.*

Proof Let $\mathbf{y} = (\mathbf{x}^\top, s)^\top, \mathbf{x} \in \mathbf{R}^n_+, s \in \mathbf{R}_+$. Define the mapping $F : \mathbf{R}^{n+1}_+ \to \mathbf{R}^{n+1}$ such that

$$F(\mathbf{y}) = \begin{pmatrix} \mathscr{A}\mathbf{x}^{m-1} + s\mathbf{q} + s\mathbf{1} \\ s \end{pmatrix}, \tag{4.45}$$

where $\mathbf{1} = (1, 1, \ldots, 1)^\top \in \mathbf{R}^n$, and $\mathbf{q} \in \mathbf{R}^n$. Let $S = \{\mathbf{y} \in \mathbf{R}^{n+1}_+ \mid \Sigma_{i=1}^{n+1} y_i = 1\}$. Since F is continuous and S is compact, from Lemma 4.4, we know that there exists a vector $\tilde{y} = (\tilde{\mathbf{x}}^\top, \tilde{s})^\top \in S$ such that

$$\mathbf{y}^\top F(\tilde{\mathbf{y}}) \geq \tilde{\mathbf{y}}^\top F(\tilde{\mathbf{y}}), \quad \forall\, \mathbf{y} \in S,$$

$$(F(\tilde{\mathbf{y}}))_k = \min_{i \in [n+1]} (F(\tilde{\mathbf{y}}))_i = \omega \;\text{ if }\; \tilde{y}_k > 0, \tag{4.46}$$

$$(F(\tilde{\mathbf{y}}))_k \geq \omega \;\text{ if }\; \tilde{y}_k = 0. \tag{4.47}$$

We claim that $\tilde{s} > 0$. If $\tilde{s} = 0$, then by (4.47), one obtain

$$\omega \leq (F(\tilde{\mathbf{y}}))_{n+1} = \tilde{s} = 0.$$

Combining this with (4.46), we obtain that

$$(F(\tilde{\mathbf{y}}))_k = (\mathscr{A}\tilde{\mathbf{x}}^{m-1})_k = \omega \;\text{ if }\; \tilde{x}_k > 0,$$

which contradicts that $\mathscr{A}$ is a strictly semi-positive tensor, i.e., there is $k \in [n]$ such that $\tilde{x}_k > 0$, $(\mathscr{A}\tilde{\mathbf{x}}^{m-1})_k > 0$. Thus $\tilde{s} > 0$.

Now, we prove that the TCP($\mathbf{q}, \mathscr{A}$) has a solution for all $\mathbf{q} \in \mathbf{R}^n$. The result hold obviously when $\mathbf{q} \geq \mathbf{0}$ since $\mathbf{0} \in \mathbf{R}^n$ is a solution of the TCP($\mathbf{q}, \mathscr{A}$). When $\mathbf{q} \notin \mathbf{R}^n_+$, by (4.45)–(4.47), we have

$$(F(\tilde{\mathbf{y}}))_{n+1} = \min_{i \in [n+1]} (F(\tilde{\mathbf{y}}))_i = \omega = \tilde{s} = \tilde{y}_{n+1} > 0,$$

and for all $i \in [n]$,

$$(F(\tilde{\mathbf{y}}))_i = (\mathscr{A}\tilde{\mathbf{x}}^{m-1})_i + \tilde{s}q_i + \tilde{s} = \omega = \tilde{s} \;\text{ if }\; \tilde{y}_i = \tilde{x}_i > 0,$$
$$(F(\tilde{\mathbf{y}}))_i = (\mathscr{A}\tilde{\mathbf{x}}^{m-1})_i + \tilde{s}q_i + \tilde{s} \geq \omega = \tilde{s} \;\text{ if }\; \tilde{y}_i = \tilde{x}_i = 0,$$

which is equivalent with

$$(F(\tilde{\mathbf{y}}))_i = (\mathscr{A}\tilde{\mathbf{x}}^{m-1})_i + \tilde{s}q_i = 0 \;\text{ if }\; \tilde{y}_i = \tilde{x}_i > 0,$$
$$(F(\tilde{\mathbf{y}}))_i = (\mathscr{A}\tilde{\mathbf{x}}^{m-1})_i + \tilde{s}q_i \geq 0 \;\text{ if }\; \tilde{y}_i = \tilde{x}_i = 0.$$

Let $\mathbf{z} = \frac{\tilde{\mathbf{x}}}{\tilde{s}^{\frac{1}{m-1}}} \in \mathbf{R}^n_+$. Then it is not difficult to see that

$$\mathbf{z} \geq \mathbf{0}, \quad \mathscr{A}\mathbf{z}^{m-1} + \mathbf{q} \geq \mathbf{0}, \quad \text{and} \quad \mathbf{z}^\top(\mathscr{A}\mathbf{z}^{m-1} + \mathbf{q}) = 0,$$

and the desired results hold. □

Next, we will show the relationships between R(R_0)-tensors and tensor complementarity problems (4.1).

Proposition 4.4 *Let $\mathscr{A} = (a_{i_1 \ldots i_m}) \in T_{m,n}$. Then the following results hold.*

(1) $\mathscr{A}$ is an R_0-tensor if and only if the problem TCP($\mathbf{0}$, $\mathscr{A}$) has a unique solution $\mathbf{0}$;

(2) $\mathscr{A}$ is an R-tensor if and only if it is an R_0-tensor and the problem TCP($\mathbf{1}$, $\mathscr{A}$) has a unique solution $\mathbf{0}$, where $\mathbf{1}$ is the all one vector.

Proof (1) It is clear that the problem TCP($\mathbf{0}$, $\mathscr{A}$) does not have nonzero vector solution if and only if the system

$$\begin{cases} 0 \neq \mathbf{x} = (x_1, \ldots, x_n)^\top \geq 0, \\ \left(\mathscr{A}\mathbf{x}^{m-1}\right)_i = 0 \text{ if } x_i > 0, \\ \left(\mathscr{A}\mathbf{x}^{m-1}\right)_i \geq 0 \text{ if } x_i = 0 \end{cases}$$

has no solution. By Definition 4.10, it is equivalent with $\mathscr{A}$ being an R_0-tensor, and the desired result holds.

(2) The necessity is obvious ($t = 1$) from Definition 4.10. Now we prove the sufficient condition by contradiction. If $\mathscr{A}$ is not an R-tensor, then there exists $\mathbf{x} \in \mathbf{R}^n_+ \setminus \{\mathbf{0}\}$ satisfying the system (4.44), which means that the problem TCP($t\mathbf{1}$, $\mathscr{A}$) has nonzero vector solution $\mathbf{x}$ for some $t \geq 0$. We have $t > 0$ since $\mathscr{A}$ is an R_0-tensor. So the problem TCP($\mathbf{1}$, $\mathscr{A}$) has nonzero vector solution $\frac{\mathbf{x}}{\sqrt[m-1]{t}}$, where we obtain a contradiction, and the desired results follow. □

Moreover, we have the following corollary [246].

Corollary 4.1 *Suppose $\mathscr{A} \in T_{m,n}$ is an R-tensor. Then $\mathscr{A}$ is a Q-tensor.*

The next theorem shows when an R_0-tensor is an R-tensor.

Theorem 4.21 *Let $\mathscr{A} \in T_{m,n}$ be an R_0-tensor. If $\mathscr{A}$ is semi-positive, then $\mathscr{A}$ is an R-tensor, and hence $\mathscr{A}$ is a Q-tensor.*

Proof By contradiction, assume that $\mathscr{A}$ is not an R-tensor. Let $\bar{\mathbf{x}} \geq 0$ be a nonzero solution of the system (4.44). Since $t = 0$ contradicts the assumption that $\mathscr{A}$ is an R_0-tensor, we have $t > 0$, and for any $i \in [n]$,

$$\left(\mathscr{A}\mathbf{x}^{m-1}\right)_i + t = 0 \ \text{ if } \ x_i > 0,$$

and hence,

$$\left(\mathscr{A}\mathbf{x}^{m-1}\right)_i = -t < 0 \ \text{if} \ x_i > 0,$$

which contradicts the assumption that $\mathscr{A}$ is semi-positive. Thus $\mathscr{A}$ is an R-tensor, and further $\mathscr{A}$ is a Q-tensor by Corollary 4.1. □

By the analysis above, the relationship between several classes of structured sensors is listed below.

		Semi-Positive R_0-Tensors
		$\Downarrow$
P-Tensors $\Rightarrow$	Strictly Semi-Positive Tensors	$\Rightarrow$ R-Tensors $\Rightarrow$ Q-Tensors
$\Downarrow$	$\Downarrow$	$\Downarrow$
P_0 − Tensors $\Rightarrow$	Semi-Positive Tensors	R_0−Tensors

Furthermore, we provide a checkable characterization for Q-tensors through diagonal entries.

Theorem 4.22 *Let $\mathscr{A} = (a_{i_1 \ldots i_m}) \in T_{m,n}$ be a nonnegative tensor. Then $\mathscr{A}$ is a Q-tensor if and only if $a_{ii\ldots i} > 0$ for all $i \in [n]$.*

Proof To prove the sufficiency, by Definition 4.9, we know that $\mathscr{A}$ is strictly semi-positive since $\mathscr{A}$ is nonnegative with positive diagonal entries, and hence $\mathscr{A}$ is a Q-tensor by Lemma 4.9.

For necessity, suppose that there exists $k \in [n]$ such that $a_{kk\ldots k} = 0$. Let $\mathbf{q} = (q_1, \ldots, q_n)^\top$ with $q_k < 0$ and $q_i > 0$ for all $i \in [n]$ and $i \neq k$. Since $\mathscr{A}$ is a Q-tensor, the problem TCP($\mathbf{q}$, $\mathscr{A}$) has at least a solution $\mathbf{z}$ such that

$$\mathbf{z} \geq \mathbf{0}, \ \mathbf{w} = \mathscr{A}\mathbf{z}^{m-1} + \mathbf{q} \geq \mathbf{0} \text{ and } \mathbf{z}^\top \mathbf{w} = 0. \tag{4.48}$$

Apparently $\mathbf{z} \neq \mathbf{0}$. Since $\mathbf{z} \geq \mathbf{0}$ and $\mathscr{A} \geq 0$, together with $q_i > 0$ for each $i \in [n]\backslash\{k\}$, we have

$$w_i = \left(\mathscr{A}\mathbf{z}^{m-1}\right)_i + q_i = \sum_{i_2,\ldots,i_m=1}^{n} a_{ii_2\ldots i_m} z_{i_2} \cdots z_{i_m} + q_i > 0.$$

By (4.48), it follows that $z_i = 0$ for $i \neq k$ and $i \in [n]$, which implies that

$$w_k = \left(\mathscr{A}\mathbf{z}^{m-1}\right)_k + q_k = \sum_{i_2,\ldots,i_m=1}^{n} a_{ki_2\ldots i_m} z_{i_2} \cdots z_{i_m} + q_k = a_{kk\ldots k} z_k^{m-1} + q_k = q_k < 0$$

since $a_{kk\ldots k} = 0$. This contradicts the fact that $\mathbf{w} \geq \mathbf{0}$. Thus $a_{ii\ldots i} > 0$ for all $i \in [n]$, and the desired results hold. □

From Definition 4.10 and Theorem 4.22, the following corollaries can be obtained readily.

Corollary 4.2 *Let $\mathscr{A}$ be a Q-tensor. If $\mathscr{A}$ is nonnegative, then the following results hold.*
(1) All principal sub-tensors of $\mathscr{A}$ are also Q-tensors.
(2) $\mathbf{x} = \mathbf{0}$ is the unique feasible solution to the problem TCP($\mathbf{q}$, $\mathscr{A}$) with $\mathbf{q} \geq \mathbf{0}$.

By Theorems 4.21, 4.22 and Definitions 4.9, 4.10, we obtain the following result.

Corollary 4.3 *Suppose $\mathscr{A} = (a_{i_1 i_2 \ldots i_m}) \in T_{m,n}$ is a nonnegative tensor. Then the following results are equivalent:*

(1) $\mathscr{A}$ is a Q-tensor;
(2) $\mathscr{A}$ is an R-tensor;
(3) $\mathscr{A}$ is a strictly semi-positive tensor.

We continue to study properties of Q-tensors, and two famous results related to Q-matrices are extended to the tensor space. Now, we give another characterization for a given tensor to be Q-tensor.

Theorem 4.23 *Let $\mathscr{A} = (a_{i_1 \ldots i_m}) \in T_{m,n}$. Suppose $\mathscr{A}_{1\ldots 1}$, $\mathscr{A}_{2\ldots 2}$ are two tensors with order m dimension $n-1$ such that*

$$\mathscr{A}_{1\ldots 1} := (a_{i_1 \ldots i_m}), \quad i_1, \ldots, i_m \in [n] \setminus \{1\},$$

$$\mathscr{A}_{2\ldots 2} := (a_{i_1 \ldots i_m}), \quad i_1, \ldots, i_m \in [n] \setminus \{2\}.$$

If $a_{1 i_2 \ldots i_m} = a_{2 i_2 \ldots i_m}$ for all $i_2, \ldots, i_m \in [n]$, and both $\mathscr{A}_{1\ldots 1}$ and $\mathscr{A}_{2\ldots 2}$ are Q-tensors, then $\mathscr{A}$ is a Q-tensor.

Proof To prove the results, the following two cases are considered for any given vector $\mathbf{q} = (q_1, \ldots, q_n)^T \in \mathbf{R}^n$.
(1) Assume $q_2 \leq q_1$. Define

$$\mathscr{N} := \{(i_2, \ldots, i_m) : i_2, \ldots, i_m \in [n] \setminus \{1\}\} \quad \text{and} \quad \mathbf{q}_{-1} := (q_2, \ldots, q_n)^\top.$$

Then, for any $\mathbf{x} = (0, \hat{\mathbf{x}}^\top)^\top \in \mathbf{R} \times \mathbf{R}^{n-1}$ with $\hat{\mathbf{x}} := (x_2, \ldots, x_n)^\top \in \mathbf{R}^{n-1}$, it follows that

$$\begin{aligned}(\mathscr{A}_{1\ldots 1}\hat{\mathbf{x}}^{m-1})_i &= \sum_{(i_2,\ldots,i_m)\in \mathscr{N}} a_{(i+1)i_2\ldots i_m} x_{i_2} \cdots x_{i_m} \\ &= \sum_{i_2,\ldots,i_m \in [n]} a_{(i+1)i_2\ldots i_m} x_{i_2} \cdots x_{i_m} = (\mathscr{A}\mathbf{x}^{m-1})_{i+1}\end{aligned} \tag{4.49}$$

for all $i \in [n-1]$, and

$$\begin{aligned}(\mathscr{A}\mathbf{x}^{m-1})_1 &= \sum_{i_2,\ldots,i_m \in [n]} a_{1 i_2\ldots i_m} x_{i_2} \cdots x_{i_m} \\ &= \sum_{i_2,\ldots,i_m \in [n]} a_{2 i_2\ldots i_m} x_{i_2} \cdots x_{i_m} = (\mathscr{A}\mathbf{x}^{m-1})_2\end{aligned} \tag{4.50}$$

since $a_{1i_2\dots i_m} = a_{2i_2\dots i_m}$ for all $i_2, \dots, i_m \in [n]$. On the other hand, $\mathscr{A}_{1\dots 1} \in T_{m,n-1}$ is a Q-tensor, which implies that TCP($\mathbf{q}_{-1}$,$\mathscr{A}_{1\dots 1}$) has a solution $\hat{\mathbf{y}} := (\bar{y}_2, \dots, \bar{y}_n)^\top$. Then, $\hat{\mathbf{y}} \in \mathbf{R}_+^{n-1}$ and for any $i \in [n-1]$, it holds that

$$0 \le (\mathscr{A}_{1\dots 1}\hat{\mathbf{y}}^{m-1} + \mathbf{q}_{-1})_i = (\mathscr{A}_{1\dots 1}\hat{\mathbf{y}}^{m-1})_i + q_{i+1}, \tag{4.51}$$

and

$$\begin{aligned} 0 &= \bar{y}_i(\mathscr{A}_{1\dots 1}\hat{\mathbf{y}}^{m-1} + \mathbf{q}_{-1})_i = \bar{y}_{i+1}(\mathscr{A}_{1\dots 1}\hat{\mathbf{y}}^{m-1})_i + \bar{y}_{i+1}(\mathbf{q}_{-1})_i \\ &= \bar{y}_{i+1}q_{i+1} + \bar{y}_{i+1} \sum_{(i_2,\dots,i_m)\in\mathscr{N}} a_{(i+1)i_2\dots i_m}\bar{y}_{i_2}\cdots\bar{y}_{i_m}. \end{aligned} \tag{4.52}$$

Let $\mathbf{y} := (0, \hat{\mathbf{y}})^\top \in \mathbf{R}_+^n$. By (4.49)–(4.52), we have that

$$\begin{aligned} (\mathscr{A}\mathbf{y}^{m-1} + \mathbf{q})_1 &= (\mathscr{A}\mathbf{y}^{m-1})_1 + q_1 \\ &= (\mathscr{A}\mathbf{y}^{m-1})_2 + q_1 \\ &\ge (\mathscr{A}\mathbf{y}^{m-1})_2 + q_2 \\ &= (\mathscr{A}_{1\dots 1}\hat{\mathbf{y}}^{m-1})_1 + q_2 \\ &\ge 0, \\ (\mathscr{A}\mathbf{y}^{m-1} + \mathbf{q})_{i+1} &= (\mathscr{A}_{1\dots 1}\hat{\mathbf{y}}^{m-1})_i + q_{i+1} \\ &\ge 0, \quad \forall\, i \in [n-1], \end{aligned}$$

and

$$\begin{aligned} y_1(\mathscr{A}\mathbf{y}^{m-1} + \mathbf{q})_1 &= 0, \\ y_{i+1}(\mathscr{A}\mathbf{y}^{m-1} + q)_{i+1} &= \bar{y}_{i+1}q_{i+1} + \bar{y}_{i+1}(\mathscr{A}_{1\dots 1}\hat{\mathbf{y}}^{m-1})_i \\ &= \bar{y}_{i+1}q_{i+1} + \bar{y}_{i+1} \sum_{(i_2,\dots,i_m)\in\mathscr{N}} a_{i+1i_2\dots i_m}\bar{y}_{i_2}\cdots\bar{y}_{i_m} \\ &= 0, \quad \forall\, i \in [n-1]. \end{aligned}$$

Thus, $\mathbf{y}$ is a solution to TCP($\mathbf{q}$, $\mathscr{A}$).
(2) Now suppose $q_1 \le q_2$. In this case, by using the condition that $\mathscr{A}_{2\dots 2} \in T_{m,n-1}$ is a Q-tensor, similar to the proof of (1), we can obtain that TCP($\mathbf{q}$, $\mathscr{A}$) has a solution, and the desired result holds. □

Next, we introduce several equivalent classes of Q-tensors within the class of strong P_0-tensors and nonnegative tensors. Thus all tensor complementarity problems related to those class of tensors always have solutions. First of all, we review a result obtained in the matrix case.

Proposition 4.5 (Agangic, Cottle 1979) *If $A \in \mathbf{R}^{n\times n}$ is a P_0-matrix, then*

$$A \text{ is an } R_0\text{-matrix} \iff A \text{ is an } R\text{-matrix} \iff A \text{ is a } Q\text{-matrix}.$$

Since the concepts of P_0-tensors and P'_0-tensors (see Sect. 4.4) are extensions of P_0-matrix, it is natural to consider whether the following result holds or not:

- Suppose $\mathscr{A} \in T_{m,n}$. If $\mathscr{A}$ is a P_0-tensor or a P'_0-tensor, then

$$\mathscr{A} \text{ is an } R_0\text{-tensor} \iff \mathscr{A} \text{ is an R-tensor} \iff A \text{ is a Q-tensor.}$$

Unfortunately, two examples show that the conjecture does not hold for both P_0-tensors and P'_0-tensors.

Example 4.1 Let $A = (a_{i_1i_2i_3i_4}) \in T_{4,2}$, where $a_{1122} = a_{2222} = 1, a_{2112} = -1$ and all other $a_{i_1i_2i_3i_4} = 0$. Then the following results hold:

(1) $\mathscr{A}$ is a P_0-tensor;
(2) $\mathscr{A}$ is a Q-tensor;
(3) $\mathscr{A}$ is not an R_0-tensor.

To show that the results in Example 4.1 *hold, obviously, for any* $\mathbf{x} \in \mathbf{R}^2$,

$$\mathscr{A}\mathbf{x}^3 = \begin{pmatrix} x_1x_2^2 \\ x_2^3 - x_1^2x_2 \end{pmatrix},$$

and hence,

$$x_1(\mathscr{A}\mathbf{x}^3)_1 = x_1^2x_2^2 \text{ and } x_2(\mathscr{A}\mathbf{x}^3)_2 = x_2^4 - x_1^2x_2^2. \tag{4.53}$$

Therefore, for any $\mathbf{x} \in \mathbf{R}^2$ *with* $\mathbf{x} \neq \mathbf{0}$, *if* $x_1 \neq 0$, *then* $x_1(\mathscr{A}\mathbf{x}^3)_1 = x_1^2x_2^2 \geq 0$; *and if* $x_1 = 0$, *then* $x_2 \neq 0$ *since* $\mathbf{x} \neq \mathbf{0}$, *and* $x_2(\mathscr{A}\mathbf{x}^3)_2 = x_2^4 \geq 0$. *Thus,* $\mathscr{A}$ *is a* $P_0 - tensor$.

Then, it is not difficult to see from (4.53) *that* $(1,0)^\top \in \mathbf{R}^2$ *is a solution to* TCP$(\mathbf{0}, \mathscr{A})$, *which, together with the definition of* R_0*-tensor, implies that* $\mathscr{A}$ *is not an* R_0 *-tensor. Finally, we will show that* $\mathscr{A}$ *is a Q-tensor. Let a and b be two nonnegative real numbers, we consider the following four cases:*

(1) Let $\mathbf{q} = (a^3, b^3)^\top$, *then* $\mathbf{z} = (0,0)^\top$ *is a solution of* TCP$(\mathbf{q}, \mathscr{A})$.

(2) Let $\mathbf{q} = (a^3, -b^3)^\top$, *then* $\mathbf{z} = (0,b)^\top$ *is a solution of* TCP$(\mathbf{q}, \mathscr{A})$.

(3) Let $\mathbf{q} = (-a^3, -b^3)^\top$ *with* $(a,b) \neq (0,0)$, *we show that* TCP$(\mathbf{q}, \mathscr{A})$ *has a solution. In this case, in order to ensure that* $(\mathscr{A}\mathbf{x}^3)_i + q_i \geq 0$ *for* $i \in \{1,2\}$, *it must hold that* $x_1 \neq 0$ *and* $x_2 \neq 0$. *So we need to show that the system of equations*

$$0 = \mathscr{A}\mathbf{x}^3 + \mathbf{q} = \begin{pmatrix} x_1x_2^2 - a^3 \\ x_2^3 - x_1^2x_2 - b^3 \end{pmatrix} \tag{4.54}$$

has a nonnegative solution. From the first equation in (4.54) *it follows that* $x_1 = \frac{a^3}{x_2^2}$; *and hence, the second equation in* (4.54) *becomes*

$$(x_2^3)^2 - b^3x_2^3 - a^6 = 0.$$

It is easy to see that the above equation has a solution $x_2^* := [(b^3 + \sqrt{b^6 + 4a^6})/2]^{1/3} > 0$. *Furthermore,* $(a^3/(x_2^*)^2, x_2^*)^\top$ *is a solution to TCP(*$\mathbf{q}, \mathscr{A}$*).*
(4) Let $\mathbf{q} = (-a^3, b^3)^\top$. *Similar to the proof given in the case (3), we can obtain that TCP(*$\mathbf{q}, \mathscr{A}$*) has a solution in this case.*

Above all, from (1)–(4), we obtain that $\mathscr{A}$ *is a Q-tensor and the desired results hold.* □

Example 4.2 Suppose $\mathscr{A} = (a_{i_1 i_2 i_3}) \in T_{3,2}$, where $a_{122} = a_{222} = 1$, $a_{212} = -1$ and all other $a_{i_1 i_2 i_3} = 0$. Then it holds that

(1) $\mathscr{A}$ is a P_0'-tensor;
(2) $\mathscr{A}$ is a Q-tensor;
(3) $\mathscr{A}$ is not an R_0-tensor.

To prove the results in Example 4.2*, we first consider the problem TCP(*$\mathbf{0}, \mathscr{A}$*), i.e., finding* $\mathbf{x} \in \mathbf{R}^2$ *satisfying*

$$\mathbf{x} \geq \mathbf{0}, \quad \mathscr{A}\mathbf{x}^2 = \begin{pmatrix} x_2^2 \\ x_2^2 - x_1 x_2 \end{pmatrix} \geq \mathbf{0}, \quad \mathbf{x}^\top \mathscr{A}\mathbf{x}^2 = 0.$$

It is easy to see that $(1, 0)^\top$ *is a solution to TCP*$(0, \mathscr{A})$*, and hence* $\mathscr{A}$ *is not an* $R_0 - tensor$*.*

Then, for any $\mathbf{x} \in \mathbf{R}^2, \mathbf{x} \neq \mathbf{0}$*, by Definition* 4.8*, it is not difficult to know that* $\mathscr{A}$ *is a* $P_0' - tensor$ *since*

$$x_1^2(\mathscr{A}\mathbf{x}^2)_1 = x_1^2 x_2^2 \text{ and } x_2^2(\mathscr{A}\mathbf{x}^2)_2 = x_2^4.$$

On the other hand, let a and b be two nonnegative real numbers.
(1) Let $\mathbf{q} = (a^2, b^2)^\top$. *Obviously,* $(0, 0)^\top$ *is a solution to TCP(*$\mathbf{q}, \mathscr{A}$*).*
(2) Let $\mathbf{q} = (-a^2, b^2)^\top$ *with* $a \neq 0$. *Take* $\mathbf{z} := ((a^2 + b^2)/a, a)^\top$*, then*

$$\mathbf{z} \geq \mathbf{0}, \quad \mathscr{A}\mathbf{z}^2 + \mathbf{q} = \begin{pmatrix} a^2 - a^2 \\ a^2 - a \times \frac{a^2+b^2}{a} + b^2 \end{pmatrix} = \mathbf{0}, \quad \mathbf{z}^\top(\mathscr{A}\mathbf{z}^2 + \mathbf{q}) = 0.$$

Thus, $\mathbf{z}$ *solves TCP(*$\mathbf{q}, \mathscr{A}$*) in this case.*
(3) Let $\mathbf{q} = (a^2, -b^2)^\top$. *Take* $\mathbf{z} := (0, b)^\top$*, then*

$$\mathbf{z} \geq \mathbf{0}, \quad \mathscr{A}\mathbf{z}^2 + \mathbf{q} = \begin{pmatrix} b^2 + a^2 \\ 0 \end{pmatrix} \geq \mathbf{0}, \quad \mathbf{z}^\top(\mathscr{A}\mathbf{z}^2 + \mathbf{q}) = 0.$$

Thus,$\mathbf{z}$ *solves TCP(*$\mathbf{q}, \mathscr{A}$*) in this case.*
(4) Let $\mathbf{q} = (-a^2, -b^2)^\top$ *with* $a \leq b$. *Take* $\mathbf{z} := (0, b)^\top$*, then*

$$\mathbf{z} \geq \mathbf{0}, \quad \mathscr{A}\mathbf{z}^2 + \mathbf{q} = \begin{pmatrix} b^2 - a^2 \\ 0 \end{pmatrix} \geq \mathbf{0}, \quad \mathbf{z}^\top(\mathscr{A}\mathbf{z}^2 + \mathbf{q}) = 0.$$

Thus, $\mathbf{z}$ *solves TCP*$(\mathbf{q}, \mathscr{A})$ *in this case.*
(5) Let $\mathbf{q} = (-a^2, -b^2)^\top$ *with* $0 \neq a \geq b$. *Take* $\mathbf{z} := ((a^2 - b^2)/a, a)^\top$, *then*

$$\mathbf{z} \geq 0, \quad \mathscr{A}\mathbf{z}^2 + \mathbf{q} = \begin{pmatrix} a^2 - a^2 \\ a^2 - a \times \frac{a^2-b^2}{a} - b^2 \end{pmatrix} = 0, \quad \mathbf{z}^\top(\mathscr{A}\mathbf{z}^2 + \mathbf{q}) = 0.$$

Thus, $\mathbf{z}$ *solves TCP*$(\mathbf{q}, \mathscr{A})$ *in this case.*

Therefore, it follows from (1)–(5) that the problem TCP$(\mathbf{q}, \mathscr{A})$ *has a solution for each* $\mathbf{q} \in \mathbf{R}^2$. *Thus,* $\mathscr{A}$ *is a Q-tensor, and the desired result follows.* □

Now we present a new class of tensors named strong P_0-tensors, based on which one can extend Proposition 4.5 to the tensor case. Recall that a function $f : \mathbf{R}^n \to \mathbf{R}^n$ is called a P_0-function, if for all $\mathbf{x}, \mathbf{y} \in \mathbf{R}^n$ with $\mathbf{x} \neq \mathbf{y}$, there is an index $i \in [n]$ such that

$$x_i \neq y_i \quad \text{and} \quad (x_i - y_i)[f_i(\mathbf{x}) - f_i(\mathbf{y})] \geq 0.$$

It is not difficult to verify that the mapping $f(\mathbf{x}) := A\mathbf{x} + \mathbf{q}$ with $\mathbf{q} \in \mathbf{R}^n$ is a P_0-function if and only if $A \in \mathbf{R}^{n\times n}$ is a P_0-matrix. We now present the definition of strong P_0-tensors in a similar way.

Definition 4.11 Let $\mathscr{A} \in T_{m,n}$. If the mapping $f(\mathbf{x}) := \mathscr{A}\mathbf{x}^{m-1} + \mathbf{q}$ with $\mathbf{q} \in \mathbf{R}^n$ is a **P_0-function**, we call $\mathscr{A}$ is a **strong P_0-tensor**, abbreviated as **SP_0-tensor**.

Obviously, when $m = 2$, an SP_0-tensor reduces to a P_0-matrix, and all SP_0-tensors are P_0-tensors. Furthermore, we have the following results.

Theorem 4.24 *If* $\mathscr{A} \in T_{m,n}$ *is an* SP_0*-tensor, then the three conditions are equivalent to each other:*

(1) $\mathscr{A}$ *is an* R_0*-tensor;*
(2) $\mathscr{A}$ *is an R-tensor;*
(3) $\mathscr{A}$ *is a Q-tensor.*

Proof Since an SP_0-tensor is a P_0-tensor and every P_0-tensor is semi-positive, by Definition 4.10 and Theorem 4.21, it follows that

$$\mathscr{A} \text{ is an } \mathrm{R}_0\text{-tensor} \quad \Leftrightarrow \quad \mathscr{A} \text{ is an R-tensor.}$$

Furthermore, as discussed before, an R-tensor is a Q-tensor. It remains to prove (3)⇒(1).

Suppose that $\mathscr{A}$ is a Q-tensor but not an R_0-tensor. Then there exists a vector $\bar{\mathbf{x}} \in \mathbf{R}^n_+ \setminus \{\mathbf{0}\}$ such that

$$\begin{cases} (\mathscr{A}\bar{\mathbf{x}}^{m-1})_i = 0, & \text{if } \bar{x}_i > 0, \\ (\mathscr{A}\bar{\mathbf{x}}^{m-1})_i \geq 0, & \text{if } \bar{x}_i = 0. \end{cases}$$

Denote $J_1 = \{i \in [n] : \bar{x}_i = 0\}$ and $J_2 = \{i \in [n] : \bar{x}_i > 0\}$. Let $\mathbf{q} \in \mathbf{R}^n$ with $q_i > 0$ for any $i \in J_1$ and $q_i < 0$ for any $i \in J_2$. Since $\mathscr{A}$ is a Q-tensor, we can assume that $\bar{\mathbf{y}}$ is a solution of TCP($\mathbf{q}$, $\mathscr{A}$). It is obvious that $\bar{\mathbf{x}} \neq \bar{\mathbf{y}}$. Let λ be a positive real number. For any $i \in [n]$ such that $\bar{x}_i \neq \bar{y}_i$, we consider the following two cases.
(1) If $i \in J_1$, then $\bar{x}_i = 0$ and $\bar{y}_i > 0$. Hence, $(\mathscr{A}\bar{\mathbf{x}}^{m-1})_i \geq 0$ and $(\mathscr{A}\bar{\mathbf{y}}^{m-1} + \mathbf{q})_i = 0$. The above equality implies that $(\mathscr{A}\bar{\mathbf{y}}^{m-1})_i < 0$ since $q_i > 0$ for any $i \in J_1$. This further yields that $(\mathscr{A}(\lambda\bar{\mathbf{y}})^{m-1})_i = \lambda^{m-1}(\mathscr{A}\bar{\mathbf{y}}^{m-1})_i < 0$ for any $i \in J_1$ since $\lambda > 0$. Thus, for any $i \in J_1$,

$$(\bar{x}_i - \lambda\bar{y}_i)[(\mathscr{A}\bar{\mathbf{x}}^{m-1})_i - (\mathscr{A}(\lambda\bar{\mathbf{y}})^{m-1})_i] < 0$$

holds for any $\lambda > 0$.
(2) If $i \in J_2$, then $(\mathscr{A}\bar{\mathbf{x}}^{m-1})_i = 0$ and $(\mathscr{A}\bar{\mathbf{y}}^{m-1} + \mathbf{q})_i \geq 0$. The above inequality implies that $(\mathscr{A}\bar{\mathbf{y}}^{m-1})_i > 0$ since $q_i < 0$ for any $i \in J_2$, which yields that $(\mathscr{A}(\lambda\bar{\mathbf{y}})^{m-1})_i = \lambda^{m-1}(\mathscr{A}\bar{\mathbf{y}}^{m-1})_i > 0$ for any $\lambda > 0$. Now, we can choose sufficiently small $\lambda > 0$ such that $(\bar{\mathbf{x}} - \lambda\bar{\mathbf{y}})_i > 0$. Hence, for any $i \in J_2$,

$$(\bar{x}_i - \lambda\bar{y}_i)[(\mathscr{A}\bar{\mathbf{x}}^{m-1})_i - (\mathscr{A}(\lambda\bar{\mathbf{y}})^{m-1})_i] < 0.$$

Thus, by picking sufficiently small $\lambda > 0$, we obtain that for any $i \in [n]$ with $\bar{x}_i \neq \bar{y}_i$,

$$(\bar{x}_i - \lambda\bar{y}_i)[(\mathscr{A}\bar{\mathbf{x}}^{m-1})_i - (\mathscr{A}(\lambda\bar{\mathbf{y}})^{m-1})_i] < 0,$$

which contradicts the condition that $\mathscr{A}$ is an SP_0-tensor. Therefore, $\mathscr{A}$ is an R_0-tensor, and the desired results are obtained. □

Now, we address the relationships among SP_0-tensors, P_0-tensors and P'_0-tensors.

Example 4.3 Let $\mathscr{A} = (a_{i_1 \ldots i_m}) \in T_{m,n}$ with entries $a_{122\ldots2} = 1$ and $a_{i_1\ldots i_m} = 0$ otherwise. Then $\mathscr{A}$ is an SP_0-tensor; $\mathscr{A}$ is a P_0-tensor; $\mathscr{A}$ is a P'_0-tensor.

To prove the Example 4.3, for any $\mathbf{x} \in \mathbf{R}^n$, *we have*

$$\mathscr{A}\mathbf{x}^{m-1} = \begin{pmatrix} x_2^{m-1} \\ 0 \\ \vdots \\ 0 \end{pmatrix} \in \mathbf{R}^n.$$

On one hand, for any $\mathbf{x}, \mathbf{y} \in \mathbf{R}^n$ *with* $\mathbf{x} \neq \mathbf{y}$, *if there exists an index* $i_0 \in [n]$ *such that* $x_{i_0} \neq y_{i_0}$, *then* $(x_{i_0} - y_{i_0})[(\mathscr{A}\mathbf{x}^{m-1})_{i_0} - (\mathscr{A}\mathbf{y}^{m-1})_{i_0}] = 0$. *Otherwise, we have that* $x_i = y_i$ *for all* $i \in [n]$, *and hence,* $x_1 \neq y_1$ *and* $x_2 = y_2$. *Furthermore,* $(x_1 - y_1)$ $[(\mathscr{A}\mathbf{x}^{m-1})_1 - (\mathscr{A}\mathbf{y}^{m-1})_1] = (x_1 - y_1)(x_2^{m-1} - y_2^{m-1}) = 0$. *Thus* $\mathscr{A}$ *is an* SP_0*-tensor.*

On the other hand, for any $\mathbf{x} \in \mathbf{R}^n$ *with* $\mathbf{x} \neq 0$, *if there exists an index* $i_0 \in \{2, \ldots, n\}$ *such that* $x_{i_0} \neq 0$, *then* $x_{i_0}(\mathscr{A}\mathbf{x}^{m-1})_{i_0} = 0$ *and* $x_{i_0}^{m-1}(\mathscr{A}\mathbf{x}^{m-1})_{i_0} = 0$. *Otherwise, we have that* $x_i = 0$ *for all* $i \in \{2, \ldots, n\}$, *and hence,* $x_1 \neq 0$ *and* $x_2 = 0$.

Furthermore, $x_1(\mathscr{A}\mathbf{x}^{m-1})_1 = x_1 x_2^{m-1} = 0$ *and* $x_1^{m-1}(\mathscr{A}\mathbf{x}^{m-1})_1 = x_1^{m-1}x_2^{m-1} = 0$. *Thus* $\mathscr{A}$ *is a* P_0*-tensor as well as a* P_0'*-tensor.* □

Example 4.4 Let $\mathscr{A} = (a_{i_1i_2i_3i_4}) \in T_{4,2}$ with entries $a_{1122} = a_{2122} = 1$ and $a_{i_1i_2i_3i_4} = 0$ elsewhere. Then $\mathscr{A}$ is a P_0-tensor, but not an SP_0-tensor.

Obviously, for any $\mathbf{x} \in \mathbf{R}^2$ with $\mathbf{x} \neq 0$, *we have*

$$\mathscr{A}\mathbf{x}^3 = \begin{pmatrix} x_1x_2^2 \\ x_1x_2^2 \end{pmatrix}.$$

If $x_1 \neq 0$, *then* $x_1(\mathscr{A}\mathbf{x}^3)_1 = x_1^2x_2^2 \geq 0$. *If* $x_1 = 0$, *then* $x_2 \neq 0$ *and* $x_2(\mathscr{A}\mathbf{x}^3)_2 = x_1x_2^3 = 0$. *We obtain that* $\mathscr{A}$ *is a* P_0*-tensor.*

In addition, for any given $\mathbf{q} \in \mathbf{R}^2$, *let* $f(\mathbf{x}) = \mathscr{A}\mathbf{x}^3 + \mathbf{q}$ *for any* $\mathbf{x} \in \mathbf{R}^2$. *Take* $\bar{\mathbf{x}} = (1, 1)^\top$ and $\bar{\mathbf{y}} = (1, -2)^\top$. *It is easy to see that*

$$\bar{x}_1 = \bar{y}_1, \quad \text{and} \quad \bar{x}_2 \neq \bar{y}_2, \ (\bar{x}_2 - \bar{y}_2)(f(\bar{\mathbf{x}}) - f(\bar{\mathbf{y}}))_2 = -9 < 0.$$

These demonstrate that $\mathscr{A}$ *is not an* SP_0*-tensor.* □

For abbreviation, denote the set of all SP_0-tensors, P_0-tensors and P_0'-tensors by $\mathbb{SP}_0$, $\mathbb{P}_0$ and $\mathbb{P}_0'$ respectively. Conclusions from Examples 4.3 and 4.4 can be drawn as below.

Proposition 4.6 *Let* $\mathscr{A} \in T_{m,n}$, *then the following conditions hold:*

(1) $T_{m,n} \bigcap \mathbb{SP}_0 \bigcap \mathbb{P}_0 \bigcap \mathbb{P}_0' \neq \emptyset$;
(2) if $\mathscr{A}$ *is an* SP_0*-tensor, then it is a* P_0*-tensor, but the converse is not true.*

The relationships of P_0-tensors and P_0'-tensors have been studied in Sect. 4.4. Next, we study further connections between P_0-tensors, P_0'-tensors and SP_0-tensors.

Example 4.5 Let $\mathscr{A} = (a_{i_1i_2i_3}) \in T_{3,2}$ with entries $a_{121} = 1$, $a_{211} = -1$ and $a_{i_1i_2i_3} = 0$ elsewhere. Then $\mathscr{A} \in T_{3,2} \bigcap \mathbb{P}_0$, but $\mathscr{A} \notin T_{3,2} \bigcap \mathbb{P}_0'$.

To show the results of Example 4.5, *for any* $\mathbf{x} \in \mathbf{R}^2$, *we have*

$$\mathscr{A}\mathbf{x}^2 = \begin{pmatrix} x_2x_1 \\ -x_1^2 \end{pmatrix}.$$

On one hand, from $x_1(\mathscr{A}\mathbf{x}^2)_1 = x_2x_1^2$ *and* $x_2(\mathscr{A}\mathbf{x}^2)_2 = -x_2x_1^2$, *it is easy to see that for any* $\mathbf{x} \in \mathbf{R}^2\backslash\{\mathbf{0}\}$, *there exists an index* $i \in [2]$ *such that* $x_i \neq 0$ *and* $x_i(\mathscr{A}\mathbf{x}^2)_i \geq 0$. *Thus* $\mathscr{A} \in T_{3,2} \bigcap \mathbb{P}_0$. *On the other hand, for any* $\alpha > 0$ *and* $\beta < 0$, *by taking* $(x_1, x_2) := (\alpha, \beta)$, *we have*

$$x_1^2(\mathscr{A}\mathbf{x}^2)_1 = x_2x_1^3 = \beta\alpha^3 < 0 \quad \text{and} \quad x_2^2(\mathscr{A}\mathbf{x}^2)_2 = -x_2^2x_1^2 = -\beta^2\alpha^2 < 0,$$

and hence $\mathscr{A} \notin T_{3,2} \bigcap \mathbb{P}_0'$. □

Example 4.6 Let $\mathscr{A} = (a_{i_1i_2i_3}) \in T_{3,2}$ with entries $a_{122} = 1, a_{211} = -1$ and other $a_{i_1i_2i_3} = 0$. Then $\mathscr{A} \in T_{3,2} \bigcap \mathbb{P}_0'$, but $\mathscr{A} \notin T_{3,2} \bigcap \mathbb{P}_0$.

Obviously, for any $\mathbf{x} \in \mathbf{R}^2$, *we have*

$$\mathscr{A}\mathbf{x}^2 = \begin{pmatrix} x_2^2 \\ -x_1^2 \end{pmatrix}.$$

On one hand, from $x_1^2(\mathscr{A}\mathbf{x}^2)_1 = x_1^2x_2^2$ *and* $x_2^2(\mathscr{A}\mathbf{x}^2)_2 = -x_2^2x_1^2$, *it is clear that for any* $\mathbf{x} \in \mathbf{R}^2$ *with* $\mathbf{x} \neq 0$, *there exists an index* $i \in \{1, 2\}$ *such that* $x_i \neq 0$ *and* $x_i^2(\mathscr{A}\mathbf{x}^2)_i \geq 0$. *Thus* $\mathscr{A} \in T_{3,2} \bigcap \mathbb{P}_0'$. *On the other hand, for any* $\alpha < 0$ and $\beta > 0$, *by taking* $(x_1, x_2) := (\alpha, \beta)$, *we have*

$$x_1(\mathscr{A}\mathbf{x}^2)_1 = x_1x_2^2 = \alpha\beta^2 < 0 \quad \text{and} \quad x_2(\mathscr{A}\mathbf{x}^2)_2 = -x_2x_1^2 = -\beta\alpha^2 < 0,$$

and hence $\mathscr{A} \notin T_{3,2} \bigcap \mathbb{P}_0$. □

By Definitions 4.7, 4.8 and Examples 4.5, 4.6, we obtain the following conclusions.

Proposition 4.7 *Suppose* $\mathscr{A} \in T_{m,n}$.

(1) If m is even, then $T_{m,n} \bigcap \mathbb{P}_0' = T_{m,n} \bigcap \mathbb{P}_0$;
(2) There is no inclusion relation between $\mathbb{P}_0$ *and* $\mathbb{P}_0'$;
(3) When m is even, if $\mathscr{A}$ *is an* SP_0*-tensor then* $\mathscr{A}$ *is a* P_0'*-tensor, but the converse is not true.*

In the odd order case, we have more results about SP_0-tensors and P_0'-tensors.

Lemma 4.10 *Let* $\mathscr{A} \in T_{m,n} \bigcap \mathbb{SP}_0$ *with m being odd. Then, for any* $i \in [n]$, *either* $(\mathscr{A}\mathbf{x}^{m-1})_i \equiv 0$, *or* $(\mathscr{A}\mathbf{x}^{m-1})_i$ *is a function in variables* $x_1, \ldots, x_{i-1}, x_{i+1}, \ldots, x_n$, *but independent of the variable* x_i.

Proof Let $\mathbf{x} = (x_1, \ldots, x_n)^\top \in \mathbf{R}^n$. Suppose that a is an arbitrary fixed real number and $i_0 \in [n]$, we take $\mathbf{y} = (x_1, \ldots, x_{i_0-1}, a, x_{i_0+1}, \ldots, x_n)^\top$. Then for each

$$i \in J := \{1, \ldots, i_0 - 1, i_0 + 1, \ldots, n\},$$

one has $x_i = y_i$. For any $x_{i_0} \in \mathbf{R} \setminus \{a\}$, $x_{i_0} \neq y_{i_0}$, which together with $\mathscr{A} \in \mathbb{SP}_0$, implies that

$$(x_{i_0} - a)[(\mathscr{A}\mathbf{x}^{m-1})_{i_0} - (\mathscr{A}\mathbf{y}^{m-1})_{i_0}] \geq 0. \tag{4.55}$$

For $-\mathbf{x}$ and $-\mathbf{y}$, we have $\mathscr{A}(-\mathbf{x})^{m-1} = \mathscr{A}\mathbf{x}^{m-1}$ since m is odd. Moreover, $-x_{i_0} \neq -y_{i_0}$ and $-x_i = -y_i$ for any $i \in J$. These and $\mathscr{A} \in \mathbb{SP}_0$ imply that

$$(a - x_{i_0})[(\mathscr{A}\mathbf{x}^{m-1})_{i_0} - (\mathscr{A}\mathbf{y}^{m-1})_{i_0}] \geq 0. \tag{4.56}$$

Combining (4.55) with (4.56), we obtain that for any $x_{i_0} \in \mathbf{R} \setminus \{a\}$,

$$(\mathscr{A}\mathbf{x}^{m-1})_{i_0} = (\mathscr{A}\mathbf{y}^{m-1})_{i_0}.$$

By the arbitrariness of x_{i_0}, the above equality means either $(\mathscr{A}\mathbf{x}^{m-1})_{i_0} \equiv 0$ or $(\mathscr{A}\mathbf{x}^{m-1})_{i_0}$ is independent of the variable x_{i_0}. The desired result follows from the arbitrariness of i_0. □

Proposition 4.8 *Let m be odd. Then* $T_{m,2}\bigcap \mathbb{SP}_0 \subseteq T_{m,2}\bigcap \mathbb{P}'_0$.

Proof It is clear that if $\mathscr{A} \in T_{m,2}\bigcap \mathbb{SP}_0$, then $\mathscr{A} \in T_{m,2}\bigcap \mathbb{P}_0$.

First, we show that for any $\mathbf{x} \in \mathbf{R}^2$, there exists an index $i \in \{1, 2\}$ such that $(\mathscr{A}\mathbf{x}^{m-1})_i \equiv 0$. Assume that $(\mathscr{A}\mathbf{x}^{m-1})_i \not\equiv 0$ for all $i \in \{1, 2\}$. Since $\mathscr{A} \in T_{m,2}\bigcap \mathbb{SP}_0$, it follows from Lemma 4.10 that

$$\mathscr{A}\mathbf{x}^{m-1} = \begin{pmatrix} \alpha x_2^{m-1} \\ \beta x_1^{m-1} \end{pmatrix},$$

where $\alpha, \beta \in \mathbf{R} \setminus \{0\}$. Without loss of generality, assume that $\bar{\mathbf{x}} = (\bar{x}_1, \bar{x}_2)^\top \in \mathbf{R}^2_+$ and $(\mathscr{A}\mathbf{x}^{m-1})_i \neq 0$ for all $i \in \{1, 2\}$. Let

$$\hat{\mathbf{x}} = \begin{cases} (-\bar{x}_1, -\bar{x}_2)^\top & \text{if } (\mathscr{A}\bar{\mathbf{x}}^{m-1})_1 > 0 \text{ and } (\mathscr{A}\bar{\mathbf{x}}^{m-1})_2 > 0, \\ (-\bar{x}_1, \bar{x}_2)^\top & \text{if } (\mathscr{A}\bar{\mathbf{x}}^{m-1})_1 > 0 \text{ and } (\mathscr{A}\bar{\mathbf{x}}^{m-1})_2 < 0, \\ (\bar{x}_1, -\bar{x}_2)^\top & \text{if } (\mathscr{A}\bar{\mathbf{x}}^{m-1})_1 < 0 \text{ and } (\mathscr{A}\bar{\mathbf{x}}^{m-1})_2 > 0, \\ (\bar{x}_1, \bar{x}_2)^\top & \text{if } (\mathscr{A}\bar{\mathbf{x}}^{m-1})_1 < 0 \text{ and } (\mathscr{A}\bar{\mathbf{x}}^{m-1})_2 < 0. \end{cases}$$

Then $\mathscr{A}\hat{\mathbf{x}}^{m-1} = \mathscr{A}\bar{\mathbf{x}}^{m-1}$ since m is odd. Hence, $\hat{x}_1(\mathscr{A}\hat{\mathbf{x}}^{m-1})_1 < 0$ and $\hat{x}_2(\mathscr{A}\hat{\mathbf{x}}^{m-1})_2 < 0$, which is a contradiction with $\mathscr{A} \in \mathbb{P}_0$.

Second, we show that $\mathscr{A} \in T_{m,2}\bigcap \mathbb{P}'_0$. Without loss of generality, assume that $(\mathscr{A}\mathbf{x}^{m-1})_2 \equiv 0$ for any $\mathbf{x} \in \mathbf{R}^2$. Then for $\mathbf{x} \in \mathbf{R}^2$ with $\mathbf{x} \neq 0$,

- if $x_2 \neq 0$, then $x_2^{m-1}(\mathscr{A}\mathbf{x}^{m-1})_2 = 0$; and
- if $x_2 = 0$, then $x_1 \neq 0$ and $x_1^{m-1}(\mathscr{A}\mathbf{x}^{m-1})_1 = x_1^{m-1}(\alpha x_2^{m-1}) = 0$,

thus $\mathscr{A} \in \mathbb{P}'_0$. Therefore, the desired result holds. □

4.7 Z-Tensor Complementarity Problems

In this section, we mainly pay attention to two issues. First, we try to find the sparsest solution to a tensor complementarity problem, which can be modeled as

$$(\mathbf{P_0}) \quad \begin{cases} \min \ \|\mathbf{x}\|_0 \\ \text{s.t.} \ \ \mathscr{A}\mathbf{x}^{m-1} + \mathbf{q} \geq \mathbf{0}, \ \mathbf{x} \geq \mathbf{0}, \ \mathbf{x}^\top(\mathscr{A}\mathbf{x}^{m-1} + \mathbf{q}) = 0. \end{cases}$$

By the nonconvexity and noncontinuity of the involved ℓ_0-norm in (P_0), it is generally NP-hard to find its sparsest solutions. Stimulated by the scheme of the most popular convex relaxation and inspired by the least element theory in nonlinear complementarity problems, we will show that the sparsest solution of (P_0) can be obtained under conditions such that the involved $\mathbf{q}$ is nonnegative and $\mathscr{A}$ is a Z-tensor. Therefore, the problem (P_0) can be achieved by solving the following polynomial programming problem:

$$(\mathbf{P}_1) \quad \begin{cases} \min \ \mathbf{1}^\top \mathbf{x} \\ \text{s.t.} \ \ \mathscr{A}\mathbf{x}^{m-1} + \mathbf{q} = \mathbf{0}, \mathbf{x} \geq \mathbf{0}. \end{cases}$$

Second, we focus on tensor complementarity problems associated with Z-tensors and describe various equivalent conditions for a Z-tensor to have the global solvability property. These conditions/properties include the strong M-tensor property, the S-property, positive stable property, strict semi-monotonicity property, etc.

Next, the partial Z-tensor will be introduced and analyzed in order to explore the Z-property of homogeneous polynomial functions

Definition 4.12 Let $\mathscr{A} = (a_{i_1 \ldots i_m}) \in T_{m,n}$. $\mathscr{A}$ is called a partial Z-tensor if for any $i_1 \in [n]$, $a_{i_1 \ldots i_m} \leq 0$ for all $i_2, \ldots, i_m$ satisfying $i_1 \notin \{i_2, i_3, \ldots, i_m\}$.

Recall the notion of Z-tensor(see Chap. 2) i.e. all off-diagonal entries are nonpositive. Therefore, it is obvious that a Z-tensor is a partial Z-tensor, but not vice versa. Moreover, recall the definition of Z-functions and related properties in Sect. 4.1, both Z-tensors and partial Z-tensors of order $m = 2$ are exactly Z-matrices. Properties on these two kinds of tensors will be discussed first which will play an essential role in the following analysis.

Theorem 4.25 *Let $\mathscr{A} \in T_{m,n}$ be a partial Z-tensor. Then the implication (4.4) holds with $F(\mathbf{x}) = \mathscr{A}\mathbf{x}^{m-1}$; if $\mathscr{A}$ is a Z-tensor, then $F(\mathbf{x}) = \mathscr{A}\mathbf{x}^{m-1}$ is a Z-function.*

Proof For any $\mathbf{x}, \mathbf{y} \in \mathbf{R}^n_+$ satisfying $\mathbf{x}^\top \mathbf{y} = 0$, it implies that

$$x_i \geq 0, \ \ y_i \geq 0, \ \ x_i y_i = 0, \ \ \forall\, i \in [n]. \tag{4.57}$$

Hence, it follows that

$$\begin{aligned} & \mathbf{x}^\top (F(\mathbf{y}) - F(\mathbf{0})) \\ = & \ \mathbf{x}^\top \mathscr{A}\mathbf{y}^{m-1} = \sum_{i=1}^n x_i \left(\mathscr{A}\mathbf{y}^{m-1}\right)_i \\ = & \sum_{i=1}^n x_i \sum_{i_2,\ldots,i_m=1}^n a_{i i_2 \ldots i_m} y_{i_2} \cdots y_{i_m} \\ = & \sum_{i=1}^n \left(\sum_{\substack{i_2,\ldots,i_m=1 \\ i \notin \{i_2,\ldots,i_m\}}}^n a_{i i_2 \ldots i_m} x_i y_{i_2} \cdots y_{i_m} + \sum_{\substack{i_2,\ldots,i_m=1 \\ i \in \{i_2,\ldots,i_m\}}}^n a_{i i_2 \ldots i_m} x_i y_{i_2} \cdots y_{i_m} \right) \end{aligned}$$

$$= \sum_{i=1}^{n} \sum_{\substack{i_2,\ldots,i_m=1 \\ i \notin \{i_2,\ldots,i_m\}}}^{n} a_{ii_2\ldots i_m} x_i y_{i_2} \cdots y_{i_m}$$
$$\le 0,$$

where the second term of the fourth formula is zero owing to (4.57) and the last inequality follows from Definition 4.12.

(2) According to Proposition 4.5, it suffices to show that for any $\mathbf{x} \in \mathbf{R}^n_+$, $\nabla_{\mathbf{x}} \left(\mathscr{A}\mathbf{x}^{m-1}\right)$ is a Z-matrix. Combining with Lemma 4.5, it only needs to show the implication (4.5) holds with $A := \nabla_{\mathbf{x}} \left(\mathscr{A}\mathbf{x}^{m-1}\right)$ for any given $\mathbf{x} \in \mathbf{R}^n_+$. Assume $\mathbf{y}, \mathbf{z}$ are any two arbitrary nonnegative vectors, and $\mathbf{y}^\top \mathbf{z} = 0$. Then

$$\mathbf{y}^\top \nabla_{\mathbf{x}} \left(\mathscr{A}\mathbf{x}^{m-1}\right) \mathbf{z}$$
$$= \sum_{i=1}^{n} y_i \sum_{i_2,\ldots,i_m=1}^{n} \left(a_{ii_2\ldots i_m} + a_{i_2 i\ldots i_m} + \cdots + a_{i_2\ldots i_m i}\right) x_{i_2} \cdots x_{i_{m-1}} z_{i_m}$$
$$= \sum_{i=1}^{n} \left(\sum_{\substack{i_2,\ldots,i_m=1 \\ i_m \ne i}}^{n} \left(a_{ii_2\ldots i_m} + a_{i_2 i\ldots i_m} + \cdots + a_{i_2\ldots i_m i}\right) x_{i_2} \cdots x_{i_{m-1}} z_{i_m} y_i \right)$$
$$\le 0,$$

where the last inequality is from the fact that $\mathscr{A}$ is a Z-tensor. Thus $F(\mathbf{x})$ is a Z-function. □

To characterize the TCP problem which has a unique least solution, the following theorem is needed, which is a nice property possessed by general nonlinear complementarity problems with Z-functions.

Theorem 4.26 (Facchinei, Pang 2003) *Let $F : \mathbf{R}^n \to \mathbf{R}^n$ be a continuous Z-function. Suppose that the following nonlinear complementarity problem*

$$(NCP(F)) \quad \mathbf{x} \ge \mathbf{0}, \; F(\mathbf{x}) \ge \mathbf{0}, \; \mathbf{x}^\top F(\mathbf{x}) = 0$$

is feasible, i.e., $\mathbb{F} := \{\mathbf{x} \in \mathbf{R}^n : \mathbf{x} \ge \mathbf{0}, F(\mathbf{x}) \ge \mathbf{0}\} \ne \emptyset$. Then $\mathbb{F}$ has a unique least element $\mathbf{x}^*$ which is also a solution to $(NCP(F))$.

By the relation between Z-tensors and Z-functions, one can get the following properties for Z-tensor complementarity problems.

Corollary 4.4 *Let $\mathscr{A}$ be a Z-tensor and $\mathbf{q} \in \mathbf{R}^n$. Suppose that TCP($\mathbf{q}, \mathscr{A}$) is feasible, i.e., $\mathbb{F} := \{\mathbf{x} \in \mathbf{R}^n : \mathbf{x} \ge \mathbf{0}, \mathscr{A}\mathbf{x}^{m-1} + \mathbf{q} \ge \mathbf{0}\} \ne \emptyset$. Then $\mathbb{F}$ has a unique least element $\mathbf{x}^*$ which is also a solution to TCP($\mathbf{q}, \mathscr{A}$).*

Proof Since $\mathscr{A}$ is a Z-tensor, by Theorem 4.25, we know that $\mathscr{A}\mathbf{x}^{m-1}$ is a Z-function. By Proposition 4.5, it is not difficult to verify that $\mathscr{A}\mathbf{x}^{m-1} + \mathbf{q}$ is also a Z-function for any $\mathbf{q} \in \mathbf{R}^n$. Thus, the desired result follows directly from Theorem 4.26. □

Assume that the tensor complementarity problem is defined with respect to a nonpositive vector $\mathbf{q}$ and a partial Z-tensor $\mathscr{A}$, the following proposition shows the equivalence between the proposed TCP and a multilinear equation with nonnegative constraints.

Proposition 4.9 *Let $\mathscr{A}$ be a partial Z-tensor. Assume $\mathbf{q} \leq \mathbf{0}$ (all entries of $\mathbf{q}$ are nonpositive). Then the two systems below are equivalent:*

(1) $\mathbf{x} \in \mathbf{R}^n_+,\ \mathscr{A}\mathbf{x}^{m-1} + \mathbf{q} \in \mathbf{R}^n_+,\ \mathbf{x}^\top(\mathscr{A}\mathbf{x}^{m-1} + \mathbf{q}) = 0;$
(2) $\mathbf{x} \in \mathbf{R}^n_+,\ \mathscr{A}\mathbf{x}^{m-1} + \mathbf{q} = \mathbf{0}.$

Proof On one hand, it is obvious that any solution of system (2) is a solution to system (1).

On the other hand, suppose $\mathbf{y}$ is a solution of system (1). Since $\mathscr{A}$ is a partial Z-tensor and $\mathbf{q} \leq \mathbf{0}$, by Theorem 4.25, it follows that

$$\begin{aligned} 0 &\geq (\mathscr{A}\mathbf{y}^{m-1} + \mathbf{q})^\top(\mathscr{A}\mathbf{y}^{m-1}) \\ &= (\mathscr{A}\mathbf{y}^{m-1} + \mathbf{q})^\top(\mathscr{A}\mathbf{y}^{m-1} + \mathbf{q}) - (\mathscr{A}\mathbf{y}^{m-1} + \mathbf{q})^\top\mathbf{q} \\ &\geq \|\mathscr{A}\mathbf{y}^{m-1} + \mathbf{q}\|_2^2. \end{aligned}$$

This indicates that $\mathscr{A}\mathbf{y}^{m-1} + \mathbf{q} = \mathbf{0}$, which implies that $\mathbf{y}$ is a solution of (2) and the desired results hold. □

Note that Z-tensors are partial Z-tensors. Thus the results of Proposition 4.9 also hold for Z-tensors. Furthermore, from Proposition 4.9, one can characterize the feasibility of TCP($\mathbf{q}$, $\mathscr{A}$) in terms of the consistency of the corresponding nonnegative constrained multilinear equation. To move on, the definition of M-tensors is needed, which have been studied in Chap. 2.

Suppose $\mathscr{A} \in T_{m,n}$ is a Z-tensor with $\mathscr{A} = s\mathscr{I} - \mathscr{B}$, where $\mathscr{I}$ is the identity tensor, $\mathscr{B}$ is a nonnegative tensor and $s \in \mathbf{R}_+$. If $s \geq \rho(\mathscr{B})$, then $\mathscr{A}$ is called an M-tensor. If $s > \rho(\mathscr{B})$, then $\mathscr{A}$ is called a strong M-tensor. Here $\rho(\mathscr{B})$ stands for the spectral radius of $\mathscr{B}$. Combining this with Definition 4.9, we can get the following results.

Lemma 4.11 *A Z-tensor $\mathscr{A}$ is a strong M-tensor if and only if there is a vector $\mathbf{x} > \mathbf{0}$ such that $\mathscr{A}\mathbf{x}^{m-1} > \mathbf{0}$.*

Proposition 4.10 *Suppose TCP($\mathbf{q}$, $\mathscr{A}$) is a problem with $\mathscr{A}$ being a strong M-tensor and $\mathbf{q} \leq \mathbf{0}$. Then TCP($\mathbf{q}$, $\mathscr{A}$) is feasible.*

Proof Let $\mathscr{A} = s\mathscr{I} - \mathscr{B}$ be a strong M-tensor with $\mathscr{B} \geq \mathscr{O}$ and $s > \rho(\mathscr{B})$. Since M-tensors are partial Z-tensors, by Proposition 4.9, it suffices to show that there exists some nonnegative $\mathbf{x}$ such that $\mathscr{A}\mathbf{x}^{m-1} = -\mathbf{q}$. Let $T_{s,\mathscr{B},\mathbf{q}} : \mathbf{R}^n_+ \to \mathbf{R}^n_+$ be the mapping defined as follows:

$$T_{s,\mathscr{B},\mathbf{q}}(\mathbf{x}) := \left(s^{-1}\mathscr{B}\mathbf{x}^{m-1} - s^{-1}\mathbf{q}\right)^{[\frac{1}{m-1}]}, \quad \forall\, \mathbf{x} \in \mathbf{R}^n.$$

Thus, it is enough to find a fixed point of this mapping $T_{s,\mathscr{B},\mathbf{q}}$. Besides, since $\mathscr{A}$ is a strong M-tensor, from Lemma 4.11, there always exists a positive $\mathbf{z}$ such that $\mathscr{A}\mathbf{z}^{m-1} > \mathbf{0}$. Denote

$$\alpha := \min_{i\in[n]} \frac{-q_i}{\left(\mathscr{A}\mathbf{z}^{m-1}\right)_i}, \text{ and } \beta := \max_{i\in[n]} \frac{-q_i}{\left(\mathscr{A}\mathbf{z}^{m-1}\right)_i}.$$

Obviously,

$$\mathbf{0} \le \alpha\mathscr{A}\mathbf{z}^{m-1} \le -\mathbf{q} \le \beta\mathscr{A}\mathbf{z}^{m-1}.$$

Set $\mathbf{v} := \mathscr{A}\left(\alpha^{\frac{1}{m-1}}\mathbf{z}\right)^{m-1}$ and $\mathbf{w} := \mathscr{A}\left(\beta^{\frac{1}{m-1}}\mathbf{z}\right)^{m-1}$. Therefore,

$$\alpha^{\frac{1}{m-1}}\mathbf{z} = T_{s,\mathscr{B},\mathbf{v}}\left(\alpha^{\frac{1}{m-1}}\mathbf{z}\right) \le T_{s,\mathscr{B},\mathbf{q}}\left(\alpha^{\frac{1}{m-1}}\mathbf{z}\right),$$

$$\beta^{\frac{1}{m-1}}\mathbf{z} = T_{s,\mathscr{B},\mathbf{w}}\left(\alpha^{\frac{1}{m-1}}\mathbf{z}\right) \ge T_{s,\mathscr{B},\mathbf{q}}\left(\beta^{\frac{1}{m-1}}\mathbf{z}\right)$$

Note that $T_{s,\mathscr{B},\mathbf{q}}$ is an increasing continuous mapping on $\mathbf{R}^n_+$. By Theorem 2.1 in Chap. 2, there exists at least one fixed point $\mathbf{x}$ of $T_{s,\mathscr{B},\mathbf{q}}$ such that $\mathbf{0} \le \alpha^{\frac{1}{m-1}}\mathbf{z} \le \mathbf{x} \le \beta^{\frac{1}{m-1}}\mathbf{z}$. This completes the proof. □

Now we show the exact relaxation theorem for the ℓ_0 norm minimization problem (P_0).

Theorem 4.27 *Suppose $\mathscr{A}$ is a Z-tensor and $\mathbf{q} \le \mathbf{0}$. If the problem (P_0) is feasible, then (P_0) has a solution $\mathbf{x}^*$. Furthermore, $\mathbf{x}^*$ is also the unique solution for problem (P_1).*

Proof By Theorem 4.26, we know that $\mathscr{A}\mathbf{x}^{m-1}$ is a Z-function. Combining this with Proposition 4.1, it is easy to verify that $\mathscr{A}\mathbf{x}^{m-1} + \mathbf{q}$ is also a Z-function. Thus, by Corollary 4.4 and Theorem 4.26, there exists a unique least element $\mathbf{x}^* \in \mathbb{F}$ which is also a solution of the tensor complementarity problem. Since the constraint is non-negative, it yields that $\mathbf{x}^*$ is one of the sparsest solutions of (P_0). By Proposition 4.9, $\mathbf{x}^*$ should be the least element in $\{\mathbf{x} \in \mathbf{R}^n : \mathbf{x} \ge \mathbf{0}, \mathscr{A}\mathbf{x}^{m-1} = -\mathbf{q}\}$, which implies that $\mathbf{x}^*$ is definitely the unique solution of (P_1), and the desired results hold. □

From Proposition 4.10 and Theorem 4.27, one can obtain the following results directly.

Corollary 4.5 *Suppose $\mathscr{A}$ is a strong M-tensor and $\mathbf{q} \le \mathbf{0}$. Then the problem (P_1) can be uniquely solved. Furthermore, the unique solution is also an optimal solution to problem (P_0).*

In addition, some examples are raised below to illustrate the exact relaxation theorem for pursuing the sparsest solution to strong M-tensor complementarity problems.

Example 4.7 Suppose $\mathscr{A} = (a_{i_1 i_2 i_3 i_4}) \in T_{4,2}$ is a tensor with entries $a_{1111} = a_{2222} = 1$, $a_{1112} = -2$, and other entries 0. Let $\mathbf{q} = (0, -1)^\top$. By a direct computation, one can get that $\mathscr{A}$ is a strong M-tensor, and the solution set of the corresponding tensor complementarity problem TCP($\mathbf{q}, \mathscr{A}$) is $\{(0, 1)^\top, (2, 1)^\top\}$. Furthermore, it follows that $(0, 1)^\top$ is the unique common solution to both problems (P_0) and (P_1).

Example 4.8 Assume $\mathscr{A} = (a_{ijk}) \in T_{3,n}$ is a strictly diagonally dominated Z-tensor (and hence strong M-tensor as shown in Chap. 5 of [228]) with entries $a_{ijj} = 0$ for any $i \neq j, i, j \in \{1, \ldots, n\}$. Let $\mathbf{q} \leq \mathbf{0}$ with $\|\mathbf{q}\|_0 = 1$. In this case, it is not hard to verify that $\mathbf{x}^* = (0, \ldots, 0, \sqrt{\frac{b_i}{a_{iii}}}, 0, \ldots, 0)^\top$ is the unique solution to problem (P_0), where i is the index such that $q_i \neq 0$. This optimal solution is also the unique solution to problem (P_1) by direct verification or by applying the Gauss–Seidel iteration method in [87] for the system $\mathscr{A}\mathbf{x}^2 = -\mathbf{q}$ with the zero vector as the initial point.

In the following analysis, we discuss some other conditions for a Z-tensor to be Q-tensor. Different with the analysis above, the following results are mainly presented based on the degree theory. For the sake of completeness, we first review some preliminaries of the degree theory. More information about degree theory can be found in [91].

For any continuous function $f : \mathbf{R}^n \to \mathbf{R}^n$, we assume that $f(\mathbf{x}) = \mathbf{0}$ implies $\mathbf{x} = \mathbf{0}$. Then, the local degree of f at the origin (which equals the degree of f relative to any bounded open set containing zero) is well defined and is denoted by $\deg(f, \mathbf{0})$. When this degree is nonzero, the equation $f(\mathbf{x}) = \mathbf{p}$ will have solutions for all $\mathbf{p}$ near the origin. We now apply this idea to tensor complementarity problems. Given a tensor $\mathscr{A}$, let

$$\Phi(\mathbf{x}) := \min\{\mathbf{x}, F(\mathbf{x})\},$$

where $F(\mathbf{x}) = \mathscr{A}\mathbf{x}^{m-1}$ and $\min\{\mathbf{x}, \mathbf{y}\}$ stands for a vector whose ith entry is $\min\{x_i, y_i\}$.

Theorem 4.28 *Suppose that*

$$\Phi(\mathbf{x}) = \mathbf{0} \Rightarrow \mathbf{x} = \mathbf{0} \quad \textit{and} \quad \deg(\Phi, \mathbf{0}) \neq \mathbf{0}.$$

Then, $\mathscr{A}$ is a Q-tensor and TCP($\mathbf{q}, \mathscr{A}$) has a nonempty compact solution for any $\mathbf{q} \in \mathbf{R}^n$.

By the degree theory, several properties of Q-tensors are established in the following results.

Corollary 4.6 *Let $\mathscr{A} \in T_{m,n}$. If one of the following conditions hold, then $\mathscr{A}$ is a Q-tensor and the corresponding TCP($\mathbf{q}, \mathscr{A}$) has a nonempty compact solution for any $\mathbf{q} \in \mathbf{R}^n$.*

(1) There exists a vector $\mathbf{d} > 0$ such that for TCP($\mathbf{0}, \mathscr{A}$) and TCP($\mathbf{d}, \mathscr{A}$), zero (vector) is the only solution.
(2) $\mathscr{A}$ is a strictly semi-monotone (or strictly semi-positive) tensor, which means that, for each nonzero $\mathbf{x} \geq \mathbf{0}$, $\max_{i \in [n]} x_i (\mathscr{A}\mathbf{x}^{m-1})_i > 0$.

(3) $\mathscr{A}$ *is a strictly copositive tensor, which means that, for all* $\mathbf{0} \neq \mathbf{x} \geq \mathbf{0}$, $\mathscr{A}\mathbf{x}^m := \mathbf{x}^\top \mathscr{A}\mathbf{x}^{m-1} > 0$.
(4) $\mathscr{A}$ *is a positive definite tensor, which means that, for all* $\mathbf{x} \neq \mathbf{0}$, $\mathscr{A}\mathbf{x}^m := \mathbf{x}^\top \mathscr{A}\mathbf{x}^{m-1} > 0$.

Furthermore, the following results hold.

Theorem 4.29 *Let* $\mathscr{A} \in T_{m,n}$. *Consider the following statements:*

(1) $\mathscr{A}$ *is a Q-tensor.*
(2) *For every* $\mathbf{q} \in \mathbf{R}^n$, *TCP*$(\mathbf{q}, \mathscr{A})$ *is feasible.*
(3) *There exists* $\mathbf{d} > \mathbf{0}$ *such that* $\mathscr{A}\mathbf{d}^{m-1} > \mathbf{0}$.

Then, (1) $\Rightarrow$ (2) $\Leftrightarrow$ (3). *Furthermore, these statements are equivalent when* $\mathscr{A}$ *is a Z-tensor.*

Analogous to Theorem 4.29, we have the following results in the even order Z-tensor case.

Theorem 4.30 *Let m be even. If* $\mathscr{A}$ *is a Z-tensor, then the following conditions are equivalent:*

(1) $\mathscr{A}$ *is a strong M-tensor.*
(2) *If* $F(\mathbf{x}) = \mathbf{0}$ *then* $\mathbf{x} = \mathbf{0}$, *and* $\deg(F, \mathbf{0}) = 1$.
(3) $F(\mathbf{x})$ *is surjective.*

Theorem 4.31 *Let m be even. Suppose* $\mathscr{A} \in T_{m,n}$ *is a Z-tensor. Then* $\mathscr{A}$ *is a strong M-tensor if and only if for every principal sub-tensor* $\widetilde{\mathscr{A}}$ *of* $\mathscr{A}$, *the corresponding function* $\widetilde{F}(\mathbf{x}) := \widetilde{\mathscr{A}}\mathbf{x}^{m-1}$ *satisfies the conditions*

$$\widetilde{F}(\mathbf{x}) = \mathbf{0} \Rightarrow \mathbf{x} = \mathbf{0} \quad \textit{and} \quad \deg(\widetilde{F}, \mathbf{0}) = 1.$$

In the result below, we offer an easily checkable sufficient condition for a strong M-tensor to have the global uniqueness solution property.

Theorem 4.32 *Let* $\mathscr{A} = (a_{i_1 \ldots i_m}) \in T_{m,n}$ *be a strong M-tensor such that for each index i,*

$$a_{i\, i_2 \ldots i_m} = 0 \quad \textit{whenever } i_j \neq i_k \ \textit{ for some } j \neq k.$$

Then, for any $\mathbf{q} \in \mathbf{R}^n$, *TCP*$(\mathbf{q}, \mathscr{A})$ *has a unique solution.*

4.8 Solution Boundedness of Tensor Complementarity Problems

In this section, the properties of the solution set of the tensor complementarity problem are studied. The notion of S-tensor is introduced first, and then its two equivalent definitions by means of the tensor complementarity problem are presented. Then it is proved that the solution sets of R_0-tensor, P-tensor and strictly semi-positive tensor complementarity problems are bounded. Furthermore, several new upper bounds of the solution set for tensor complementarity problems with respect to strictly semi-positive tensors are established.

Definition 4.13 Let $\mathscr{A} = (a_{i_1 \ldots i_m}) \in T_{m,n}$. Then

(1) $\mathscr{A}$ is said to be an **S-tensor** if the system $\mathscr{A}\mathbf{x}^{m-1} > \mathbf{0},\ \mathbf{x} > \mathbf{0}$ has a solution;
(2) $\mathscr{A}$ is said to be an **S_0-tensor** if the system $\mathscr{A}\mathbf{x}^{m-1} \geq \mathbf{0},\ \mathbf{x} \geq \mathbf{0},\ \mathbf{x} \neq \mathbf{0}$ has a solution.

Next we present a sufficient and necessary condition for a given tensor to be S-tensor

Proposition 4.11 *Let $\mathscr{A} \in T_{m,n}$. Then $\mathscr{A}$ is an S-tensor if and only if the system*

$$\mathscr{A}\mathbf{x}^{m-1} > \mathbf{0},\ \mathbf{x} \geq \mathbf{0} \tag{4.58}$$

has a solution.

Proof By Definition 4.13, It is enough to prove the sufficient condition. Suppose that there exists $\mathbf{y} \geq \mathbf{0}$ satisfying $\mathscr{A}\mathbf{y}^{m-1} > \mathbf{0}$. Hence, it is easy to know that $\mathbf{y} \neq \mathbf{0}$. Since $\mathscr{A}\mathbf{y}^{m-1}$ is continuous on $\mathbf{y}$, it follows that $\mathscr{A}(\mathbf{y} + t\mathbf{1})^{m-1} > 0$ for some small enough $t > 0$, where $\mathbf{1} = (1, 1, \ldots, 1)^\top$. By Definition 4.13 and $\mathbf{y} + t\mathbf{1} > 0$, we obtain $\mathscr{A}$ is an S-tensor and the desired result holds. □

Based on the system of tensor complementarity problem, a necessary and sufficient condition for a given tensor to be an S-tensor is stated below.

Theorem 4.33 *Let $\mathscr{A} \in T_{m,n}$. Then $\mathscr{A}$ is an S-tensor if and only if the TCP($\mathbf{q}$, $\mathscr{A}$) is feasible for any $\mathbf{q} \in \mathbf{R}^n$.*

Proof On one hand, if $\mathscr{A}$ is an S-tensor, from Definition 4.13, there exists $\mathbf{y}$ such that

$$\mathscr{A}\mathbf{y}^{m-1} > \mathbf{0},\ \mathbf{y} > \mathbf{0}.$$

Hence for any $\mathbf{q} \in \mathbf{R}^n$, there exists $t > 0$ such that

$$\mathscr{A}(\sqrt[m-1]{t}\mathbf{y})^{m-1} = t\mathscr{A}\mathbf{y}^{m-1} \geq -\mathbf{q}.$$

Clearly, $\sqrt[m-1]{t}\mathbf{y} > \mathbf{0}$, then $\sqrt[m-1]{t}\mathbf{y}$ is a feasible vector of the TCP($\mathbf{q}$, $\mathscr{A}$).

On the other hand, if the TCP($\mathbf{q}$, $\mathscr{A}$) is feasible for all $\mathbf{q} \in \mathbf{R}^n$, let $\mathbf{q} < \mathbf{0}$ and $\mathbf{z}$ be a feasible solution of the TCP($\mathbf{q}$, $\mathscr{A}$). Thus

$$\mathbf{z} \geq \mathbf{0} \text{ and } \mathbf{q} + \mathscr{A}\mathbf{z}^{m-1} \geq \mathbf{0}.$$

Therefore

$$\mathscr{A}\mathbf{z}^{m-1} \geq -\mathbf{q} > \mathbf{0},$$

which implies that $\mathbf{z}$ is a solution of the system (4.58). By Proposition 4.11, it follows that $\mathscr{A}$ is an S-tensor. □

Recall the R_0-tensor in Definition 4.10, we have the following results.

Theorem 4.34 *Suppose $\mathscr{A} \in T_{m,n}$ is a given tensor. Then the following three conclusions are equivalent to each other:*

(1) $\mathscr{A}$ is an R_0-tensor;
(2) The set

$$\Gamma(\mathbf{q}, s, t) = \{\mathbf{x} \geq \mathbf{0} \;:\; \mathbf{q} + \mathscr{A}\mathbf{x}^{m-1} \geq \mathbf{0} \textit{ and } \mathbf{x}^\top \mathbf{q} + t\mathscr{A}\mathbf{x}^m \leq s\}$$

is bounded for any $\mathbf{q} \in \mathbf{R}^n$ and each $t, s \in \mathbf{R}$ with $t > 0$.
(3) The solution set of the TCP $(\mathbf{q}, \mathscr{A})$ is bounded for any $\mathbf{q} \in \mathbf{R}^n$.

Proof (1) $\Rightarrow$ (2) Let $\mathscr{A}$ be an R_0-tensor. Assume that there exist $\mathbf{q}' \in \mathbf{R}^n$, $s' \in \mathbf{R}$ and $t' > 0$ such that the set $\Gamma(\mathbf{q}', s', t')$ is not bounded i.e. there is an unbounded sequence $\{\mathbf{x}^k\} \subset \Gamma(\mathbf{q}', s', t')$. Then the sequence $\left\{\frac{\mathbf{x}^k}{\|\mathbf{x}^k\|}\right\}$ is bounded, which implies that there are a $\mathbf{x}' \in \mathbf{R}^n$ and a subsequence $\left\{\frac{\mathbf{x}^{k_j}}{\|\mathbf{x}^{k_j}\|}\right\}$ such that

$$\lim_{j\to\infty} \frac{\mathbf{x}^{k_j}}{\|\mathbf{x}^{k_j}\|} = \mathbf{x}' \neq \mathbf{0} \;\text{ and }\; \lim_{j\to\infty} \|\mathbf{x}^{k_j}\| = \infty.$$

Thus, it follows that

$$\frac{\mathbf{q}'}{\|\mathbf{x}^{k_j}\|^{m-1}} + \mathscr{A}\left(\frac{\mathbf{x}^{k_j}}{\|\mathbf{x}^{k_j}\|}\right)^{m-1} \geq \mathbf{0} \;\text{ and }\; \frac{\mathbf{x}^\top \mathbf{q}'}{\|\mathbf{x}^{k_j}\|^m} + t'\mathscr{A}\left(\frac{\mathbf{x}^{k_j}}{\|\mathbf{x}^{k_j}\|}\right)^m \leq \frac{s'}{\|\mathbf{x}^{k_j}\|^m}. \tag{4.59}$$

Let $j \to \infty$. By the continuity of $\mathscr{A}\mathbf{x}^m$ and $\mathscr{A}\mathbf{x}^{m-1}$, one has

$$\mathscr{A}(\mathbf{x}')^{m-1} \geq \mathbf{0} \;\text{ and }\; \mathscr{A}(\mathbf{x}')^m \leq 0.$$

Since $\mathbf{x}' \geq \mathbf{0}$, we obtain

$$\mathscr{A}(\mathbf{x}')^m = (\mathbf{x}')^\top \mathscr{A}(\mathbf{x}')^{m-1} \geq \mathbf{0}.$$

Thus, $\mathscr{A}(\mathbf{x}')^m = 0$, and $\mathbf{x}'$ is a nonzero solution of the TCP($\mathbf{0}$, $\mathscr{A}$). This contradicts the assumption that $\mathscr{A}$ is R_0-tensor.
(2) $\Rightarrow$ (3) By conditions, let $t = 1, s = 0$, it holds that

$$\Gamma(\mathbf{q}, 0, 1) = \{\mathbf{x} \geq \mathbf{0} \;:\; \mathbf{q} + \mathscr{A}\mathbf{x}^{m-1} \geq \mathbf{0} \text{ and } \mathbf{x}^\top(\mathbf{q} + \mathscr{A}\mathbf{x}^{m-1}) = 0\},$$

which implies that $\Gamma(\mathbf{q}, 0, 1)$ is the solution set of the TCP($\mathbf{q}$, $\mathscr{A}$), and the result (3) follows.
(3) $\Rightarrow$ (1) Assume $\mathscr{A}$ is not R_0-tensor. Then the problem TCP($\mathbf{0}$, $\mathscr{A}$) has a nonzero solution $\mathbf{x}^*$, i.e., $\mathbf{x}^* \in \Gamma(\mathbf{0}, 0, 1)$. Since $\mathbf{x}^* \neq \mathbf{0}$, $\tau\mathbf{x}^* \in \Gamma(\mathbf{0}, 0, 1)$ for all $\tau > 0$. Therefore, the set $\Gamma(\mathbf{0}, 0, 1)$ is not bounded. This contradicts the assumption (3). Hence $\mathscr{A}$ is R_0-tensor and the desired result holds. □

The study of previous sections invokes that a strictly semi-positive tensor is an R_0-tensor and each P_0-tensor is a strictly semi-positive tensor. Thus, if $\mathscr{A}$ is a strictly semi-positive tensor, the solution set of the TCP($\mathbf{q}$, $\mathscr{A}$) is bounded for each $\mathbf{q} \in \mathbf{R}^n$; if $\mathscr{A}$ is a P-tensor, the solution set of the TCP($\mathbf{q}$, $\mathscr{A}$) is bounded for each $\mathbf{q} \in \mathbf{R}^n$.

In the following analysis, we discuss the global upper bound for solutions of TCP($\mathbf{q}$, $\mathscr{A}$) with symmetric strictly semi-positive tensors.

Theorem 4.35 *Suppose $\mathscr{A} = (a_{i_1 \ldots i_m}) \in S_{m,n}$ is strictly semi-positive. If $\mathbf{x}$ is a solution of the TCP($\mathbf{q}$, $\mathscr{A}$), then*

$$\|\mathbf{x}\|_m^{m-1} \leq \frac{\|(-\mathbf{q})_+\|_{\frac{m}{m-1}}}{\lambda(\mathscr{A})}, \tag{4.60}$$

where $\mathbf{x}_+ := (\max\{x_1, 0\}, \max\{x_2, 0\}, \ldots, \max\{x_n, 0\})^\top$, *and*

$$\lambda(\mathscr{A}) := \min_{\mathbf{x} \geq \mathbf{0}, \|\mathbf{x}\|_m = 1} \mathscr{A}\mathbf{x}^m.$$

Proof Since $\mathscr{A}$ is strictly semi-positive, by Theorem 4.18, we know that $\mathscr{A}$ is strictly copositive. From Proposition 5.8, it follows that $\mathscr{A}$ has a Pareto H-eigenvalue $\lambda(\mathscr{A}) = \min_{\mathbf{y} \geq \mathbf{0}, \|\mathbf{y}\|_m = 1} \mathscr{A}\mathbf{y}^m > 0$. Since $\mathbf{x}$ is a solution of the TCP($\mathbf{q}$, $\mathscr{A}$), one has

$$\mathbf{x} \geq \mathbf{0},\ \mathscr{A}\mathbf{x}^{m-1} + \mathbf{q} \geq \mathbf{0} \text{ and } \mathscr{A}\mathbf{x}^m - \mathbf{x}^\top(-\mathbf{q}) = \mathbf{x}^\top(\mathscr{A}\mathbf{x}^{m-1} + \mathbf{q}) = 0.$$

When $\mathbf{q} \geq \mathbf{0}$, by Theorem 4.17, we know that $\mathbf{x} = \mathbf{0}$, and the conclusion holds. If $\mathbf{q} \notin \mathbf{R}^n_+$, then $\mathbf{x} \neq \mathbf{0}$ (suppose not, $\mathbf{x} = \mathbf{0}$, $\mathscr{A}\mathbf{x}^{m-1} + \mathbf{q} = \mathbf{q}$, which contradicts to the fact that $\mathscr{A}\mathbf{x}^{m-1} + \mathbf{q} \geq \mathbf{0}$). Therefore, one has

$$\frac{\mathscr{A}\mathbf{x}^m}{\|\mathbf{x}\|_m^m} = \mathscr{A}\left(\frac{\mathbf{x}}{\|\mathbf{x}\|_m}\right)^m \geq \lambda(\mathscr{A}) > 0,$$

which implies that

$$0 < \|\mathbf{x}\|_m^m \lambda(\mathscr{A}) \leq \mathscr{A}\mathbf{x}^m = \mathbf{x}^\top(-\mathbf{q}) \leq \mathbf{x}^\top(-\mathbf{q})_+ \leq \|\mathbf{x}\|_m \|(-\mathbf{q})_+\|_m,$$

i.e.,

$$\|\mathbf{x}\|_m^{m-1} \leq \frac{\|(-\mathbf{q})_+\|_{\frac{m}{m-1}}}{\lambda(\mathscr{A})},$$

and the desired conclusion follows. □

Theorem 4.36 *Suppose* $\mathscr{A} = (a_{i_1 \ldots i_m}) \in S_{m,n}$ *is strictly semi-positive. If* $\mathbf{x}$ *is a solution of the TCP*($\mathbf{q}$, $\mathscr{A}$)*, then it holds that*

$$\|\mathbf{x}\|_2^{m-1} \leq \frac{\|(-\mathbf{q})_+\|_2}{\mu(\mathscr{A})},$$

where $\mu(\mathscr{A}) := \min_{\mathbf{x}\geq \mathbf{0}, \|\mathbf{x}\|_2=1} \mathscr{A}\mathbf{x}^m$.

Proof From Proposition 5.9, we know that $\mathscr{A}$ has a Pareto Z-eigenvalue such that $\mu(\mathscr{A}) = \min_{\substack{\mathbf{y}\geq \mathbf{0} \\ \|\mathbf{y}\|_2=1}} \mathscr{A}\mathbf{y}^m > 0$. Similar to the proof of Theorem 4.35, we only need to prove the case that $\mathbf{q} \notin \mathbf{R}^n_+$, then $\mathbf{x} \neq \mathbf{0}$. It holds that

$$\frac{\mathscr{A}\mathbf{x}^m}{\|\mathbf{x}\|_2^m} = \mathscr{A}\left(\frac{\mathbf{x}}{\|\mathbf{x}\|_2}\right)^m \geq \mu(\mathscr{A}) > 0,$$

which implies that

$$0 < \|\mathbf{x}\|_2^m \mu(\mathscr{A}) \leq \mathscr{A}\mathbf{x}^m = \mathbf{x}^\top(-\mathbf{q}) \leq \mathbf{x}^\top(-\mathbf{q})_+ \leq \|\mathbf{x}\|_2\|(-\mathbf{q})_+\|_2,$$

the desired conclusion follows. □

Recall the quantity (4.42), which is defined for a strictly semi-positive tensor $\mathscr{A}$ such that

$$\beta(\mathscr{A}) := \min_{\substack{\mathbf{x}\geq 0 \\ \|\mathbf{x}\|_\infty=1}} \max_{i\in[n]} x_i(\mathscr{A}\mathbf{x}^{m-1})_i,$$

where $\|\mathbf{x}\|_\infty := \max\{|x_i| \ : \ i \in [n]\}$. It follows from the definition of strictly semi-positive tensor that $\beta(\mathscr{A}) > 0$. By the quantity $\beta(\mathscr{A})$, we have the following results.

Theorem 4.37 *Suppose* $\mathscr{A} = (a_{i_1 \ldots i_m}) \in T_{m,n}$ *is strictly semi-positive. If* $\mathbf{x}$ *is a solution of the TCP*($\mathbf{q}$, $\mathscr{A}$)*, then*

$$\|\mathbf{x}\|_\infty^{m-1} \leq \frac{\|(-\mathbf{q})_+\|_\infty}{\beta(\mathscr{A})}.$$

Proof By the proof of Theorems 4.35 and 4.36, the conclusion is obvious when $\mathbf{q} \geq \mathbf{0}$. Assume that $\mathbf{q}$ is not nonnegative, similarly to the proof technique of Theorem 4.35, we have $\mathbf{0} \leq \mathbf{x} \neq \mathbf{0}$. From the notion of $\beta(\mathscr{A})$ and the tensor complementarity problem, it holds that

$$\begin{aligned} 0 < \|\mathbf{x}\|_\infty^m \beta(\mathscr{A}) &\leq \max_{i\in[n]} x_i(\mathscr{A}\mathbf{x}^{m-1})_i = \max_{i\in[n]} x_i(-q)_i \\ &\leq \max_{i\in[n]} x_i((-\mathbf{q})_+)_i \leq \|x\|_\infty\|(-\mathbf{q})_+\|_\infty, \end{aligned}$$

and the desired conclusion follows. □

Based on the mappings $T_{\mathscr{A}}$ and $F_{\mathscr{A}}$ in Sect. 4.3, one has the following result.

Theorem 4.38 *Let $\mathscr{A} \in T_{m,n}$ $(m \geq 2)$ be a strictly semi-positive tensor. If* $\mathbf{x}$ *is a solution of TCP*$(\mathbf{q}, \mathscr{A})$*, then the following inequalities hold.*

(1) $\frac{\|(-\mathbf{q})_+\|_\infty}{n^{\frac{m-2}{2}}\|T_{\mathscr{A}}\|_\infty} \leq \|\mathbf{x}\|_\infty^{m-1}$;
(2) if m is even, then $\frac{\|(-\mathbf{q})_+\|_\infty}{\|F_{\mathscr{A}}\|_\infty^{m-1}} \leq \|\mathbf{x}\|_\infty^{m-1}$;
(3) $\frac{\|(-\mathbf{q})_+\|_2}{\|T_{\mathscr{A}}\|_2} \leq \|\mathbf{x}\|_2^{m-1}$;
(4) if m is even, then $\frac{\|(-\mathbf{q})_+\|_m}{\|F_{\mathscr{A}}\|_m^{m-1}} \leq \|\mathbf{x}\|_m^{m-1}$;

where $T_{\mathscr{A}}$ and $F_{\mathscr{A}}$ are defined as in (4.30) and (4.31).

Proof For $\mathbf{q} \geq \mathbf{0}$, by Theorem 4.17, we know that $\mathbf{x} = \mathbf{0}$. Hence the conclusion holds automatically since $\|(-\mathbf{q})_+\| = 0$. It is enough to assume that $\mathbf{x} \neq \mathbf{0}$, or equivalently, that $\mathbf{q}$ is not nonnegative.
(1) If $\mathbf{x}$ is a solution of TCP($\mathbf{q}$, $\mathscr{A}$), we have

$$(\mathscr{A}\mathbf{x}^{m-1})_i \geq (-\mathbf{q})_i \quad \text{for all } i \in [n].$$

Particularly, it holds that

$$|(\mathscr{A}\mathbf{x}^{m-1})_i| \geq ((\mathscr{A}\mathbf{x}^{m-1})_+)_i \geq ((-\mathbf{q})_+)_i, \quad \forall\, i \in [n].$$

Thus,

$$\|\mathscr{A}\mathbf{x}^{m-1}\|_\infty \geq \|(-\mathbf{q})_+\|_\infty. \tag{4.61}$$

Combining (4.61) with Proposition 4.3, we obtain that

$$\begin{aligned}\|(-\mathbf{q})_+\|_\infty &\leq \|\mathbf{x}\|_2^{m-2}\|\|\mathbf{x}\|_2^{2-m}\mathscr{A}\mathbf{x}^{m-1}\|_\infty \\ &= \|\mathbf{x}\|_2^{m-2}\|T_{\mathscr{A}}(\mathbf{x})\|_\infty \\ &\leq \|\mathbf{x}\|_2^{m-2}\|\mathbf{x}\|_\infty\|T_{\mathscr{A}}\|_\infty \\ &\leq n^{\frac{m-2}{2}}\|\mathbf{x}\|_\infty^{m-1}\|T_{\mathscr{A}}\|_\infty,\end{aligned}$$

where the last inequality is obtained by $\|\mathbf{x}\|_2 \leq \sqrt{n}\|\mathbf{x}\|_\infty$, and the desired result holds.

(2) Similarly, using Proposition 4.3 and (4.61), we also have

$$\begin{aligned}\|(-\mathbf{q})_+\|_\infty &\leq \|(\mathscr{A}\mathbf{x}^{m-1})^{[\frac{1}{m-1}]}\|_\infty^{m-1} \\ &= \|F_{\mathscr{A}}(\mathbf{x})\|_\infty^{m-1} \\ &\leq \|\mathbf{x}\|_\infty^{m-1}\|F_{\mathscr{A}}\|_\infty^{m-1}.\end{aligned}$$

(3) By (4.61) and Proposition 4.3 again, it follows that

$$\begin{aligned}\|(-\mathbf{q})_+\|_2 &\le \|\mathscr{A}\mathbf{x}^{m-1}\|_2 = \|\mathbf{x}\|_2^{m-2}\|\|\mathbf{x}\|_2^{2-m}\mathscr{A}\mathbf{x}^{m-1}\|_2 \\ &= \|\mathbf{x}\|_2^{m-2}\|T_{\mathscr{A}}(\mathbf{x})\|_2 \\ &\le \|\mathbf{x}\|_2^{m-2}\|\mathbf{x}\|_2\|T_{\mathscr{A}}\|_2 \\ &\le \|\mathbf{x}\|_2^{m-1}\|T_{\mathscr{A}}\|_2.\end{aligned}$$

(4) It follows from (4.61) and Proposition 4.3 that

$$\begin{aligned}\|(-\mathbf{q})_+\|_m &\le \|\mathscr{A}\mathbf{x}^{m-1}\|_m \le \|\mathscr{A}\mathbf{x}^{m-1}\|_{\frac{m}{m-1}} = \|(\mathscr{A}\mathbf{x}^{m-1})^{[\frac{1}{m-1}]}\|_m^{m-1} \\ &= \|F_{\mathscr{A}}(\mathbf{x})\|_m^{m-1} \\ &\le \|\mathbf{x}\|_m^{m-1}\|F_{\mathscr{A}}\|_m^{m-1},\end{aligned}$$

and the desired inequality follows. □

According to Lemma 4.7, Theorems 4.35–4.38, the following theorem can be easily proved.

Theorem 4.39 *Let* $\mathscr{A} = (a_{i_1 i_2 \ldots i_m}) \in T_{m,n}$ $(m \ge 2)$ *be strictly semi-positive. Suppose* $\mathbf{x}$ *is a solution of TCP*$(\mathbf{q}, \mathscr{A})$. *Then the following results hold.*
(1)

$$\frac{\|(-\mathbf{q})_+\|_\infty}{n^{\frac{m-2}{2}}\max\limits_{i\in[n]}\left(\sum\limits_{i_2,\ldots,i_m=1}^{n}|a_{ii_2\ldots i_m}|\right)} \le \|\mathbf{x}\|_\infty^{m-1} \le \frac{\|(-\mathbf{q})_+\|_\infty}{\beta(\mathscr{A})}.$$

(2) When m is even, it follows that

$$\frac{\|(-\mathbf{q})_+\|_\infty}{\max\limits_{i\in[n]}\left(\sum\limits_{i_2,\ldots,i_m=1}^{n}|a_{ii_2\ldots i_m}|\right)} \le \|\mathbf{x}\|_\infty^{m-1} \le \frac{\|(-\mathbf{q})_+\|_\infty}{\beta(\mathscr{A})}.$$

4.9 Global Uniqueness and Solvability

A famous property for linear complementarity problem is the **global uniqueness and solvability** (GUS-property). The equivalent condition for the property is that the matrix involved is a P-matrix. Unfortunately, the GUS-property generally does not hold for nonlinear complementarity problem. Thus, one may ask whether such a result can be generalized to the tensor complementarity problem or not? One may wonder whether P-tensors imply the GUS-property?

In this section, we will present answers to the above questions by constructing two counterexamples. Furthermore, we will prove that when the GUS-property holds for the tensor complementarity problem.

First of all, we introduce an example, that illustrates a TCP($\mathbf{q}$, $\mathscr{A}$) problem with a P-tensor may not have the GUS-property.

Example 4.9 Suppose $\mathscr{A} = (a_{i_1i_2i_3i_4}) \in T_{4,2}$ is a tensor with entries such that $a_{1111} = 1, a_{1112} = -2, a_{1122} = 1, a_{2222} = 1$ and $a_{i_1i_2i_3i_4} = 0$ otherwise. Then,

$$\mathscr{A}\mathbf{x}^3 = \begin{pmatrix} x_1^3 - 2x_1^2x_2 + x_1x_2^2 \\ x_2^3 \end{pmatrix},$$

and $x_1(\mathscr{A}\mathbf{x}^3)_1 = x_1^4 - 2x_1^3x_2 + x_1^2x_2^2, x_2(\mathscr{A}\mathbf{x}^3)_2 = x_2^4$. For any $\mathbf{x} \in \mathbf{R}^2\backslash\{\mathbf{0}\}$, it is not difficult to see that

(1) $x_2(\mathscr{A}\mathbf{x}^3)_2 > 0$, when $x_2 \neq 0$;
(2) $x_1(\mathscr{A}\mathbf{x}^3)_1 > 0$, when $x_1 \neq 0, x_2 = 0$.

From Definition 4.7, we know that $\mathscr{A}$ is a P-tensor. Let $\mathbf{q} = (0, -1)^\top$. It is obvious that $\mathbf{x} = (0, 1)^\top$ and $\mathbf{x} = (1, 1)^\top$ are two solutions of TCP($\mathbf{q}$, $\mathscr{A}$).

By Theorem 4.18, a given tensor is a P-tensor if and only if it is strictly semi-positive. Hence the corresponding TCP problem has a solution for every $\mathbf{q} \in \mathbf{R}^n$ from Lemma 4.9. Next, we show the compactness of the solution set for TCP($\mathbf{q}$, $\mathscr{A}$) with $\mathscr{A}$ being P-tensor.

Theorem 4.40 *Suppose $\mathbf{q} \in \mathbf{R}^n$ and $\mathscr{A} \in T_{m,n}$ is a P-tensor. Then the solution set of TCP($\mathbf{q}$, $\mathscr{A}$) is nonempty and compact.*

Proof By the conditions and the above analysis, we know that TCP($\mathbf{q}$, $\mathscr{A}$) has a solution for every $\mathbf{q} \in \mathbf{R}^n$. So it is enough to show that the solution set of TCP($\mathbf{q}$, $\mathscr{A}$) is compact. We divide the proof into two parts.
(1) To show the boundedness of the solution set, we first prove that if there is a sequence $\{\mathbf{x}^{(k)}\} \subset \mathbf{R}^n_+$ such that

$$\|\mathbf{x}^{(k)}\| \to \infty \text{ and } \frac{[-\mathscr{A}(\mathbf{x}^{(k)})^{m-1} - \mathbf{q}]_+}{\|\mathbf{x}^{(k)}\|} \to \infty \text{ as } k \to \infty, \tag{4.62}$$

then there exists an $i \in [n]$ satisfying $x_i^{(k)}[\mathscr{A}(\mathbf{x}^{(k)})^{m-1} + \mathbf{q}]_i > 0$ holds for some $k \geq 0$. We suppose conversely that there is a sequence $\{\mathbf{x}^{(k)}\} \subset \mathbf{R}^n_+$ satisfying (4.62). However

$$x_i^{(k)}[\mathscr{A}(\mathbf{x}^{(k)})^{m-1} + \mathbf{q}]_i \leq 0, \quad \forall\, i \in [n], \ \forall\, k \geq 0. \tag{4.63}$$

Without loss of generality, suppose $\lim_{k\to\infty} \frac{\mathbf{x}^{(k)}}{\|\mathbf{x}^{(k)}\|} = \bar{\mathbf{x}} \in \mathbf{R}^n$ since $\left\{\frac{\mathbf{x}^{(k)}}{\|\mathbf{x}^{(k)}\|}\right\}$ is bounded. Combining this with the fact that $\{\mathbf{x}^{(k)}\} \subset \mathbf{R}^n_+$ and $\|\mathbf{x}^{(k)}\| \to \infty, k \to \infty$, we obtain

$$\bar{\mathbf{x}} \geq \mathbf{0}, \quad \bar{\mathbf{x}} \neq \mathbf{0}. \tag{4.64}$$

On the one hand, if $(\mathscr{A}(\mathbf{x}^{(k)})^{m-1}+\mathbf{q})_i \leq 0$, then

$$\left(-(\mathscr{A}(\mathbf{x}^{(k)})^{m-1}+\mathbf{q})_i\right)_+ = -(\mathscr{A}(\mathbf{x}^{(k)})^{m-1}+\mathbf{q})_i.$$

Since $\lim_{k\to\infty}\frac{q_i}{\|\mathbf{x}^{(k)}\|}=0$ for all $i \in [n]$, one has

$$\begin{aligned} 0 &= \lim_{k\to\infty}\frac{\left(-(\mathscr{A}(\mathbf{x}^{(k)})^{m-1}+\mathbf{q})_i\right)_+}{\|\mathbf{x}^{(k)}\|} = \lim_{k\to\infty}\frac{-\left(\mathscr{A}(\mathbf{x}^{(k)})^{m-1}\right)_i-\mathbf{q}_i}{\|\mathbf{x}^{(k)}\|^{m-1}} \\ &= \lim_{k\to\infty}\frac{-\left(\mathscr{A}(\mathbf{x}^{(k)})^{m-1}\right)_i}{\|\mathbf{x}^{(k)}\|^{m-1}} = -\left(\mathscr{A}\bar{\mathbf{x}}^{m-1}\right)_i. \end{aligned}$$

On the other hand, if $\left(\mathscr{A}(\mathbf{x}^{(k)})^{m-1}+\mathbf{q}\right)_i \geq 0$, then

$$0 \leq \lim_{k\to\infty}\frac{\left(\mathscr{A}(\mathbf{x}^{(k)})^{m-1}+\mathbf{q}\right)_i}{\|\mathbf{x}^{(k)}\|^{m-1}} = \lim_{k\to\infty}\frac{\left(\mathscr{A}(\mathbf{x}^{(k)})^{m-1}\right)_i}{\|\mathbf{x}^{(k)}\|^{m-1}} = \left(\mathscr{A}\bar{\mathbf{x}}^{m-1}\right)_i.$$

Combining these two situations, it follows that

$$\left(\mathscr{A}\bar{\mathbf{x}}^{m-1}\right)_i \geq 0, \quad \forall\, i \in [n]. \tag{4.65}$$

By (4.63), we obtain

$$\bar{x}_i\left(\mathscr{A}\bar{\mathbf{x}}^{m-1}\right)_i = \lim_{k\to\infty}\frac{x_i^k}{\|\mathbf{x}^{(k)}\|}\frac{\left(\mathscr{A}(\mathbf{x}^{(k)})^{m-1}\right)_i}{\|\mathbf{x}^{(k)}\|^{m-1}} = \lim_{k\to\infty}\frac{x_i^k}{\|\mathbf{x}^{(k)}\|}\frac{\left(\mathscr{A}(\mathbf{x}^{(k)})^{m-1}+\mathbf{q}\right)_i}{\|\mathbf{x}^{(k)}\|^{m-1}} \leq 0.$$

Combining this with (4.64) and (4.65), it follows that

$$\bar{x}_i\left(\mathscr{A}\bar{\mathbf{x}}^{m-1}\right)_i = 0, \quad \forall\, i \in [n]. \tag{4.66}$$

Furthermore, by (4.64)–(4.66), we know that $\bar{\mathbf{x}}$ is a nonzero solution of TCP($\mathbf{0}$, $\mathscr{A}$). By the fact that $\mathscr{A}$ is a P-tensor, hence it is a strictly semi-positive tensor. From Theorem 4.17, we obtain that TCP($\mathbf{0}$, $\mathscr{A}$) has a unique solution $\mathbf{0}$, which contradicts $\bar{\mathbf{x}} \neq \mathbf{0}$. Thus, when (4.62) holds, there exists an $i \in [n]$ satisfying $x_i^{(k)}(\mathscr{A}(\mathbf{x}^{(k)})^{m-1}+\mathbf{q})_i > 0$ for some $k \geq 0$.

Now, we prove the main conclusion of the first part, i.e. the solution set of TCP(q, $\mathscr{A}$) is bounded. Conversely, we suppose that the solution set of TCP($\mathbf{q}$, $\mathscr{A}$) is unbounded, which means that there exists an unbounded solution sequence $\{\mathbf{x}^{(k)}\}$ of TCP($\mathbf{q}$, $\mathscr{A}$) such that $\|\mathbf{x}^{(k)}\| \to \infty$ as $k \to \infty$, and for all $i \in [n], k \geq 0$, it follows that

$$\mathbf{x}^{(k)} \geq \mathbf{0}, \quad \mathscr{A}(\mathbf{x}^{(k)})^{m-1}+\mathbf{q} \geq \mathbf{0}, \quad (\mathbf{x}^{(k)})^\top(\mathscr{A}(\mathbf{x}^{(k)})^{m-1}+\mathbf{q}) = 0. \tag{4.67}$$

It is obvious that $\left(\mathscr{A}(\mathbf{x}^{(k)})^{m-1}+\mathbf{q}\right)_i \geq 0$, which implies that

$$\frac{\left(-(\mathscr{A}(\mathbf{x}^{(k)})^{m-1}+\mathbf{q})_i\right)_+}{\|\mathbf{x}^{(k)}\|} \to 0 \quad \text{as } k \to \infty.$$

Hence the sequence $\{\mathbf{x}^{(k)}\}$ satisfies (4.62), and there must be an index i_0 and a positive integer $k^* > 0$ such that $x_{i_0}^{(k^*)}\left(\mathscr{A}(\mathbf{x}^{(k^*)})^{m-1}+\mathbf{q}\right)_{i_0} > 0$, which is contrary to that $x_i^k\left(\mathscr{A}(\mathbf{x}^{(k)})^{m-1}+\mathbf{q}\right)_i = 0$ for all $i \in [n]$ and $k \geq 0$. Thus, the solution set of TCP($\mathbf{q}$, $\mathscr{A}$) is bounded.

(2) Now we prove that the solution set of TCP($\mathbf{q}$, $\mathscr{A}$) is a closed set. Suppose that $\{\mathbf{x}^{(k)}\}$ is a solution sequence of TCP($\mathbf{q}$, $\mathscr{A}$) and

$$\lim_{k\to\infty} \mathbf{x}^{(k)} = \bar{\mathbf{x}}. \tag{4.68}$$

It is sufficient to show that $\bar{\mathbf{x}}$ is a solution of TCP($\mathbf{q}$, $\mathscr{A}$).

By (4.68) and the continuity of $\mathscr{A}\mathbf{x}^{m-1}+\mathbf{q}$, we have

$$\lim_{k\to\infty}\left((\mathscr{A}(\mathbf{x}^{(k)})^{m-1}+\mathbf{q}\right) = \mathscr{A}\left(\lim_{k\to\infty}\mathbf{x}^{(k)}\right)^{m-1}+\mathbf{q} = \mathscr{A}\bar{\mathbf{x}}^{m-1}+\mathbf{q}. \tag{4.69}$$

Since $\{\mathbf{x}^{(k)}\}$ is a solution sequence of TCP($\mathbf{q}$, $\mathscr{A}$), we have that (4.67) holds. By (4.67)–(4.69), we obtain that

$$\bar{\mathbf{x}} \geq \mathbf{0}, \quad \mathscr{A}\bar{\mathbf{x}}^{m-1}+\mathbf{q} \geq \mathbf{0}, \quad \bar{\mathbf{x}}^\top(\mathscr{A}\bar{\mathbf{x}}^{m-1}+\mathbf{q}) = 0.$$

So $\bar{\mathbf{x}}$ is a solution of TCP($\mathbf{q}$, $\mathscr{A}$). Therefore, the solution set is closed.

Combining (1) with (2), we know that the solution set is compact. □

As stated at the beginning of this section, the GUS-property of TCP($\mathbf{q}$, $\mathscr{A}$) may not hold for P-tensor $\mathscr{A}$. Then it is natural to ask that for which kind of tensors, TCP($\mathbf{q}$, $\mathscr{A}$) has the GUS-property. In the following sequel, we will study the TCP with respect to strong P-tensors.

Now we introduce the definition of strong P-tensors, which is a strengthened version of the notion of strong P_0-tensors in Definition 4.11.

Definition 4.14 Let $\mathscr{A} = (a_{i_1\ldots i_m}) \in T_{m,n}$. We say that $\mathscr{A}$ is a **strong P-tensor** if $\mathscr{A}\mathbf{x}^{m-1}$ is a P-function.

It is not difficult to check that all strong P-tensors are P-tensors. Many properties of P-tensors have been studied in the previous sections. Inspired by the results before, we may obtain the following properties of strong P-tensor:

Proposition 4.12 *If $\mathscr{A} \in T_{m,n}$ is a strong P-tensor, then*

(1) $\mathscr{A}$ is strictly semi-positive;
(2) $\mathscr{A}$ is an R-tensor;
(3) all of its H-eigenvalues and Z-eigenvalues are positive;

(4) all the diagonal entries of $\mathscr{A}$ are positive;
(5) every principal sub-tensor of $\mathscr{A}$ is a strong P-tensor.

Proof We only need to prove the result (5). Let $\mathscr{A}^J \in T_{m,r}$ be an principal sub-tensor of the strong P-tensor $\mathscr{A} \in T_{m,n}$. We choose any $\mathbf{x} = (x_{j_1}, x_{j_2}, \ldots, x_{j_r}) \in \mathbf{R}^r \setminus \{\mathbf{0}\}$ and $\mathbf{y} = (y_{j_1}, y_{j_2}, \ldots, y_{j_r}) \in \mathbf{R}^r \setminus \{\mathbf{0}\}$ with $\mathbf{x} \neq y$. Then let $\bar{\mathbf{x}} = (\bar{x}_1, \bar{x}_2, \ldots, \bar{x}_n) \in \mathbf{R}^n$ where $\bar{x}_i = x_{j_i}$ for $i \in J$ and $\bar{x}_i = 0$ for $i \notin J$. In a similar way, let $\bar{\mathbf{y}} = (\bar{y}_1, \bar{y}_2, \ldots, \bar{y}_n) \in \mathbf{R}^n$ where $\bar{y}_i = y_{j_i}$ for $i \in J$ and $\bar{y}_i = 0$ for $i \notin J$. Since $\mathscr{A}$ is a strong P-tensor, then there exists an index $k \in [n]$ such that

$$\begin{aligned} 0 &< \max_{k \in [n]} (\bar{x}_k - \bar{y}_k)((\mathscr{A}\bar{\mathbf{x}}^{m-1})_k - (\mathscr{A}\bar{\mathbf{y}}^{m-1})_k) \\ &= \max_{k \in J} (x_k - y_k)((\mathscr{A}^J\mathbf{x}^{m-1})_k - (\mathscr{A}^J\mathbf{y}^{m-1})_k). \end{aligned}$$

Thus, $\mathscr{A}^J$ is a strong P-tensor. □

We now present the main result of this section.

Theorem 4.41 *Let $\mathscr{A} \in T_{m,n}$ be a strong P-tensor, then TCP($\mathbf{q}$, $\mathscr{A}$) has the GUS-property.*

Proof Since $\mathscr{A}$ is a strong P-tensor, by Definition 4.7, it follows that $\mathscr{A}$ is a P-tensor. Thus $\mathscr{A}$ is a strictly semi-positive tensor from Theorem 4.18. Furthermore, by Lemma 4.9, we know that TCP($\mathbf{q}$, $\mathscr{A}$) has a solution for every $\mathbf{q} \in \mathbf{R}^n$. Moreover since $\mathscr{A}$ is a strong P-tensor, it is easy to see that $\mathscr{A}\mathbf{x}^{m-1} + \mathbf{q}$ is a P-function. Hence, from Lemma 4.6, it follows that TCP($\mathbf{q}$, $\mathscr{A}$) has no more than one solution. Therefore, TCP($\mathbf{q}$, $\mathscr{A}$) has a unique solution for every $\mathbf{q} \in \mathbf{R}^n$. □

4.10 Exceptional Regular Tensors and Tensor Complementarity Problems

For nonlinear complementarity problems, the concept of *exceptional family of elements* is a powerful tool, and many good theoretical results for the NCP have been obtained by applying this tool [136, 144–146]. In this section, inspired by the notion of exceptional family of elements, a new class of tensors named *exceptionally regular tensor* is introduced. It is shown that this is a wide tensor class, which includes many important structured tensors as its subcases. Furthermore, the main result for this section is to study the tensor complementarity problem with an exceptionally regular tensor. The nonempty and compactness for its solution set will be considered.

We first introduce the definitions of exceptional regular(ER) function and exceptional regular tensors(ER-tensor).

Definition 4.15 (*Zhao, Isac 2000*) The function $g(\mathbf{x}) = f(\mathbf{x}) - f(\mathbf{0})$ is **exceptionally regular** if there is no $(\mathbf{x}, \alpha) \in \mathbf{R}^n_+ \times \mathbf{R}_+$ with $\|\mathbf{x}\|_2 = 1$ such that

$$\begin{cases} g_i(\mathbf{x})/x_i = -\alpha, & \text{if } x_i > 0, \\ g_i(\mathbf{x}) \geq 0, & \text{if } x_i = 0. \end{cases} \tag{4.70}$$

Now, we define the exceptionally regular tensor (ER-tensor for short) below.

Definition 4.16 $\mathscr{A} \in T_{m,n}$ is called an **ER-tensor**, if there exists no $(\mathbf{x}, t) \in (\mathbf{R}^n_+ \setminus \{\mathbf{0}\}) \times \mathbf{R}_+$ such that

$$\begin{cases} (\mathscr{A}\mathbf{x}^{m-1})_i + tx_i = 0, & \text{if } x_i > 0, \\ (\mathscr{A}\mathbf{x}^{m-1})_i \geq 0, & \text{if } x_i = 0. \end{cases} \tag{4.71}$$

Recall Definition 4.9, one can obtain that all strictly semi-positive tensors form a subset of the class of ER-tensors.

Next, we show the relationship between P′-tensor and ER-tensor, and several examples are raised to verify the corresponding conclusions.

Theorem 4.42 *Assume $\mathscr{A}$ is a P′-tensor defined by Definition 4.8, then $\mathscr{A}$ is an ER-tensor.*

Proof Since $\mathscr{A}$ is a P′-tensor, by Definition 4.8, it holds that for any $\mathbf{x} \in \mathbf{R}^n \setminus \{\mathbf{0}\}$, there exists an index $i \in [n]$ such that $x_i^{m-1}(\mathscr{A}\mathbf{x}^{m-1})_i > 0$. Without loss of generality, let $i_0 \in [n]$ such that

$$x_{i_0}^{m-1}(\mathscr{A}\mathbf{x}^{m-1})_{i_0} > 0.$$

Here if we choose $\mathbf{x} \in \mathbf{R}^n_+ \setminus \{\mathbf{0}\}$, then it implies that $x_{i_0} \neq 0$, and then $x_{i_0} > 0$. Furthermore, we get $(\mathscr{A}\mathbf{x}^{m-1})_{i_0} > 0$. That is to say, for any $\mathbf{x} \in \mathbf{R}^n_+ \setminus \{\mathbf{0}\}$, there exists an index i_0 such that $x_{i_0} > 0$ and $(\mathscr{A}x^{m-1})_{i_0} > 0$, which implies that the system (4.71) has no solution. So, $\mathscr{A}$ is an ER-tensor. □

Example 4.10 Suppose $\mathscr{A} = (a_{ijk}) \in T_{3,2}$ with entries such that $a_{111} = 1, a_{122} = -1, a_{211} = 2, a_{222} = -1$ and $a_{i_1 i_2 i_3} = 0$ otherwise. Then $\mathscr{A}$ is an ER-tensor but not a P′-tensor, which shows that the P′-tensor is a proper subset of the ER-tensor.

Here, we only prove that $\mathscr{A}$ is not a P′-tensor. The proof for $\mathscr{A}$ being an ER-tensor will be presented in Example 4.12. *In fact, let $\bar{\mathbf{x}} = (0, 1)^\top$ and we have that*

$$\begin{cases} \bar{x}_1 = 0, & \bar{x}_1^2(\mathscr{A}\bar{\mathbf{x}}^2)_1 = 0; \\ \bar{x}_2 = 1 > 0, & \bar{x}_2^2(\mathscr{A}\bar{\mathbf{x}}^2)_2 = -1 < 0. \end{cases}$$

Thus, for given $\bar{\mathbf{x}} = (0, 1)^\top$, there exists no index $i \in \{1, 2\}$ such that $\bar{x}_i^2(\mathscr{A}\bar{\mathbf{x}}^2)_i > 0$. Thereby, the tensor $\mathscr{A}$ is not a P′-tensor. □

Remark 4.1 In [81], it has proved that many classes of structured tensors are subclasses of P′-tensors, including positive definite tensors, strongly completely positive tensors, nonsingular H-tensors with all positive diagonal entries, strictly diagonally dominant tensors with positive diagonal entries, Cauchy tensors with mutually distinct entries of generating vector, addition tensors of P′-tensors and completely positive tensors, odd-order B-tensors or symmetric even-order B-tensors, and so on. Therefore, all kinds of tensors mentioned above are ER-tensors by Theorem 4.42.

Invoking Definitions 4.10 and 4.16, it seems that the definitions of ER-tensors and R-tensors have some similarities. However, two kinds of structured tensors are different. Two examples below illustrate their difference.

Example 4.11 Assume $\mathscr{A} = (a_{ijk}) \in T_{3,2}$, where $a_{111} = -16, a_{122} = 1, a_{211} = -17, a_{222} = 1$ and all other elements $a_{i_1i_2i_3} = 0$, then $\mathscr{A}$ is an R-tensor, but not an ER-tensor.

First, we show that $\mathscr{A}$ is an R-tensor. For any $\mathbf{x} \in \mathbf{R}^2_+$, it is obvious that

$$\mathscr{A}\mathbf{x}^2 = \begin{pmatrix} -16x_1^2 + x_2^2 \\ -17x_1^2 + x_2^2 \end{pmatrix}.$$

(1) If $x_1 > 0$, then $(\mathscr{A}\mathbf{x}^2)_1 + t = -16x_1^2 + x_2^2 + t = 0$, i.e., $x_2^2 = 16x_1^2 - t$, but

$$(\mathscr{A}\mathbf{x}^2)_2 + t = -17x_1^2 + x_2^2 + t = -x_1^2 < 0.$$

(2) If $x_2 > 0$, then *$(\mathscr{A}\mathbf{x}^2)_2 + t = -17x_1^2 + x_2^2 + t = 0$, i.e., $17x_1^2 = x_2^2 + t > 0$, but*

$$(\mathscr{A}\mathbf{x}^2)_1 + t = -16x_1^2 + x_2^2 + t = -16x_1^2 + 17x_1^2 - t + t = x_1^2 > 0.$$

Therefore, by (1)–(2) we know that $\mathscr{A}$ is an R-tensor.

On the other hand, we show that $\mathscr{A}$ is not an ER-tensor. Consider the following system

$$x_1 > 0, \quad (\mathscr{A}\mathbf{x}^2)_1 + tx_1 = -16x_1^2 + x_2^2 + tx_1 = 0; \tag{4.72}$$

$$x_2 > 0, \quad (\mathscr{A}\mathbf{x}^2)_2 + tx_2 = -17x_1^2 + x_2^2 + tx_2 = 0. \tag{4.73}$$

By (4.72), we have $x_2^2 = x_1(16x_1 - t)$ and $16x_1 \geq t$. Combining this with (4.73), we obtain

$$(\mathscr{A}\mathbf{x}^2)_2 + tx_2 = -17x_1^2 + x_2^2 + tx_2 = -x_1^2 - tx_1 + t\sqrt{16x_1^2 - tx_1} = 0.$$

Since $x_1 > 0$, we further have

$$x_1^3 + 2tx_1^2 - 15t^2x_1 + t^3 = 0. \tag{4.74}$$

Let $f(z)$ be defined as the zeros of

$$f(z) := z^3 + 2z^2 - 15z + 1. \tag{4.75}$$

It is obvious that $f(z) \to +\infty$ as *$z \to +\infty$. Let the derivation of $f(z)$ equal to zero, i.e.,*

$$3z^2 + 4z - 15 = 0,$$

one has $z = \frac{5}{3} > \frac{1}{16}$ *and* $f(\frac{5}{3}) = -\frac{373}{27} < 0$. *Hence, the Eq.* (4.75) *has a root* $z^* \in (\frac{5}{3}, +\infty) \subset (\frac{1}{16}, +\infty)$. *Thus,* $(x_1, t) := (z^*, 1)$ *solves the Eq.* (4.74). *Furthermore, if we take* $x_2 = \sqrt{16(z^*)^2 - z^*}$, *then* $(\bar{\mathbf{x}}, \bar{t}) \in (\mathbf{R}^2_+ \setminus \{\mathbf{0}\}) \times \mathbf{R}_+$, *with* $\bar{\mathbf{x}} = (z^*, \sqrt{16(z^*)^2 - z^*})^\top$ *and* $\bar{t} = 1$, *solves the system* (4.70), *which implies that* $\mathscr{A}$ *is not an ER-tensor.* □

Example 4.12 Suppose $\mathscr{A}$ is defined as in Example 4.10, then $\mathscr{A}$ is an ER-tensor, but not an R-tensor.

We first show that $\mathscr{A}$ *is an ER-tensor. By Definition* 4.16, *for any* $\mathbf{x} \in \mathbf{R}^2_+$,

$$\mathscr{A}\mathbf{x}^2 = \begin{pmatrix} x_1^2 - x_2^2 \\ 2x_1^2 - x_2^2 \end{pmatrix}.$$

Consider the following two cases.
(1) When $x_1 > 0$, $(\mathscr{A}\mathbf{x}^2)_1 + tx_1 = x_1^2 - x_2^2 + tx_1 = 0$, *it holds that*

$$x_2^2 = x_1^2 + tx_1 > 0 \quad \text{and} \quad x_2 - x_1 = \frac{tx_1}{x_2 + x_1}.$$

Thus, $x_2 > 0$. *Now we state that the equation*

$$(\mathscr{A}\mathbf{x}^2)_2 + tx_2 = 2x_1^2 - x_2^2 + tx_2 = x_1^2 + t(x_2 - x_1) = x_1^2 + t\frac{tx_1}{x_2 + x_1} = 0 \tag{4.76}$$

has no solution. Otherwise, if (4.76) *has a solution, then* $x_1^3 + x_1^2 x_2 + t^2 x_1 = 0$, *which is impossible because* $x_1 > 0$ and $x_2 > 0$.
(2) When $x_2 > 0$, $(\mathscr{A}\mathbf{x}^2)_2 + tx_2 = 2x_1^2 - x_2^2 + tx_2 = 0$. *Then*

$$x_1^2 = \frac{x_2^2 - tx_2}{2} \geq 0. \tag{4.77}$$

If $x_1 = 0$, *then*

$$(\mathscr{A}\mathbf{x}^2)_1 + tx_1 = x_1^2 - x_2^2 + tx_1 = -x_2^2 \leq 0,$$

which contradicts the condition that $x_2 > 0$. *If* $x_1 > 0$, *by* (4.77), *we know that* $x_2 > t$ *and*

$$(\mathscr{A}\mathbf{x}^2)_1 + tx_1 = x_1^2 - x_2^2 + tx_1 = \frac{x_2^2 - tx_2}{2} - x_2^2 + t\sqrt{\frac{x_2^2 - tx_2}{2}}.$$

Let $(\mathscr{A}\mathbf{x}^2)_1 + tx_1 = 0$, *then we derive a contradiction. Since* $(\mathscr{A}\mathbf{x}^2)_1 + tx_1 = 0$, *it follows that*

$$\frac{x_2^2}{2} + \frac{tx_2}{2} = t\sqrt{\frac{x_2^2 - tx_2}{2}},$$

and hence,

$$x_2^2(x_2+t)^2 = t^2(2x_2^2 - 2tx_2). \tag{4.78}$$

By the fact that $x_2 > 0$, the Eq. (4.78) can be simplified as

$$x_2^3 + 2tx_2^2 - t^2x_2 + 2t^3 = 0,$$

which does not hold since $x_2^3 > t^2x_2$ from $x_2 > t$ and $t \geq 0$.

Therefore, by the analysis of (1)–(2), we know that the system (4.71) has no solution in $(\mathbf{R}_+^n \setminus \{\mathbf{0}\}) \times \mathbf{R}_+$, which demonstrates that $\mathscr{A}$ is an ER-tensor.

On the other hand, we show that $\mathscr{A}$ is not an R-tensor. In fact, take $\bar{x}_1 = 0, \bar{x}_2 = a > 0$ and $\bar{t} = a^2$. Then, it is easy to check that $(\bar{\mathbf{x}}, \bar{t}) \in (\mathbf{R}_+^2 \setminus \{\mathbf{0}\}) \times \mathbf{R}_+$ is a solution of (4.44) in Definition 4.10. Therefore, $\mathscr{A}$ is not an R-tensor. □

Several more properties of ER-tensors are listed below.

Proposition 4.13 *Let $\mathscr{A} \in T_{m,n}$ be an ER-tensor. Then $\mathscr{A}$ is an R_0-tensor and every principal sub-tensor of $\mathscr{A}$ is also an ER-tensor. Furthermore, $\mathscr{A}$ has no nonpositive Z-eigenvalue associated with a nonnegative Z-eigenvector.*

Proof (1) Suppose $\mathscr{A}$ is an ER-tensor. Then any point in $(\mathbf{R}_+^n \setminus \{\mathbf{0}\}) \times \mathbf{R}_+$ is not a solution of the system (4.71). Hence, the system (4.71) has no nonzero solution when $t = 0$, that is, the system (4.44) has no solution $\mathbf{x} \in \mathbf{R}_+^n \setminus \{\mathbf{0}\}$ when $t = 0$. Therefore, $\mathscr{A}$ is an R_0-tensor.
(2) Suppose $\mathscr{A}^J$ is one of the principal subtensors of $\mathscr{A}$, where $J \subset [n]$ and $|J| = r$ $(1 \leq r \leq n)$. If $\mathscr{A}^J$ is not an ER-tensor, then there exists a point $(\mathbf{x}_J, t) \in (\mathbf{R}_+^r \setminus \{\mathbf{0}\}) \times \mathbf{R}_+$ satisfying the system (4.71). Define $\bar{\mathbf{x}} \in \mathbf{R}_+^n \setminus \{\mathbf{0}\}$ by

$$\bar{x}_i = \begin{cases} (\mathbf{x}_J)_i, & \text{if } i \in J, \\ 0, & \text{if } i \notin J. \end{cases}$$

Then, the point $(\bar{\mathbf{x}}, t) \in (\mathbf{R}_+^n \setminus \{\mathbf{0}\}) \times \mathbf{R}_+$ solves the system (4.71). Therefore, $\mathscr{A}$ is not an ER-tensor, which causes a contradiction.
(3) Suppose $\mathbf{x} \geq \mathbf{0}$ is a Z-eigenvector of $\mathscr{A}$ and λ is the corresponding Z-eigenvalue, i.e., $\mathscr{A}\mathbf{x}^{m-1} = \lambda\mathbf{x}$. By contradiction, if $\lambda \leq 0$ and let $t = -\lambda \geq 0$, then we have

$$\begin{cases} (\mathscr{A}\mathbf{x}^{m-1})_i + tx_i = 0, & \text{if } x_i > 0, \\ (\mathscr{A}\mathbf{x}^{m-1})_i \geq 0, & \text{if } x_i = 0. \end{cases}$$

That is, the point $(\mathbf{x}, t) \in (\mathbf{R}_+^n \setminus \{\mathbf{0}\}) \times \mathbf{R}_+$ solves the system (4.71), which contradicts that $\mathscr{A}$ is an ER-tensor. Hence, we have $\lambda > 0$. □

Recall the definition of semi-positive tensors in Definition 4.9, we have the following result.

Theorem 4.43 *Suppose $\mathscr{A} \in T_{m,n}$ is semi-positive, then the following results are equivalent.*

(1) $\mathscr{A}$ *is an* R_0*-tensor;*
(2) $\mathscr{A}$ *is an ER-tensor;*
(3) $\mathscr{A}$ *is an R-tensor.*

Proof From Definition 4.10 and Theorem 4.21, we know that the condition (1) is equivalent to (3).

On the other hand, from Proposition 4.13, we have that every ER-tensor is an R_0-tensor. Suppose $\mathscr{A}$ is an R_0-tensor. We process by contradiction. If $\mathscr{A}$ is not an ER-tensor, then there exists a point $(\bar{\mathbf{x}}, \bar{t}) \in (\mathbf{R}^n_+ \setminus \{\mathbf{0}\}) \times \mathbf{R}_+$ satisfying the system (4.71). By the notion of R_0-tensor, we have $\bar{t} > 0$. Thus, it holds that

$$\begin{cases} (\mathscr{A}\bar{\mathbf{x}}^{m-1})_i + \bar{t}\bar{x}_i = 0, & \text{if } \bar{x}_i > 0, \\ (\mathscr{A}\bar{\mathbf{x}}^{m-1})_i \qquad\quad \geq 0, & \text{if } \bar{x}_i = 0, \end{cases}$$

i.e.,

$$\begin{cases} (\mathscr{A}\bar{\mathbf{x}}^{m-1})_i = -\bar{t}\bar{x}_i < 0, & \text{if } \bar{x}_i > 0, \\ (\mathscr{A}\bar{\mathbf{x}}^{m-1})_i \geq 0, & \text{if } \bar{x}_i = 0. \end{cases}$$

Therefore, it follows that

$$\bar{x}_i \left(\mathscr{A}\bar{\mathbf{x}}^{m-1}\right)_i = -\bar{t}\bar{x}_i^2 < 0, \quad \forall\, i \in \{j \in [n] : \bar{x}_j > 0\},$$

which contradicts the condition that $\mathscr{A}$ is a semi-positive tensor. Hence, $\mathscr{A}$ is an ER-tensor, and the desired results hold. □

In what follows, we study properties of the solution set of TCP$(\mathbf{q}, \mathscr{A})$. For any $\mathbf{x} \in \mathbf{R}^n$, denote $\mathbf{x}_+$ as defined in previous sections such that

$$\mathbf{x}_+ := (\max\{x_1, 0\}, \ldots, \max\{x_n, 0\})^\top.$$

To move on, we list several results below, which are useful for our sequel analysis.

Definition 4.17 (*Isac, Bulavski, Kalashnikov 1997*) A set of points $\{\mathbf{x}^{(k)}\} \subset \mathbf{R}^n_+$ is an exceptional family of elements for the continuous function f if $\|\mathbf{x}^{(k)}\| \to \infty$ as $k \to \infty$ and, for each $k > 0$, there exists a scalar $\mu_k > 0$ such that

$$\begin{cases} f_i(\mathbf{x}^{(k)}) = -\mu_k x_i^{(k)}, & \text{if } x_i^{(k)} > 0, \\ f_i(\mathbf{x}^{(k)}) \geq 0, & \text{if } x_i^{(k)} = 0. \end{cases}$$

Theorem 4.44 (Isac, Bulavski, Kalashnikov 1997) *For any continuous function* $f : \mathbf{R}^n_+ \to \mathbf{R}^n$*, there exists either a solution to NCP*(f) *or an exceptional family of elements for* f*, where the NCP*(f) *is defined as below:*

$$\mathbf{x} \geq \mathbf{0}, \quad f(\mathbf{x}) \geq \mathbf{0}, \quad \mathbf{x}^\top f(\mathbf{x}) = 0.$$

The following result is the main result of this section.

Theorem 4.45 *Let $\mathscr{A} \in T_{m,n}$ be an ER-tensor and $\mathbf{q} \in \mathbf{R}^n$. Then the solution set of TCP($\mathbf{q}$, $\mathscr{A}$) is nonempty and compact.*

Proof We first show that the solution set of TCP($\mathbf{q}$, $\mathscr{A}$) is nonempty. By contradiction, if TCP($\mathbf{q}$, $\mathscr{A}$) has no solution, we know that there exists an exceptional family of elements for $f(\mathbf{x}) = \mathscr{A}\mathbf{x}^{m-1} + \mathbf{q}$, i.e., there exists a sequence $\{\mathbf{x}^{(k)}\} \subset \mathbf{R}^n_+$ satisfying $\|\mathbf{x}^{(k)}\| \to \infty$ as $k \to \infty$ and, for each $k > 0$, there exists a scalar $\mu_k > 0$ such that

$$(\mathscr{A}(\mathbf{x}^{(k)})^{m-1})_i + q_i = -\mu_k x_i^{(k)}, \quad \text{if } x_i^{(k)} > 0, \tag{4.79}$$

$$(\mathscr{A}(\mathbf{x}^{(k)})^{m-1})_i + q_i \geq 0, \qquad \text{if } x_i^{(k)} = 0. \tag{4.80}$$

Without loss of generality, assume that $\frac{\mathbf{x}^{(k)}}{\|\mathbf{x}^{(k)}\|} \to \mathbf{x}^*$. Then we have

$$\mathbf{x}^* \in \mathbf{R}^n_+ \quad \text{and} \quad \mathbf{x}^* \neq \mathbf{0}. \tag{4.81}$$

We now consider the following two cases.

(1) If $x_i^* > 0$, then we have $\frac{x_i^{(k)}}{\|\mathbf{x}^{(k)}\|} \to x_i^* > 0$, $k \to \infty$. By (4.79), it holds that

$$\left(\mathscr{A}\left(\frac{\mathbf{x}^{(k)}}{\|\mathbf{x}^{(k)}\|}\right)^{m-1}\right)_i + \frac{q_i}{\|\mathbf{x}^{(k)}\|^{m-1}} = -\frac{\mu_k}{\|\mathbf{x}^{(k)}\|^{m-2}} \cdot \frac{x_i^{(k)}}{\|\mathbf{x}^{(k)}\|},$$

i.e.,

$$\frac{\mu_k}{\|\mathbf{x}^{(k)}\|^{m-2}} = -\frac{\left(\mathscr{A}\left(\frac{\mathbf{x}^{(k)}}{\|\mathbf{x}^{(k)}\|}\right)^{m-1}\right)_i + \frac{q_i}{\|\mathbf{x}^{(k)}\|^{m-1}}}{\frac{x_i^{(k)}}{\|\mathbf{x}^{(k)}\|}}.$$

It is obvious that the limit of the right-hand side of the above equality exists, which implies that $\frac{\mu_k}{\|x^{(k)}\|^{m-2}}$ as $k \to \infty$ exists. Denote $t^* := \lim_{k\to\infty} \frac{\mu_k}{\|\mathbf{x}^{(k)}\|^{m-2}}$, then we have $t^* \geq 0$ and

$$t^* = \lim_{k\to\infty} \frac{\mu_k}{\|\mathbf{x}^{(k)}\|^{m-2}} = -\lim_{k\to\infty} \frac{\left(\mathscr{A}\left(\frac{\mathbf{x}^{(k)}}{\|\mathbf{x}^{(k)}\|}\right)^{m-1}\right)_i + \frac{q_i}{\|\mathbf{x}^{(k)}\|^{m-1}}}{\frac{x_i^{(k)}}{\|\mathbf{x}^{(k)}\|}} = -\frac{\left(\mathscr{A}(\mathbf{x}^*)^{m-1}\right)_i}{x_i^*},$$

which yields that $(\mathscr{A}(\mathbf{x}^*)^{m-1})_i + t^* x_i^* = 0$ for any $i \in \{j \in [n] : x_j^* > 0\}$.

(2) If $x_i^* = 0$, then $x_i^{(k)}/\|x^{(k)}\| \to 0$, $k \to \infty$. Then one has

$$\mu_k x_i^{(k)}/\|\mathbf{x}^{(k)}\| \to 0, \quad k \to \infty.$$

Furthermore, it holds that

$$(\mathscr{A}(\mathbf{x}^*)^{m-1})_i = \lim_{k\to\infty}\left(\mathscr{A}\left(\frac{\mathbf{x}^{(k)}}{\|\mathbf{x}^{(k)}\|}\right)^{m-1}\right)_i = \lim_{k\to\infty}\frac{(\mathscr{A}(\mathbf{x}^{(k)})^{m-1})_i + q_i}{\|\mathbf{x}^{(k)}\|^{m-1}}$$

$$= \begin{cases} \lim\limits_{k\to\infty}\frac{(\mathscr{A}(\mathbf{x}^{(k)})^{m-1})_i+q_i}{\|\mathbf{x}^{(k)}\|^{m-1}} \ge 0, \text{ if } x_i^{(k)} = 0, & \text{(by (4.80))} \\ \lim\limits_{k\to\infty}\frac{-\mu_k x_i^{(k)}}{\|\mathbf{x}^{(k)}\|^{m-1}} = 0, \qquad \text{if } x_i^{(k)} > 0, & \text{(by (4.79))} \end{cases}$$

which yields that $(\mathscr{A}(\mathbf{x}^*)^{m-1})_i \ge 0$ for any $i \in \{j \in [n] : x_j^* = 0\}$.

Combining the proof of (1)–(2) with the Eq. (4.81), we know that $(\mathbf{x}^*, t^*) \in (\mathbf{R}_+^n \setminus \{\mathbf{0}\}) \times \mathbf{R}_+$ satisfies the system (4.71), which contradicts the condition that $\mathscr{A}$ is an ER-tensor. Thus, the solution set of TCP($\mathbf{q}$, $\mathscr{A}$) is nonempty.

On the other hand, we prove that the solution set of TCP($\mathbf{q}$, $\mathscr{A}$) is compact. Since $\mathscr{A}$ is an ER-tensor, by Proposition 4.13(1), it follows that $\mathscr{A}$ is an R_0-tensor. Thus, it is not difficult to prove that the Eq. (4.62) in Theorem 4.40 holds. Thus, with a similar proof with that of Theorem 4.40, it follows that the solution set of TCP($\mathbf{q}$, $\mathscr{A}$) is bounded and closed, which implies that the solution set of TCP($\mathbf{q}$, $\mathscr{A}$) is compact, and the desired results hold. □

4.11 Notes

In this chapter, a comprehensive study on the solution of tensor complementarity problem is presented such as existence, uniqueness, stability, sparsity, computation, compactness, etc. For recent advance in this area, we refer to [107, 278].

Section 4.1 The lemma about the uniqueness for the solution of nonlinear complementarity problem was given by Cottle in [62]. The definition for P-function was defined by Moré in [196]. More details about P-function and uniform P-function can be found in [91] written by Facchinei and Pang. The existence for the solution of the nonlinear complementarity problem with P-functions was proved by Moré in [196]. Furthermore, Harker and Pang showed more details about copositive mapping, strictly copositive mapping, strongly copositive mapping and **d**-regular mapping in [118]. The nonlinear complementarity problem with respect to Z-functions is given by Isac in [143].

Section 4.2 Huang and Qi studied the relationship between the m person game and the tensor complementarity problem in [137]. Then convergence of the Algorithm 4.1 can be obtained in [135].

Section 4.3 The content for this section is mainly given by Che, Qi, and Wei [37]. Copositive tensors was first defined by Qi in [223]. One can check more properties about copositive tensors and strictly copositive tensors in book [228].

Section 4.4 The P-matrix was first introduced by Fiedler and Ptak [97]. Stimulated by the notion of P-matrix, Song and Qi [247] extended the concepts to P-tensors and P_0-tensors. However, it was pointed out by Yuan and You [300], there are no odd order P-tensors according to the definition of Song and Qi [247]. Then, the deficiency was solved by Ding, Luo, and Qi in [81], and their applications in tensor complementarity problems and others are also discussed.

Section 4.5 The results of this section were originally given by Song and Qi in [250, 251].

Section 4.6 The definition of Q-tensors was first given by Song and Qi in [246]. More results about Q-tensors were originally given by Huang, Suo, and Wang [139]. Proposition 4.5 was first proved by Agangic and Cottle in [3].

Section 4.7 In [305], Zhang, Qi and Zhou first introduced the definition of Z-tensors. Here, we mainly studied the complementarity problems with respect to Z-tensors, and its applications, which was originally given by Luo, Qi and Xiu in [189]. The Theorem 4.26 was proved by Facchinei and Pang in [91]. The Theorems 4.28–4.32 were presented by Gowda, Luo, Qi, and Xiu in [108].

Section 4.8 Most conclusions of this section come from papers written by Song and Yu [253]. Song and Qi proved the Theorems 4.38–4.39 in [251].

Section 4.9 The contents of this section mainly studied the global uniqueness and solvability for tensor complementarity problems. The original of these contents is established by Bai, Huang and Wang in [6].

Section 4.10 Most results of this part were originally given by Wang, Huang and bai in [277]. Definition 4.17 and Theorem 4.44 were cited from [145] by Isac, Bulavski, Kalashnikov.

4.12 Exercises

1 Let $\mathscr{A} = (a_{i_1 i_2 \ldots i_m})$ with $a_{i_1 i_2 \ldots i_m} = 1$ for all $i_1, i_2, \ldots, i_m \in [n]$. Please check that $\mathscr{A}$ is strictly semi-positive, but not a P-tensor.

2 Let $\mathscr{A}$ be an SP_0-tensor(strong P_0-tensor). Prove that $\mathscr{A}$ is an ER-tensor if and only if it is a Q-tensor.

3 Suppose $\mathscr{A} \in T_{m,n}$ is nonnegative. Prove that $\mathscr{A}$ is an ER-tensor if and only if it is a Q-tensor.

4 Suppose $\mathscr{A} \in T_{m,n}$ is an ER-tensor, where m is even. Prove that $\mathscr{A}$ has no non-positive Z-eigenvalue associated with a nonpositive Z-eigenvector.

Chapter 5
Tensor Eigenvalue Complementarity Problems

This chapter is a companion chapter of Chap. 4. In this chapter, we mainly discuss *tensor eigenvalue complementarity problems* (TEiCP). It is a generalization of the matrix eigenvalue complementarity problem (EiCP), which possess a broad range of interesting applications.

A tensor eigenvalue complementarity problem can be formulated as follows: finding $\lambda \in \mathbf{R}$ and $\mathbf{x} \in \mathbf{R}^n_+ \setminus \{\mathbf{0}\}$ such that

$$\mathbf{x} \geq \mathbf{0}, \ \lambda \mathscr{B}\mathbf{x}^{m-1} - \mathscr{A}\mathbf{x}^{m-1} \geq \mathbf{0}, \ \text{and} \ \mathbf{x}^\top(\lambda \mathscr{B}\mathbf{x}^{m-1} - \mathscr{A}\mathbf{x}^{m-1}) = 0, \tag{5.1}$$

where $\mathscr{A}, \mathscr{B} \in T_{m,n}$.

At the beginning section, we shall first study the tensor eigenvalue complementarity problem in a more general way, which is called the tensor generalized eigenvalue complementarity problem (TGEiCP): $\lambda \in \mathbf{R}$ and nonzero vector $\mathbf{x} \in \mathbf{R}^n$ such that finding

$$\mathbf{x} \in \mathbb{K}, \ \lambda \mathscr{B}\mathbf{x}^{m-1} - \mathscr{A}\mathbf{x}^{m-1} \in \mathbb{K}^*, \ \text{and} \ \mathbf{x}^\top(\lambda \mathscr{B}\mathbf{x}^{m-1} - \mathscr{A}\mathbf{x}^{m-1}) = 0, \tag{5.2}$$

where $\mathbb{K} \subseteq \mathbf{R}^n$ is a closed convex cone, and $\mathbb{K}^*$ is the dual cone of $\mathbb{K}$, i.e.,

$$\mathbb{K}^* := \{\omega \in \mathbf{R}^n \ : \ \omega^\top \mathbf{k} \geq 0, \ \forall\, \mathbf{k} \in \mathbb{K}\}.$$

Actually, the TGEiCP (5.2) reduces to TEiCP (5.1) when we choose $\mathbb{K} = \mathbf{R}^n_+$. Particularly, a simple case is also studied such that $\mathscr{B} = \mathscr{I}$ for TEiCP (5.1).

First of all, the definitions of K-eigenvalue and Pareto-eigenvalue are introduced, which are related to TGEiCP and TEiCP respectively. Then, it is proved that the TGEiCP has at least one solution under reasonable conditions. And several more properties about the related eigenvalues for TGEiCP are given. Along the line of tensor complementarity problem, it is showed the TEiCP is solvable and has a nonempty

L. Qi et al., *Tensor Eigenvalues and Their Applications*, Advances in Mechanics and Mathematics 39, https://doi.org/10.1007/978-981-10-8058-6_5

solution set under reasonable assumptions. Actually, this is related to Pareto eigenvalues of tensors. In addition, two optimization reformulations of TEiCP are introduced, which beneficially establish an upper bound on Pareto eigenvalues of tensors. Moreover, some results concerning the bounds on the number of Pareto eigenvalues will further enrich the theory of TEiCP.

The concepts of Pareto H-eigenvalue and Pareto Z-eigenvalue of a tensor are introduced, and a comprehensive study for these two kinds of eigenvalues will present in the following section. Necessary and sufficient conditions of existence of such eigenvalues are given. It is proved that a symmetric tensor has at least one Pareto H-eigenvalue (Pareto Z-eigenvalue). Furthermore, the minimum Pareto H-eigenvalue (or Pareto Z-eigenvalue) of a symmetric tensor is exactly equal to the minimum value of a constrained minimization problem of a homogeneous polynomial deduced by such a tensor, which gives an alternative method for solving the constrained minimization problem. In particular, a symmetric tensor $\mathscr{A}$ is strictly copositive if and only if every Pareto H-eigenvalue (Z-eigenvalue) of $\mathscr{A}$ is positive, and $\mathscr{A}$ is copositive if and only if every Pareto H-eigenvalue (Z-eigenvalue) of $\mathscr{A}$ is nonnegative.

With the numerical computation aspects, on one hand, it can be easily seen that the tensor eigenvalue complementarity problem is closely related to the optimality conditions for polynomial optimization. By introducing an NCP-function, the tensor eigenvalue complementarity problem will be reformulated as a system of nonlinear equations. It is showed that this function is strongly semi-smooth but not differentiable, in which case the classical smooth methods cannot apply. Furthermore, a damped semi-smooth Newton method is proposed for tensor eigenvalue complementarity problem. A new procedure to evaluate an element of the generalized Jacobian is given, which turns out to be an element of the B-subdifferential under mild assumptions. As a result, the convergence of the damped semi-smooth Newton method is guaranteed by existing results.

On the other hand, tensor eigenvalue complementarity problems can also be reformulated as constrained polynomial optimization. When one tensor is strictly copositive, the Pareto eigenvalues can be computed by solving polynomial optimization with normalization by strict copositivity. When no tensor is strictly copositive, the tensor eigenvalue complementarity problem can be equivalently reformulated as polynomial optimization by a randomization process. The Pareto eigenvalues can be computed sequentially in this case. The formulated polynomial optimization can be solved by Lasserre's hierarchy of semidefinite relaxations. With standard polynomial optimization techniques, it can be shown that the proposed method has finite convergence for generic tensors.

Finally, a unified framework of *tensor higher-degree eigenvalue complementarity problem* (THDEiCP) is presented, which goes beyond the framework of the classical quadratic eigenvalue complementarity problem for matrices. We will study some topological properties of higher-degree *cone eigenvalues* of tensors. Based upon the symmetry assumptions on the underlying tensors, the THDEiCP is reformulated as a weakly coupled homogeneous polynomial optimization problem, and some results concerning existence of solutions of THDEiCP without symmetry conditions are presented.

5.1 Tensor Eigenvalue Complementarity Problems

In this section, some preliminaries and the definition of Pareto-eigenvalue related to the TGEiCP and TEiCP will be given first. Then, we study the existence of solutions for tensor generalized eigenvalue complementarity problems under reasonable conditions. Then, two optimization reformulations of TEiCP are introduced, thereby beneficially establishing an upper bound on Pareto-eigenvalues of tensors. Moreover, some new results concerning the bounds on the number of K-eigenvalues of TGEiCP further enrich the theory of TGEiCP.

For the TGEiCP, the scalar $\lambda \in \mathbf{R}$ and the nonzero vector $\mathbf{x} \in \mathbf{R}^n$ are called a **K-eigenvalue** of $(\mathscr{A}, \mathscr{B})$ and an associated **K-eigenvector** if they satisfy the related system (5.2). Denote $\sigma_K(\mathscr{A}, \mathscr{B})$ the set of all K-eigenvalues, i.e., **K-spectrum** of $(\mathscr{A}, \mathscr{B})$. Furthermore, if $\lambda \in \sigma_K(\mathscr{A}, \mathscr{B})$ and the related K-eigenvector $\mathbf{x} \in \text{int}(\mathbb{K})$, then λ is called the **strict K-eigenvalue** of $(\mathscr{A}, \mathscr{B})$, where $\text{int}(\mathbb{K})$ denotes the interior of $\mathbb{K}$.

For the TEiCP (5.1), the scalar $\lambda \in \mathbf{R}$ and the nonzero vector $\mathbf{x} \in \mathbf{R}^n_+$ are called a **Pareto-eigenvalue** of $(\mathscr{A}, \mathscr{B})$ and an associated **Pareto-eigenvector**, respectively. The set of all Pareto-eigenvalues, defined by $\sigma(\mathscr{A}, \mathscr{B})$, is called the **Pareto-spectrum** of $(\mathscr{A}, \mathscr{B})$. Similarly, if $\lambda \in \sigma(\mathscr{A}, \mathscr{B})$ with the associated Pareto-eigenvector $\mathbf{x} > \mathbf{0}$, then λ is called a **strict Pareto-eigenvalue**. Particularly, if $\mathscr{B} = \mathscr{I}$, then the K(Pareto)-eigenvalue/eigenvector of $(\mathscr{A}, \mathscr{B})$ is called the K(Pareto)-eigenvalue/eigenvector of $\mathscr{A}$ for simple, and the K (Pareto)-spectrum of $(\mathscr{A}, \mathscr{B})$ is called the K(Pareto)-spectrum of $\mathscr{A}$.

To move on, several preliminaries should be recalled. If a nonempty set $S \subseteq \mathbf{R}^n$ generates a cone $\mathbb{K}$, then it can be written by $\mathbb{K} := \text{cone}(S)$ i.e. $\mathbb{K} := \{\, t\mathbf{s} \;:\; \mathbf{s} \in S,\ t \in \mathbf{R}_+ \,\}$. If $\mathbf{0} \notin S$ and for any $\mathbf{k} \in \mathbb{K}\backslash\{\mathbf{0}\}$, and if there exist unique $\mathbf{s} \in S$ and $t \in \mathbf{R}_+$ such that $\mathbf{k} = t\mathbf{s}$, then S is called a basis for $\mathbb{K}$. The cone(conv(S)) is called a polyhedral cone whenever S is a finite set, where conv(S) is the convex hull of S.

First of all, we want to study the existence of solution for TGEiCP. The assumptions below are needed.

Assumption 5.1 Suppose $\mathbb{K} := \text{cone}(S)$. It holds that $\mathscr{B}\mathbf{x}^m \neq 0$ for every vector $\mathbf{x} \in S$.

By (5.2), let $\bar{\lambda} \in \sigma_K(\mathscr{A}, \mathscr{B})$, with associate K-eigenvector $\bar{\mathbf{x}} \in \mathbf{R}^n\backslash\{\mathbf{0}\}$ under Assumption 5.1. Then one has

$$\bar{\lambda} = \frac{\mathscr{A}\bar{\mathbf{x}}^m}{\mathscr{B}\bar{\mathbf{x}}^m}.$$

Therefore, by the second expression of (5.2), it holds that

$$\frac{\mathscr{A}\bar{\mathbf{x}}^m}{\mathscr{B}\bar{\mathbf{x}}^m}\mathscr{B}\bar{\mathbf{x}}^{m-1} - \mathscr{A}\bar{\mathbf{x}}^{m-1} \in \mathbb{K}^*.$$

The following lemma is cited from [4].

Lemma 5.1 *Let $S \subseteq \mathbf{R}^n$ be a compact convex subset. Suppose $f(\mathbf{x}, \mathbf{y}) : S \times S \to \mathbf{R}$ is a given function. If $f(\mathbf{x}, \mathbf{y})$ is lower semicontinuous in $\mathbf{x}$, and concave in $\mathbf{y}$, then there exists $\mathbf{x}^* \in S$ such that $\sup_{\mathbf{y}\in S} f(\mathbf{x}^*, \mathbf{y}) \leq \sup_{\mathbf{y}\in S} f(\mathbf{y}, \mathbf{y})$.*

Now, we give the conclusion for the existence of solutions for TGEiCP under Assumption 5.1.

Theorem 5.1 *Suppose $\mathbb{K} := cone(S)$, and S is a convex compact basis for $\mathbb{K}$. Then TGEiCP (5.2) has at least one solution under Assumption 5.1.*

Proof Define $F : S \times S \to \mathbf{R}$ by

$$F(\mathbf{x}, \mathbf{y}) = \mathbf{y}^\top \mathscr{A}\mathbf{x}^{m-1} - \frac{\mathscr{A}\mathbf{x}^m}{\mathscr{B}\mathbf{x}^m}\mathbf{y}^\top \mathscr{B}\mathbf{x}^{m-1}. \tag{5.3}$$

By Assumption 5.1, it holds that $\mathscr{B}\mathbf{x}^m \neq 0$ for any $\mathbf{x} \in S$, which implies that $F(\cdot, \mathbf{y})$ is lower-semicontinuous on S for any fixed $\mathbf{y} \in S$. Since $F(\mathbf{x}, \mathbf{y})$ is a linear function for any fixed $\mathbf{x}$, we obtain $F(\mathbf{x}, \cdot)$ is concave on S. Therefore, by Lemma 5.1, there exists a vector $\bar{\mathbf{x}} \in S$ such that

$$\sup_{\mathbf{y}\in S} F(\bar{\mathbf{x}}, \mathbf{y}) \leq \sup_{\mathbf{y}\in S} F(\mathbf{y}, \mathbf{y}). \tag{5.4}$$

On the other hand, since $F(\mathbf{y}, \mathbf{y}) = 0$, for all $\mathbf{y} \in S$, by (5.4), it follows that $F(\bar{\mathbf{x}}, \mathbf{y}) \leq 0$ for any $\mathbf{y} \in S$. Then we know that $\mathbf{y}^\top(\bar{\lambda}\mathscr{B}\bar{\mathbf{x}}^{m-1} - \mathscr{A}\bar{\mathbf{x}}^{m-1}) \geq 0$ for all $\mathbf{y} \in S$ from (5.3), where

$$\bar{\lambda} = \frac{\mathscr{A}\bar{\mathbf{x}}^m}{\mathscr{B}\bar{\mathbf{x}}^m}.$$

Hence, it follows that

$$\bar{\lambda}\mathscr{B}\bar{\mathbf{x}}^{m-1} - \mathscr{A}\bar{\mathbf{x}}^{m-1} \in \mathbb{K}^*. \tag{5.5}$$

Moreover, one can observe that

$$\bar{\mathbf{x}}^\top(\bar{\lambda}\mathscr{B}\bar{\mathbf{x}}^{m-1} - \mathscr{A}\bar{\mathbf{x}}^{m-1}) = 0,$$

which means, together with (5.5) and the fact $\bar{\mathbf{x}} \neq \mathbf{0}$, that $(\bar{\lambda}, \bar{\mathbf{x}})$ is a solution of (5.2). Thus the desired result holds. □

The following example shows that the condition $\mathscr{B}\mathbf{x}^m \neq 0$, for all $\mathbf{x} \in S$, is necessary to ensure the existence for the solution of TGEiCP related some special cone.

Example 5.1 Let $m = 2$. Consider the TGEiCP with $S = \mathbf{R}^n \backslash \{\mathbf{0}\}$ with matrices such that

$$\mathscr{A} = \begin{pmatrix} 1 & 3 \\ 4 & 1 \end{pmatrix} \text{ and } \mathscr{B} = \begin{pmatrix} 1 & 0 \\ 0 & -1 \end{pmatrix}.$$

By a direct computation, for any $\mathbf{x} = (x_0, x_0)^\top \in S$, it holds that

$$\mathbf{x}^\top \mathscr{B}\mathbf{x} = x_0^2 - x_0^2 = 0.$$

Since $\det(\lambda\mathscr{B} - \mathscr{A}) = -\lambda^2 - 11 \neq 0$ for any $\lambda \in \mathbf{R}$, it follows that the system $(\lambda\mathscr{B} - \mathscr{A})\mathbf{x} = 0$ has only one unique solution $\mathbf{x} = \mathbf{0}$ for any $\lambda \in \mathbf{R}$. Moreover, one may check that $(\lambda\mathscr{B} - \mathscr{A})\mathbf{x} \geq 0$ does not hold for any $(\lambda, \mathbf{x}) \in \mathbf{R} \times (\mathbf{R}^2_+ \backslash \{\mathbf{0}\})$ with $\mathbf{x} = (\mathbf{x}_1, 0)^\top$ or $\mathbf{x} = (0, \mathbf{x}_2)^\top$. Therefore, problem (5.1) has no solution.

In the following discussion, we discuss the existence of solutions for TEiCP. A simple corollary is listed below, which is given based on Definition 4.5 and Theorem 5.1.

Corollary 5.1 *When tensor $\mathscr{B}$ is strictly copositive, then problem (5.1) has at least one solution.*

Proof Let S be the standard simplex in $\mathbf{R}^n$, i.e.,

$$S := \left\{ \mathbf{x} \in \mathbf{R}^n_+ \ : \ \sum_{i=1}^{n} x_i = 1 \right\}. \tag{5.6}$$

It is apparent that S is a convex compact basis of $\mathbf{R}^n_+$. Thus, by Theorem 5.1, the desired result holds. □

To move on, let

$$\lambda(\mathbf{x}) = \frac{\mathscr{A}\mathbf{x}^m}{\mathscr{B}\mathbf{x}^m}. \tag{5.7}$$

Therefore $\mathscr{B}\mathbf{x}^m \neq 0$ under Assumption 5.1. When $\mathscr{A}$ and $\mathscr{B}$ are symmetric tensors, the gradient of $\lambda(\mathbf{x})$ is

$$\nabla\lambda(\mathbf{x}) = \frac{m}{\mathscr{B}\mathbf{x}^m}\left[\mathscr{A}\mathbf{x}^{m-1} - \lambda(\mathbf{x})\mathscr{B}\mathbf{x}^{m-1}\right]. \tag{5.8}$$

In this case, one stationary point of $\lambda(\mathbf{x})$ corresponds to a solution of (5.1). However, if either $\mathscr{A}$ or $\mathscr{B}$ is not symmetric, (5.8) is incorrect, and the relationship between stationary points and solutions of the (5.1) ceases to hold.

The following lemma presents two fundamental properties of $\lambda(\mathbf{x})$ in (5.8).

Lemma 5.2 *The function $\lambda(\mathbf{x})$ in (5.7) is a zero order homogeneous function, i.e., $\lambda(\tau\mathbf{x}) = \lambda(\mathbf{x})$ for all $\mathbf{x} \in \mathbf{R}^n \backslash \{\mathbf{0}\}$ and $\tau > 0$. Furthermore, since $\lambda(\mathbf{x})$ is continuously differentiable, we have $\mathbf{x}^\top \nabla\lambda(\mathbf{x}) = 0$.*

We now have to introduce two optimization reformulations for (5.1). The first optimization problem is such that

$$\rho(\mathscr{A}, \mathscr{B}) := \max\{\lambda(\mathbf{x}) \ : \ \mathbf{x} \in S\}, \tag{5.9}$$

where the constraint set S is determined by (5.6).

Proposition 5.1 *Suppose $\mathscr{A}$, $\mathscr{B} \in S_{m,n}$ and $\mathscr{B}$ is strictly copositive. If $\bar{\mathbf{x}} \in S$ is a stationary point of (5.9), then $(\lambda(\bar{\mathbf{x}}), \bar{\mathbf{x}})$ is a solution of the TEiCP (5.1).*

Proof By the definition of S, the optimization problem (5.9) is equivalent with

$$\max\left\{\lambda(\mathbf{x}) \mid \mathbf{x} \geq \mathbf{0},\ \sum_{i=1}^{n} x_i = 1\right\}.$$

Since $\bar{\mathbf{x}}$ is a stationary point of (5.9), there exist $\bar{\alpha} \in \mathbf{R}^n$ and $\bar{\beta} \in \mathbf{R}$, such that

$$\begin{cases} -\nabla\lambda(\bar{\mathbf{x}}) = \bar{\alpha} + \bar{\beta}\mathbf{1}, \\ \bar{\alpha} \geq \mathbf{0},\ \bar{\mathbf{x}} \geq \mathbf{0}, \\ \bar{\alpha}^\top \bar{\mathbf{x}} = 0, \\ \mathbf{1}^\top \bar{\mathbf{x}} = 1, \end{cases} \tag{5.10}$$

where $\mathbf{1} \in \mathbf{R}^n$ is the all one vector. By (5.10) and Lemma 5.2, we obtain that $\bar{\beta} = 0$. Hence, combining (5.8), (5.10) with the fact that $\mathscr{B}\bar{\mathbf{x}}^m > 0$, it holds that

$$\lambda(\bar{\mathbf{x}})\mathscr{B}\bar{\mathbf{x}}^{m-1} - \mathscr{A}\bar{\mathbf{x}}^{m-1} \geq \mathbf{0}.$$

Since $\bar{\mathbf{x}} \geq \mathbf{0}$ and

$$\bar{\mathbf{x}}^\top(\lambda(\bar{\mathbf{x}})\mathscr{B}\bar{\mathbf{x}}^{m-1} - \mathscr{A}\bar{\mathbf{x}}^{m-1}) = 0,$$

it follows that $(\lambda(\bar{\mathbf{x}}), \bar{\mathbf{x}})$ is a solution of (5.1) and the desired results hold. □

The next theorem is raised on the following scalar

$$\lambda_{max}(\mathscr{A}, \mathscr{B}) = \max\{\lambda\ :\ \exists\, \mathbf{x} \in \mathbf{R}^n_+ \backslash \{\mathbf{0}\} \text{ such that } (\lambda, \mathbf{x}) \text{ is a solution of } (5.1)\},$$

which denotes the largest Pareto-eigenvalue of $(\mathscr{A}, \mathscr{B})$. Using this, the following theorem characterizes the relationship between optimization problem (5.9) and TEiCP (5.1).

Theorem 5.2 *Suppose $\mathscr{A}$, $\mathscr{B} \in S_{m,n}$. Then $\lambda_{max}(\mathscr{A}, \mathscr{B}) = \rho(\mathscr{A}, \mathscr{B})$ if $\mathscr{B}$ is strictly copositive.*

Proof Since the constraint set S of (5.9) is compact, there exists a vector $\bar{\mathbf{x}} \in S$ such that $\rho(\mathscr{A}, \mathscr{B}) = \lambda(\bar{\mathbf{x}})$. Denote $I(\bar{\mathbf{x}}) := \{i \in N\ :\ \bar{x}_i = 0\}$ since $\bar{\mathbf{x}} \neq \mathbf{0}$. We know that $\{\mathbf{1}\} \cup \{\mathbf{1}_i\ :\ i \in I(\bar{\mathbf{x}})\}$ is linearly independent. Thus the first order optimality condition of (5.9) holds, i.e., $\bar{\mathbf{x}}$ is a stationary point of (5.9). By Proposition 5.1, one obtain that $(\lambda(\bar{\mathbf{x}}), \bar{\mathbf{x}})$ is a solution of (5.1), which implies that $\rho(\mathscr{A}, \mathscr{B}) \leq \lambda_{max}(\mathscr{A}, \mathscr{B})$.

On the other hand, suppose $\lambda \in \sigma(\mathscr{A}, \mathscr{B})$ with associated Pareto-eigenvector $\mathbf{x}$ for TEiCP (5.1). Then it holds that $\lambda = \frac{\mathscr{A}\mathbf{x}^m}{\mathscr{B}\mathbf{x}^m}$. Denote $\mathbf{y} = \frac{\mathbf{x}}{\mathbf{1}^\top \mathbf{x}}$, and it follows that

$$\mathbf{y} \in S \text{ and } \lambda = \frac{\mathscr{A}\mathbf{y}^m}{\mathscr{B}\mathbf{y}^m},$$

which implies that $\lambda \le \rho(\mathscr{A}, \mathscr{B})$. By the arbitrariness of λ, we have $\lambda_{max}(\mathscr{A}, \mathscr{B}) \le \rho(\mathscr{A}, \mathscr{B})$, and hence the desired results hold. □

We now study another optimization reformulation of (5.1) such that

$$\varphi(\mathscr{A}, \mathscr{B}) = \max\{\mathscr{A}\mathbf{x}^m \ : \ \mathbf{x} \in \Sigma\}, \tag{5.11}$$

where $\Sigma := \{\mathbf{x} \in \mathbf{R}^n_+ \ : \ \mathscr{B}\mathbf{x}^m = 1\}$ is assumed to be compact. Similar to Theorem 5.2, we have the following theorem.

Theorem 5.3 *Suppose $\mathscr{A}$, $\mathscr{B} \in S_{m,n}$ and $\mathscr{B}$ is strictly copositive. then λ_{max} $(\mathscr{A}, \mathscr{B}) = \varphi(\mathscr{A}, \mathscr{B})$.*

Proof On one hand, let $\lambda \in \sigma(\mathscr{A}, \mathscr{B})$ with associated Pareto-eigenvector $\mathbf{x} \in \mathbf{R}^n_+ \backslash \{\mathbf{0}\}$. By (5.1), it holds that $\lambda = \frac{\mathscr{A}\mathbf{x}^m}{\mathscr{B}\mathbf{x}^m}$. Since $\mathscr{B}$ is strictly copositive, it follows that $\mathscr{B}\mathbf{x}^m > 0$ for any $\mathbf{x} \in \mathbf{R}^n_+ \backslash \{\mathbf{0}\}$. Let $\mathbf{y} = \frac{\mathbf{x}}{(\mathscr{B}\mathbf{x}^m)^{\frac{1}{m}}} \in \mathbf{R}^n_+ \backslash \{\mathbf{0}\}$. Then one has that

$$\mathscr{B}\mathbf{y}^m = 1 \quad \lambda = \mathscr{A}\mathbf{y}^m,$$

which implies that $\mathbf{y} \in \Sigma$ and $\lambda \le \varphi(\mathscr{A}, \mathscr{B})$. From the arbitrariness of λ, we know that $\lambda_{max}(\mathscr{A}, \mathscr{B}) \le \varphi(\mathscr{A}, \mathscr{B})$.

On the other hand, there exists a vector $\bar{\mathbf{x}} \in \Sigma$ satisfying $\varphi(\mathscr{A}, \mathscr{B}) = \mathscr{A}\bar{\mathbf{x}}^m$ since Σ is compact. Then, we obtain that

$$\varphi(\mathscr{A}, \mathscr{B}) = \mathscr{A}\bar{\mathbf{x}}^m = \frac{\mathscr{A}\bar{\mathbf{x}}^m}{\mathscr{B}\bar{\mathbf{x}}^m},$$

which is equivalent to

$$\varphi(\mathscr{A}, \mathscr{B})\mathscr{B}\bar{\mathbf{x}}^m - \mathscr{A}\bar{\mathbf{x}}^m = 0.$$

Using this equation, we know that $\bar{\mathbf{x}}^\top(\varphi(\mathscr{A}, \mathscr{B})\mathscr{B}\bar{\mathbf{x}}^{m-1} - \mathscr{A}\bar{\mathbf{x}}^{m-1}) = 0$. It implies that $\varphi(\mathscr{A}, \mathscr{B}) \in \sigma(\mathscr{A}, \mathscr{B})$ and $\varphi(\mathscr{A}, \mathscr{B}) \le \lambda_{max}(\mathscr{A}, \mathscr{B})$, and the desired result follows. □

In the sequel, we estimate the value of K-eigenvalues for TGEiCP and Pareto-eigenvalues for TEiCP. For the sake of simplicity, based upon Theorem 5.3, we first study the bound on Pareto-eigenvalue of TEiCP with $\mathscr{B} = \mathscr{I}$. Let Ω^* denote the solution set of (5.1) with respect to $\mathscr{B} = \mathscr{I}$. Define $\rho_{max}(A)$ such as

$$\rho_{max}(A) = \max\{|\lambda| \ : \ \exists\, \mathbf{x} \in \mathbf{R}^n_+ \backslash \{\mathbf{0}\} \text{ such that } (\lambda, \mathbf{x}) \in \Omega^*\}.$$

Theorem 5.4 *For TEiCP with $\mathscr{B} := \mathscr{I}$ in (5.1), it holds that*

$$\rho_{max}(A) \leq \min\left\{n^{\frac{m-2}{2}}\|\mathscr{A}\|_F,\ \bar{a}n^{m-1}\right\},$$

where $\bar{a} := \max\{|a_{i_1 i_2 \ldots i_m}| \ :\ 1 \leq i_1, i_2, \ldots, i_m \leq n\}$.

Proof Let $(\lambda, \mathbf{x})$ be an arbitrary solution of (5.1) with $\mathscr{B} := \mathscr{I}$. Since $\mathscr{I}\mathbf{x}^m = \sum_{i=1}^n x_i^m$, it follows that

$$\lambda = \frac{\mathscr{A}\mathbf{x}^m}{\sum_{i=1}^n x_i^m},$$

which implies that

$$|\lambda| = \frac{|\mathscr{A}\mathbf{x}^m|}{\sum_{i=1}^n x_i^m} \leq \frac{\|\mathscr{A}\|_F \|\mathbf{x}^{\otimes m}\|_F}{\sum_{i=1}^n x_i^m},$$

where $\mathbf{x}^{\otimes m} := (x_{i_1} x_{i_2} \ldots x_{i_m})_{1 \leq i_1, \ldots, i_m \leq n}$ is a symmetric rank-1 tensor as discussed in Chap. 1. Since

$$\|\mathbf{x}^{\otimes m}\|_F^2 = \sum_{i_1, i_2, \ldots, i_m = 1}^{n} (x_{i_1} x_{i_2} \ldots x_{i_m})^2 = \left(\sum_{i=1}^n x_i^2\right)^m \leq n^{m-2}\left(\sum_{i=1}^n x_i^m\right)^2,$$

where the last inequality uses the Hölder inequality, hence we obtain

$$|\lambda| \leq n^{\frac{m-2}{2}}\|\mathscr{A}\|_F.$$

On the other hand, by the definition of $\bar{a}$, it follows that

$$|\lambda| = \frac{|\mathscr{A}\mathbf{x}^m|}{\sum_{i=1}^n x_i^m} \leq \frac{\bar{a}(\sum_{i=1}^n x_i)^m}{\sum_{i=1}^n x_i^m} \leq \bar{a}n^{m-1}.$$

So, we obtain

$$\rho_{max}(A) \leq \min\left\{n^{\frac{m-2}{2}}\|\mathscr{A}\|_F,\ \bar{a} \cdot n^{m-1}\right\},$$

Since λ is arbitrary, we obtain the desired result and complete the proof. □

With a similar proof of Theorem 5.4, we have the following conclusion.

Corollary 5.2 *If $\mathscr{B}$ is strictly copositive, then for TEiCP in (5.1), it holds that*

$$|\lambda_{max}(\mathscr{A}, \mathscr{B})| \leq \frac{1}{N_{\min}(\mathscr{B})}\min\left\{n^{\frac{m-2}{2}}\|\mathscr{A}\|_F,\ \bar{a} \cdot n^{m-1}\right\},$$

where $N_{\min}(\mathscr{B}) := \min\{\mathscr{B}\mathbf{x}^m \ :\ \mathbf{x} \in \mathbf{R}_+^n,\ \sum_{i=1}^n x_i^m = 1\} > 0$.

To study the estimation of the numbers of Pareto-eigenvalues of $(\mathscr{A}, \mathscr{B})$, the following basic concept and property are needed.

For given tensors $\mathscr{A}$ and $\mathscr{B}$ with the same size, we say that $(\mathscr{A}, \mathscr{B})$ is an **identical singular pair**, if

$$\left\{\mathbf{x} \in \mathbf{C}^n \backslash \{\mathbf{0}\} \ : \ \mathscr{A}\mathbf{x}^{m-1} = \mathbf{0}, \ \mathscr{B}\mathbf{x}^{m-1} = \mathbf{0}\right\} \neq \emptyset.$$

On the other hand, we know that

$$(\mathscr{A} - \lambda\mathscr{B})\mathbf{x}^{m-1} = \mathbf{0}, \tag{5.12}$$

is indeed a set of n homogeneous polynomials with n variables of degree $(m-1)$. We denote the n homogeneous polynomials by $P_1, \ldots, P_n$. By the resultant theory as discussed in Chap. 1, applying it to the current system (5.12), we have the following conclusion.

Proposition 5.2 *The following results hold for system (5.12):*

(1) $\mathrm{RES}(P_1, \ldots, P_n) = 0$ *if and only if there exists* $(\lambda, \mathbf{x}) \in \mathbf{C} \times (\mathbf{C}^n \backslash \{\mathbf{0}\})$ *satisfying (5.12);*
(2) The degree of λ *in* $\mathrm{RES}(P_1, \ldots, P_n)$ *is at most* $n(m-1)^{n-1}$.

The following proposition presents a sufficient and necessary condition for a given $\lambda \in \mathbf{R}$ to be a Pareto-eigenvalue of $(\mathscr{A}, \mathscr{B})$.

Theorem 5.5 *Let* $\mathscr{A}, \mathscr{B} \in T_{m,n}$. *Then* $\lambda \in \mathbf{R}$ *is a Pareto-eigenvalue of* $(\mathscr{A}, \mathscr{B})$, *if and only if there exist a nonempty subset* $J \subseteq [n]$ *and a vector* $\mathbf{w} \in \mathbf{R}_{++}^{|J|}$ *such that*

$$\mathscr{A}^J \mathbf{w}^{m-1} = \lambda \mathscr{B}^J \mathbf{w}^{m-1} \tag{5.13}$$

and

$$\sum_{i_2,\ldots,i_m \in J} (\lambda b_{ii_2\ldots i_m} - a_{ii_2\ldots i_m}) w_{i_2} \ldots w_{i_m} \geq 0, \quad \forall\, i \in [n] \backslash J.$$

In such a case, the vector $\mathbf{x} \in \mathbf{R}_+^n$ *defined by*

$$x_i = \begin{cases} w_i, & i \in J \\ 0, & i \in [n] \backslash J \end{cases} \tag{5.14}$$

is a Pareto-eigenvector of $(\mathscr{A}, \mathscr{B})$, *associated with the real number* λ.

Proof Suppose $\lambda \in \mathbf{R}$ be a Pareto-eigenvalue of $(\mathscr{A}, \mathscr{B})$ with a corresponding Pareto-eigenvector $\mathbf{y} \in \mathbf{R}_+^n$. Then we know that

$$\begin{cases} \mathbf{y}^\top(\lambda\mathscr{B}\mathbf{y}^{m-1} - \mathscr{A}\mathbf{y}^{m-1}) = 0, \\ \lambda\mathscr{B}\mathbf{y}^{m-1} - \mathscr{A}\mathbf{y}^{m-1} \geq \mathbf{0}, \\ \mathbf{y} \geq \mathbf{0}, \end{cases} \tag{5.15}$$

and hence

$$\begin{cases} \sum_{i=1}^{n} y_i(\lambda \mathscr{B}\mathbf{y}^{m-1} - \mathscr{A}\mathbf{y}^{m-1})_i = 0, \\ \left(\lambda \mathscr{B}\mathbf{y}^{m-1} - \mathscr{A}\mathbf{y}^{m-1}\right)_i \geq 0, \quad \text{for } i \in [n], \\ y_i \geq 0, \quad \text{for } i \in [n]. \end{cases} \tag{5.16}$$

By (5.16), one has that

$$y_i \left(\lambda \mathscr{B}\mathbf{y}^{m-1} - \mathscr{A}\mathbf{y}^{m-1}\right)_i = 0, \quad \text{for all } \; i \in [n]. \tag{5.17}$$

Take $J = \{i \in [n] \; : \; y_i > 0\}$. Let $\omega \in \mathbf{R}^{|J|}$ be defined by

$$\omega_i = y_i \text{ for all } i \in J.$$

Clearly, $\omega \in \mathbf{R}^{|J|}_{++}$. Combining (5.17) with the fact that $\omega_i > 0$ for all $i \in J$, we have

$$\left(\lambda \mathscr{B}\mathbf{y}^{m-1} - \mathscr{A}\mathbf{y}^{m-1}\right)_i = 0, \quad \text{for all } \; i \in J,$$

and so

$$\mathscr{A}^J \omega^{m-1} = \lambda \mathscr{B}^J \omega^{[m-1]}, \quad \omega \in \mathbf{R}^{|J|}_{++}.$$

By the middle inequality in (5.16), we directly know that

$$\sum_{i_2,\ldots,i_m \in J} (\lambda b_{ii_2\ldots i_m} - a_{ii_2\ldots i_m}) w_{i_2} \ldots w_{i_m} \geq 0, \quad \forall\, i \in N \backslash J.$$

which implies that necessary condition holds.

For sufficiency, suppose that there exists a nonempty subset $J \subseteq [n]$ and a vector $\omega \in \mathbf{R}^{|J|}$ satisfying (5.13). Then the vector $\mathbf{x}$ defined by (5.14) is a nonzero vector in $\mathbf{R}^n_+$ such that $(\lambda, \mathbf{x})$ satisfying (5.15). Then the desired conclusions hold. □

A direct conclusion based on Theorem 5.5 is that if λ is a Pareto-eigenvalue of $(\mathscr{A}, \mathscr{B})$, then there exists a nonempty subset $J \subseteq [n]$ such that λ is a strict Pareto-eigenvalue of $(\mathscr{A}^J, \mathscr{B}^J)$.

Theorem 5.6 *Suppose $\mathscr{A}, \mathscr{B} \in T_{m,n}$, and assume $(\mathscr{A}, \mathscr{B})$ is not an identical singular pair. Then there are at most $\rho_{m,n} := nm^{n-1}$ Pareto-eigenvalues of $(\mathscr{A}, \mathscr{B})$.*

Proof For any $k = 0, 1, \ldots, n-1$, there are $\binom{n}{n-k}$ corresponding principal subtensors pair of order m dimension $n-k$. By Proposition 5.2, we know that every principal subtensor with order m dimension $n-k$ has at most $(n-k)(m-1)^{n-k-1}$ Pareto-eigenvalues. By Theorem 5.5, we obtain the upper bound

$$\rho_{m,n} = \sum_{k=0}^{n-1} \binom{n}{n-k} (n-k)(m-1)^{n-k-1} = nm^{n-1},$$

and the desired result holds. □

We now study the estimation for the number of K-eigenvalues of TGEiCP in case that $\mathbb{K}$ is a polyhedral convex cone. Recall that a closed convex cone $\mathbb{K}$ in $\mathbf{R}^n$ is said to be finitely generated if there is a linearly independent collection $\{\mathbf{c}_1, \mathbf{c}_2, \ldots, \mathbf{c}_p\}$ of vectors in $\mathbf{R}^n$ such that

$$\mathbb{K} = cone\{\mathbf{c}_1, \mathbf{c}_2, \ldots, \mathbf{c}_p\} = \left\{ \sum_{i=1}^{p} \alpha_j \mathbf{c}_j \ : \ \alpha = (\alpha_1, \alpha_2, \ldots, \alpha_p)^\top \in \mathbf{R}_+^p \right\}. \quad (5.18)$$

Apparently, $\mathbb{K} = \{C^\top \alpha \ : \ \alpha \in \mathbf{R}_+^p\}$, where $C = (c_1, c_2, \ldots, c_p)^\top \in \mathbf{R}^{p\times n}$. Moreover, we can see that the dual cone of $\mathbb{K}$, denoted by $\mathbb{K}^*$, is equivalent to $\{\mathbf{w} \in \mathbf{R}^n \ : \ C\mathbf{w} \geq 0\}$.

Theorem 5.7 *Suppose $\mathscr{A}$, $\mathscr{B} \in T_{m,n}$, then $(\mathscr{A}, \mathscr{B})$ has at most $\rho_{m,p} := pm^{p-1}$ K-eigenvalues if the closed convex cone $\mathbb{K}$ admits representation (5.18).*

Proof We first prove that TGEiCP (5.2) with $\mathbb{K}$ defined by (5.18) is equivalent to finding a vector $\bar{\alpha} \in \mathbf{R}^p\backslash\{\mathbf{0}\}$ and $\bar{\lambda} \in \mathbf{R}$ such that

$$\bar{\alpha} \geq 0, \ \bar{\lambda}\mathscr{D}\bar{\alpha}^{m-1} - \mathscr{G}\bar{\alpha}^{m-1} \geq 0, \ \bar{\alpha}^\top \left(\bar{\lambda}\mathscr{D}\bar{\alpha}^{m-1} - \mathscr{G}\bar{\alpha}^{m-1}\right) = 0, \quad (5.19)$$

where $\mathscr{D}, \mathscr{G} \in T_{m,p}$ are defined by

$$d_{i_1 i_2 \ldots i_m} = \sum_{j_1, j_2, \ldots, j_m}^{n} b_{j_1 j_2 \ldots j_m} c_{i_1 j_1} c_{i_2 j_2} \cdots c_{i_m j_m}$$

and

$$g_{i_1 i_2 \ldots i_m} = \sum_{j_1, j_2, \ldots, j_m}^{n} a_{j_1 j_2 \ldots j_m} c_{i_1 j_1} c_{i_2 j_2} \cdots c_{i_m j_m},$$

for $i_1, i_2, \ldots, i_m \in [p]$, respectively.

Without loss of generality, suppose $(\bar{\lambda}, \bar{\mathbf{x}}) \in \mathbf{R} \times (\mathbf{R}^n\backslash\{\mathbf{0}\})$ is a solution of (5.2) with $\mathbb{K}$ in (5.18). Since $\bar{\mathbf{x}} \in \mathbb{K}$, there exists a nonzero vector $\bar{\alpha} \in \mathbf{R}_+^p$ such that $\bar{\mathbf{x}} = C^\top\bar{\alpha}$. Consequently, from $\bar{\lambda}\mathscr{B}\bar{\mathbf{x}}^{m-1} - \mathscr{A}\bar{\mathbf{x}}^{m-1} \in K^*$ and the expression of K^*, it holds that $C(\bar{\lambda}\mathscr{B}\bar{\mathbf{x}}^{m-1} - \mathscr{A}\bar{\mathbf{x}}^{m-1}) \geq \mathbf{0}$, which implies that

$$C(\bar{\lambda}\mathscr{B}(C^\top\bar{\alpha})^{m-1} - \mathscr{A}(C^\top\bar{\alpha})^{m-1}) \geq \mathbf{0}. \quad (5.20)$$

By the definitions of $\mathscr{D}$ and $\mathscr{G}$, we know that (5.20) can be equivalently written as

$$\bar{\lambda}\mathscr{D}\bar{\alpha}^{m-1} - \mathscr{G}\bar{\alpha}^{m-1} \geq \mathbf{0}.$$

Moreover, it is easy to verify that $\bar{\alpha}^\top(\bar{\lambda}\mathscr{D}\bar{\alpha}^{m-1} - \mathscr{G}\bar{\alpha}^{m-1}) = 0$. Conversely, if $(\bar{\lambda}, \bar{\alpha}) \in \mathbf{R} \times (\mathbf{R}^p\backslash\{\mathbf{0}\})$ satisfies (5.19), then we can prove that $(\bar{\lambda}, \bar{\mathbf{x}})$ with $\bar{\mathbf{x}} = C^\top\bar{\alpha}$ satisfies (5.2) in a similar way.

Above all, applying Theorem 5.6 to (5.19), it follows that $(\mathscr{A}, \mathscr{B})$ has at most $\rho_{m,p} = pm^{p-1}$ K-eigenvalues, and the desired result holds. □

At last, we discuss the TEiCP with $\mathscr{B} := \mathscr{I}$, and several interesting conclusions will be presented.

Lemma 5.3 *Assume $\mathscr{A} \in T_{m,n}$ is a nonnegative tensor. If $\mathscr{A}$ has two Pareto-eigenvectors in $\mathbf{R}^n_{++}$, then the corresponding Pareto-eigenvalues are equal.*

Proof Suppose λ_1 and λ_2 are two Pareto-eigenvalues of $\mathscr{A}$ with Pareto-eigenvectors $\mathbf{x} \in \mathbf{R}^n_{++}$ and $\mathbf{y} \in \mathbf{R}^n_{++}$ respectively. Then it holds that

$$\mathscr{A}\mathbf{x}^{m-1} = \lambda_1\mathbf{x}^{[m-1]} \text{ and } \mathscr{A}\mathbf{y}^{m-1} = \lambda_2\mathbf{y}^{[m-1]}. \tag{5.21}$$

It is clear that λ_1, λ_2 are nonnegative since $\mathscr{A}$ is nonnegative. Without loss of generality, assume $\lambda_1 \geq \lambda_2$. If $\lambda_1 = 0$, then the first equation in (5.21) implies that $\mathscr{A}$ is a zero tensor. Thus, $\lambda_2 = 0$. If $\lambda_1 > 0$, denote

$$t_0 = \min\{t > 0 \ : \ t\mathbf{y} - \mathbf{x} \in \mathbf{R}^n_+\}. \tag{5.22}$$

Hence t_0 is well defined since $\mathbf{y} \in \mathbf{R}^n_{++}$, and we know that $t_0\mathbf{y} - \mathbf{x} \in \mathbf{R}^n_+$, which immediately implies that $t_0 y_i \geq x_i$ for all $i \in [n]$. By (5.21) and the fact that $\mathscr{A}$ is nonnegative, we obtain that

$$t_0^{m-1}\lambda_2\mathbf{y}^{[m-1]} - \lambda_1\mathbf{x}^{[m-1]} = \mathscr{A}(t_0\mathbf{y})^{m-1} - \mathscr{A}\mathbf{x}^{m-1} \in \mathbf{R}^n_+,$$

which implies

$$t_0\left(\frac{\lambda_2}{\lambda_1}\right)^{\frac{1}{m-1}}\mathbf{y} - \mathbf{x} \in \mathbf{R}^n_+.$$

By (5.22), we know that $t_0 \leq t_0\left(\frac{\lambda_2}{\lambda_1}\right)^{\frac{1}{m-1}}$, which means $\lambda_1 \geq \lambda_2$. Thus, we know that $\lambda_1 = \lambda_2$ and complete the proof. □

Recall the definition of Z-tensor discussed in Chap. 2, we say that $\mathscr{A}$ is a Z-tensor, if all off-diagonal entries of $\mathscr{A}$ are nonpositive. Then, we have the following lemma.

Lemma 5.4 *Suppose $\mathscr{A} \in T_{m,n}$ is a given tensor. If $-\mathscr{A}$ is a Z-tensor or $\mathscr{A}$ is a Z-tensor, then $\mathscr{A}$ has at most one strict Pareto-eigenvalue.*

Proof If $-\mathscr{A}$ is a Z-tensor, and suppose $\lambda_1, \lambda_2 \in \mathbf{R}$ are two strict Pareto-eigenvalues of $\mathscr{A}$. Let $\mathbf{x}, \mathbf{y} \in \mathbf{R}^n_{++}$ be associate strict Pareto-eigenvectors corresponding λ_1 and $\lambda_2 \in \mathbf{R}$ respectively. Then it holds that

$$\mathscr{A}\mathbf{x}^{m-1} = \lambda_1 \mathbf{x}^{[m-1]} \text{ and } \mathscr{A}\mathbf{y}^{m-1} = \lambda_2 \mathbf{y}^{[m-1]}.$$

Hence

$$(\mathscr{A} + \mu\mathscr{I})\mathbf{x}^{m-1} = (\lambda_1 + \mu)\mathbf{x}^{[m-1]} \text{ and } (\mathscr{A} + \mu\mathscr{I})\mathbf{y}^{m-1} = (\lambda_2 + \mu)\mathbf{y}^{[m-1]},$$

for any $\mu \in \mathbf{R}$. Since $-\mathscr{A}$ is a Z-tensor, $\mathscr{A} + \mu\mathscr{I}$ is nonnegative for enough large μ. By Lemma 5.3, we obtain that $\lambda_1 + \mu = \lambda_2 + \mu$, which implies the desired conclusion.

For the case $\mathscr{A}$ is a Z-tensor, the conclusion can be proved similarly. □

We have the following result on counting of Pareto-eigenvalues.

Proposition 5.3 *Suppose $\mathscr{A} \in T_{m,n}$ is a given tensor. If $-\mathscr{A}$ is a Z-tensor or $\mathscr{A}$ is a Z-tensor, then $\mathscr{A}$ has at most $\rho_n := 2^n - 1$ Pareto-eigenvalues.*

Proof If $-\mathscr{A}$ is a Z-tensor, for every $k = 0, 1, \ldots, n-1$, there are $\binom{n}{n-k}$ principal sub-tensors of order m dimension $n-k$. Since $-\mathscr{A}$ is a Z-tensor, it is clear that any principal subtensors of $-\mathscr{A}$ are also Z-tensors. By Lemma 5.4, it follows that every principal subtensor has at most one strict eigenvalue. Hence, by Theorem 5.5, one has the upper bound

$$\rho_n = \sum_{k=0}^{n-1} \binom{n}{n-k} \cdot 1 = 2^n - 1.$$

We can prove similarly when $\mathscr{A}$ is a Z-tensor, and the desired results hold. □

5.2 Pareto H(Z)-Eigenvalues of Tensors

In this section, we give an comprehensive study on Pareto H-eigenvalue and Pareto Z-eigenvalue for symmetric tensors, which are introduced from studying constrained minimization problems. Necessary and sufficient conditions for a given scalar to be a Pareto H(Z)-eigenvalues are presented. Furthermore, it is proved that a symmetric tensor has at least one Pareto H-eigenvalue (Pareto Z-eigenvalue).

First of all, we give definitions of Pareto H-eigenvalue and Pareto Z-eigenvalue below.

Definition 5.1 Let $\mathscr{A} \in T_{m,n}$ be a given tensor.

(1) The real number λ is called a **Pareto H-eigenvalue** of $\mathscr{A}$ if there exists a non-zero vector $\mathbf{x} \in \mathbf{R}^n$ satisfying the the following system.

$$\begin{cases} \mathscr{A}\mathbf{x}^m = \lambda \mathbf{x}^\top \mathbf{x}^{[m-1]}, \\ \mathscr{A}\mathbf{x}^{m-1} - \lambda \mathbf{x}^{[m-1]} \geq \mathbf{0}, \\ \mathbf{x} \geq \mathbf{0}. \end{cases} \tag{5.23}$$

The non-zero vector $\mathbf{x}$ is called a **Pareto H-eigenvector** of $\mathscr{A}$ associated to λ.

(2) The real number μ is called a **Pareto Z-eigenvalue** of $\mathscr{A}$ if there is a non-zero vector $\mathbf{x} \in \mathbf{R}^n$ satisfying the following system

$$\begin{cases} \mathscr{A}\mathbf{x}^m = \mu(\mathbf{x}^\top \mathbf{x})^{\frac{m}{2}}, \\ \mathscr{A}\mathbf{x}^{m-1} - \mu(\mathbf{x}^\top \mathbf{x})^{\frac{m}{2}-1} \geq \mathbf{0}, \\ \mathbf{x} \geq \mathbf{0}. \end{cases} \tag{5.24}$$

The non-zero vector $\mathbf{x}$ is called a **Pareto Z-eigenvector** of $\mathscr{A}$ associated to μ.

By Definition 5.1, we can see that the Pareto H-eigenvalue is a Pareto-eigenvalue when we replace $\mathscr{A}$ and $\mathscr{B}$ in (5.1) by $-\mathscr{A}$ and $-\mathscr{I}$ respectively. Hence, we can see that the Pareto H-eigenvalue is a special case for the Pareto-eigenvalue.

It should be noted that it is an interesting work to compute the Pareto H-eigenvalue (Z-eigenvalue) for a higher order tensor. Since then, one can solve the following constrained optimization problems by means of the Pareto H-eigenvalue and Pareto Z-eigenvalue of the corresponding symmetric tensors:

$$\begin{cases} \min & \dfrac{1}{m}\mathscr{A}\mathbf{x}^m \\ \text{s.t.} & \mathbf{x}^\top \mathbf{x}^{[m-1]} = 1, \\ & \mathbf{x} \in \mathbf{R}^n_+, \end{cases} \qquad \begin{cases} \min & \dfrac{1}{m}\mathscr{A}\mathbf{x}^m \\ \text{s.t.} & \mathbf{x}^\top \mathbf{x} = 1, \\ & \mathbf{x} \in \mathbf{R}^n_+. \end{cases}$$

To move on, we now recall the definitions of H^+-eigenvalues and Z^+-eigenvalue for tensors. An H-eigenvalue λ of tensor $\mathscr{A}$ is called an $\mathbf{H^+}$**-eigenvalue** of $\mathscr{A}$, if its H-eigenvector $\mathbf{x} \in \mathbf{R}^n_+$; λ is called an $\mathbf{H^{++}}$**-eigenvalue** of $\mathscr{A}$, if its H-eigenvector $\mathbf{x} \in \mathbf{R}^n_{++}$. Similarly, we have the concepts of Z^+-eigenvalue and Z^{++}-eigenvalue. More details about these eigenvalues can be found in [228].

By the concepts above, we have the following conclusions.

Proposition 5.4 *Let $\mathscr{A} \in T_{m,n}$. If $\lambda \in \mathbf{R}$ is an H^+-eigenvalue (Z^+-eigenvalue) of $\mathscr{A}$, then λ is its Pareto H-eigenvalue (Z-eigenvalue, respectively).*

Similar to the proof of Theorem 5.5, one can obtain the following conclusions about Pareto H-eigenvalues and Pareto Z-eigenvalues of $\mathscr{A}$.

Proposition 5.5 *Let $\mathscr{A} \in T_{m,n}$. A real number λ is a Pareto H-eigenvalue of $\mathscr{A}$ if and only if there exist a nonempty subset $J \subseteq [n]$ and a vector $\omega \in \mathbf{R}^{|J|}$ such that*

$$\mathscr{A}^J \omega^{m-1} = \lambda \omega^{[m-1]}, \quad \omega \in \mathbf{R}^{|J|}_{++}, \tag{5.25}$$

$$\sum_{i_2,\ldots,i_m \in J} a_{ii_2\ldots i_m} \omega_{i_2} \omega_{i_3} \ldots \omega_{i_m} \geq \mathbf{0} \ \text{for} \ i \in [n] \backslash J. \tag{5.26}$$

In such a case, the vector $\mathbf{y} \in \mathbf{R}^n_+$ defined by

$$y_i = \begin{cases} \omega_i, & i \in J, \\ 0, & i \in [n]\backslash J, \end{cases} \tag{5.27}$$

is a Pareto H-eigenvalue of $\mathscr{A}$ associated to the real number λ.

Proposition 5.6 *Let $\mathscr{A} \in T_{m,n}$. Then $\mu \in \mathbf{R}$ is a Pareto Z-eigenvalue of $\mathscr{A}$ if and only if there exist a nonempty subset $J \subseteq [n]$ and a vector $\omega \in \mathbf{R}^{|J|}{}_{++}$ such that*

$$\mathscr{A}^J \omega^{m-1} = \mu(\omega^\top \omega)^{\frac{m-2}{2}} \omega, \tag{5.28}$$

$$\sum_{i_2,\ldots,i_m \in J} a_{ii_2\ldots i_m} \omega_{i_2} \omega_{i_3} \cdots \omega_{i_m} \geq \mathbf{0} \text{ for } i \in [n]\backslash J. \tag{5.29}$$

In such a case, the vector $\mathbf{y} \in \mathbf{R}^n_+$ defined by

$$y_i = \begin{cases} \omega_i, & i \in J, \\ 0, & i \in [n]\backslash J, \end{cases} \tag{5.30}$$

is a Pareto Z-eigenvector of $\mathscr{A}$ associated to the real number μ.

From Propositions 5.5 and 5.6, we can obtain the following corollary directly.

Corollary 5.3 *Let $\mathscr{A} \in T_{m,n}$. Then, the following results hold.*

(1) If $\lambda \in \mathbf{R}$ is a Pareto H-eigenvalue (Z-eigenvalue) of $\mathscr{A}$, then λ is an H^{++}-eigenvalue (Z^{++}-eigenvalue, respectively) for some principal subtensors of $\mathscr{A}$.
(2) If $\mathscr{A}$ is a diagonal tensor, then Pareto H-eigenvalues (Z-eigenvalues) of $\mathscr{A}$ coincide with its diagonal entries.
(3) A diagonal tensor with dimension n may have at most n distinct Pareto H-eigenvalues (Z-eigenvalues).

Similar to Theorem 5.3, we consider the constrained minimization problem

$$\gamma(\mathscr{A}) = \min\{\mathscr{A}\mathbf{x}^m \ : \ \mathbf{x} \geq \mathbf{0} \text{ and } \|\mathbf{x}\|_m = 1\}. \tag{5.31}$$

For a given tensor $\mathscr{A}$, let $\lambda_{min}(\mathscr{A})$ denote its smallest Pareto H-eigenvalue. Then we have the following result.

Proposition 5.7 *Let $\mathscr{A} \in S_{m,n}$. If $\mathscr{A}$ has Pareto H-eigenvalues, then $\gamma(\mathscr{A}) = \lambda_{min}(\mathscr{A})$.*

By Proposition 5.7, we know that $\gamma(\mathscr{A}) \geq \lambda(\mathscr{A})$ is always a Pareto H-eigenvalue for symmetric tensor $\mathscr{A}$. Then, it is clear the following result holds.

Proposition 5.8 *If $\mathscr{A} \in T_{m,n}$ is symmetric, then $\mathscr{A}$ has at least one Pareto H-eigenvalue*

$$\gamma(\mathscr{A}) = \min_{\mathbf{x} \geq 0, \|\mathbf{x}\|_m = 1} \mathscr{A}\mathbf{x}^m.$$

Similar to Proposition 5.7, we can obtain the following conclusions for the Pareto Z-eigenvalue of a symmetric tensor $\mathscr{A}$.

Proposition 5.9 *Let $\mathscr{A} \in S_{m,n}$. Then $\mathscr{A}$ has at least one Pareto Z-eigenvalue $\mu(\mathscr{A}) = \min_{\mathbf{x}\geq \mathbf{0}, \|\mathbf{x}\|_2=1} \mathscr{A}\mathbf{x}^m$. It further follows that $\mu(\mathscr{A})$ is the smallest Pareto Z-eigenvalue.*

At the end of this section, we will discuss the relationship between copositive tensors and Pareto H-eigenvalues or Pareto Z-eigenvalues. By Definition 4.5, we first give the following conclusions.

Lemma 5.5 *Let $\mathscr{A} \in S_{m,n}$. Then we have*

(1) $\mathscr{A}$ is copositive if and only if $\mathscr{A}\mathbf{x}^m \geq 0$ for all $\mathbf{x} \in \mathbf{R}^n_+$ with $\|\mathbf{x}\| = 1$;
(2) $\mathscr{A}$ is strictly copositive if and only if $\mathscr{A}\mathbf{x}^m > 0$ for all $\mathbf{x} \in \mathbf{R}^n_+$ with $\|\mathbf{x}\| = 1$.

Proof (1) The necessary condition can be easily proved by Definition 4.5. On the other hand, if $\|x\| = 0$, then it follows that $\mathbf{x} = \mathbf{0}$, and hence $\mathscr{A}\mathbf{x}^m = 0$. If $\|x\| > 0$, $\mathbf{x} \geq \mathbf{0}$, then let $\mathbf{y} = \frac{\mathbf{x}}{\|\mathbf{x}\|}$. It holds that $\|\mathbf{y}\| = 1$ and $\mathbf{x} = \|\mathbf{x}\|\mathbf{y}$. Hence, we obtain

$$\mathscr{A}\mathbf{x}^m = \mathscr{A}(\|\mathbf{x}\|\mathbf{y})^m = \|\mathbf{x}\|^m \mathscr{A}\mathbf{y}^m \geq 0,$$

which implies that $\mathscr{A}\mathbf{x}^m \geq 0$ for all $\mathbf{x} \in \mathbf{R}^n_+$.

Similarly, we can prove (2), and the desired results hold. □

According to Lemma 5.5 and Definition 5.1, we have the following conclusion.

Corollary 5.4 *Let $\mathscr{A} \in S_{m,n}$. Then*

(1) $\mathscr{A}$ is copositive (strictly copositive) if and only if all of its Pareto H-eigenvalues are nonnegative (positive, respectively).
(2) $\mathscr{A}$ is copositive (strictly copositive) if and only if all of its Pareto Z-eigenvalues are nonnegative (positive, respectively).

5.3 Computational Methods for Tensor Eigenvalue Complementarity Problems

In this section, we mainly study computational algorithms for tensor eigenvalue complementarity problem. It contains a kind of semismooth Newton method and a kind of implementable projection algorithm.

1 The Semismooth Newton Method

For the semismooth Newton method, some basic definitions and properties in nonsmooth analysis and nonlinear complementarity problems should be presented first. Suppose that $F : U \subseteq \mathbf{R}^{n_1} \to \mathbf{R}^{n_2}$ is a locally Lipschitz function, where U is nonempty and open. By Rademacher's Theorem, F is differentiable almost everywhere. Let $D_F \subseteq \mathbf{R}^{n_1}$ denote the set of points at which F is differentiable. For any $\mathbf{x} \in D_F$, we write $\nabla F(\mathbf{x})$ for the usual $n_2 \times n_1$ Jacobian matrix of partial derivatives. The B-subdifferential of F at $\mathbf{x} \in U$ is the set defined by

$$\partial_B F(\mathbf{x}) := \{V \in \mathbf{R}^{n_2 \times n_1} \,:\, \exists \{\mathbf{x}^k\} \subseteq D_F \text{ with } \mathbf{x}^k \to \mathbf{x}, \nabla F(\mathbf{x}^k) \to V\}.$$

The Clark's generalized Jacobian of F at $\mathbf{x}$ is the set defined by

$$\partial F(\mathbf{x}) = \mathrm{conv}(\partial_B F(\mathbf{x})),$$

where "conv" denotes the convex hull. In the case of $n_2 = 1$, $\partial F(\mathbf{x})$ is called the generalized gradient. Some fundamental properties about generalized Jacobian are given below.

Proposition 5.10 (Clark 1983) *Suppose that the function $F : U \subseteq \mathbf{R}^{n_1} \to \mathbf{R}^{n_2}$ is locally Lipschitz, where U is nonempty and open. Then for any $\mathbf{x} \in U$, we have*

(1) $\partial F(\mathbf{x})$ is a nonempty convex compact subset of $\mathbf{R}^{n_2 \times n_1}$;
(2) $\partial F(\mathbf{x}) = \partial_B F(\mathbf{x}) = \{\nabla F(\mathbf{x})\}$ if F is continuously differentiable at $\mathbf{x}$;
(3) $\partial F(\mathbf{x}) \subseteq \partial f^1(\mathbf{x}) \times \partial f^2(\mathbf{x}) \times \cdots \times \partial f^m(\mathbf{x})$, where $F(\mathbf{x}) = [f^1(\mathbf{x}), f^2(\mathbf{x}), \ldots, f^m(\mathbf{x})]$ and the latter denotes the set of all matrices whose ith row belongs to $\partial f^i(\mathbf{x})$ for each i.

The following definition is useful in the study of the semismooth Newton method.

Definition 5.2 Let $U \subseteq \mathbf{R}^{n_1}$ be nonempty and open.

(1) If F is locally Lipschitz at $\mathbf{x}$, and if the following limit exists for all $\mathbf{d} \in \mathbf{R}^n$:

$$\lim_{V \in \partial F(\mathbf{x}+t\tilde{\mathbf{d}}), \tilde{\mathbf{d}} \to \mathbf{d}, t \to 0} V\tilde{\mathbf{d}}$$

Then F is called semismooth at $\mathbf{x}$. If F is semismooth at all $\mathbf{x} \in U$, we call F semismooth on U.

(2) The function F is called strongly semismooth if it is semismooth and for any $\mathbf{x} \in U$ and $V \in \partial F(\mathbf{x} + \mathbf{d})$,

$$V\mathbf{d} - F'(\mathbf{x}, \mathbf{d}) = o(\|\mathbf{d}\|^2), \;\; \mathbf{d} \to \mathbf{0},$$

where $F'(\mathbf{x}, \mathbf{d})$ denotes the directional derivative of F at $\mathbf{x}$ in direction $\mathbf{d}$, i.e.,

$$F'(\mathbf{x}, \mathbf{d}) = \lim_{t \to 0} \frac{F(\mathbf{x} + t\mathbf{d}) - F(\mathbf{x})}{t}.$$

It is worth mentioning that if the function F is semismooth, the directional derivative $F'(\mathbf{x}, \mathbf{d})$ exists for all $\mathbf{d} \in \mathbf{R}^n$ and $F'(\mathbf{x}, \mathbf{d}) = \lim_{V \in \partial F(\mathbf{x}+t\tilde{\mathbf{d}}), \tilde{\mathbf{d}} \to \mathbf{d}, t \to 0} V\tilde{\mathbf{d}}$.

Furthermore, Mifflin in [195] showed that the composition of strongly semismooth functions is again strongly semismooth.

We have recalled the basic model for nonlinear complementarity problem (NCP) in Chap. 4. Moreover, many solution methods developed for NCP or related problems are based on the so-called NCP-functions. Here, a function $\phi : \mathbf{R}^2 \to \mathbf{R}$ is called an **NCP-function** if

$$\phi(a, b) = 0 \Leftrightarrow ab = 0, \ a \succeq 0, \ b \succeq 0.$$

Given an NCP-function ϕ, let us define $\Phi(\mathbf{x}) = [\phi(x_i, F_i(\mathbf{x}))]_{i=1}^n$. Thus, for a given function $F(\mathbf{x})$, $\mathbf{x} \in \mathbf{R}^n_+$ is a solution of NCP(F) if and only if $\Phi(\mathbf{x}) = \mathbf{0}$.

Now, for the sake of simplicity, some NCP-functions are listed below, which are widely used in nonlinear complementarity problems.

- The min function $\phi_{min}(a, b) := a - (a - b)_+$.
- The Fischer–Burmeister function $\phi_{FB}(a, b) := (a + b) - \sqrt{a^2 + b^2}$.
- The penalized Fischer–Burmeister function $\phi_\tau(a, b) := \tau\phi_{FB}(a, b) + (1 - \tau)a_+b_+$, where $\tau \in (0, 1)$.

Here, $t_+ = \max\{t, 0\}$ for $t \in \mathbf{R}$. For the three kinds of NCP-functions, their generalized gradients are given as follows.

Proposition 5.11 (Chen, Chen, Kanzow 2000) *Let $\phi_{min}(a, b), \phi_{FB}(a, b)$ and $\phi_\tau(a, b)$ be defined as above. Then*

(1) The generalized gradient $\partial\phi_{min}(a, b)$ is equal to the set of all (v_a, v_b) such that

$$(v_a, v_b) = \begin{cases} (1, \ 0) & \text{if } a < b, \\ (1 - v, \ v) & \text{if } a = b, \\ (0, \ 1) & \text{if } a > b, \end{cases}$$

where v is any scalar in the interval $[0, \ 1]$.

(2) The generalized gradient $\partial\phi_{FB}(a, b)$ is equal to the set of all (v_a, v_b) such that

$$(v_a, v_b) = \begin{cases} (1 - \frac{a}{\|(a,b)\|}, 1 - \frac{b}{\|(a,b)\|}) & \text{if } (a, b) \neq (0, 0), \\ (1 - \sigma, 1 - \eta) & \text{if } (a, b) = (0, 0), \end{cases}$$

where (σ, η) is any vector satisfying $\|(\sigma, \eta)\| \leq 1$.

(3) For any $\tau \in (0, 1)$, the generalized gradient $\partial\phi_\tau(a, b)$ is equal to the set of all (v_a, v_b) such that

$$(v_a, v_b) = \begin{cases} \tau\left(1 - \frac{a}{\|(a,b)\|}, 1 - \frac{b}{\|(a,b)\|}\right) + (1 - \tau)(b_+\partial a_+, a_+\partial b_+) & \text{if } (a, b) \neq (0, 0), \\ \tau(1 - \sigma, 1 - \eta) & \text{if } (a, b) = (0, 0), \end{cases}$$

where (σ, η) is any vector satisfying $\|(\sigma, \eta)\| \leq 1$ and

$$\partial x_+ = \begin{cases} 0 & \text{if } x < 0, \\ [0, 1] & \text{if } x = 0, \\ 1 & \text{if } x > 0. \end{cases}$$

Recall the TEiCP defined as in (5.1), $\sigma(\mathscr{A}, \mathscr{B})$ denotes the solution set for the TEiCP with respect to eigenvalues $\mathscr{A}, \mathscr{B} \in T_{m,n}$. Notice that if $(\lambda, \mathbf{x}) \in \sigma(\mathscr{A}, \mathscr{B})$, then $(\lambda, s\mathbf{x}) \in \sigma(\mathscr{A}, \mathscr{B})$ for any $s > 0$. Therefore, we can always consider solutions of TEiCP such that $\|\mathbf{x}\|_2 = 1$.

By introducing a new variable $t \in \mathbf{R}$, denote

$$F(\mathbf{x}, t) := (t^2\mathscr{B} - \mathscr{A})\mathbf{x}^{m-1}, \text{ for } \mathbf{x} \in \mathbf{R}^n, \ t \in \mathbf{R}. \tag{5.32}$$

Then we obtain a parameterized nonlinear complementarity problem, i.e.,

$$\mathbf{x} \geq \mathbf{0}, \ \ F(\mathbf{x}, t) \geq \mathbf{0}, \ \ \mathbf{x}^\top F(\mathbf{x}, t) = 0, \tag{5.33}$$

with the constraint $\mathbf{x}^\top\mathbf{x} = 1$. Define $H : \mathbf{R}^{n+1} \to \mathbf{R}^{n+1}$ by

$$H(\mathbf{z}) = \begin{pmatrix} \Phi(\mathbf{z}) \\ \mathbf{x}^\top\mathbf{x} - 1 \end{pmatrix}, \tag{5.34}$$

where $\Phi : \mathbf{R}^{n+1} \to \mathbf{R}^n$ is given by

$$\Phi(\mathbf{z}) = [\phi(x_i, F_i(\mathbf{x}, t))]_{i=1}^n, \tag{5.35}$$

and $\phi(a, b)$ is an NCP-function.

By the discussion above, it is clear that the following conclusions hold.

Proposition 5.12 *Let $\mathscr{A}, \mathscr{B} \in S_{m,n}$. Then, the following results hold.*

(1) If $(\lambda, \mathbf{x})$ is a solution of TEiCP (5.1) with $\|\mathbf{x}\| = 1$, then $H(\mathbf{z}) = 0$ with $\mathbf{z} = (\mathbf{x}, \pm\sqrt{|\lambda|}) \in \mathbf{R}^{n+1}$.
(2) If $H(\mathbf{z}) = 0$ with $\mathbf{z} = (\mathbf{x}, t) \in \mathbf{R}^{n+1}$ and $t \neq 0$, then $(t^2, \mathbf{x})$ is a solution of (5.1) with $\|\mathbf{x}\| = 1$.

Since there are three kinds of NCP-functions listed above, let $H_{min}(\mathbf{z})$, $H_{FB}(\mathbf{z})$ and $H_\tau(\mathbf{z})$ be defined as (5.34) with respect to ϕ_{min}, ϕ_{FB} and ϕ_τ, respectively. Then we will show that these functions are all strongly semismooth. Before that, the following lemma is needed [257].

Lemma 5.6 *Suppose H and F are defined by (5.34) and (5.32). If ∇F is locally Lipschitz continuous around a point $\mathbf{x} \in \mathbf{R}^n$, then H is strongly semismooth at $\mathbf{x}$.*

Lemma 5.7 *Suppose $\Phi_{min}(\mathbf{z})$, $\Phi_{FB}(\mathbf{z})$ and $\Phi_\tau(\mathbf{z})$ are defined by (5.35). Then the following results hold.*

(1) The functions $\Phi_{min}(\mathbf{z})$, $\Phi_{FB}(\mathbf{z})$ and $\Phi_\tau(\mathbf{z})$ are strongly semismooth.
(2) The functions $H_{min}(\mathbf{z})$, $H_{FB}(\mathbf{z})$ and $H_\tau(\mathbf{z})$ are strongly semismooth.

Proof (1) For any $\mathbf{z} = (\mathbf{x}, t) \in \mathbf{R}^{n+1}$, the functions $F(\mathbf{z}) = (t^2\mathscr{B} - \mathscr{A})\mathbf{x}^{m-1}$ is continuously differentiable and its Jacobian $\nabla F(\mathbf{z})$ is locally Lipschitz continuous.

By Lemma 5.6 and the fact that the composition of strongly semismooth functions is again strongly semismooth, we know that $\Phi_{FB}(\mathbf{z})$, $\Phi_\lambda(\mathbf{z})$ and $\Phi_{min}(\mathbf{z})$ is strongly semismooth, and the desired results hold.

(2) It is clear that $H_{n+1}(\mathbf{z}) = \mathbf{x}^\top\mathbf{x} - 1$ is continuously differentiable. By (1), the functions $\Phi_{min}(\mathbf{z})$, $\Phi_{FB}(\mathbf{z})$ and $\Phi_\tau(\mathbf{z})$ are strongly semismooth. It follows that $H_{min}(\mathbf{z})$, $H_{FB}(\mathbf{z})$ and $H_\tau(\mathbf{z})$ are strongly semismooth since all their components are strongly semismooth. □

In [60], Clark presented results about the Jacobian chain rule, upon which we have the following result.

Proposition 5.13 *Let $F : \mathbf{R}^{n_1+n_2} \to \mathbf{R}^{n_1}$ be continuous differentiable. Suppose $\phi : \mathbf{R}^2 \to \mathbf{R}$ is locally Lipschitz. Define $\Phi(\mathbf{z}) = [\phi(x_i, F_i(\mathbf{z}))]_{i=1}^{n_1}$, $\mathbf{z} = (\mathbf{x}, \mathbf{y}) \in \mathbf{R}^{n_1+n_2}$. Then we have*

$$\partial\Phi(\mathbf{z}) \subseteq (D_a(\mathbf{z}), \mathbf{0}_{n_1\times n_2}) + D_b(\mathbf{z})\nabla F(\mathbf{z}),$$

where $D_a(\mathbf{z}) = \operatorname{diag}\{a_i(\mathbf{z})\}$ and $D_b(\mathbf{z}) = \operatorname{diag}\{b_i(z)\}$ are diagonal matrices in $\mathbf{R}^{n_1\times n_1}$ with entries $(a_i(\mathbf{z}), b_i(\mathbf{z})) \in \partial\phi(x_i, F_i(\mathbf{z}))$.

Let $F(\mathbf{z})$ be defined in (5.32) with $\mathscr{A}, \mathscr{B}$ being symmetric tensors. By a direct computation, the Jacobian of F is given by

$$\nabla F(\mathbf{z}) = \left[(m-1)(t^2\mathscr{B} - \mathscr{A})\mathbf{x}^{m-2}, 2t\mathscr{B}\mathbf{x}^{m-1}\right] \in \mathbf{R}^{n\times(n+1)}.$$

Here, for a given tensor $\mathscr{T} = (t_{i_1i_2\ldots i_m}) \in T_{m,n}$ and a vector $\mathbf{x} \in \mathbf{R}^n$, let $\mathscr{T}\mathbf{x}^{m-2} \in \mathbf{R}^{n\times n}$ be a matrix such that

$$(\mathscr{T}\mathbf{x}^{m-2})_{ij} = \sum_{i_3,\ldots,i_m=1}^{n} t_{iji_3\ldots i_m}x_{i_3}\ldots x_{i_m}.$$

By Propositions 5.11 and 5.13, one can obtain the overestimation of $\partial\Phi_{FB}(\mathbf{z})$ and $\partial\Phi_\tau$, respectively. To move on, an algorithm for obtaining the element of $\partial\Phi_\tau(\mathbf{z})$ for any $\mathbf{z} \in \mathbf{R}^{n+1}$, $\tau \in (0, 1]$ is given. Note that $\partial\Phi_1(\mathbf{z}) = \partial\Phi_{FB}(\mathbf{z})$.

Algorithm 1

Step 0. Given $\tau \in (0, 1]$, $\mathbf{z} = (\mathbf{x}, t) \in \mathbf{R}^{n+1}$ and let V_i be the ith row of a matrix $V \in \mathbf{R}^{n\times(n+1)}$.

Step 1. Set $S_1 = \{i \in [n] : x_i = 0, F_i(\mathbf{x}, t) = 0\}$, $S_2 = \{i \in [n] : x_i = 0,$

$F_i(\mathbf{x}, t) > 0\}$, $S_3 = \{i \in [n] : x_i > 0, F_i(\mathbf{x}, t) = 0\}$,

$S_4 = \{i \in [n] : x_i > 0, F_i(\mathbf{x}, t) > 0\}$

Step 2. Let $\mathbf{c} \in \mathbf{R}^n$ such that $c_i = 1$ for $i \in S_1 \cup S_2 \cup S_3$ and 0 otherwise.

Step 3. For $i \in S_1$, set

$$V_i = \tau\left(1 + \frac{c_i}{\|(c_i, \nabla_{\mathbf{x}} F_i(\mathbf{z})^\top \mathbf{c})\|}\right)(\mathbf{1}_i^\top, 0) + \tau\left(1 + \frac{\nabla_{\mathbf{x}} F_i(\mathbf{z})^\top \mathbf{c}}{\|(c_i, \nabla_{\mathbf{x}} F_i(\mathbf{z})^\top \mathbf{c})\|}\right)\nabla_{\mathbf{x}} F_i(\mathbf{z})^\top.$$

Step 4. For $i \in S_3$, set $V_i = \begin{cases} (\tau + (1-\tau)x_i)\nabla F_i(\mathbf{z})^\top & \text{if } \nabla_{\mathbf{x}} F_i(\mathbf{z})^\top \mathbf{c} < 0, \\ \tau \nabla F_i(\mathbf{z})^\top & \textit{otherwise.} \end{cases}$

Step 5. For $i \in S_4$, set

$$V_i = \left[\tau\left(1 - \frac{x_i}{\|(x_i, F_i(\mathbf{z}))\|}\right) + (1-\tau)F_i(\mathbf{z})\right](\mathbf{1}_i^\top, 0) + \left[\tau\left(1 - \frac{F_i(\mathbf{z})}{\|(x_i, F_i(\mathbf{z}))\|}\right) + (1-\tau)x_i\right]\nabla F_i(\mathbf{z})^\top.$$

Step 6. For $i \notin S_1 \cup S_3 \cup S_4$, set

$$V_i = \tau\left(1 - \frac{x_i}{\|(x_i, F_i(\mathbf{z}))\|}\right)(\mathbf{1}_i^\top, 0) + \tau\left(1 - \frac{F_i(\mathbf{z})}{\|(x_i, F_i(\mathbf{z}))\|}\right)\nabla F_i(\mathbf{z})^\top.$$

From Algorithm 1, we have the following conclusion.

Theorem 5.8 *Suppose $V \in \mathbf{R}^{n\times(n+1)}$ is generated by Algorithm 1 with $\tau \in (0, 1)$. Let $\mathbf{z} = (\mathbf{x}, t) \in \mathbf{R}^{n+1}$ and $\mathbf{c}$ be defined as in Algorithm 1. If $\nabla_{\mathbf{x}} F_i(\mathbf{z})^\top \mathbf{c} \neq 0$ for all $i \in S_3$. Then $G = \begin{pmatrix} V \\ \bar{\mathbf{x}} \end{pmatrix}$, $\bar{\mathbf{x}} = (2\mathbf{x}^\top, 0) \in \mathbf{R}^{n+1}$ is an element of $\partial_B H_\tau(\mathbf{z})$.*

Proof First of all, it is clear that $H_\tau(\mathbf{z})$ is differentiable everywhere except on the set

$$\Omega := \{\mathbf{z} = (\mathbf{x}, t) \in \mathbf{R}^{n+1} \ : \ x_i \geq 0,\ F_i(\mathbf{z}) \geq 0,\ x_i F_i(\mathbf{z}) \geq 0 \text{ for some } i \in [n]\}.$$

Then, we prove that there is a sequence $\{\mathbf{z}^{(k)}\}_{k=1}^{\infty} \subseteq \mathbf{R}^{n+1}\backslash\Omega$ such that $\nabla H(\mathbf{z}^{(k)})$ tends to the matrix G. Then the conclusion follows immediately by the definition of B-subdifferential.

It is clear the result holds if $\mathbf{z} \notin \Omega$, i.e., $S_1 \cup S_2 \cup S_3 = \emptyset$. Hence, let $\mathbf{z} \in \Omega$, i.e., $S_1 \cup S_2 \cup S_3 \neq \emptyset$ in the following analysis. Denote $\mathbf{z}^{(k)} = \mathbf{z} - \frac{1}{k}(\mathbf{c}^\top, 0)$. Then we know that $z_i^{(k)} < 0$, $i \in S_1 \cup S_2$. For $i \in S_1 \cup S_3$, by Taylor expansion, it holds that

$$F_i(\mathbf{z}^{(k)}) = F_i(\mathbf{z}) + \nabla F_i(\xi^{(k)})^\top(\mathbf{z}^{(k)} - \mathbf{z}) = -\frac{1}{k}\nabla_{\mathbf{x}} F_i(\xi^{(k)})^\top \mathbf{c}, \tag{5.36}$$

where $\xi^{(k)} \to \mathbf{z}$ as $k \to \infty$. Since $\nabla_{\mathbf{x}} F_i(\mathbf{z})^\top \mathbf{c} \neq 0$ for all $i \in S_3$, by continuity, we have that for all $i \in S_3$, $F_i(\mathbf{z}^{(k)}) \neq 0$ when k is large enough. Hence, there exists $N > 0$ such that $H_\tau(\mathbf{z}^k)$ is differentiable for all $k > N$.

For $i \in [n+1]$, let $\nabla H(\mathbf{z}^{(k)})_i$ be the ith row of $\nabla H(\mathbf{z}^{(k)})$. If $i \notin S_1 \cup S_2 \cup S_3$ or $i = n+1$, by continuity, it is obvious that $\nabla H(\mathbf{z}^{(k)})_i$ tends to the ith row of G. For $i \in S_1 \cup S_2$, we have

$$\nabla H(\mathbf{z}^{(k)})_i = \tau \left(1 - \frac{x_i^{(k)}}{\|(x_i^{(k)}, F_i(\mathbf{z}^{(k)}))\|}\right)(\mathbf{1}_i^\top, 0) + \tau \left(1 - \frac{F_i(\mathbf{z}^{(k)})}{\|(x_i^{(k)}, F_i(\mathbf{z}^{(k)}))\|}\right)\nabla F_i(\mathbf{z}^{(k)})^\top.$$

For $i \in S_3$, it is not difficult to show that

$$\nabla H(\mathbf{z}^{(k)})_i = \begin{cases} \left[\tau \left(1 - \frac{x_i^{(k)}}{\|(x_i^{(k)}, F_i(\mathbf{z}^{(k)}))\|}\right) + (1-\tau)F_i(\mathbf{z}^{(k)})\right](\mathbf{1}_i^\top, 0) & \\ + \left[\tau \left(1 - \frac{F_i(\mathbf{z}^{(k)})}{\|(x_i^{(k)}, F_i(\mathbf{z}^{(k)}))\|}\right) + (1-\tau)x_i^{(k)}\right]\nabla F_i(\mathbf{z}^{(k)})^\top & \text{if } \nabla F_i(\mathbf{z}^{(k)})^\top \mathbf{c} < 0, \\ \tau \left(1 - \frac{x_i^{(k)}}{\|(x_i^{(k)}, F_i(\mathbf{z}^{(k)}))\|}\right)(\mathbf{1}_i^\top, 0) + & \\ \tau \left(1 - \frac{F_i(\mathbf{z}^{(k)})}{\|(x_i^{(k)}, F_i(\mathbf{z}^{(k)}))\|}\right)\nabla F_i(\mathbf{z}^{(k)})^\top & \text{if } \nabla_{\mathbf{x}} F_i(\mathbf{z}^{(k)})^\top \mathbf{c} > 0. \end{cases}$$

Note that for $i \in S_1$, by substituting (5.36), it follows that

$$\lim_{k\to\infty} \frac{x_i^{(k)}}{\|(x_i^{(k)}, F_i(\mathbf{z}^{(k)}))\|} = \lim_{k\to\infty} \frac{-1/k}{\sqrt{(1/k)^2 + F_i(\mathbf{z}^{(k)})^2}} = \frac{1}{\sqrt{1 + (\nabla_{\mathbf{x}} F_i(\mathbf{z})^\top \mathbf{c})^2}}.$$

Similarly, $\lim_{k\to\infty} \frac{F_i(\mathbf{z}^{(k)})}{\|(x_i^{(k)}, F_i(\mathbf{z}^{(k)}))\|} = \frac{-\nabla_{\mathbf{x}} F_i(\mathbf{z})^\top \mathbf{c}}{\sqrt{1+(\nabla_{\mathbf{x}} F_i(\mathbf{z})^\top \mathbf{c})^2}}$. It follows that for $i \in S_1 \cup S_2 \cup S_3$, $\nabla H(\mathbf{z}^{(k)})_i$ tends to the ith row of the matrix G. □

With a similar way to the proof of Theorem 5.8, we know that $H_{FB}(\mathbf{z})$ is differentiable everywhere except on the set

$$\{\mathbf{z} = (\mathbf{x}, t) \in \mathbf{R}^{n+1} \;:\; x_i = 0,\, F_i(\mathbf{z}) = 0 \text{ for some } i \in [n]\}.$$

Hence $H_{FB}(\mathbf{z})_i$ is differentiable for all $i \in S_3$, and the following conclusion can be proved similarly.

Theorem 5.9 *Let $\mathbf{z} = (\mathbf{x}, t) \in \mathbf{R}^{n+1}$ be given and let $V \in \mathbf{R}^{n\times(n+1)}$ be the matrix generated by Algorithm 1 with $\tau = 1$. Then $G = \begin{pmatrix} V \\ \bar{\mathbf{x}} \end{pmatrix}$, $\bar{\mathbf{x}} = (2\mathbf{x}^\top, 0) \in \mathbf{R}^{n+1}$ is an element of $\partial_B H_{FB}(\mathbf{z})$.*

Now we present a damped Newton method for tensor eigenvalue complementarity problem. Here, we only take the NCP-functions ϕ_{FB} and ϕ_τ. Moreover, a natural metric function of $H(\mathbf{z})$ is defined by

$$\Psi(\mathbf{z}) = \frac{1}{2} H(\mathbf{z})^\top H(\mathbf{z}). \tag{5.37}$$

Algorithm 2 Damped semismooth Newton method for TEiCP
Step 0. Given $\varepsilon > 0$, $\rho > 0$, $p > 2$, $\beta \in (0, \frac{1}{2})$ and choose $\mathbf{z}^0 = (\mathbf{x}^0, t^0) \in \mathbf{R}^{n+1}$. Set $k = 0$.
Step 1. If $\|H(\mathbf{z}^{(k)})\| \leq \varepsilon$, stop. Otherwise, go to Step 2.
Step 2. Compute an element $G_k = \begin{pmatrix} V_k \\ \bar{\mathbf{x}}^{(k)} \end{pmatrix}$ of $\partial \Phi(\mathbf{z}^k)$, where $V_k \in \mathbf{R}^{n\times(n+1)}$ and $\bar{\mathbf{x}}^{(k)} = (2(\mathbf{x}^k)^\top, 0)$ are generalized by Algorithm 1. Find the solution $\mathbf{d}^{(k)}$ of the system

$$G_k \mathbf{d} = -H(\mathbf{z}^{(k)}). \tag{5.38}$$

If G_k in (5.38) is ill-conditioned or if the condition $\nabla\Psi(\mathbf{z}^{(k)})^\top \mathbf{d}^{(k)} \leq -\rho\|\mathbf{d}^{(k)}\|^p$ is not satisfied, set $\mathbf{d}^{(k)} = -\nabla\Psi(\mathbf{z}^{(k)})$.
Step 3. Find the smallest $i_k = 0, 1, \ldots,$ such that $\alpha_k = 2^{-i_k}$ and

$$\Psi(\mathbf{z}^{(k)} + \alpha_k \mathbf{d}^{(k)}) \leq \Psi(\mathbf{z}^{(k)}) + \beta\alpha_k \nabla\Psi(\mathbf{z}^{(k)})^\top \mathbf{d}^{(k)}.$$

Set $\mathbf{z}^{(k+1)} = \mathbf{z}^{(k)} + \alpha_k \mathbf{d}^{(k)}$.
Step 4. Set $k = k + 1$ and go back to Step 1.

The convergence of Algorithm 2 is guaranteed by the following theorem, where one can find the proof in references [76, 216, 220].

Theorem 5.10 *Let $\sigma(\mathcal{A}, \mathcal{B})$ be nonempty. Suppose $\{\mathbf{z}^{(k)}\} \subseteq \mathbf{R}^{n+1}$ is generated by Algorithm 2. Assume that $H(\mathbf{z}^{(k)}) \neq 0$ for all k. Then the following conclusions hold:*

(1) $\|H(\mathbf{z}^{(k+1)})\| \leq \|H(\mathbf{z}^{(k)})\|$;
(2) Each accumulation point $\mathbf{z}^*$ of the sequence $\{\mathbf{z}^{(k)}\}$ is a stationary point of Ψ. Furthermore, if $H(\mathbf{z})$ is strongly BD-regular $\mathbf{z}^*$, then $\mathbf{z}^*$ is a zero of $H(\mathbf{z})$ if and only if $\{\mathbf{z}^{(k)}\}$ converges to $\mathbf{z}^*$ quadratically and α_k eventually becomes 1. On the other hand, $\mathbf{z}^*$ is not a zero of $H(\mathbf{z})$ if and only if $\{\mathbf{z}^{(k)}\}$ diverges or $\lim_{k\to\infty} \alpha_k = 0$.

2 The Scaling and Projection Algorithm

To end this section, we introduce an implementable algorithm for solving the TGEiCP (5.2). Hence, it is can be used for TEiCP (5.1) too.

As we all know that the general NCP can be transformed into a system of equations. Thus, it is can be solved by the semismooth and smoothing Newton methods. Unfortunately, due to the high-dimensional structure of tensors, TGEiCP is more complicated than the classical eigenvalue complementarity problem for the matrix case. So it is difficult to implement for making such a second-order algorithm.

Now, we study a so-called **scaling-and-projection algorithm** (SPA) to solve TGEiCP (5.2). Here, $\mathcal{B}$ is assumed to be strictly K-positive, i.e., $\mathcal{B}\mathbf{x}^m > 0$ for any $\mathbf{x} \in \mathbb{K}\backslash\{\mathbf{0}\}$, where $\mathbb{K}$ is a closed convex cone.

Algorithm 3 (SPA)
Step 1. Take any starting point $\mathbf{u}^{(0)} \in \mathbb{K}\backslash\{\mathbf{0}\}$, and define $\mathbf{x}^{(0)} = \frac{\mathbf{u}^0}{\sqrt[m]{\mathcal{B}(\mathbf{u}^0)^m}}$.

Step 2. for $k = 0, 1, 2, \ldots$ do
Step 3. One has a current point $\mathbf{x}^{(k)} \in \mathbb{K}\backslash\{\mathbf{0}\}$. Compute $\lambda_k = \frac{\mathscr{A}(\mathbf{x}^{(k)})^m}{\mathscr{B}(\mathbf{x}^{(k)})^m}$ and $\mathbf{y}^{(k)} = \mathscr{A}(\mathbf{x}^{(k)})^{m-1} - \lambda_k \mathscr{B}(\mathbf{x}^{(k)})^{m-1}$.
Step 4. If $\|\mathbf{y}^{(k)}\| = 0$, then stop. Otherwise, let $s_k := \|\mathbf{y}^{(k)}\|$, and compute $\mathbf{u}^{(k)} = \Pi_k[\mathbf{x}^{(k)} + s_k \mathbf{y}^{(k)}]$ and $\mathbf{x}^{k+1} = \frac{\mathbf{u}^{(k)}}{\sqrt[m]{\mathscr{B}(\mathbf{u}^{(k)})^m}}$
Step 5. end for

It is clear that Step 3 always ensures that

$$(\mathbf{x}^{(k)})^\top \mathbf{y}^{(k)}) = 0.$$

Therefore, $\mathbf{y}^{(k)} \in \mathbb{K}^*$, which means that $(\mathbf{x}^{(k)}, \mathbf{y}^{(k)})$ is a solution of TGEiCP (5.2). However, for the sake of simplicity, $\|\mathbf{y}^{(k)}\| = 0$ can be used as the stoping condition in algorithmic framework instead of $\mathbf{y}^{(k)} \in \mathbb{K}^*$.

The convergence of Algorithm 2 is listed below, and the proof is skipped. The interested reader is referred to [73] for a similar proof.

Theorem 5.11 *Let the sequence* $\{\mathbf{x}^{(k)}\}$ *be generated by Algorithm 2 and further satisfy* $\mathscr{B}(\mathbf{x}^{(k)})^m = 1$. *Assume convergence of* $\{\mathbf{x}^{(k)}\}$ *toward some limit that one denotes by* $\bar{\mathbf{x}}$. *Then*

$$\lim_{k\to\infty} \lambda_k = \bar{\lambda} := \frac{\mathscr{A}\bar{\mathbf{x}}^m}{\mathscr{B}\bar{\mathbf{x}}^m}, \lim_{k\to\infty} \mathbf{y}^{(k)} = \bar{\mathbf{y}} := \mathscr{A}\bar{\mathbf{x}}^{m-1} - \bar{\lambda}\mathscr{B}\bar{\mathbf{x}}^{m-1},$$

and $(\bar{\lambda}, \bar{\mathbf{x}})$ *is a solution of (5.2).*

5.4 A Unified Framework of Tensor Higher-Degree Eigenvalue Complementarity Problems

In this section, we mainly study eigenvalue complementarity problems in a higher degree case, i.e. the Tensor Higher-Degree Eigenvalue Complementarity Problem (THDEiCP). Suppose $\mathbb{K} \subseteq \mathbf{R}^n$ is a closed convex cone. Let $\mathbb{K}^*$ denote the dual cone of $\mathbb{K}$. The THDEiCP is to find a scalar $\lambda \in \mathbf{R}^n$ and a vector $\mathbf{x} \in \mathbf{R}^n\backslash\{\mathbf{0}\}$ such that

$$\mathbf{x} \in \mathbb{K}, (\lambda^m \mathscr{A} + \lambda\mathscr{B} + \mathscr{C})\mathbf{x}^{m-1} \in \mathbb{K}^*, \text{ and } (\lambda^m \mathscr{A} + \lambda\mathscr{B} + \mathscr{C})\mathbf{x}^m = 0, \quad (5.39)$$

where $\mathscr{A}, \mathscr{B}, \mathscr{C} \in T_{m,n}$. Denote $F_{m,n} = T_{m,n} \times T_{m,n} \times T_{m,n}$ based on (5.39). Then we present the following definition.

Definition 5.3 Suppose $\mathscr{Q} = (\mathscr{A}, \mathscr{B}, \mathscr{C}) \in F_{m,n}$.

(1) If (5.39) holds, then $\lambda \in \mathbf{R}$ and $\mathbf{x} \in \mathbf{R}^n\backslash\{\mathbf{0}\}$ are called an **m-degree K-eigenvalue** and an **m-degree K-eigenvector** respectively for the tensor triplet $\mathscr{Q}$. In this

case, $(\lambda, \mathbf{x})$ is called an **m-degree K-eigenpair**, and let $\sigma(\mathcal{Q}, \mathbb{K})$ be the set of all m-degree K-eigenvalues:

$$\sigma(\mathcal{Q}, \mathbb{K}) := \{\lambda \in \mathbf{R} \mid \exists\, \mathbf{x} \in \mathbf{R}^n \backslash \{\mathbf{0}\} \text{ such that } \mathbf{x} \in \mathbb{K}, \\ (\lambda^m \mathscr{A} + \lambda \mathscr{B} + \mathscr{C})\mathbf{x}^{m-1} \in \mathbb{K}^*, \text{ and } (\lambda^m \mathscr{A} + \lambda \mathscr{B} + \mathscr{C})\mathbf{x}^m = 0\}. \tag{5.40}$$

(2) If $\mathbb{K} = \mathbf{R}^n_+$, the m-degree K-eigenvalue (m-degree K-eigenvector) can be called the **m-degree Pareto-eigenvalue** (**m-degree Pareto-eigenvector**) of $\mathcal{Q}$.
(3) If $\lambda \in \sigma(\mathcal{Q}, \mathbb{K})$ ($\mathbb{K} = \mathbf{R}^n_+$), then λ can be called a **strict m-degree K-eigenvalue** (strict m-degree Pareto-eigenvalue) of $\mathcal{Q}$ when $\mathbf{x} \in \text{int}\mathbb{K}$ ($\mathbf{x} \in \mathbf{R}^n_{++}$).

To present the main result, we review some basic concepts. The next definition was first given in [248].

Definition 5.4 Suppose $\mathbb{K} \subseteq \mathbf{R}^n$ is a closed convex cone and $\mathscr{G} \in T_{m,n}$. If $\mathscr{G}\mathbf{x}^m \geq 0$ (resp. > 0) for any $\mathbf{x} \in \mathbb{K}\backslash\{\mathbf{0}\}$, then $\mathscr{G}$ is called a (resp. strictly) K-positive tensor.

When $\mathbb{K} = \mathbf{R}^n_+$, the (strictly) K-positive tensor $\mathscr{G}$ reduces to the (strictly) copositive tensor.

Let $\mathscr{C}(\mathbf{R}^n)$ be the set of nonzero closed convex cones in $\mathbf{R}^n$. We introduce the K-regular tensor.

Definition 5.5 Suppose $\mathcal{Q} = (\mathscr{A}, \mathscr{B}, \mathscr{C}) \in F_{m,n}$ and $\mathbb{K} \in \mathscr{C}(\mathbf{R}^n)$. Then $\mathcal{Q}$ is **K-regular** if the leading tensor $\mathscr{A}$ satisfies $\mathscr{A}\mathbf{x}^m \neq 0$ for any $\mathbf{x} \in \mathbb{K}\backslash\{\mathbf{0}\}$. Also, if $\mathcal{Q}$ is K-regular, either $\mathscr{A}$ in $\mathcal{Q}$ or$-\mathscr{A}$ is K-positive.

Let $\Sigma := \{(\mathcal{Q}, \mathbb{K}, \lambda) \in F_{m,n} \times \mathscr{C}(\mathbf{R}^n) \times \mathbf{R} \mid \lambda \in \sigma(\mathcal{Q}, \mathbb{K})\}$. Then several basic properties are listed below.

Proposition 5.14 (Ling, He, Qi 2016)

(1) The set Σ is closed in the product space $F_{m,n} \times \mathscr{C}(\mathbf{R}^n) \times \mathbf{R}$. Particularly, $\sigma(\mathcal{Q}, \mathbb{K})$ is a closed subset of $\mathbf{R}$, for any $(\mathcal{Q}, \mathbb{K}) \in F_{m,n} \times \mathscr{C}(\mathbf{R}^n)$;
(2) Assume $(\mathcal{Q}, \mathbb{K}) \in F_{m,n} \times \mathscr{C}(\mathbf{R}^n)$. Then $\sigma(\mathcal{Q}, \mathbb{K})$ is compact if $\mathcal{Q}$ is K-regular.

We now give the following conclusion, which is useful to obtain the estimation on the numbers of m-degree Pareto-eigenvalues.

Proposition 5.15 *Suppose $\mathcal{Q} = (\mathscr{A}, \mathscr{B}, \mathscr{C}) \in F_{m,n}$. Then $\lambda \in \mathbf{R}$ is an m-degree Pareto-eigenvalue of $\mathcal{Q}$ if and only if there exist a nonempty subset $J \subseteq [n]$ and a vector $\mathbf{w} \in \mathbf{R}^{|J|}_{++}$ satisfying*

$$(\lambda^m \mathscr{A}^J + \lambda \mathscr{B}^J + \mathscr{C}^J)\mathbf{w}^{m-1} = 0 \tag{5.41}$$

and

$$\sum_{i_2,\ldots,i_m \in J} (\lambda^m a_{ii_2\ldots i_m} + \lambda b_{ii_2\ldots i_m} + c_{ii_2\ldots i_m}) w_{i_2} \ldots w_{i_m} \geq 0, \quad \forall\, i \in [n]\backslash J.$$

Furthermore, the vector $\mathbf{x} \in \mathbf{R}^n_+$ *with its elements being*

$$x_i = \begin{cases} w_i, & i \in J, \\ 0, & i \in [n]\backslash J \end{cases}$$

is an m*-degree Pareto-eigenvector of* $\mathcal{Q}$*, associated to the* m*-degree Pareto-eigenvalue* λ*.*

Proof The results can be proved with a similar process of Theorem 5.5. □

Under some assumptions, the following theorem shows an upper bound of the number of m-degree Pareto-eigenvalues.

Theorem 5.12 *Let* $\mathcal{Q} = (\mathcal{A}, \mathcal{B}, \mathcal{C}) \in F_{m,n}$*. Suppose* $\mathcal{Q}$ *is* $\mathbf{R}^n_+$*-regular. Then* $\mathcal{Q}$ *has at most* $\tau_{m,n} := nm^n$ m*-degree Pareto-eigenvalues.*

Proof For any $k \in [n-1]$, there are $\binom{n}{n-k}$ mth order $(n-k)$-dimensional principal subtensor triplets, each of which can have at most $m(n-k)(m-1)^{n-k-1}$ strict m-degree Pareto-eigenvalues. Furthermore, by Proposition 5.15, we have

$$\tau_{m,n} = \sum_{k=0}^{n-1} \binom{n}{n-k} m(n-k)(m-1)^{n-k-1} = nm^n$$

and the desired result holds. □

Similar to the proof of Theorem 5.7, we have the following conclusion with the cone $\mathbb{K}$ is finitely generated as follows.

Theorem 5.13 *Suppose* $\mathcal{Q} = (\mathcal{A}, \mathcal{B}, \mathcal{C}) \in F_{m,n}$*. Let* $\mathbb{K}$ *be defined as in (5.18). If* $\mathcal{Q}$ *is K-regular, then* $\mathcal{Q}$ *has at most* $\tau_{m,p} := pm^p$ m*-degree K-eigenvalues.*

When $\mathcal{A}$ and $\mathcal{B}$ are symmetric tensors, $\mathcal{C} = -\mathcal{I}$ and $\mathbb{K} := \mathbf{R}^n_+$, we transform the THDEiCP into the following weakly coupled homogeneous polynomial optimization problem.

$$\begin{cases} \max & \varphi_0(\mathbf{u}, \mathbf{v}) := m(m-1)^{\frac{1}{m}-1} \mathbf{v}^\top \mathbf{u}^{[m-1]} - \mathcal{B}\mathbf{u}^m \\ \text{s.t.} & \mathcal{A}\mathbf{u}^m + \mathbf{v}^\top \mathbf{v}^{[m-1]} = 1, \\ & \mathbf{u} \geq \mathbf{0}, \quad \mathbf{v} \geq \mathbf{0}. \end{cases} \tag{5.42}$$

Define $\phi_0(\mathbf{u}, \mathbf{v}) = \mathcal{A}\mathbf{u}^m + \mathbf{v}^\top \mathbf{v}^{[m-1]} - 1$. We obtain

$$\nabla_{\mathbf{u}}\varphi_0(\mathbf{u},\mathbf{v}) = m(m-1)^{\frac{1}{m}}\mathrm{diag}(\mathbf{v})\mathbf{u}^{[m-2]} - m\mathscr{B}\mathbf{u}^{m-1} \tag{5.43a}$$
$$\nabla_{\mathbf{v}}\varphi_0(\mathbf{u},\mathbf{v}) = m(m-1)^{\frac{1}{m}-1}\mathbf{u}^{[m-1]} \tag{5.43b}$$
$$\nabla_{\mathbf{u}}\phi_0(\mathbf{u},\mathbf{v}) = m\mathscr{A}\mathbf{u}^{m-1} \tag{5.43c}$$
$$\nabla_{\mathbf{v}}\phi_0(\mathbf{u},\mathbf{v}) = m\mathbf{v}^{m-1}. \tag{5.43d}$$

The next theorem shows the relationship between (5.39) and (5.42).

Theorem 5.14 *Let $\mathscr{Q} = (\mathscr{A}, \mathscr{B}, -\mathscr{I}) \in F_{m,n}$ with symmetric $\mathscr{A}$ and $\mathscr{B}$. Assume $(\bar{\mathbf{u}}, \bar{\mathbf{v}})$ with $\bar{\mathbf{u}} \neq \mathbf{0}$ is a stationary point of (5.42). Then $(\bar{\lambda}, \bar{\mathbf{u}})$ is an m-degree Pareto-eigenpair of $\mathscr{Q}$, in which $\bar{\lambda} = (\varphi_0(\bar{\mathbf{u}}, \bar{\mathbf{v}}))^{\frac{1}{m-1}}$.*

Proof Since $(\bar{\mathbf{u}}, \bar{\mathbf{v}})$ is a stationary point (5.42), by (5.43), we know that there are three multipliers $\bar{\alpha}, \bar{\beta} \in \mathbf{R}^n$ and $\bar{\gamma} \in \mathbf{R}$ satisfying

$$m\mathscr{B}\bar{\mathbf{u}}^{m-1} - m(m-1)^{\frac{1}{m}}\mathrm{diag}(\bar{\mathbf{v}})\bar{\mathbf{u}}^{[m-2]} = \bar{\alpha} + \bar{\gamma}m\mathscr{A}\bar{\mathbf{u}}^{m-1}, \tag{5.44a}$$
$$-m(m-1)^{\frac{1}{m}-1}\bar{\mathbf{u}}^{[m-1]} = \bar{\beta} + \bar{\gamma}m\bar{v}^{[m-1]}, \tag{5.44b}$$
$$\bar{\alpha} \geq 0,\ \ \bar{\mathbf{u}} \geq 0,\ \ \bar{\alpha}^{\top}\mathbf{u} = 0, \tag{5.44c}$$
$$\bar{\beta} \geq 0,\ \ \bar{\mathbf{v}} \geq 0,\ \ \bar{\beta}^{\top}\mathbf{v} = 0, \tag{5.44d}$$
$$\mathscr{A}\bar{\mathbf{u}}^m + \bar{\mathbf{v}}^{\top}\bar{\mathbf{v}}^{m-1} = 1. \tag{5.44e}$$

By (5.44b), we have

$$-(m-1)^{\frac{1}{m}-1}\bar{\mathbf{u}}^{[m-1]} - \bar{\gamma}\bar{v}^{[m-1]} = \frac{\bar{\beta}}{m}. \tag{5.45}$$

We now prove $\bar{\beta} = 0$. If $\bar{\beta} \neq 0$, then there is $i_0 \in [n]$ such that $\bar{\beta}_{i_0} > 0$, which implies $\bar{\mathbf{v}}_{i_0} = 0$ from (5.44d). Then we have $-(m-1)^{\frac{1}{m}-1}\bar{\mathbf{u}}_{i_0}^{[m-1]} - \frac{\bar{\beta}_{i_0}}{m} > 0$, which is impossible. Based on (5.45) we obtain

$$-(m-1)^{\frac{1}{m}-1}\bar{\mathbf{u}}^{[m-1]} = \bar{\gamma}\bar{v}^{[m-1]}. \tag{5.46}$$

By (5.46) and the facts that $\bar{\mathbf{u}} \neq \mathbf{0}$, we have $\bar{\gamma} < 0$ and

$$\bar{\mathbf{v}} = (-\bar{\gamma})^{-\frac{1}{m-1}}(m-1)^{-\frac{1}{m}}\bar{\mathbf{u}}. \tag{5.47}$$

By (5.44a) and (5.47), it holds that

$$\mathscr{B}\bar{\mathbf{u}}^{m-1} - (-\bar{\gamma})^{-\frac{1}{m-1}}\bar{\mathbf{u}}^{[m-1]} - \bar{\gamma}\mathscr{A}\bar{\mathbf{u}}^{m-1} = \frac{\bar{\alpha}}{m} \geq \mathbf{0},$$

which implies that

$$(-\bar{\gamma})^{\frac{m}{m-1}}\mathscr{A}\bar{\mathbf{u}}^{m-1} + (-\bar{\gamma})^{\frac{1}{m-1}}\mathscr{B}\bar{\mathbf{u}}^{m-1} - \bar{\mathbf{u}}^{[m-1]} \geq \mathbf{0}. \tag{5.48}$$

Furthermore, by (5.47), (5.44a) and (5.44c), it is clear that

$$\bar{\mathbf{u}}^\top\left((-\bar{\gamma})^{\frac{m}{m-1}}\mathscr{A}\bar{\mathbf{u}}^{m-1}+(-\bar{\gamma})^{\frac{1}{m-1}}\mathscr{B}\bar{\mathbf{u}}^{m-1}-\bar{\mathbf{u}}^{[m-1]}\right)=0. \tag{5.49}$$

On the other hand, from (5.44a) and (5.44c), we have

$$m\mathscr{B}\bar{\mathbf{u}}^m-m\bar{\gamma}\mathscr{A}\bar{\mathbf{u}}^m-m(m-1)^{\frac{1}{m}}\bar{\mathbf{v}}^\top\bar{\mathbf{u}}^{[m-1]}=0.$$

Thus, by (5.44e) and (5.46), we obtain

$$\begin{aligned}-m\varphi_0(\bar{\mathbf{u}},\bar{\mathbf{v}})&=m\bar{\gamma}\mathscr{A}\bar{\mathbf{u}}^m-m(m-1)^{\frac{1}{m}-1}\bar{\mathbf{v}}^\top\bar{\mathbf{u}}^{[m-1]}\\&=m\bar{\gamma}.\end{aligned} \tag{5.50}$$

It follows from (5.50) that $\varphi_0(\bar{\mathbf{u}},\bar{\mathbf{v}})=-\bar{\gamma}>0$, and (5.48) and (5.49) imply that $\bar{\mathbf{u}}$ is an m-degree Pareto-eigenvector of (5.42) associated to the m-degree Pareto-eigenvalue $\bar{\lambda}$. □

If the leading tensor is copositive, we have the following conclusion.

Theorem 5.15 *Suppose $\mathscr{Q}=(\mathscr{A},\mathscr{B},-\mathscr{I})\in F_{m,n}$. Let $\mathscr{A},\mathscr{B}\in T_{m,n}$ be symmetric tensors, and $(\bar{\lambda},\bar{\mathbf{x}})$ be an m-degree Pareto-eigenpair of $\mathscr{Q}$. If $\bar{\lambda}>0$ and $\mathscr{A}$ is copositive, the vector $(\bar{\mathbf{u}},\bar{\mathbf{v}})$ being defined as*

$$(\bar{\mathbf{u}},\bar{\mathbf{v}})=\frac{1}{(\mathscr{A}\bar{\mathbf{x}}^m+\bar{\mathbf{y}}^\top\bar{\mathbf{y}}^{[m-1]})^{\frac{1}{m}}}(\bar{\mathbf{x}},\bar{\mathbf{y}}) \tag{5.51}$$

with $\bar{\mathbf{y}}=(m-1)^{-\frac{1}{m}}(\bar{\lambda})^{-1}\bar{\mathbf{x}}$, is a stationary point of (5.42).

Proof If $\mathscr{A}$ is copositive, then $\mathscr{A}\bar{\mathbf{x}}^m+\bar{\mathbf{y}}^\top\bar{\mathbf{y}}^{[m-1]}>0$ for any $\bar{\mathbf{y}}\in\mathbf{R}^n_+\backslash\{\mathbf{0}\}$. Furthermore, it is not difficult to see that $(\bar{\mathbf{u}},\bar{\mathbf{v}})$ in (5.51) is feasible for (5.42). Let

$$\bar{\alpha}=\frac{m}{\bar{\lambda}(\mathscr{A}\bar{\mathbf{x}}^m+\bar{\mathbf{y}}^\top\bar{\mathbf{y}}^{[m-1]})^{\frac{m-1}{m}}}(\bar{\lambda}^m\mathscr{A}\bar{\mathbf{x}}^{m-1}+\bar{\lambda}\mathscr{B}\bar{\mathbf{x}}^{m-1}-\bar{\mathbf{x}}^{[m-1]}),$$

$\bar{\beta}=0$ and $\bar{\gamma}=-\bar{\lambda}^{m-1}$. Since $(\bar{\lambda},\bar{\mathbf{x}})$ satisfies (5.39) and $\bar{\lambda}>0$, we have $\bar{\alpha}\geq 0$. By a direct computation, we obtain $\bar{\alpha}^\top\bar{\mathbf{u}}=0$ and $\bar{\beta}^\top\bar{\mathbf{v}}=0$. Then we have(5.44a) and (5.44c) hold from (5.51), and $(\bar{\mathbf{u}},\bar{\mathbf{v}})$ is a stationary point of (5.42). □

When $\mathscr{A}$ is strictly copositive, the feasible set of (5.42) is compact, and (5.42) has global optimization solution. Suppose $\varphi_0^{\max}=\varphi(\bar{\mathbf{w}})$, where $\bar{\mathbf{w}}:=(\bar{\mathbf{u}}^\top,\bar{\mathbf{v}}^\top)^\top$. Denote

$$\bar{\mathbf{d}}=((\mathbf{1}_{I(\bar{\mathbf{w}})})^\top,-\bar{t}(\bar{\mathbf{w}}_{I^c(\bar{\mathbf{w}})})^\top)^\top\ \text{ with }\ \bar{t}=\sum_{i\in I(\bar{\mathbf{w}})}(\mathscr{A}\bar{\mathbf{u}}^{m-1})_i,$$

where $I(\bar{\mathbf{w}})=\{i\in[2n]\mid\bar{\mathbf{w}}_i=0\}$ and $I^c(\bar{\mathbf{w}})=[n]\backslash I(\bar{\mathbf{w}})$. Since ϕ_0 is homogeneous and $\phi_0(\bar{\mathbf{w}})=1$, it holds that $\bar{\mathbf{w}}^\top\nabla\phi_0(\bar{\mathbf{w}})=m\phi_0(\bar{\mathbf{w}})=m\neq 0$, which indicates that

$\nabla\phi_0(\bar{\mathbf{w}}) \neq 0$ and hence $\{\nabla\phi_0(\bar{\mathbf{w}})\}$ is linearly independent. Furthermore, it is easy to show that

$$\bar{\mathbf{d}}^\top \nabla\phi_0(\bar{\mathbf{w}}) = m\left(\sum_{i\in I(\bar{\mathbf{w}})} (\mathscr{A}\bar{\mathbf{u}}^{m-1})_i - \bar{t}\right) = 0$$

and $\bar{\mathbf{d}}^\top \nabla\phi_i(\bar{\mathbf{w}}) = 1 > 0$, for any $i \in I(\bar{\mathbf{w}})$. Therefore the Mangasarian Fromovitz constraint qualification (MFCQ) holds at $\bar{\mathbf{w}}$, and $\bar{\mathbf{w}}$ is a stationary point of (5.42). Also, it holds that $\bar{\mathbf{u}} \neq 0$. Take

$$\mathbf{u}_t = t\mathbf{1}, \mathbf{v}_t = \left(\frac{1-\bar{a}t^m}{n}\right)^{1/m}\mathbf{1}$$

in which $\bar{a} = \sum_{i_1,\ldots,i_m=1}^n a_{i_1\ldots i_m}$. Then $\mathbf{w}_t = (\mathbf{u}_t, \mathbf{v}_t)$ is a feasible solution of (5.42) with the corresponding objective value being

$$\varphi_0(\mathbf{u}_t, \mathbf{v}_t) = t^m\left(nm(m-1)^{\frac{1}{m}-1}\left(\frac{1-\bar{a}t^m}{n}\right)^{1/m} - \bar{b}t\right),$$

where $\bar{b} = \sum_{i_1,\ldots,i_m=1}^n b_{i_1\ldots i_m}$. When $t > 0$ is small enough, we have $\varphi_0(\mathbf{u}_t, \mathbf{v}_t) > 0$. Hence $\varphi_0(\bar{\mathbf{u}}, \bar{\mathbf{v}}) > 0$ because $(\bar{\mathbf{u}}, \bar{\mathbf{v}})$ is an optimal point of (5.42). Consequently, the inequality $\bar{\mathbf{u}} \neq 0$ is gained.

Furthermore, it follows from Theorem 5.14 that $(\bar{\lambda}, \bar{\mathbf{u}})$ with $\bar{\lambda} = (\varphi_0(\bar{\mathbf{u}}, \bar{\mathbf{v}}))^{\frac{1}{m-1}}$ is a solution of (5.39). Then (5.39) has at least a positive m-degree Pareto-eigenvalue. Hence, we have $\varphi_0^{\max} \leq \lambda_{\max}^{m-1}$, where

$$\lambda_{\max} = \max\{\lambda \in \mathbf{R} \mid \exists\, \mathbf{x} \in \mathbf{R}^n, (\lambda, \mathbf{x}) \text{ is an } m\text{-degree Pareto-eigenpair of } \mathscr{Q}\}.$$

By the analysis above, we have the following theorem.

Theorem 5.16 *Suppose* $\mathscr{Q} = (\mathscr{A}, \mathscr{B}, -\mathscr{I}) \in F_{m,n}$. *Let* $\mathscr{A}, \mathscr{B} \in T_{m,n}$ *be symmetric tensors. If* $\mathscr{A}$ *is strictly copositive, it holds that* $\lambda_{\max}^{m-1} = \varphi_0^{\max}$.

Proof Suppose $(\bar{\lambda}, \bar{\mathbf{x}})$ is an m-degree Pareto-eigenpair of $\mathscr{Q}$ with $\bar{\lambda} > 0$. Without loss of generality, let $\bar{\mathbf{x}} \in \mathbf{R}_+^n$ satisfy $\mathbf{1}^\top\bar{\mathbf{x}} = 1$. Since $\mathscr{A}$ is strictly copositive, we have $\mathscr{A}\bar{\mathbf{x}}^m > 0$. Let

$$\bar{\mathbf{y}} = \frac{(m-1)^{-\frac{1}{m}}}{\bar{\lambda}}\bar{\mathbf{x}} \text{ and } (\bar{\mathbf{u}}, \bar{\mathbf{v}}) = \frac{(\bar{\mathbf{x}}, \bar{\mathbf{y}})}{(\mathscr{A}\bar{\mathbf{x}}^m + \bar{\mathbf{y}}^\top\bar{\mathbf{y}}^{[m-1]})^{\frac{1}{m}}}. \tag{5.52}$$

Since $(\bar{\mathbf{u}}, \bar{\mathbf{v}}) \in \mathbf{R}_+^n \times \mathbf{R}_+^n$ and

$$\mathscr{A}\bar{\mathbf{u}}^m = \frac{1}{\mathscr{A}\bar{\mathbf{x}}^m + \bar{\mathbf{y}}^\top\bar{\mathbf{y}}^{[m-1]}}\mathscr{A}\bar{\mathbf{x}}^m, \quad \bar{\mathbf{v}}^\top\bar{\mathbf{v}}^{[m-1]} = \frac{1}{\mathscr{A}\bar{\mathbf{x}}^m + \bar{\mathbf{y}}^\top\bar{\mathbf{y}}^{[m-1]}}\bar{\mathbf{y}}^\top\bar{\mathbf{y}}^{[m-1]},$$

we have $\mathscr{A}\bar{\mathbf{u}}^m + \bar{\mathbf{v}}^\top \bar{\mathbf{v}}^{[m-1]} = 1$. Therefore, $(\bar{\mathbf{u}}, \bar{\mathbf{v}})$ is feasible for (5.42) and $\varphi_0(\bar{\mathbf{u}}, \bar{\mathbf{v}}) \le \varphi_0^{\max}$.

On the other hand, by Definition 5.3, it holds that $\bar{\lambda}^m \mathscr{A}\bar{\mathbf{x}}^m + \bar{\lambda}\mathscr{B}\bar{\mathbf{x}}^m - \bar{\mathbf{x}}\bar{\mathbf{x}}^{m-1} = 0$. Substituting $(\bar{\mathbf{u}}, \bar{\mathbf{v}})$ into $\varphi_0(\mathbf{u}, \mathbf{v})$, it follows from (5.52) that

$$\begin{aligned}\varphi_0(\bar{\mathbf{u}}, \bar{\mathbf{v}}) =& m(m-1)^{\frac{1}{m}-1}\bar{\mathbf{v}}^\top \bar{\mathbf{u}}^{[m-1]} - \mathscr{B}\bar{\mathbf{u}}^m \\ =& \frac{m\bar{\mathbf{x}}^\top \bar{\mathbf{x}}^{[m-1]} - (m-1)\bar{\lambda}\mathscr{B}\bar{\mathbf{x}}^m}{(m-1)\bar{\lambda}^m \mathscr{A}\bar{\mathbf{x}}^m + \bar{\mathbf{x}}^\top \bar{\mathbf{x}}^{[m-1]}}\bar{\lambda}^{m-1} \\ =& \bar{\lambda}^{m-1}.\end{aligned}$$

Thus we have $\bar{\lambda}^{m-1} \le \varphi_0^{\max}$, and $\lambda_{\max}^{m-1} \le \varphi_0^{\max}$. The proof is completed. □

In the following analysis, we study the existence of solutions of THDEiCP. For the symmetric case $\mathscr{A}, \mathscr{B} \in S_{m,n}$, if $\mathscr{A}$ is strictly copositive, then $\mathscr{Q} = (\mathscr{A}, \mathscr{B}, -\mathscr{I}) \in F_{m,n}$ has at least an m-degree Pareto-eigenpair. However, the assumption of $\mathscr{A}$ being strictly copositive is not enough for nonsymmetric cases. The following theorem give a sufficient condition for the existence of solutions of THDEiCP under the condition that no symmetric assumption is posed on $\mathscr{A}, \mathscr{B}$.

Theorem 5.17 *Let $\mathscr{Q} = (\mathscr{A}, \mathscr{B}, -\mathscr{I}) \in F_{m,n}$, where $\mathscr{A}$ be strictly copositive. If $\mathscr{A} = (a_{i_1 i_2 \ldots i_m})$ and $\mathscr{B} = (b_{i_1 i_2 \ldots i_m})$ satisfy that*

$$(a_{ii\ldots i} + 1 - m)(m-1)^{\frac{1}{m}-1} - b_{ii\ldots i} > 0, \quad \forall\, i \in [n], \tag{5.53}$$

there exists at least an m-degree Pareto-eigenpair of $\mathscr{Q}$.

Proof Let $S = \{(\mathbf{x}, \mathbf{y}) \in \mathbf{R}_+^n \times \mathbf{R}_+^n \; : \; \mathbf{x} \ge \mathbf{0}, \mathbf{1}^\top \mathbf{x} = 1, \mathbf{y} \ge \mathbf{0}\}$ and $S_0 = \{(\mathbf{x}, \mathbf{y}) \in S \; : \; \|\mathbf{y}\| \le 1\}$. Then $S_0 \subseteq S$ is compact and convex. Define $F : S \times S \to \mathbf{R}$ by

$$\begin{aligned}F(\mathbf{x}, \mathbf{y}; \mathbf{z}, \mathbf{w}) = \mathbf{z}^\top & \left(-\mathscr{B}\mathbf{x}^{m-1} - f(\mathbf{x}, \mathbf{y})\mathscr{A}\mathbf{x}^{m-1} + (m-1)^{\frac{1}{m}}\mathrm{diag}(\mathbf{y})\mathbf{x}^{[m-2]}\right) \\ \mathbf{w}^\top & \left((m-1)^{\frac{1}{m}-1}\mathbf{x}^{[m-1]} - f(\mathbf{x}, \mathbf{y})\mathbf{y}^{[m-1]}\right),\end{aligned} \tag{5.54}$$

where

$$f(\mathbf{x}, \mathbf{y}) = \frac{m(m-1)^{\frac{1}{m}-1}\mathbf{y}^\top \mathbf{x}^{[m-1]} - \mathscr{B}\mathbf{x}^m}{\mathscr{A}\mathbf{x}^m + \mathbf{y}^\top \mathbf{y}^{[m-1]}}.$$

Apparently, $F(\mathbf{x}, \mathbf{y}; \mathbf{x}, \mathbf{y}) = 0$ for any $(\mathbf{x}, \mathbf{y}) \in S$. Furthermore, $F(\cdot, \cdot; \mathbf{z}, \mathbf{w})$ is lower-semicontinuous on S for any fixed $(\mathbf{z}, \mathbf{w}) \in S$, while $F(\mathbf{x}, \mathbf{y}; \cdot, \cdot)$ is concave on S for any fixed $(\mathbf{x}, \mathbf{y}) \in S$. It can be seen from (5.53) that

$$\Omega := \{(\mathbf{z}, \mathbf{w}) \in S \mid F(\mathbf{x}, \mathbf{y}; \mathbf{z}, \mathbf{w}) \le 0, \forall\, (\mathbf{x}, \mathbf{y}) \in S_0\}$$

is compact. Otherwise, there is a sequence $\{(\mathbf{z}^{(k)}, \mathbf{w}^{(k)})\}$ of Ω such that

$$\|(\mathbf{z}^{(k)}, \mathbf{w}^{(k)})\| \to +\infty \text{ as } k \to +\infty.$$

Since $\{\mathbf{z}^{(k)}\}$ is bounded, we claim $\|\mathbf{w}^{(k)}\| \to +\infty$ without loss of generality. Consequently, there exists $i_0 \in [n]$ such that $\mathbf{w}_{i_0}^{(k)} \to +\infty$. Take $\mathbf{x}^{(k)} = \mathbf{y}^{(k)} = \mathbf{1}_{i_0} \in S_0$ with $\mathbf{1}_{i_0}$ being the i_0th unit vector in $\mathbf{R}^n$. We obtain

$$F(\mathbf{x}^{(k)}, \mathbf{y}^{(k)}; \mathbf{z}^{(k)}, \mathbf{w}^{(k)}) = \theta_k + \frac{(a_{i_0 i_0 \ldots i_0} + 1 - m)(m-1)^{\frac{1}{m}-1} - b_{i_0 i_0 \ldots i_0}}{a_{i_0 i_0 \ldots i_0} + 1} \mathbf{w}_{i_0}^{(k)},$$

where

$$\theta_k = (\mathbf{z}^{(k)})^\top \left(-\mathscr{B}(\mathbf{x}^{(k)})^{m-1} - f(\mathbf{x}^{(k)}, \mathbf{y}^{(k)}) \mathscr{A}(\mathbf{x}^{(k)})^{m-1} + (m-1)^{\frac{1}{m}} \operatorname{diag}(\mathbf{y}^{(k)})(\mathbf{x}^{(k)})^{[m-2]} \right).$$

and $\{\theta_k\}$ is bounded. By condition (5.53), it holds that $F(\mathbf{x}^{(k)}, \mathbf{y}^{(k)}; \mathbf{z}^{(k)}, \mathbf{w}^{(k)}) > 0$ when k is large enough, which contradicts the fact $(\mathbf{z}^{(k)}, \mathbf{w}^{(k)}) \in \Omega$. Therefore, there exists $(\bar{\mathbf{x}}, \bar{\mathbf{y}}) \in S$ such that

$$F(\bar{\mathbf{x}}, \bar{\mathbf{y}}; \mathbf{z}, \mathbf{w}) \le 0, \ \forall\, (\mathbf{z}, \mathbf{w}) \in S. \tag{5.55}$$

Let $\mathbf{w} = 0$ in (5.55). For any $\mathbf{z} \in D := \{\mathbf{z} \in \mathbf{R}^n \mid \mathbf{z} \ge \mathbf{0}, \mathbf{1}^\top \mathbf{z} = 1\}$, we have

$$F(\bar{\mathbf{x}}, \bar{\mathbf{y}}; \mathbf{z}, \mathbf{0}) = \mathbf{z}^\top \left(-\mathscr{B}\bar{\mathbf{x}}^{m-1} - \bar{f} \mathscr{A}\bar{\mathbf{x}}^{m-1} + (m-1)^{\frac{1}{m}} \operatorname{diag}(\bar{\mathbf{y}})\bar{\mathbf{x}}^{[m-2]} \right) \le 0,$$

where $\bar{f} = f(\bar{\mathbf{x}}, \bar{\mathbf{y}})$. Since D is a basis of $\mathbf{R}^n_+$, we obtain

$$\mathscr{B}\bar{\mathbf{x}}^{m-1} + \bar{f}\mathscr{A}\bar{\mathbf{x}}^{m-1} - (m-1)^{\frac{1}{m}} \operatorname{diag}(\bar{\mathbf{y}})\bar{\mathbf{x}}^{[m-2]} \ge \mathbf{0}. \tag{5.56}$$

For any $\mathbf{w} \in \mathbf{R}^n_+$, it is obvious that $(\bar{\mathbf{x}}, \bar{\mathbf{y}} + \mathbf{w}) \in S$. Thus, we have

$$\begin{aligned} F(\bar{\mathbf{x}}, \bar{\mathbf{y}}; \bar{\mathbf{x}}, \bar{\mathbf{y}} + \mathbf{w}) &= F(\bar{\mathbf{x}}, \bar{\mathbf{y}}; \bar{\mathbf{x}}, \bar{\mathbf{y}}) + \mathbf{w}^\top \left((m-1)^{\frac{1}{m}-1} \bar{\mathbf{x}}^{[m-1]} - \bar{f}\bar{\mathbf{y}}^{[m-1]}, \right) \\ &= \mathbf{w}^\top \left((m-1)^{\frac{1}{m}-1} \bar{\mathbf{x}}^{[m-1]} - \bar{f}\bar{\mathbf{y}}^{[m-1]} \right) \\ &\le 0, \end{aligned}$$

in which the second equality holds because $F(\bar{\mathbf{x}}, \bar{\mathbf{y}}; \bar{\mathbf{x}}, \bar{\mathbf{y}}) = 0$. Then we have

$$\bar{f}\bar{\mathbf{y}}^{[m-1]} - (m-1)^{\frac{1}{m}-1} \bar{\mathbf{x}}^{[m-1]} \ge \mathbf{0}. \tag{5.57}$$

Due to $\bar{\mathbf{x}} \ge \mathbf{0}$ and $\bar{\mathbf{x}} \ne \mathbf{0}$, there exists $i_0 \in [n]$ such that $\bar{\mathbf{x}}_{i_0} > 0$. Immediately we get

$$\bar{f}\bar{y}_{i_0}^{m-1} \ge (m-1)^{\frac{1}{m}-1} \bar{x}_{i_0}^{m-1} > 0,$$

which means $\bar{f} > 0$. Let $I(\bar{\mathbf{y}}) := \{i \in [n] \mid \bar{y}_i = 0\}$. Since $I(\bar{\mathbf{y}}) \subset [n]$, it follows from (5.57) that $\bar{x}_i = 0$ for every $i \in I(\bar{\mathbf{y}})$. Therefore

$$\bar{f}\bar{y}_i^{m-1} = (m-1)^{\frac{1}{m}-1}\bar{x}_i^{m-1}, \quad \forall\ i \in I(\bar{\mathbf{y}}). \tag{5.58}$$

For any $i \in [n]\backslash I(\bar{\mathbf{y}})$ and any real number t with $|t|$ being small enough, by taking $\mathbf{w} = t\mathbf{1}_i$ with $t \in \mathbf{R}$ we have

$$(\bar{\mathbf{x}}, \bar{\mathbf{y}} + \mathbf{w}) \in S$$

from $\bar{y}_i > 0$. It follows from (5.55) that

$$F(\bar{\mathbf{x}}, \bar{\mathbf{y}}; \bar{\mathbf{x}}, \bar{\mathbf{y}} + \mathbf{w}) \leq 0,$$

which indicates that

$$\left((m-1)^{\frac{1}{m}-1}\bar{x}_i^{m-1} - \bar{f}\bar{y}_i^{m-1}\right)t \leq 0$$

for any $i \in [n]\backslash I[\bar{y}]$ and any real number t with $|t|$ being small enough. Then, we have

$$(m-1)^{\frac{1}{m}-1}\bar{x}_i^{m-1} = \bar{f}\bar{y}_i^{m-1}, \quad \forall\, i \in [n]\backslash I[\bar{y}]. \tag{5.59}$$

From (5.58) and (5.59), we obtain

$$\bar{f}\bar{\mathbf{y}}^{[m-1]} = (m-1)^{\frac{1}{m}-1}\bar{\mathbf{x}}^{[m-1]}, \tag{5.60}$$

which is equivalent with

$$\bar{f}^{\frac{1}{m-1}}\bar{\mathbf{y}} = (m-1)^{-\frac{1}{m}}\bar{\mathbf{x}}. \tag{5.61}$$

By (5.56) and (5.60), we have

$$\begin{aligned} 0 \leq &\mathscr{B}\bar{\mathbf{x}}^{m-1} + \bar{f}\mathscr{A}\bar{\mathbf{x}}^{m-1} - \bar{f}^{-\frac{1}{m-1}}\bar{\mathbf{x}}^{[m-1]} \\ =&\bar{f}^{-\frac{1}{m-1}}\left\{\bar{f}^{\frac{1}{m-1}}\mathscr{B}\bar{\mathbf{x}}^{m-1} + \bar{f}^{\frac{m}{m-1}}\mathscr{A}\bar{\mathbf{x}}^{m-1} - \bar{\mathbf{x}}^{[m-1]}\right\}, \end{aligned}$$

which means

$$\bar{\lambda}^m\mathscr{A}\bar{\mathbf{x}}^{m-1} + \bar{\lambda}\mathscr{B}\bar{\mathbf{x}}^{m-1} - \bar{\mathbf{x}}^{[m-1]} \geq \mathbf{0} \tag{5.62}$$

with $\bar{\lambda} = \bar{f}^{\frac{1}{m-1}}$.

Next, we prove that

$$\bar{\mathbf{x}}^\top\left(\bar{\lambda}^m\mathscr{A}\bar{\mathbf{x}}^{m-1} + \bar{\lambda}\mathscr{B}\bar{\mathbf{x}}^{m-1} - \bar{\mathbf{x}}^{[m-1]}\right) = 0.$$

We only need to verify that

$$\bar{f}\mathscr{A}\bar{\mathbf{x}}^m + \mathscr{B}\bar{\mathbf{x}}^m - \bar{f}^{-\frac{1}{m-1}}\sum_{i=1}^n \bar{x}_i^m = 0.$$

Due to $F(\bar{\mathbf{x}}, \bar{\mathbf{y}}; \bar{\mathbf{x}}, \bar{\mathbf{y}}) = 0$ and

$$\bar{f}\mathscr{A}\bar{\mathbf{x}}^m + \mathscr{B}\bar{\mathbf{x}}^m = m(m-1)^{\frac{1}{m}-1}\bar{\mathbf{y}}^\top\bar{\mathbf{x}}^{[m-1]} - \bar{f}\sum_{i=1}^n \bar{y}_i^m,$$

what we should prove is that

$$m(m-1)^{\frac{1}{m}-1}\bar{\mathbf{y}}^\top\bar{\mathbf{x}}^{[m-1]} - \bar{f}\sum_{i=1}^n \bar{y}_i^m - \bar{f}^{-\frac{1}{m-1}}\sum_{i=1}^n \bar{x}_i^m = 0. \tag{5.63}$$

The left-hand side of (5.63) is equal to

$$\begin{aligned}
& m(m-1)^{\frac{1}{m}-1}\bar{\mathbf{y}}^\top\bar{\mathbf{x}}^{[m-1]} - \bar{f}\bar{\mathbf{y}}^\top\bar{\mathbf{y}}^{m-1} - \bar{f}^{-\frac{1}{m-1}}\sum_{i=1}^n \bar{x}_i^m \\
&= m(m-1)^{\frac{1}{m}-1}\bar{\mathbf{y}}^\top\bar{\mathbf{x}}^{[m-1]} - (m-1)^{\frac{1}{m}-1}\bar{\mathbf{y}}^\top\bar{\mathbf{x}}^{[m-1]} - \bar{f}^{-\frac{1}{m-1}}\sum_{i=1}^n \bar{x}_i^m \\
&= (m-1)^{\frac{1}{m}-1}\bar{\mathbf{y}}^\top\bar{\mathbf{x}}^{[m-1]} - \bar{f}^{-\frac{1}{m-1}}\bar{\mathbf{x}}^\top\bar{\mathbf{x}}^{[m-1]} \\
&= (m-1)^{\frac{1}{m}-1}\bar{\mathbf{y}}^\top\bar{\mathbf{x}}^{[m-1]} - \bar{f}^{-\frac{1}{m-1}}(m-1)^{\frac{1}{m}}\bar{f}^{\frac{1}{m-1}}\bar{\mathbf{y}}^\top\bar{\mathbf{x}}^{[m-1]} \\
&= 0,
\end{aligned}$$

where the first equality comes from (5.60), while the second equality is based on (5.61). Hence, $(\bar{\lambda}, \bar{\mathbf{x}})$ is an m-degree Pareto-eigenpair of $\mathscr{Q}$, and the desired results hold. □

For the problem of (5.42), we rewrite it as a standard minimization problem:

$$\begin{cases} \min & \mathscr{B}\mathbf{u}^m + a\mathbf{v}^\top\mathbf{u}^{[m-1]}, \\ \text{s.t.} & \mathscr{A}\mathbf{u}^m + \mathscr{I}\mathbf{v}^m = 1, \\ & \mathbf{u} \geq \mathbf{0},\ \mathbf{v} \geq \mathbf{0}, \end{cases} \tag{5.64}$$

where a is a constant being $a := -m(m-1)^{\frac{1}{m}-1}$. The powerful semismooth and smoothing Newton methods is feasible for this problem. Here we introduce a first-order structure-exploiting algorithm, which is much easier to be implemented than the second-order type methods. By fully exploiting the weakly coupled structure of the optimization model of THDEICP, an implementable splitting algorithm based on the augmented Lagrangian method of (5.64) is proposed [180].

The augmented Lagrangian function of (5.64) is as follows

$$L(\mathbf{u},\mathbf{v},\xi) = \mathscr{B}\mathbf{u}^m + a\mathbf{v}^\top\mathbf{u}^{[m-1]} - \xi(\mathscr{A}\mathbf{u}^m + \mathscr{I}^m - 1) + \frac{\beta}{2}(\mathscr{A}\mathbf{u}^m + \mathscr{I}^m - 1)^2, \tag{5.65}$$

where $\xi \in \mathbf{R}$ is the Lagrangian multiplier associated to the equality constraint and $\beta > 0$ is the penalty parameter. For a given $\xi^{(k)} \in \mathbf{R}$, the iterative scheme of Augmented Lagrangian Method (ALM) of (5.66) is:

$$(\mathbf{u}^{(k+1)}, \mathbf{v}^{(k+1)}) = \arg\min_{\mathbf{u},\mathbf{v}}\{L(\mathbf{u},\mathbf{v},\xi^{(k)}) \mid \mathbf{u} \geq \mathbf{0},\ \mathbf{v} \geq \mathbf{0}\}; \tag{5.66}$$

$$\xi^{(k+1)} = \xi^{(k)} - \beta(\mathscr{A}(\mathbf{u}^{(k+1)})^m + \mathscr{I}(\mathbf{v}^{(k+1)})^m - 1). \tag{5.67}$$

However, the above iterative technique is not easy to be carried out due to the coupled structure and high nonlinearity of the objective function and constraint. By updating the variables in an alternating (Gauss–Seidel) order, the so-called Alternating Direction Method of Multipliers (ADMM) was developed for separable convex minimization to improve its practicability and numerical performance. For given $(\mathbf{v}^{(k)}, \xi^{(k)})$, the ADMM scheme of the subproblem (5.66) is:

$$\mathbf{u}^{(k+1)} = \arg\min_{\mathbf{u}}\left\{L(\mathbf{u},\mathbf{v}^{(k)},\xi^{(k)}) \mid \mathbf{u} \geq \mathbf{0}\right\}; \tag{5.68}$$

$$\mathbf{v}^{(k+1)} = \arg\min_{\mathbf{v}}\left\{L(\mathbf{u}^{(k+1)},\mathbf{v},\xi^{(k)}) \mid \mathbf{v} \geq \mathbf{0}\right\}; \tag{5.69}$$

$$\xi^{(k+1)} = \xi^{(k)} - \beta(\mathscr{A}(\mathbf{u}^{(k+1)})^m + \mathscr{I}(\mathbf{v}^{(k+1)})^m - 1). \tag{5.70}$$

Since constraints of (5.68) and (5.69) are simple convex sets, we solve subproblems by computing the projection onto these sets. On the other hand, owing to the multilinearity of tensors $\mathscr{A}$ and $\mathscr{B}$, it is not easy to find the closed-form solutions of subproblems (5.68) and (5.69). Therefore, we linearize (5.65) so that each subproblem has closed form representation. Furthermore, in order to avoid nonconvexity of $L(\mathbf{u},\mathbf{v},\xi)$ and make both subproblems well-posed, we attach two proximal terms $\frac{\gamma_1}{2}\|\mathbf{u}-\mathbf{u}^{(k)}\|^2$ and $\frac{\gamma_2}{2}\|\mathbf{v}-\mathbf{v}^{(k)}\|^2$ to (5.68) and (5.69), respectively. Here γ_1 and γ_2 are positive constants. Then, we obtain the linearized ADMM as follows:

$$\mathbf{u}^{(k+1)} = \Pi_{\mathbf{R}^n_+}\left[\mathbf{u}^{(k)} - \frac{\Phi^{(k)}}{\gamma_1}\right], \tag{5.71}$$

$$\mathbf{v}^{(k+1)} = \Pi_{\mathbf{R}^n_+}\left[\mathbf{v}^{(k)} - \frac{a(\mathbf{u}^{(k+1)})^{[m-1]} + \Upsilon^{(k)}}{\gamma_2}\right], \tag{5.72}$$

$$\xi^{(k+1)} = \xi^{(k)} - \beta(\mathscr{A}(\mathbf{u}^{(k+1)})^m + \mathscr{I}(\mathbf{v}^{(k+1)})^m - 1), \tag{5.73}$$

where $\Pi_{\mathbf{R}^n_+}$ is the projection onto $\mathbf{R}^n_+$ and

$$\Phi^{(k)} = m\mathscr{B}(\mathbf{u}^{(k)})^{(m-1)} + a(m-1)\mathrm{diag}(\mathbf{v}^{(k)})(\mathbf{u}^{(k)})^{[m-2]} + \beta m q^{(k)}\mathscr{A}(\mathbf{u}^{(k)})^{m-1},$$

with $q^{(k)} = \mathscr{A}(\mathbf{u}^{(k)})^m + \mathscr{I}(\mathbf{v}^{(k)})^m - 1 - \frac{\xi^{(k)}}{\beta}$ and

$$\Upsilon^{(k)} = \beta m\left(\mathscr{A}(\mathbf{u}^{(k+1)})^m + \mathscr{I}(\mathbf{v}^{(k)})^m - 1 - \frac{\xi^{(k)}}{\beta}\right)\mathscr{I}(\mathbf{v}^{(k)})^{m-1}.$$

Although the convergence result of such a linearized ADMM of the nonconvex model is still open, it is illustrated in [180] that (5.71) is numerically convergent for model (5.64) in many cases.

5.5 The Semidefinite Relaxation Method*

In the following analysis, the TEiCP will be formulated as constrained polynomial optimization. And the formulated polynomial optimization can be solved by Lasserre's hierarchy of semidefinite relaxations.

To move on, some basic symbols often used in polynomial optimization theory should be recalled first. For $\mathbf{x} = (x_1, \ldots, x_n)$ and $\alpha = (\alpha_1, \ldots, \alpha_n)$, let $\mathbf{x}^\alpha := x_1^{\alpha_1} \ldots x_n^{\alpha_n}$ denote the monomial power. And let $[\mathbf{x}]_d$ denote the following vector of monomials

$$[\mathbf{x}]_d^\top = (1, x_1, \ldots, x_n, x_1^2, x_1x_2, \ldots\ldots, x_1^d, x_1^{d-1}x_2, \ldots\ldots, x_n^d).$$

Let $\mathbf{R}[\mathbf{x}] := \mathbf{R}[x_1, \ldots, x_n]$ be the ring of polynomials in $\mathbf{x}$ and with real coefficients.

For any $f(\mathbf{x}) \in \mathbf{R}[\mathbf{x}]$, let $\deg(f)$ denote its degree. For $t \in \mathbf{R}$, $\lceil t \rceil$ (resp., $\lfloor t \rfloor$) denotes the smallest (resp. largest) integer not smaller (resp. bigger) than t. Let $\mathbb{I}$ be an ideal in $\mathbf{R}[\mathbf{x}]$ such that $\mathbb{I} \cdot \mathbf{R}[\mathbf{x}] \subseteq \mathbb{I}$ and $\mathbb{I} + \mathbb{I} \subseteq \mathbb{I}$. For a tuple $h = (h_1, \ldots, h_m)$ in $\mathbf{R}[\mathbf{x}]$, denote the ideal

$$\mathbb{I}(h) := h_1 \cdot \mathbf{R}[\mathbf{x}] + \cdots + h_m \cdot \mathbf{R}[\mathbf{x}].$$

Let $\mathbb{I}_k(h) \subseteq \mathbb{I}(h)$ denote the set of the kth truncation of the ideal such that

$$h_1 \cdot \mathbf{R}[\mathbf{x}]_{k\text{-}deg(h_1)} + \cdots + h_m \cdot \mathbf{R}[\mathbf{x}]_{k\text{-}deg(h_m)}, \tag{5.74}$$

where

$$\mathbf{R}[\mathbf{x}]_t := \{f \in \mathbf{R}[\mathbf{x}] \mid \deg(f) \le t\}.$$

The function ψ is called a sum of squares (SOS) if $\psi = q_1^2 + \cdots + q_k^2$ for some $q_1, \ldots, q_k \in \mathbf{R}[\mathbf{x}]$, and $\Sigma[\mathbf{x}]$ denotes the set of all SOS polynomials. Hence, $\Sigma[\mathbf{x}]_m := \Sigma[\mathbf{x}] \cap \mathbf{R}[\mathbf{x}]_m$ denotes the truncation of degree m.

For a tuple $g = (g_1, \dots, g_t)$, its quadratic module is the set

$$Q(g) := \Sigma[\mathbf{x}] + g_1 \cdot \Sigma[\mathbf{x}] + \cdots + g_t \cdot \Sigma[\mathbf{x}].$$

The kth truncation of $Q(g)$ is the set

$$Q_k(g) := \Sigma[\mathbf{x}]_{2k} + g_1 \cdot \Sigma[\mathbf{x}]_{d_1} + \cdots + g_t \cdot \Sigma[\mathbf{x}]_{d_t}, \tag{5.75}$$

where each $d_i = 2k - deg(g_i)$. Furthermore, $\mathbb{I}(h) + Q(g)$ is called **archimedean** if there exists $a > 0$ such that $a - \|\mathbf{x}\|^2 \in \mathbb{I}(h) + Q(g)$. Define

$$E(h) := \{\mathbf{x} \in \mathbf{R}^n \mid h(\mathbf{x}) = \mathbf{0}\}, \quad S(g) := \{\mathbf{x} \in \mathbf{R}^n \mid g(\mathbf{x}) \geq \mathbf{0}\}, \tag{5.76}$$

where h, g are tuples defined as above. A known result is that the set $E(h) \cap S(g)$ is compact if $\mathbb{I}(h) + Q(g)$ is archimedean. Conversely, by adding the polynomial $M - \|\mathbf{x}\|^2$ to the tuple g for M being large enough, then $\mathbb{I}(h) + Q(g)$ can be forced to be archimedean if $\mathbb{I}(h) + Q(g)$ is compact.

Suppose $\alpha = (\alpha_1, \dots, \alpha_n) \in \mathbf{N}^n$. Then, denote $\mathbf{N}_d^n$ such that

$$\mathbf{N}_d^n := \{\alpha \in \mathbf{N}^n \ : \ |\alpha| := \alpha_1 + \cdots + \alpha_n \leq d\}.$$

Let $\mathbf{R}^{\mathbf{N}_d^n}$ be the space of real vectors indexed by $\alpha \in \mathbf{N}_d^n$. Let $\mathbf{y} \in \mathbf{R}^{\mathbf{N}_d^n}$. Then $\mathbf{y}$ is called a truncated multi-sequence (tms) of degree d. Using those vectors, define the operation

$$\langle \sum_{\alpha \in \mathbf{N}_d^n} p_\alpha x_1^{\alpha_1} \dots x_n^{\alpha_n}, \ \mathbf{y} \rangle := \sum_{\alpha \in \mathbf{N}_d^n} p_\alpha \mathbf{y}_\alpha, \tag{5.77}$$

where each $p_\alpha \in \mathbf{R}$ is a coefficient. For a polynomial $q \in \mathbf{R}[\mathbf{x}]_{2k}$, the kth localizing matrix of q, i.e., $L_q^{(k)}(\mathbf{y})$ is a symmetric matrix such that

$$vec(p_1)^\top (L_q^{(k)}(\mathbf{y})) vec(p_2) = \langle q p_1 p_2, \mathbf{y} \rangle,$$

where $\mathbf{y} \in \mathbf{N}_{2k}^n$ is a tms, and $p_1, p_2 \in \mathbf{R}[\mathbf{x}]$ with $\deg(p_1), \deg(p_2) \leq k - \lceil \deg(q)/2 \rceil$. Besides, $vec(p_i)$ denotes the coefficient vector of the polynomial p_i. The matrix $L_q^{(k)}(\mathbf{y})$ becomes a moment matrix if $q = 1$, and is denoted by

$$M_k(\mathbf{y}) := L_1^{(k)}(\mathbf{y}). \tag{5.78}$$

When $q = (q_1, \dots, q_r)$ is a tuple of r polynomials, then we denote

$$L_q^{(k)}(\mathbf{y}) := (L_{q_1}^{(k)}(\mathbf{y}), \dots, L_{q_r}^{(k)}(\mathbf{y})). \tag{5.79}$$

Let $h = (h_1, \dots, h_m)$ and $g = (g_1, \dots, g_t)$ be two polynomial tuples. In applications, people are often interested in whether or not a tms $\mathbf{y} \in \mathbf{R}^{\mathbf{N}_{2k}^n}$ admits a representing measure whose support is contained in $E(h) \cap S(g)$, as in (5.76). For this to be true, a necessary condition (cf. [72, 120]) is that

$$M_k(\mathbf{y}) \succeq 0, \quad L_g^{(k)}(\mathbf{y}) \succeq 0, \quad L_h^{(k)}(\mathbf{y}) = 0. \tag{5.80}$$

However, the above is typically not sufficient. Let

$$d_0 = \max\{1, \lceil \deg(h)/2 \rceil, \lceil \deg(g)/2 \rceil\}.$$

If $\mathbf{y}$ satisfies (5.80) and the rank condition

$$\text{rank} M_{k-d_0}(\mathbf{y}) = \text{rank} M_k(\mathbf{y}), \tag{5.81}$$

then $\mathbf{y}$ admits a measure supported in $E(h) \cap S(g)$ (cf. [72]). In such case, $\mathbf{y}$ admits a unique finitely atomic measure on $E(h) \cap S(g)$. For convenience, we just call that $\mathbf{y}$ is flat with respect to $h = 0$ and $g \geq 0$ if (5.80) and (5.81) are both satisfied.

For $t \leq d$ and $\mathbf{w} \in \mathbf{R}^{\mathbf{N}_d^n}$, denote the truncation of w:

$$\mathbf{w}|_t = (w_\alpha)_{\alpha \in \mathbf{N}_t^n}.$$

For two tms' $\mathbf{y} \in \mathbf{R}^{\mathbf{N}_{2k}^n}$ and $\mathbf{z} \in \mathbf{R}^{\mathbf{N}_{2l}^n}$ with $k < l$, we say that $\mathbf{y}$ is a truncation of $\mathbf{z}$ (equivalently, $\mathbf{z}$ is an extension of $\mathbf{y}$), if $\mathbf{y} = \mathbf{z}|_{2k}$. For such case, $\mathbf{y}$ is called a flat truncation of $\mathbf{z}$ if $\mathbf{y}$ is flat, and $\mathbf{z}$ is a flat extension of $\mathbf{y}$ if $\mathbf{z}$ is flat. Flat extensions and flat truncations are very useful in solving polynomial optimization and truncated moment problems.

In the sequel, we want to present the definition of complementarity eigenvalues, which is useful in the study of TEiCP (5.1). Before that, we have to recall the generalized tensor eigenvalues. For two given complex tensors $\mathscr{A}, \mathscr{B} \in T_{m,n}$, a number $\lambda \in \mathbf{C}$ is called a generalized eigenvalue of the pair $(\mathscr{A}, \mathscr{B})$ if there exists a vector $\mathbf{x} \in \mathbf{C}^n \backslash \{\mathbf{0}\}$ such that

$$\mathscr{A}\mathbf{x}^{m-1} - \lambda \mathscr{B}\mathbf{x}^{m-1} = 0.$$

And $\mathbf{x}$ is called a generalized eigenvector associated with λ. Generalized tensor eigenvalues are closely related to the notion of resultant. Recall the discussion in Chapter 1, for $\mathscr{A} \in T_{m,n}$, its resultant is denoted by

$$\text{RES}(\mathscr{A}) := \text{RES}(\mathscr{A}\mathbf{x}^{m-1}).$$

Clearly, λ is a generalized eigenvalue of $(\mathscr{A}, \mathscr{B})$ if and only if $\text{RES}(\mathscr{A} - \lambda\mathscr{B}) = 0$. Note that $\text{RES}(\mathscr{A} - \lambda\mathscr{B})$ is a polynomial in λ and its degree is $n(m-1)^{n-1}$. An known results about the generalized tensor eigenvalue is listed below.

Theorem 5.18 (Fan, Nie, Zhou 2017) *Suppose $\mathscr{A}, \mathscr{B} \in T_{m,n}$ are tensors defined in complex field. Then the following results hold.*

(1) *If RES*$(\mathscr{B}) = 0$, *then* $(\mathscr{A}, \mathscr{B})$ *has* $n(m-1)^{n-1}$ *generalized eigenvalues by counting multiplicities.*
(2) *Generally,* $(\mathscr{A}, \mathscr{B})$ *has* $n(m-1)^{n-1}$ *distinct generalized eigenvalues, and each generalized eigenvalue corresponds a unique eigenvector in the sense of scaling.*

For two vectors $\mathbf{x}, \mathbf{y} \in \mathbf{R}^n$, $\mathbf{x} \circ \mathbf{y}$ denotes the **Hadmard product** of $\mathbf{x}$ and $\mathbf{y}$, i.e., the product is defined componentwise. We now present the definition of combinatorial eigenvalue for tensor pairs below.

Definition 5.6 Let $\mathscr{A}, \mathscr{B} \in T_{m,n}$ be tensors in complex field. If there exist a number $\lambda \in \mathbf{C}$ and a vector $\mathbf{x} \in \mathbf{C}^n \backslash \{\mathbf{0}\}$ such that

$$\mathbf{x} \circ (\mathscr{A}\mathbf{x}^{m-1} - \lambda \mathscr{B}\mathbf{x}^{m-1}) = 0,$$

then λ (resp., $\mathbf{x}$) is called a **combinatorial eigenvalue** (resp., combinatorial eigenvector) of the pair $(\mathscr{A}, \mathscr{B})$. Such $(\lambda, \mathbf{x})$ is called a **combinatorial eigenpair**.

For convenience, the combinatorial eigenvalue (resp., eigenvector, eigenpair) are called **CB-eigenvalue** (resp., **CB-eigenvector**, **CB-eigenpair**) in this section. Apparently, Pareto-eigenvalues and Pareto H-eigenvalues are special cases of CB-eigenvalues.

For any subset $J = \{i_1, \ldots, i_k\} \subseteq [n]$, denote $\mathbf{x}_J = (x_{i_1}, \ldots, x_{i_k})$. For a tensor $\mathscr{F} \in T_{m,n}$, we know that

$$(\mathscr{F}^J(\mathbf{x}_J)^{m-1})_j = \sum_{i_2,\ldots,i_m \in J} \mathscr{F}_{j,i_2,\ldots,i_m} x_{i_2} \ldots x_{i_k}, \quad \forall\, j \in J,$$

where $\mathscr{F}^J$ is the principal subtensor with respect to J. Let $\mathrm{RES}_J(\mathscr{F})$ denote the resultant of $\mathscr{F}^J(\mathbf{x}_J)^{m-1}$

$$\mathrm{RES}_J(\mathscr{F}) := \mathrm{RES}(\mathscr{F}^J(\mathbf{x}_J)^{m-1}). \tag{5.82}$$

Similar to the results about Pareto-eigenvalues in Sect. 5.1, we present a similar result for CB-eigenvalues.

Theorem 5.19 *Let* $\mathscr{A}, \mathscr{B} \in T_{m,n}$ *be tensors in complex field. Then the following results hold.*

(1) *Suppose* $J \subseteq [n]$ *is nonempty. If* $RES_J(\mathscr{B}) \neq 0$, *then* $(\mathscr{A}, \mathscr{B})$ *has at most* nm^{n-1} *CB-eigenvalues;*
(2) *For any* $\mathscr{A}, \mathscr{B} \in T_{m,n}$, *each CB-eigenvalue corresponds a unique CB-eigenvector (up to scaling).*

Proof (1) Suppose λ is a CB-eigenvalue of $(\mathscr{A}, \mathscr{B})$ with CB-eigenvector $\mathbf{x} \neq 0$ such that

$$\mathbf{x} \circ (\mathscr{A}\mathbf{x}^{m-1} - \lambda \mathscr{B}\mathbf{x}^{m-1}) = 0.$$

Let $J = \{j \in [n] : x_j \neq 0\}$. Then $J \neq \emptyset$, and it further implies that

$$\mathcal{A}^J(\mathbf{x}_J)^{m-1} - \lambda\mathcal{B}^J(\mathbf{x}_J)^{m-1} = 0.$$

Hence, it means that λ is an eigenvalue of $(\mathcal{A}^J, \mathcal{B}^J)$. By Theorem 5.18(1), $(\mathcal{A}^J, \mathcal{B}^J)$ has at most $|J|(m-1)^{|J|-1}$ eigenvalues. Since there are $\binom{n}{|J|}$ subsets for a fixed cardinality subset, the number of CB-eigenvalues of $(\mathcal{A}, \mathcal{B})$ is at most such that

$$\sum_{|J|=1}^{n} \binom{n}{|J|} |J|(m-1)^{|J|-1} = nm^{n-1}.$$

(2) Generally, suppose $\mathcal{A}, \mathcal{B} \in T_{m,n}$ are generic in the space $T_{m,n}$ with principal subtensors $\mathcal{A}^J, \mathcal{B}^J \in T_{m,|J|}$, where $J \subseteq [n]$, $J \neq \emptyset$. By (2) in Theorem 5.18, $(\mathcal{A}^J, \mathcal{B}^J)$ has a unique eigenvector (up to scaling) for each eigenvalue. For any CB-eigenpair $(\lambda, \mathbf{x})$ of $(\mathcal{A}, \mathcal{B})$, by the proof of (1), we know that λ is an eigenvalue of the sub-tensor pair $(\mathcal{A}^{J_1}, \mathcal{B}^{J_1})$ with the eigenvector $\mathbf{x}_{J_1}$, with the index set $J_1 = \{j \in [n] : u_j \neq 0\}$.
Without loss of generality, let $\mathbf{v}$ be another CB-eigenvector associated to λ. Let $J_2 = \{j \in [n] : v_j \neq 0\}$. By the proof of (1) again, λ is also an eigenvalue of the sub-tensor pair $(\mathcal{A}^{J_2}, \mathcal{B}^{J_2})$. Now we prove $J_1 = J_2$. Let

$$V = \{\mathcal{C} \in T_{m,n} : R_{J_2}(\mathcal{C}_{J_2}) = R_{J_1}(\mathcal{C}_{J_1}) = 0\}.$$

The polynomial $R_{J_2}(\mathcal{C}_{J_2})$ is irreducible in the entries of the sub-tensor $\mathcal{C}_{J_2}$. The same is true for $R_{J_1}(\mathcal{C}_{J_1})$. If $J_2 \neq J_1$, the dimension of the set V is at most $\dim(T_{m,n}) - 2$. If $\mathcal{A}, \mathcal{B}$ are general tensors, then $\mathcal{L} = \{\mathcal{A} - \lambda\mathcal{B} : \lambda \in \mathbf{C}\}$ does not intersect V. Therefore, if $J_2 \neq J_1$, then λ cannot be a common eigenvalue of the two different sub-tensor pairs $(\mathcal{A}^{J_2}, \mathcal{B}^{J_2})$ and $(\mathcal{A}^{J_1}, \mathcal{B}^{J_1})$. Hence, $J_2 = J_1$ and $\mathbf{x}_{J_1}, \mathbf{v}_{J_1}$ are both eigenvectors of $(\mathcal{A}^{J_1}, \mathcal{B}^{J_1})$. From (2) of Theorem 5.18, $\mathbf{x}$ is a scaling of $\mathbf{v}$. □

Next, we discuss how to compute Pareto-eigenvalues of a given tensor pair $(\mathcal{A}, \mathcal{B})$ if $\mathcal{B}$ is strictly copositive. Recall the definition in Sect. 5.1, we know that $(\lambda, \mathbf{x})$ is a Pareto-eigenpair of $(\mathcal{A}, \mathcal{B})$ if $\mathbf{x}$ is a nonzero vector and (5.1) holds. If $\mathcal{B}$ is strictly copositive, $\mathbf{x}$ can be scaled such that $\mathcal{B}\mathbf{x}^m = 1$ since $\mathbf{x} \neq \mathbf{0}$. In this case, the Pareto-eigenpair $(\lambda, \mathbf{x})$ satisfies

$$0 = \mathbf{x}^\top(\lambda\mathcal{B}\mathbf{x}^{m-1} - \mathcal{A}\mathbf{x}^{m-1}) = \lambda\mathcal{B}\mathbf{x}^m - \mathcal{A}\mathbf{x}^m = \lambda - \mathcal{A}\mathbf{x}^m,$$

which means that $\lambda = \mathcal{A}\mathbf{x}^m$. Thus, one can compute Pareto-eigenvalues of $(\mathcal{A}, \mathcal{B})$ by solving the polynomial system

$$\begin{cases} \mathcal{B}\mathbf{x}^m = 1, & \mathbf{x} \circ ((\mathcal{A}\mathbf{x}^m)\mathcal{B}\mathbf{x}^{m-1} - \mathcal{A}\mathbf{x}^{m-1}) = 0, \\ \mathbf{x} \geq \mathbf{0}, & (\mathcal{A}\mathbf{x}^m)\mathcal{B}\mathbf{x}^{m-1} - \mathcal{A}\mathbf{x}^{m-1} \geq \mathbf{0}. \end{cases} \tag{5.83}$$

Denote $a(\mathbf{x}) := \mathbf{x} \circ \mathscr{A}\mathbf{x}^{m-1}$, and $b(\mathbf{x}) := \mathbf{x} \circ \mathscr{B}\mathbf{x}^{m-1}$. The system (5.83) is equivalent with

$$\begin{cases} \mathscr{B}\mathbf{x}^m = 1, & (\mathscr{A}\mathbf{x}^m)b(\mathbf{x}) - a(\mathbf{x}) = 0, \\ \mathbf{x} \geq \mathbf{0}, & (\mathscr{A}\mathbf{x}^m)\mathscr{B}\mathbf{x}^{m-1} - \mathscr{A}\mathbf{x}^{m-1} \geq \mathbf{0}. \end{cases} \tag{5.84}$$

Under conditions that $\mathscr{B}$ is strictly copositive, the solution set of (5.84) is compact since $\{\mathbf{x} \in \mathbf{R}^n : \mathscr{B}\mathbf{x}^m = 1, \mathbf{x} \geq \mathbf{0}\}$ is compact. By Theorems 5.1 and 5.19, we know that $(\mathscr{A}, \mathscr{B})$ has finitely many Pareto-eigenvalues. Thus, it can be written as follows

$$\lambda_1 < \lambda_2 < \cdots < \lambda_N.$$

For the sake of simplicity, let

$$\begin{cases} f_0 = \mathscr{A}\mathbf{x}^m, \\ p = \begin{pmatrix} \mathscr{B}\mathbf{x}^m - 1 \\ (\mathscr{A}\mathbf{x}^m)b(\mathbf{x}) - a(\mathbf{x}) \end{pmatrix}, \\ q = \begin{pmatrix} \mathbf{x} \\ (\mathscr{A}\mathbf{x}^m)\mathscr{B}\mathbf{x}^{m-1} - \mathscr{A}\mathbf{x}^{m-1} \end{pmatrix}. \end{cases} \tag{5.85}$$

Then we know that λ_1 can be computed from the following optimization problem

$$\lambda_1 = \begin{cases} \min & f_0(\mathbf{x}) \\ \text{s.t.} & p(\mathbf{x}) = 0, \ q(\mathbf{x}) \geq 0. \end{cases} \tag{5.86}$$

By Lasserre type semidefinite relaxations and operation (5.77), for any $k = m, m + 1, \ldots$, the kth Lasserre relaxation of (5.86) is

$$\nu_{1,k} := \begin{cases} \min & \langle f_0, \mathbf{y} \rangle \\ \text{s.t.} & \langle 1, \mathbf{y} \rangle = 1, \ L_p^{(k)}(\mathbf{y}) = 0, \\ & M_k(\mathbf{y}) \succeq 0, \ L_q^{(k)}(\mathbf{y}) \succeq 0, \ \mathbf{y} \in \mathbf{R}^{\mathbf{N}_{2k}^n}. \end{cases} \tag{5.87}$$

Here, $\langle 1, \mathbf{y} \rangle = 1$ means that the first entry of $\mathbf{y}$ is one. Matrices $M_k(\mathbf{y})$, $L_p^{(k)}(\mathbf{y})$, $L_q^{(k)}(\mathbf{y})$ are defined as in (5.78) and (5.79). The dual problem of (5.87) is

$$\tilde{\nu}_{1,k} := \begin{cases} \max & \gamma \\ \text{s.t.} & f_0 - \gamma \in I_{2k}(p) + Q_k(q). \end{cases}$$

Assume $\mathbf{y}^{1,k}$ is an optimizer of (5.87). If $\hat{\mathbf{y}} = \mathbf{y}^{1,k}|_{2t}$ for some $t \in [m, k]$, satisfies

$$\text{rank} M_{t-m}(\hat{\mathbf{y}}) = \text{rank} M_t(\hat{\mathbf{y}}), \tag{5.88}$$

then we obtain that $\nu_{1,k} = \lambda_1$, and $\text{rank} M_t(\hat{\mathbf{y}})$ is the global optimizer of (5.86).

In what follows, we discuss how to compute λ_i for $i \geq 2$. We only need to present how to compute Pareto-eigenvalue λ_i if λ_{i-1} is already computed. Consider the optimization problem

$$\begin{cases} \min & f_0(\mathbf{x}) \\ \text{s.t.} & p(\mathbf{x}) = 0,\ q(\mathbf{x}) \geq 0,\ f_0(\mathbf{x}) - \lambda_{i-1} - \delta \geq 0. \end{cases} \tag{5.89}$$

Hence, λ_i is the optimal value of (5.89) if

$$0 < \delta < \lambda_i - \lambda_{i-1}. \tag{5.90}$$

Similarly, for any $k = m, m+1, \ldots$, the kth Lasserre relaxation of (5.89) is

$$\nu_{i,k} := \begin{cases} \min & \langle f_0, z\rangle \\ \text{s.t.} & \langle 1, \mathbf{z}\rangle = 1,\ L_p^{(k)}(\mathbf{z}) = 0,\ M_k(\mathbf{z}) \succeq 0, \\ & L_q^{(k)}(\mathbf{z}) \succeq 0,\ L_{f_0-\lambda_{i-1}-\delta}^{(k)}(\mathbf{z}) \succeq 0,\ \mathbf{z} \in \mathbf{R}^{\mathbf{N}_{2k}^n}, \end{cases} \tag{5.91}$$

and its dual problem is

$$\tilde{\nu}_{i,k} := \begin{cases} \max & \gamma \\ \text{s.t.} & f_0 - \gamma \in I_{2k}(p) + Q_k(q, f_0 - \lambda_{i-1} - \delta). \end{cases} \tag{5.92}$$

Let $\mathbf{y}^{i,k}$ be an optimizer of (5.91). If $\hat{\mathbf{y}} = \mathbf{y}^{i,k}|_{2t}$ satisfies (5.88) for some $t \in [m, k]$, then we obtain the optimizer of (5.89) such that $\nu_{i,k} = \lambda_i$.

In fact, we are not sure about λ_i in advance. Thus, we can determine the value of δ satisfying (5.90) by the following optimization problem:

$$\tau := \begin{cases} \max & f_0(\mathbf{x}) \\ \text{s.t.} & p(\mathbf{x}) = 0,\ q(\mathbf{x}) \geq 0,\ f_0(\mathbf{x}) \leq \lambda_{i-1} + \delta. \end{cases} \tag{5.93}$$

The optimal value τ can also be computed by Lasserre relaxations like (5.91) and (5.92). As in Proposition 5.17, δ satisfies (5.90) if and only if $\tau = \lambda_{i-1}$. When $\tau = \lambda_{i-1}$, λ_i does not exist if and only if (5.91) is infeasible for some k.

Assume that the tensor $\mathscr{B}$ is strictly copositive. Hence the Pareto-eigenvectors can be normalized as $\mathscr{B}\mathbf{x}^m = 1$. By the analysis above, an algorithm is presented below to compute the Pareto-eigenvalues sequentially, from the smallest one λ_1 to the biggest one λ_N.

Algorithm 4

For two tensors $\mathscr{A}, \mathscr{B} \in T_{m,n}$ with $\mathscr{B}$ strictly copositive, compute a set Λ of all Pareto-eigenvalues and a set U of Pareto-eigenvectors, for the pair $(\mathscr{A}, \mathscr{B})$. Let $U := \emptyset$, $\Lambda := \emptyset$, $i := 1$, $k := m$.

Step 1 Solve (5.87) with the order k for an optimizer $\mathbf{y}^{1,k}$.

Step 2 If (5.88) is satisfied for some $t \in [m, k]$, then update $U := U \cup S$, with S a set of optimizers of (5.86); let $\lambda_1 = \nu_{1,k}$, $\Lambda := \{\lambda_1\}$, $i := i + 1$ and go to Step 3. If such t does not exist, let $k := k + 1$ and go to Step 1.
Step 3 Let $\delta = 0.05$, and compute the optimal value τ of (5.93). If $\tau > \lambda_{i-1}$, let $\delta := \delta/2$ and compute τ again. Repeat this, until we get $\tau = \lambda_{i-1}$. Let $k := m$.
Step 4 Solve (5.91) with the order k. If it is infeasible, then (5.84) has no further Pareto-eigenvalues, and stop. Otherwise, compute an optimizer $\mathbf{y}^{i,k}$ for (5.91).
Step 5 If (5.88) is satisfied for some $t \in [m, k]$, then update $U := U \cup S$ where S is a set of optimizers of (5.89); let $\lambda_i = \nu_{i,k}$, $\Lambda := \Lambda \cup \{\lambda_i\}$, $i := i + 1$ and go to Step 3. If such t does not exist, let $k := k + 1$ and go to Step 4.

For a polynomial tuple p, denote the set

$$V_{\mathbf{C}}(p) := \{\mathbf{u} \in \mathbf{C}^n \mid p(\mathbf{u}) = 0\}, \quad V_{\mathbf{R}}(p) := V_{\mathbf{C}}(p) \cap \mathbf{R}^n.$$

It is interesting that the set $V_{\mathbf{R}}(p)$ is generally finite, which can be obtained by the following property.

Proposition 5.16 *Let p be defined as in (5.85). Suppose $\mathscr{A}, \mathscr{B}$ are general tensors, then $V_{\mathbf{C}}(p)$ and $V_{\mathbf{R}}(p)$ are finite sets.*

Proof By $p(\mathbf{x}) = 0$, we know that

$$\mathscr{B}\mathbf{x}^m = 1, \; a(\mathbf{x}) - (\mathscr{A}\mathbf{x}^m)b(\mathbf{x}) = \mathbf{x} \circ (\mathscr{A}\mathbf{x}^{m-1} - (\mathscr{A}\mathbf{x}^m)\mathscr{B}\mathbf{x}^{m-1}) = 0.$$

Hence $\mathbf{x} \neq \mathbf{0}$. Let $J = \{j \in [n] \; : \; x_j \neq 0\}$. Then it follows that

$$\mathscr{A}^J(\mathbf{x}_J)^{m-1} - (\mathscr{A}\mathbf{x}^m)\mathscr{B}^J(\mathbf{x}_J)^{m-1} = 0,$$

which implies that $\mathbf{x}_J$ is an eigenvector of the subtensor pair $(\mathscr{A}^J, \mathscr{B}^J)$. By (2) of Theorem 5.19, we know that $V_{\mathbf{C}}(p)$ and $V_{\mathbf{R}}(p)$ are finite sets, and the desired results hold. □

In the following conclusion, $\sigma(\mathscr{A}, \mathscr{B})$ denotes the set of all Pareto-eigenvalues of $(\mathscr{A}, \mathscr{B})$ as defined in Sect. 5.1. It verifies the existence of λ_i and the relation (5.90) can be checked accordingly. The proof of this conclusion can be obtained in [93].

Proposition 5.17 (Fan, Nie, Zhou 2017) *Suppose $\mathscr{A}, \mathscr{B} \in T_{m,n}$ with $\mathscr{B}$ being strictly copositive. Assume $\sigma(\mathscr{A}, \mathscr{B})$ is finite. Let λ_i be the ith smallest Pareto-eigenvalue of $(\mathscr{A}, \mathscr{B})$, and $\lambda_{\max}$ be the maximum of them. For all $i \geq 2$ and all $\delta > 0$, the following properties hold:*

(1) The set $\sigma(\mathscr{A}, \mathscr{B}) \cap [\lambda_{i-1} + \delta, \infty) = \emptyset$ if (5.91) is infeasible for some k.
(2) The optimization problem (5.91) may infeasible for some k if $\sigma(\mathscr{A}, \mathscr{B}) \cap [\lambda_{i-1} + \delta, \infty) = \emptyset$ and $V_{\mathbf{R}}(p)$ is finite.
(3) If $\tau = \lambda_{i-1}$ and λ_i exists, then δ satisfies (5.90).
(4) If $\tau = \lambda_{i-1}$ and (5.91) is infeasible for some k, then λ_i does not exist.

Algorithm 3 shows how to find the Pareto-eigenvalue under conditions $\mathscr{B}$ is strictly copositive. We now study how to compute Pareto-eigenvalues for general tensors $\mathscr{A}, \mathscr{B} \in T_{m,n}$. Here, we still use

$$a(\mathbf{x}) := \mathbf{x} \circ \mathscr{A}\mathbf{x}^{m-1} \quad \text{and} \quad b(\mathbf{x}) := \mathbf{x} \circ \mathscr{B}\mathbf{x}^{m-1}.$$

Since $\mathbf{x}$ can be normalized as a unit vector, then $(\lambda, \mathbf{x})$ is Pareto-eigenpair of $(\mathscr{A}, \mathscr{B})$ if and only if it is a solution of the polynomial system

$$\begin{cases} \mathbf{x}^\top \mathbf{x} = 1, & \lambda b(\mathbf{x}) - a(\mathbf{x}) = 0, \\ \mathbf{x} \geq \mathbf{0}, & \lambda \mathscr{B}\mathbf{x}^{m-1} - \mathscr{A}\mathbf{x}^{m-1} \geq \mathbf{0}. \end{cases} \tag{5.94}$$

Suppose $b(\mathbf{x}) \neq 0$, then the equation $a(\mathbf{x}) - \lambda b(\mathbf{x}) = 0$ is equivalent with

$$\text{rank}[a(\mathbf{x}), b(\mathbf{x})] \leq 1,$$

which can be written

$$a(\mathbf{x})_i b(\mathbf{x})_j - b(\mathbf{x})_i a(\mathbf{x})_j = 0 \ \text{ for } 1 \leq i < j \leq n.$$

Assume the system (5.94) has finite real solutions. Let $\xi \in \mathbf{R}^n$ be an arbitrary vector, then it follows that $\xi^\top b(\mathbf{x}) \neq 0$ for all $\mathbf{x}$ satisfying (5.94) and

$$\lambda = \frac{\xi^\top a(\mathbf{x})}{\xi^\top b(\mathbf{x})}. \tag{5.95}$$

Thus, Pareto-eigenvalues of $(\mathscr{A}, \mathscr{B})$ can be computed in two cases.
Case I: $\xi^\top b(\mathbf{x}) > 0$. we know that the system (5.94) is equivalent to

$$\begin{cases} \mathbf{x}^\top \mathbf{x} = 1, & a(\mathbf{x})_i b(\mathbf{x})_j - b(\mathbf{x})_i a(\mathbf{x})_j = 0 \ (1 \leq i < j \leq n), \\ \mathbf{x} \geq \mathbf{0}, & \xi^\top b(\mathbf{x}) \geq 0, \ \xi^\top a(\mathbf{x}) \mathscr{B}\mathbf{x}^{m-1} - \xi^\top b(\mathbf{x}) \mathscr{A}\mathbf{x}^{m-1} \geq 0. \end{cases} \tag{5.96}$$

For generic $(\mathscr{A}, \mathscr{B})$, it is clear that (5.96) has finite solutions. If $\mathbf{x}$ is obtained, the Pareto-eigenvalue λ can be computed by (5.95).

On the other hand, for an arbitrary $f(\mathbf{x}) \in \mathbf{R}[\mathbf{x}]_{2m}$, consider the optimization problem

$$\begin{cases} \min & f(\mathbf{x}) \\ \text{s.t.} & h(\mathbf{x}) = 0, \ g(\mathbf{x}) \geq 0, \end{cases} \tag{5.97}$$

where the polynomial tuples h, g are defined satisfying

$$\begin{cases} h(\mathbf{x}) = (\mathbf{x}^\top \mathbf{x} - 1, \ (a(\mathbf{x})_i b(\mathbf{x})_j - b(\mathbf{x})_i a(\mathbf{x})_j)_{1 \leq i < j \leq n}), \\ g(\mathbf{x}) = (\mathbf{x}, \ \xi^\top b(\mathbf{x}), \ \xi^\top a(\mathbf{x}) \mathscr{B}\mathbf{x}^{m-1} - \xi^\top b(\mathbf{x}) \mathscr{A}\mathbf{x}^{m-1}). \end{cases} \tag{5.98}$$

Then, $\mathbf{x}$ satisfies (5.96) if and only if $\mathbf{x}$ is feasible for (5.97).

Case II: $\xi^\top b(\mathbf{x}) < 0$. In this case, the system (5.94) is equivalent to

$$\begin{cases} \mathbf{x}^\top \mathbf{x} = 1, & a(\mathbf{x})_i b(\mathbf{x})_j - b(\mathbf{x})_i a(\mathbf{x})_j = 0 \ (1 \le i < j \le n), \\ \mathbf{x} \ge \mathbf{0}, & -\xi^\top b(\mathbf{x}) \ge 0, \ \xi^\top b(\mathbf{x}) \mathscr{A} \mathbf{x}^{m-1} - \xi^\top a(\mathbf{x}) \mathscr{B} \mathbf{x}^{m-1} \ge 0. \end{cases} \tag{5.99}$$

By a similar discussion as in case **I**, it is clear that $\mathbf{x}$ satisfies (5.99) if and only if it is feasible for the following optimization problem

$$\begin{cases} \min & f(\mathbf{x}) \\ \text{s.t.} & h(\mathbf{x}) = 0, \ \tilde{g}(\mathbf{x}) \ge 0, \end{cases} \tag{5.100}$$

where h is the same as in (5.98) while the tuple $\tilde{g}$ is given as

$$\tilde{g}(\mathbf{x}) = (\mathbf{x}, \ -\xi^\top b(\mathbf{x}), \ \xi^\top b(\mathbf{x}) \mathscr{A} \mathbf{x}^{m-1} - \xi^\top a(\mathbf{x}) \mathscr{B} \mathbf{x}^{m-1}).$$

It should be noted that the feasible sets of (5.97) and (5.100) are compact since they are contained in the unit sphere. However, the two feasible sets maybe empty. In the following analysis, we always assume that there are finitely many Pareto-eigenvectors (normalized to have unit lengths) for the tensor pair $(\mathscr{A}, \mathscr{B})$. An algorithm will be introduced for computing them.

(1) Pareto-eigenpairs for case I.

It is enough to discuss how to find the Pareto-eigenvectors satisfying (5.96). Suppose f is generically chosen and it achieves different values at different feasible points of (5.97), i.e., they are monotonically ordered as

$$f_1^{(1)} < f_2^{(1)} < \cdots < f_{N_1}^{(1)}.$$

We aim to compute the Pareto-eigenvectors, along with the values $f_i^{(1)}$, for $i \in [N_1]$. Take l_i such that

$$f_{i-1}^{(1)} < l_i < f_i^{(1)}. \tag{5.101}$$

(if $i = 1$, we can choose $f_0^{(1)}$ being any value smaller than $f_1^{(1)}$.) Note that $f_i^{(1)}$ is equal to the optimal value of

$$\begin{cases} \min & f(\mathbf{x}) \\ \text{s.t.} & h(\mathbf{x}) = 0, \ \ g(\mathbf{x}) \ge 0, \ \ f(\mathbf{x}) - l_i \ge 0. \end{cases} \tag{5.102}$$

For the orders $k = m, m+1, \ldots$, the kth Lasserre relaxation of (5.102) is

$$\mu_{1,k} := \begin{cases} \min & \langle f, \mathbf{y} \rangle \\ \text{s.t.} & \langle 1, \mathbf{y} \rangle = 1, \ L_h^{(k)}(\mathbf{y}) = 0, \ M_k(\mathbf{y}) \succeq 0, \\ & L_g^{(k)}(\mathbf{y}) \succeq 0, \ L_{f-l_i}^{(k)}(\mathbf{y}) \succeq 0, \ \mathbf{y} \in \mathbf{R}^{\mathbf{N}_{2k}^n}, \end{cases} \tag{5.103}$$

and the dual problem of (5.103) is

$$\mu_{1,k} := \begin{cases} \max & \gamma \\ \text{s. t.} & f - \gamma \in \mathbf{I}_{2k}(h) + Q_k(g, f - l_i), \end{cases}$$

where $\mathbb{I}_{2k}(h)$ and $Q_k(g, f - l_i)$ are defined as in (5.74) and (5.75). By the weak duality, it can be shown that (cf. [162])

$$\tilde{\mu}_{1,k} \leq \mu_{1,k} \leq f_i^{(1)}, \quad \forall\, k \geq m.$$

Moreover, both $\{\mu_{1,k}\}$ and $\{\tilde{\mu}_{1,k}\}$ are monotonically increasing.

On the other hand, the semidefinite relaxation (5.103) is always feasible when (5.96) has a solution. Equation (5.103) has a solution $\mathbf{y}^{i,k}$. For some $t \in [m, k]$, if the truncation $\hat{\mathbf{y}} := \mathbf{y}^{i,k}|_{2t}$ satisfies

$$\operatorname{rank} M_{t-m}(\hat{\mathbf{y}}) = \operatorname{rank} M_t(\hat{\mathbf{y}}), \tag{5.104}$$

then one can show that $\{\mu_{1,k}\} = \{\tilde{\mu}_{1,k}\} = f_i^{(1)}$. Furthermore, we obtain that $\operatorname{rank} M_t(\hat{\mathbf{y}})$ optimizers of (5.102). Details for the proof above can be found in [121, 202].

(2) Pareto-eigenpairs for case II.
For case **II**, we only need to find the Pareto-eigenvectors satisfying (5.99). Similar to the case **I**, let its objective values be monotonically as

$$f_1^{(2)} < f_2^{(2)} < \cdots < f_{N_2}^{(2)}.$$

We aim to compute the Pareto-eigenvectors, and the value $f_i^{(2)}$ for $i \in [N_2]$. Take $\tilde{l}_i$ such that

$$f_{i-1}^{(2)} < \tilde{l}_i < f_i^{(2)}. \tag{5.105}$$

(If $i = 1$, choose $f_0^{(2)}$ being any value smaller than $f_1^{(2)}$) It is clear that $f_i^{(2)}$ is equal to the minimum value of

$$\begin{cases} \min & f(\mathbf{x}) \\ \text{s.t.} & h(\mathbf{x}) = 0, \;\; \tilde{g}(\mathbf{x}) \geq 0, \;\; f(\mathbf{x}) - \tilde{l}_i \geq 0. \end{cases} \tag{5.106}$$

For an order $k \geq m$, the kth Lasserre relaxation of (5.106) is

$$\mu_{2,k} := \begin{cases} \min & \langle f, \; \mathbf{z} \rangle \\ \text{s.t.} & \langle 1, \; \mathbf{z} \rangle = 1, \;\; L_h^{(k)}(\mathbf{z}) = 0, \;\; M_k(\mathbf{z}) \succeq 0, \\ & L_{\tilde{g}}^{(k)}(\mathbf{z}) \succeq 0, \;\; L_{f-\tilde{l}_i}^{(k)}(\mathbf{z}) \succeq 0, \;\; \mathbf{z} \in \mathbf{R}^{\mathbf{N}_{2k}^n}, \end{cases} \tag{5.107}$$

and its dual optimization problem is

$$\tilde{\mu}_{2,k} := \begin{cases} \max & \gamma \\ \text{s. t.} & f - \gamma \in \mathbb{I}_{2k}(h) + Q_k(\tilde{g}, f - \tilde{l}_i), \end{cases}$$

Similarly, it holds the relationship that

$$\tilde{\mu}_{2,k} \leq \mu_{2,k} \leq f_i^{(2)}, \quad \forall\, k \geq m. \tag{5.108}$$

Moreover, both $\{\mu_{2,k}\}$ and $\{\tilde{\mu}_{2,k}\}$ are monotonically increasing.

Let $\mathbf{z}^{i,k}$ be an optimizer of (5.107). For some $t \in [m, k]$, if the truncation $\hat{\mathbf{z}} := \mathbf{z}^{i,k}|_{2t}$ satisfies

$$\text{rank } M_{t-m}(\hat{\mathbf{z}}) = \text{rank } M_t(\hat{\mathbf{z}}), \tag{5.109}$$

then $\mu_{2,k} = \tilde{\mu}_{2,k} = f_i^{(2)}$, one can get rank $M_t(\hat{\mathbf{z}})$ is a solution of (5.106), which is proved in [202]).

(3) An algorithm for computing Pareto-eigenpairs.

In fact, the $f, l_i, \tilde{l}_i$ should be chosen properly. One way to choose f is such that

$$f = [\mathbf{x}]_m^\top (R^\top R)[\mathbf{x}]_m, \tag{5.110}$$

where R is a random square matrix. For the f above, it almost always holds that

$$f_1^{(1)} > 0, \quad f_1^{(2)} > 0.$$

Therefore, one can choose

$$f_0^{(1)} = f_0^{(2)} = -1, l_1 = \tilde{l}_1 = 0.$$

In the computation of $f_i^{(1)}, f_i^{(2)}$, suppose the values of $f_{i-1}^{(1)}, f_{i-1}^{(2)}$ are already computed. In practice, for $\delta > 0$ small enough, we can choose

$$l_i = f_{i-1}^{(1)} + \delta, \quad \tilde{l}_i = f_{i-1}^{(2)} + \delta,$$

to satisfy (5.101) and (5.105). Noted that the value δ can be obtained by solving the following problems:

$$\theta_1 = \begin{cases} \max & f(\mathbf{x}) \\ \text{s. t.} & h(\mathbf{x}) = 0, \ g(\mathbf{x}) \geq 0, \ f(\mathbf{x}) \leq f_{i-1}^{(1)} + \delta, \end{cases} \tag{5.111}$$

$$\theta_2 = \begin{cases} \max & f(\mathbf{x}) \\ \text{s. t.} & h(\mathbf{x}) = 0, \ \tilde{g}(\mathbf{x}) \geq 0, \ f(\mathbf{x}) \leq f_{i-1}^{(2)} + \delta, \end{cases} \tag{5.112}$$

If $h(\mathbf{x}) = 0$ has finite real solutions, we must have $\theta_1 = f_{i-1}^{(1)}$ and $\theta_2 = f_{i-1}^{(2)}$, for $\delta > 0$ sufficiently small. It should be noted that f achieves only finitely many values in the feasible sets of (5.97), (5.100), when $(\mathscr{A}, \mathscr{B})$ has finitely many normalized Pareto-eigenvectors. From the analysis above, the following algorithm is presented for computing Pareto-eigenvalues for general tensors $(\mathscr{A}, \mathscr{B})$.

Algorithm 5

For two given tensors $\mathscr{A}, \mathscr{B} \in T_{m,n}$, compute a set $\sigma(\mathscr{A}, \mathscr{B})$ of Pareto-eigenvalues, and a set U of Pareto-eigenvectors, for the pair $(\mathscr{A}, \mathscr{B})$.
Step 0. Choose f as in (5.110), with R a random square matrix. Choose a random vector $\xi \in \mathbf{R}^n$. Let $U = \emptyset, i = 1, k = m, l_1 = 0, \tilde{l}_1 = 0$.
Step 1. Solve (5.103) for the order k. If it is infeasible, then (5.96) has no further Pareto-eigenvectors (except those in U); let $k = m, i = 1$ and go to Step 4. Otherwise, compute an optimizer $\mathbf{y}^{i,k}$ for (5.103).
Step 2. If (5.104) is satisfied for some $t \in [m, k]$, then update $U \leftarrow U \cup V$, where V is a set of optimizers of (5.102); let $i \leftarrow i + 1$ and go to Step 3. If such t does not exist, let $k \leftarrow k + 1$ and go to Step 1.
Step 3. Let $\delta = 0.05$, and compute the optimal value θ_1 of (5.111). If $\theta_1 > f_{i-1}^{(1)}$, let $\delta \leftarrow \delta/2$ and compute θ_1 again. Repeat this process, until $\theta_1 = f_{i-1}^{(1)}$ is met. Let $l_i := f_{i-1}^{(1)} + \delta, k = m$, then go to Step 1.
Step 4. Solve (5.107) for the order k. If it is infeasible, then (5.99) has no further Pareto-eigenvectors (except those in U) and go to Step 7. Otherwise, compute an optimizer $\mathbf{z}^{i,k}$ for it.
Step 5. Check whether or not (5.109) is satisfied for some $t \in [m, k]$. If yes, update $U \leftarrow U \cup V$, where V is a set of optimizers of (5.106); let $i \leftarrow i + 1$ and go to Step 6. If no, let $k \leftarrow k + 1$ and go to Step 4.
Step 6. Let $\delta = 0.05$, and compute the optimal value θ_2 of (5.112). If $\theta_2 > f_{i-1}^{(2)}$, let $\delta \leftarrow \delta/2$ and compute θ_2 again. Repeat this process, until we get $\theta_2 = f_{i-1}^{(2)}$. Let $\tilde{l}_i = f_{i-1}^{(2)} + \delta, k = m$, and go to Step 4.
Step 7. Let $\sigma(\mathscr{A}, \mathscr{B}) = \{\xi^\top a(\mathbf{u})/\xi^\top b(\mathbf{u}) \ : \ \mathbf{u} \in U\}$.

The convergence of the algorithm, and more properties can be found in [93].

5.6 Notes

The tensor eigenvalue complementarity problem is a companion problem of the tensor complementarity problem [126]. In this chapter, we study the TEiCP from theory point of view to calculation methods. It should be noted that the topic on this direction still need to study in the future since the convergence analysis of the proposed algorithm and designing algorithms for nonsymmetric cases need further study.

Section 5.1 The contents in this section was first given by Ling, He, and Qi in [179]. For the sake of completeness, we give a detailed proof for Theorem 5.5.

Section 5.2 We study properties of Pareto H-eigenvalues and Pareto Z-eigenvalues in this section. It was originally defined by Song and Qi in [249]. Related definitions about H^+-eigenvalue, H^{++}-eigenvalue, Z^+-eigenvalue and Z^{++}-eigenvalue can check the book [228].

Section 5.3 In this section, the damped semi-smooth Newton method was first presented by Chen and Qi in [56]. Proposition 5.10 was originally proved by Clarke in [60], and Proposition 5.11 was proved by Chen, Chen, and Kanzow in [40]. The scaling-and-projection algorithm was first introduced by Costa and Seeger in [73], for solving matrix cone constrained eigenvalue problems. Motivated by this, Ling He and Qi gave a new form of the algorithm for solving the tensor case [179].

Section 5.4 The main results of this section was originally given by Ling, He and Qi in [180].

Section 5.5 Fan, Nie, and Zhou gave the semidefinite relaxation method in [93]. It is a polynomial method that can be solved by Lasserre type semidefinite relaxations, which was first given by Lasserre in [162, 163]. Nie et al. developed this method in [200, 201, 203–207]. By the way, it should be noted that the Lasserre type semidefinite relaxations (5.103) and (5.107) can be solved by the software GloptiPoly 3 [122] and SeDuMi [255].

5.7 Exercises

1 Assume $\mathscr{A} = (a_{i_1i_2i_3i_4})$ is a tensor with order 4 and dimension 2. Suppose that $a_{1111} = 1, a_{2222} = 2$, and

$$a_{1122} + a_{1212} + a_{1221} = -1, \quad a_{2121} + a_{2112} + a_{2211} = -2,$$

and $a_{i_1i_2i_3i_4} = 0$ for the others. Then, please give all the Pareto H-eigenvalues and Pareto Z-eigenvalues of $\mathscr{A}$.

2 Assume $\mathscr{A} = (a_{i_1i_2i_3})$ is a 3-order and 2-dimensional tensor with entries such that

$$a_{111} = 1, a_{222} = 2, a_{122} = a_{212} = a_{221} = \frac{1}{3}, a_{112} = a_{121} = a_{211} = -\frac{2}{3}.$$

Please verify that some Pareto H-eigenvalue (Pareto Z-eigenvalue) of the tensor $\mathscr{A}$ may not be its H^+-eigenvalue (Z^+-eigenvalue).

3 Let $\mathscr{A}$ be a tensor with order 4 and dimension 2. Suppose

$$a_{1111} = a_{2222} = 1, a_{1112} = a_{1211} = a_{1121} = a_{2111} = t$$

and $a_{i_1i_2i_3i_4} = 0$ for the others. Please discuss the copositiveness of tensor $\mathscr{A}$.

4 Prove Propositions 5.5 and 5.6.

5 Prove Proposition 5.9.

Chapter 6
Higher Order Diffusion Tensor Imaging

Diffusion tensor imaging (DTI) is one of the most promising medical imaging models, and the most applicable technique in modern clinical medicine. While, there are limitations to DTI, which becomes useless in non-isotropic materials. As a resolution, diffusion kurtosis imaging (DKI) is proposed as a new model in medical engineering, which can characterize the non-Gaussian diffusion behavior in tissues, and in which a diffusion kurtosis (DK) tensor is involved. A DK tensor is a fourth order three dimensional symmetric tensor. In this chapter, we will apply the spectral theory of tensors to this particular type of tensors arising from medical imaging and derive some applications.

In order to handle DK tensor, we will introduce D-eigenvalues, which is analogue to Z-eigenvalues. The largest, the smallest and the average D-eigenvalues of a DK tensor correspond with the largest, the smallest and the average apparent kurtosis coefficients (AKC) of a water molecule in the space, respectively. The computational methods for these quantities and their related anisotropy value of AKC will be discussed into details.

In reality, a term in the extended Stejskal and Tanner equation of DKI should be positive for an appropriate range of b-values to make sense physically. The positive definiteness of the term reflects the signal attenuation in tissues during imaging. Hence, it is essential for the validation of DKI. We will then analyze the positive definiteness of DKI. We first characterize the positive definiteness of DKI through the positive definiteness of a tensor constructed by diffusion tensor and diffusion kurtosis tensor. Then, a conic linear optimization method and its simplified version are proposed to handle the positive definiteness of DKI from the perspective of numerical computation.

On the other hand, high angular resolution diffusion imaging (HARDI) is used to characterize non-Gaussian diffusion processes to overcome the limitation of DTI. One approach to analyze HARDI data is to model the apparent diffusion coefficient (ADC) with *higher order diffusion tensors*. The diffusivity function is positive semidefinite. Along to the above version of DKI, which is proposed to preserve

L. Qi et al., *Tensor Eigenvalues and Their Applications*, Advances in Mechanics and Mathematics 39, https://doi.org/10.1007/978-981-10-8058-6_6

positive semidefiniteness of fourth order diffusion tensors, we will propose a comprehensive model to approximate the ADC profile by a positive semidefinite diffusion tensor of either second or higher order. We call this the *positive semidefinite diffusion tensor* (PSDT) model. PSDT is a convex optimization problem with a convex quadratic objective function constrained by the nonnegativity requirement on the smallest Z-eigenvalue of the diffusion tensor. The smallest Z-eigenvalue is a computable measure of the extent of positive definiteness of the diffusivity function. We also propose some other invariants for the ADC profile analysis.

Similarly, Q-Ball Imaging (QBI) is used to estimate the probability density function (PDF) of the average spin displacement of water molecules. In particular, QBI is used to obtain the diffusion orientation distribution function (ODF) of these multiple fiber crossing. As a probability distribution function, the orientation distribution function should be nonnegative which is not guaranteed in the existing methods. In a line, we will give a method to guarantee the nonnegative property of ODF by solving a convex optimization problem, which has a convex quadratic objective function and a constraint involving the nonnegativity requirement on the smallest Z-eigenvalue of the diffusivity tensor.

In the last part of this chapter, we will display an application of the D-eigenvalue to image authenticity verification. We will introduce the gradient skewness tensor which involves a three order tensor derived from the skewness statistic of gradient images. The skewness of oriented gradients can measure the directional characteristic of illumination in an image, the local illumination detection problem for an image can be abstracted as solving the largest D-eigenvalue of gradient skewness tensors. We will discuss properties of D-eigenvalues and D-characteristic polynomial for gradient skewness tensors. Some numerical experiments show its effective application in illumination detection. This method also presents excellent results in a class of image authenticity verification problems, which is to distinguish artificial flat objects in a photograph.

6.1 Diffusion Kurtosis Tensor Imaging and D-Eigenvalues

Preliminary. Diffusion-weighted magnetic resonance imaging (DW-MRI) measures the diffusion of water molecules in biological tissues. Owing to the restriction of biological tissues, the diffusion of water molecules therein is not free. For instance, a molecule inside the axon of a neuron moves principally along the axis of the neural fiber. The molecule has a low probability of crossing the myelin membrane. Hence, water molecule diffusion patterns can reveal microscopic architecture of tissues. Clinically, DW-MRI helps understand the connectivity of white matter axons in the central nervous system and is useful for diagnoses of conditions or neurological disorders, e.g., stroke.

As a special kind of DW-MRI, **diffusion tensor imaging** (DTI) is extensively used to map white matter in the brain. Given a pulsed magnetic field gradient sequence with a gradient direction $\mathbf{x} = (x_1, x_2, x_2)^\top$ and a diffusion weighted b-value b, the Stejskal–Tanner equation for spin-echo signal attenuation [254] is

$$\ln[S/S_0] = -bD_{app}, \tag{6.1}$$

where S is the measured signal, S_0 is the signal with no gradients, and D_{app} is the apparent diffusion coefficient. Measuring diffusivity in multiple directions, we approximately obtain

$$D_{app} = \mathbf{x}^\top D\mathbf{x} = \sum_{i=1}^{3}\sum_{j=1}^{3} d_{ij}x_ix_j, \tag{6.2}$$

where $D = (d_{ij})$ is called the **diffusion tensor** [9, 10, 164]. Mathematically, $D \in \mathbf{R}^{3\times 3}$ is a symmetric and positive definite matrix.

To estimate six independent elements of the diffusion tensor D, we measure signals in six or more non-collinear and non-coplanar gradients directions. By fitting these signals in (6.1) and (6.2), we get the elements of D. Once the diffusion tensor D is obtained, some rotation invariants are calculable. Let $\alpha_1 \ge \alpha_2 \ge \alpha_3 > 0$ be three eigenvalues of the diffusion tensor D. Obviously, eigenvalues of D are irrelevant to coordinate system being used. Based on these eigenvalues, several invariants of DTI are defined such as the mean diffusivity

$$\mathrm{MD} = \frac{1}{3}(\alpha_1 + \alpha_2 + \alpha_3)$$

and the fractional anisotropy (FA)

$$\mathrm{FA} = \sqrt{\frac{3}{2}}\sqrt{\frac{(\alpha_1 - \mathrm{MD})^2 + (\alpha_2 - \mathrm{MD})^2 + (\alpha_3 - \mathrm{MD})^2}{\alpha_1^2 + \alpha_2^2 + \alpha_3^2}}.$$

Here, $0 \le \mathrm{FA} \le 1$. If $\mathrm{FA} = 1$, the diffusion is anisotropic. In a certain area of cerebral white matter, such as the optic nerve, there is a single fiber [178]. Then, FA gets close to one, the eigenvector of D associated with α_1 corresponds well with the orientation of the fiber.

In a voxel, DTI is able to capture the configuration of a single fiber, whose orientation corresponds to the eigenvector of the largest eigenvalue of the diffusion tensor D. For example, the left and the median images in Fig. 6.1 illustrate fibers with a horizontal direction and a vertical direction, respectively. However, if these two fibers crossing, the profile of the diffusion tensor D seems like a doughnut and a pizza; See the right image in Fig. 6.1. The orientations of crossing fibers are unavailable.

Note that one third of the white matter regions in a human brain contains more than one nerve fibers [14]. In these complex regions, matrix-based DTI fails to capture fiber crossing and kissing. Some effective approaches based on tensors are developed.

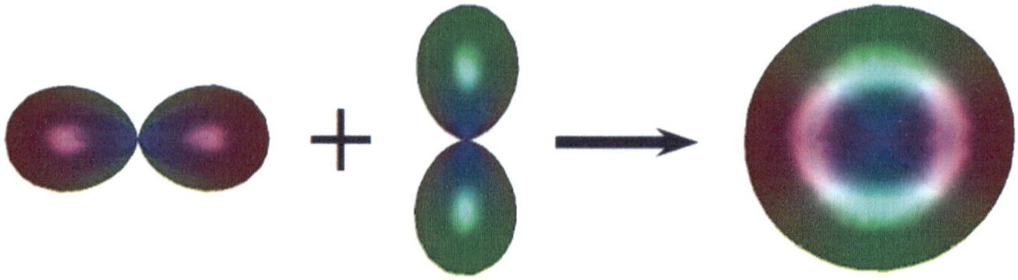

Fig. 6.1 Profiles of ADC function for DTI imaging

Diffusion kurtosis imaging. Due to the complex structure of biological tissues, which consists of various cells and their membranes, the diffusion of water through them may deviate substantially from a Gaussian form. This deviation from Gaussian behavior could be quantified conveniently by a dimensionless metric called the excess kurtosis. Hence, we regard the kurtosis as a measure of a tissue's degree of structure.

By introducing the kurtosis, diffusion kurtosis imaging (DKI) [148, 185] approximates the signal attenuation by

$$\ln[S/S_0] = -bD_{app} + \frac{1}{6}b^2 D_{app}^2 K_{app}. \tag{6.3}$$

where the apparent diffusion coefficient D_{app} is defined as in (6.2) and the apparent kurtosis coefficient (AKC)

$$K_{app} = \frac{M_D^2}{D_{app}^2} \sum_{i=1}^{3} \sum_{j=1}^{3} \sum_{k=1}^{3} \sum_{\ell=1}^{3} w_{ijk\ell} x_i x_j x_k x_\ell. \tag{6.4}$$

Here, D is a second order diffusion tensor and $\mathcal{W} = (w_{ijk\ell}) \in S_{4,3}$ is a fourth order symmetric **diffusion kurtosis tensor**. For a fixed gradient direction $\mathbf{x}$, we sample signals with multiple b-values, such as $b = 0, 500, 1000, 1500, 2000$, and $2500 s/mm^2$. Then, we estimate D_{app} and K_{app} by fitting (6.3). Taking several gradient directions into account, we could determine elements of the diffusion tensor D and the diffusion kurtosis tensor $\mathcal{W}$ [58, 141, 142].

D-eigenvalues. Let $\mathcal{W}$ be an mth order symmetric tensor and D be a second order positive definite tensor. If there exist a scalar $\lambda \in \mathbf{R}$ and a real vector $\mathbf{x} \in \mathbf{R}^3$ satisfying

$$\begin{cases} \mathcal{W}\mathbf{x}^{m-1} = \lambda D\mathbf{x}, \\ D\mathbf{x}^2 = 1, \end{cases} \tag{6.5}$$

then, λ is called a D-eigenvalue of $\mathcal{W}$, $\mathbf{x}$ is called the D-eigenvector of W associated with the D-eigenvalue λ, and $(\lambda, \mathbf{x})$ is called a D-eigenpair of $\mathcal{W}$. Here, the letter "D" means diffusion. If $D = I$, D-eigenvalues of $\mathcal{W}$ reduce to Z-eigenvalues.

Theorem 6.1 *Suppose that $\mathcal{W}$ is an mth order symmetric tensor and D is a second order positive definite tensor. Then, D-eigenvalues always exist. If $(\lambda, \mathbf{x})$ is a D-eigenpair of $\mathcal{W}$, $\lambda = \mathcal{W}\mathbf{x}^m$.*

Proof We consider an optimization problem

$$\begin{cases} \max & \mathcal{W}\mathbf{x}^m \\ \text{s.t.} & D\mathbf{x}^2 = 1. \end{cases} \tag{6.6}$$

Obviously, the objective function $\mathcal{W}\mathbf{x}^m$ is smooth. The feasible region $\{\mathbf{x} \in \mathbf{R}^3 : D\mathbf{x}^2 = 1\}$ is compact since D is positive definite. Hence, the maximum value of the objective on the feasible region is attainable. By optimality condition, the maximum point $\mathbf{x}$ satisfies (6.5) for some multiplier λ. Hence, there exist D-eigenvalues. Multiplying $\mathbf{x}$ to both sides of $\mathcal{W}\mathbf{x}^{m-1} = \lambda D\mathbf{x}$ and using $D\mathbf{x}^2 = 1$, we get $\lambda = \mathcal{W}\mathbf{x}^m$ straightforwardly. □

In the context of DKI, we study the extremal values of AKC (6.4). Obviously, maximizing K_{app} is equivalent to

$$\begin{cases} \max & \mathcal{W}\mathbf{x}^4 \\ \text{s.t.} & D\mathbf{x}^2 = 1, \mathbf{x} \in \mathbf{R}^3. \end{cases}$$

Here, $m = 4$ and the diffusion tensor D is positive definite. Let $\lambda_{D,\max}$ and $\lambda_{D,\min}$ be the largest D-eigenvalue and the smallest D-eigenvalue of the diffusion kurtosis tensor $\mathcal{W}$. Then, the largest AKC and the smallest AKC are equal to $\text{MD}^2\lambda_{D,\max}$ and $\text{MD}^2\lambda_{D,\min}$, respectively.

When $m = 4$, we consider the resultant of

$$\mathcal{W}\mathbf{x}^3 = \lambda(D\mathbf{x}^2)D\mathbf{x}, \tag{6.7}$$

which is a homogeneous representation of (6.5). By algebraic geometry [64], the resultant is a one dimensional polynomial $\phi(\lambda)$, which is defined as the D-characteristic polynomial of $\mathcal{W}$. The polynomial system (6.7) has a nonzero complex solution if and only if λ is a root of the D-characteristic polynomial $\phi(\lambda)$. Hence, we get the following theorem.

Theorem 6.2 *In DKI, a D-eigenvalue of the diffusion kurtosis tensor $\mathcal{W}$ is a real root of the D-characteristic polynomial $\phi(\lambda)$ of $\mathcal{W}$.*

Note that a real root of $\phi(\lambda)$ may not be a D-eigenvalue of $\mathcal{W}$, as such a real root may have only complex solution $\mathbf{x}$.

According to algebraic geometry [64], the degree of $\phi(\lambda)$ is at most 13. If the number ν of D-eigenvalues of $\mathcal{W}$ is finite, then $\nu \leq 13$. We order these D-eigenvalues as $\lambda_1 \geq \lambda_2 \geq \cdots \geq \lambda_\nu$. Similar to MD and FA in DTI, we define the mean kurtosis

$$\mathrm{MK} = \frac{1}{\nu}\sum_{i=1}^{\nu}\lambda_i$$

and the fractional kurtosis anisotropy

$$\mathrm{FA}_K = \sqrt{\frac{\nu}{\nu-1}}\sqrt{\frac{\sum_{i=1}^{\nu}(\lambda_i - \mathrm{MK})^2}{\sum_{i=1}^{\nu}\lambda_i^2}}.$$

Theorem 6.3 *D-eigenvalues of the diffusion kurtosis tensor* $\mathscr{W}$, MK *and* FA_K *are invariant under orthogonal transformations.*

Proof Let $P = (p_{ij})$ be an orthogonal matrix, $\tilde{D} = DP^2 = P^\top DP$, and $\tilde{\mathscr{W}} = \mathscr{W}P^4 = \mathscr{W} \times_1 P \times_2 P \times_3 P \times_4 P$. Suppose that $(\lambda, \mathbf{x})$ is a D-eigenpair of $\mathscr{W}$ and $\mathbf{y} = P^\top \mathbf{x}$. We will prove that

$$\tilde{\mathscr{W}}\mathbf{y}^3 = \lambda\tilde{D}\mathbf{y}, \quad \text{and} \quad \tilde{D}\mathbf{y}^2 = 1,$$

i.e., $(\lambda, \mathbf{y})$ is a D-eigenpair of $\tilde{\mathscr{W}}$ with $\tilde{\mathscr{D}}$.

Because P is orthogonal and $\mathbf{y} = P^\top \mathbf{x}$, we have $\mathbf{x} = P\mathbf{y}$. By some calculations, we obtain

$$\tilde{D}\mathbf{y}^2 = \mathbf{y}^\top\tilde{D}\mathbf{y} = \mathbf{y}^\top P^\top DP\mathbf{y} = \mathbf{x}^\top D\mathbf{x} = 1$$

and

$$\begin{aligned}
\tilde{\mathscr{W}}\mathbf{y}^3 &= (\mathscr{W} \times_1 P \times_2 P \times_3 P \times_4 P) \times_2 \mathbf{y} \times_3 \mathbf{y} \times_4 \mathbf{y} \\
&= P^\top(\mathscr{W} \times_2 (P\mathbf{y}) \times_3 (P\mathbf{y}) \times_4 (P\mathbf{y})) \\
&= P^\top(\mathscr{W} \times_2 \mathbf{x} \times_3 \mathbf{x} \times_4 \mathbf{x}) \\
&= P^\top(\mathscr{W}\mathbf{x}^3) \\
&= P^\top(\lambda D\mathbf{x}) \\
&= \lambda P^\top DP\mathbf{y} \\
&= \lambda\tilde{D}\mathbf{y}.
\end{aligned}$$

In addition, MK and FA_K are functions of D-eigenvalues of $\mathscr{W}$. They are also invariant under orthogonal transformations. □

6.2 Positive Definiteness of Diffusion Kurtosis Imaging

In the signal attenuation Eq. (6.3) of DKI, the term $bD_{app} - \frac{1}{6}b^2D_{app}^2K_{app}$ reflects the signal attenuation during imaging. Hence, to make sense physically, this term should be positive for all gradients $\mathbf{x}$ directions and a range of diffusion weighted b-values $[0, \bar{b}]$ [128]. That is to say, we enforce the following constraint in DKI:

$$bD\mathbf{x}^2 - \frac{1}{6}b^2\text{MD}^2\mathscr{W}\mathbf{x}^4 > 0 \qquad \forall \mathbf{x} \in \mathbb{S}^2 \text{ and } b \in (0, \bar{b}],$$

where $\mathbb{S}^2 = \{\mathbf{x} = (x_1, x_2, x_2)^\top : \mathbf{x}^\top\mathbf{x} = 1\}$. Since the diffusion tensor D is positive definite, it is sufficient to make sure

$$D\mathbf{x}^2 - \frac{1}{6}\bar{b}\text{MD}^2\mathscr{W}\mathbf{x}^4 > 0 \qquad \forall \mathbf{x} \in \mathbb{S}^2. \tag{6.8}$$

First, we recall the process of the traditional DKI. The whole biological tissues, say human brain, are divided into serval small voxels. For each voxel, DKI measures signals with multiple diffusion weighted b-values $\{b_0, b_1, \ldots, b_k\}$ and several gradient directions $\{\mathbf{x}^{(1)}, \mathbf{x}^{(2)}, \ldots, \mathbf{x}^{(r)}\}$, where $k \geq 2$, $r \geq 15$, $b_0 = 0$ usually and hence $S(b_0) = S_0$. A two-step approach is performed as follows.
Step 1 Using $k+1$ signals $\{S(b_0), S(b_1), \ldots, S(b_k)\}$ with one gradient direction $\mathbf{x}^{(j)}$, we estimate $D_{app}^{(j)}$ and $K_{app}^{(j)}$ of that gradient direction by fitting a system of nonlinear equations in two variables:

$$\begin{cases} \ln[S(b_1)/S_0] = -b_1 D_{app}^{(j)} + \frac{1}{6}b_1^2(D_{app}^{(j)})^2 K_{app}^{(j)}, \\ \quad\vdots \\ \ln[S(b_k)/S_0] = -b_k D_{app}^{(j)} + \frac{1}{6}b_k^2(D_{app}^{(j)})^2 K_{app}^{(j)}, \end{cases}$$

for $j = 1, \ldots, r$.
Step 2 The diffusion tensor D and the diffusion kurtosis tensor $\mathscr{W}$ are estimated from m estimated D_{app}'s and K_{app}'s, respectively. Under a basis $\mathbf{u}(\mathbf{x}) = (x_1^2, x_1x_2, x_1x_3, x_2^2, x_2x_3, x_3^2)^\top$, the ADC is

$$D_{app} = D\mathbf{x}^2 = \mathbf{u}(\mathbf{x})^\top\mathbf{d},$$

where $\mathbf{d} = (d_{11}, d_{12}, d_{13}, d_{22}, d_{23}, d_{33})^\top$ is a vector collecting all independent elements of the diffusion tensor D. Taking m gradient directions into account, we get

$$\begin{pmatrix} \mathbf{u}(\mathbf{x}^{(1)}) \\ \vdots \\ \mathbf{u}(\mathbf{x}^{(r)}) \end{pmatrix} \mathbf{d} = \begin{pmatrix} D_{app}^{(1)} \\ \vdots \\ D_{app}^{(r)} \end{pmatrix}.$$

Since $r \geq 6$, we determine $\mathbf{d}$ by the linear least squares method. Similarly, we can obtain 15 independent elements of the diffusion kurtosis tensor $\mathscr{W}$ since $r \geq 15$.

Obviously, the traditional two-step approach does not enforce the positive definiteness of DKI.

The positive definite DKI combines these two steps into one mathematical program and ensures the diffusion tensor and the diffusion kurtosis tensor positive definite. At the beginning, we introduce some notations. Let D be a matrix. Its trace is $\text{trace}(D) = \langle D, I\rangle$, where $\langle\cdot,\cdot\rangle$ is the inner product. $D \succ 0$ means the matrix D is positive definite. We use $\otimes$ to stand for the outer product. For symmetric matrices I and D, $I \otimes D$ generates a fourth order tensor, which may not be symmetric. For any vector $\mathbf{x}$, $(I \otimes D)\mathbf{x}^4$ is a homogeneous polynomial with degree 4. From this homogeneous polynomial, there is a unique symmetric tensor $\mathrm{Sym}(I \otimes D)$ corresponding to it, i.e., $(I \otimes D)\mathbf{x}^4 = \mathrm{Sym}(I \otimes D)\mathbf{x}^4$. Let $\mathscr{W}$ be a fourth order tensor. We also use $\mathscr{W} \succ 0$ to denote that $\mathscr{W}$ is positive definite, i.e., $\mathscr{W}\mathbf{x}^4 > 0$ for all $\mathbf{x} \neq 0$. Let $\mathbb{Q}^{n+1} = \{(\alpha, \mathbf{x}) \in \mathbf{R}^{n+1} : \alpha \geq \|\mathbf{x}\|_2\}$ be a second order cone, which is closed and convex. Then, we present the conic optimization model formally

$$
\begin{cases}
\min \ \alpha \\
\text{s.t.} \ \ D \succ 0, \\
\qquad \mathrm{Sym}(I \otimes D) - \dfrac{\bar{b}}{6}\left(\dfrac{\mathrm{trace}(D)}{3}\right)^2 \mathscr{W} \succ 0, \\
\qquad (\alpha, \mathbf{y}) \in \mathbb{Q}^{kr+1}, \\
\qquad y_{(i-1)r+j} = -\ln[S(b_i, \mathbf{x}^{(j)})/S(b_0)] - b_i\langle D, \mathbf{x}^{(j)} \otimes \mathbf{x}^{(j)}\rangle \\
\qquad \quad + \dfrac{1}{6}b_i^2\left(\dfrac{\mathrm{trace}(D)}{3}\right)^2 \langle \mathscr{W}, \mathbf{x}^{(j)} \otimes \mathbf{x}^{(j)} \otimes \mathbf{x}^{(j)} \otimes \mathbf{x}^{(j)}\rangle, \\
\qquad \qquad \text{for } i = 1, \ldots, k \text{ and } j = 1, \ldots, r,
\end{cases}
\tag{6.9}
$$

where α, $\mathbf{y}$ are auxiliary variables. Obviously, the solution of (6.9) satisfies the positive definiteness constraint (6.8).

We note here that the conic optimization (6.9) without the first two positive definiteness constraints is exactly a combination of the two-step approach.

Next, we transform the conic nonlinear programming into a conic linear programming. Let $\mathscr{K} = \mathrm{MD}^2\mathscr{W}$. If the positive definite matrix D is available, we could compute $\mathscr{K}$ and $\mathscr{W}$ from each other. The second constraint of (6.9) on the positive definite tensor reduces to $\mathrm{Sym}(I \otimes D) - \frac{\bar{b}}{6}\mathscr{K} \succ 0$. We argue that there is a one-to-one correspondence between a fourth order three dimensional symmetric tensor and a ternary quartic, which is a homogeneous polynomial of degree 4 in three variables. Hilbert [124] showed that every non-negative real ternary quartic form is a sum of three squares of quadratic forms. Hence, $\mathrm{Sym}(I \otimes D) - \frac{\bar{b}}{6}\mathscr{K} \succ 0$ if and only if there exists a positive definite matrix K such that $(\mathrm{Sym}(I \otimes D) - \frac{\bar{b}}{6}\mathscr{K})\mathbf{x}^4 = \mathbf{u}(\mathbf{x})^\top K\mathbf{u}(\mathbf{x})$. In this way, we could define a linear operator P which maps the 6×6 symmetric matrix S into a fourth order three dimensional symmetric tensor $\mathscr{T} = \mathrm{P}(S)$ such that $\mathbf{u}(\mathbf{x})^\top S\mathbf{u}(\mathbf{x}) = \mathscr{T}\mathbf{x}^4$ for all $x \in \mathbf{R}^3$.

Using the new undetermined tensor $\mathscr{K}$ and the linear operator P, we get the conic linear optimization model:

$$\begin{cases} \min & \alpha \\ \text{s.t.} & D \succ 0, \\ & S \succ 0, \\ & \mathrm{P}(S) = \mathrm{Sym}(I \otimes D) - \frac{\bar{b}}{6}\mathscr{K}, \\ & (\alpha, \mathbf{y}) \in \mathbb{Q}^{kr+1}, \\ & y_{(i-1)r+j} = -\ln[S(b_i, \mathbf{x}^{(j)})/S(b_0)] - b_i\langle D, \mathbf{x}^{(j)} \otimes \mathbf{x}^{(j)}\rangle \\ & \quad + \frac{1}{6}b_i^2\langle \mathscr{K}, \mathbf{x}^{(j)} \otimes \mathbf{x}^{(j)} \otimes \mathbf{x}^{(j)} \otimes \mathbf{x}^{(j)}\rangle, \\ & \quad\quad \text{for } i = 1, \ldots, k \text{ and } j = 1, \ldots, r. \end{cases} \tag{6.10}$$

Some sophisticated optimization softwares, such as SeDuMi [255] and SDPT3 [264], are developed for solving the conic linear optimization problem (6.10).

6.3 Positive Semidefinite Diffusion Tensor Imaging

Diffusion tensor imaging (DTI) resolves the fiber orientation within a voxel of cerebral white matter if fibers are strongly aligned. However, in more than 30% of the voxels in a brain image, there are at least two different neural fibers traveling in different directions that pass through each other. To resolve orientations of these crossing fibers, high angular resolution diffusion imaging (HARDI) [267] is proposed to look into the voxel from a large number of different gradient directions $\mathbf{x}^{(j)}$ for $j = 1, 2, \ldots, r$ (typical $r \geq 40$) with a higher diffusion weighted b-value b. The strategy is different from DKI, where multiple b-values are equipped.

From the Stejskal–Tanner equation for signal attenuation (6.1), Özarslan and Mareci [215] proposed to replace the original diffusion tensor by a higher order ($m \geq 4$) **generalized diffusion tensor** $\mathscr{D} = (d_{i_1 i_2 \ldots i_m})$, i.e.,

$$D_{app}(\mathbf{x}) = \mathscr{D}\mathbf{x}^m = \sum_{i_1=1}^{3}\sum_{i_2=1}^{3}\cdots\sum_{i_m=1}^{3} d_{i_1 i_2 \ldots i_m} x_{i_1} x_{i_2} \cdots x_{i_m}, \tag{6.11}$$

where $\mathscr{D}$ is mth order three dimensional symmetric tensor, which contains

$$N = \binom{m+2}{2} = \frac{1}{2}(m+1)(m+2)$$

independent elements. By collecting like terms, we define

$$d_{i,j,m-i-j} := \frac{m!}{i!j!(m-i-j)!} D_{\underbrace{1\ldots1}_{i}\underbrace{2\ldots2}_{j}\underbrace{3\ldots3}_{m-i-j}}.$$

Then, we get a vector that collects all the independent elements of the higher order diffusion tensor $\mathscr{D}$

$$\mathbf{d} := (d_{m,0,0}, d_{m-1,1,0}, \ldots, d_{0,0,m})^\top \in \mathbf{R}^N.$$

This vector $\mathbf{d}$ is called the representation vector of the symmetric tensor $\mathscr{D}$. Let $\mathbf{u}(\mathbf{x}) := (x_1^m, x_1^{m-1}x_2, \ldots, x_3^m)^\top$ be a basis of homogeneous polynomials with mth degree in three dimension. We could rewrite (6.11) as

$$D_{app}(\mathbf{x}) = \mathbf{u}(\mathbf{x})^\top \mathbf{d}.$$

Owing to the physical inherence of MRI, the ADC function $D_{app}(\mathbf{x})$ should be positive in all gradient directions $\mathbf{x}$, i.e.,

$$\mathscr{D}\mathbf{x}^m \geq 0, \qquad \forall \mathbf{x} \in \mathbb{S}^2. \tag{6.12}$$

If so, we call the symmetric tensor $\mathscr{D}$ positive semidefinite, and the representation vector $\mathbf{d}$ of $\mathscr{D}$ is called a positive semidefinite vector. Clearly, m should be even such that there are nonzero positive semidefinite vectors. The special case of $m = 4$ was studied in [8, 286]. Denote the set of all positive semidefinite vectors as $\mathbb{S}_m$, or simply $\mathbb{S}$ when m is fixed.

Theorem 6.4 *$\mathbb{S}$ is a closed convex cone in $\mathbf{R}^N$.*

Proof Let $\mathbf{d}^{(1)}, \mathbf{d}^{(2)} \in \mathbb{S}$ and $\alpha, \beta \geq 0$. Denote $\mathbf{d} = \alpha\mathbf{d}^{(1)} + \beta\mathbf{d}^{(2)}$. For any $\mathbf{x} \in \mathbf{R}^3$,

$$\mathbf{u}(\mathbf{x})^\top \mathbf{d} = \alpha\mathbf{u}(\mathbf{x})^\top \mathbf{d}^{(1)} + \beta\mathbf{u}(\mathbf{x})^\top \mathbf{d}^{(2)} \geq 0.$$

Hence $\mathbf{d} \in \mathbb{S}$. This proves that $\mathbb{S}$ is a convex cone. Let $\{\mathbf{d}^{(k)}\} \subset \mathbb{S}$ and $\lim_{k\to\infty} \mathbf{d}^{(k)} = \mathbf{d}$. For any $\mathbf{x} \in \mathbf{R}^3$,

$$\mathbf{u}(\mathbf{x})^\top \mathbf{d} = \lim_{k\to\infty} \mathbf{u}(\mathbf{x})^\top \mathbf{d}^{(k)} \geq 0.$$

This shows that $\mathbb{S}$ is closed. □

By measuring HARDI signals in $r(\geq N)$ gradient directions $\{\mathbf{x}^{(j)} : j = 1, \ldots, r\}$, we obtain the corresponding ADC values that was collected as a vector $\mathbf{f} = (f_j) \in \mathbf{R}^r$. We assume that $\{\mathbf{u}(\mathbf{x}^{(j)}) \in \mathbf{R}^N : j = 1, \ldots, r\}$ spans $\mathbf{R}^N$, i.e., their rank is N. We call this assumption the *full rank assumption*. The full rank assumption always holds in HARDI. Let A be an $N \times r$ matrix, whose column vectors are $\{\mathbf{u}(\mathbf{x}^{(j)}) \in \mathbf{R}^N : j = 1, \ldots, r\}$. Let $B = AA^\top$. Then B is an $N \times N$ symmetric and positive definite matrix.

To find the generalized diffusion tensor, we may fit a linear least squares

$$\min \|A^\top \mathbf{d} - \mathbf{f}\|^2 \tag{6.13}$$

whose solution is denoted as

$$\bar{\mathbf{d}} = B^{-1}A\mathbf{f}.$$

Because $\bar{\mathbf{d}}$ may not be positive semidefinite, we add the positive semidefinite constraint to the linear least squares fitting

$$\min_{\mathbf{d}\in\mathbb{S}} L(\mathbf{d}) = (\mathbf{d}-\bar{\mathbf{d}})^\top B(\mathbf{d}-\bar{\mathbf{d}}). \tag{6.14}$$

The objective function L is a convex quadratic function and the feasible region $\mathbb{S}$ is a closed convex cone. Now, we prove some important theoretical properties of the convex optimization problem (6.14), which is useful for calculating a positive definite generalized diffusion tensor.

Theorem 6.5 *The convex optimization problem (6.14) has a unique global solution* $\mathbf{d}^*$. *If* $\bar{\mathbf{d}} \in \mathbb{S}$, $\mathbf{d}^* = \bar{\mathbf{d}}$. *Otherwise,* $\bar{\mathbf{d}} \notin \mathbb{S}$, *then* $\mathbf{d}^*$ *is on* $\partial\mathbb{S}$, *the boundary of* $\mathbb{S}$.

Proof Using $\bar{\mathbf{d}}$, we could rewrite (6.13) as

$$\|A^\top\mathbf{d}-\mathbf{f}\|^2 = (\mathbf{d}-\bar{\mathbf{d}})^\top B(\mathbf{d}-\bar{\mathbf{d}}) + \|(I - A^\top(AA^\top)^{-1}A)\mathbf{f}\|^2.$$

Since the second term is a constant, the global solutions of min $\|A^\top\mathbf{d}-\mathbf{f}\|^2$ and min $L(\mathbf{d})$ are equivalent. Hence, If $\bar{\mathbf{d}} \in \mathbb{S}$, $\bar{\mathbf{d}}$ is a global solution of (6.14).

Since the matrix B is positive definite, $L(\mathbf{d})$ is a strictly convex function. Moreover, $\mathbb{S}$ is convex, $L(\mathbf{d})$ has one global solution in $\mathbb{S}$ at most. Since $\mathbb{S}$ is closed and convex and $L(\mathbf{d}) \to \infty$ as $|\mathbf{d}| \to \infty$, $L(\mathbf{d})$ has at least one global solution in $\mathbb{S}$. Hence, The convex optimization (6.14) has a unique global solution. If $\bar{\mathbf{d}} \in \mathbb{S}$, it is straightforward to see that the global solution $\mathbf{d}^* = \bar{\mathbf{d}}$.

If $\bar{\mathbf{d}} \notin \mathbb{S}$, we have $\mathbf{d}^* \neq \bar{\mathbf{d}}$ since the global solution $\mathbf{d}^* \in \mathbb{S}$. Then, there exists a scalar $t_0 \in [0, 1)$ such that $\mathbf{d}^* + t_0(\bar{\mathbf{d}} - \mathbf{d}^*)$ is on $\partial\mathbb{S}$. Recalling $\mathbf{d}^*$ is the unique global minimizer of (6.14), we have

$$L(\mathbf{d}^*) \le L(\mathbf{d}^* + t_0(\bar{\mathbf{d}} - \mathbf{d}^*)),$$

i.e.,

$$(\mathbf{d}^*-\bar{\mathbf{d}})^\top B(\mathbf{d}^*-\bar{\mathbf{d}}) \le (1-t_0)^2(\mathbf{d}^*-\bar{\mathbf{d}})^\top B(\mathbf{d}^*-\bar{\mathbf{d}}).$$

Since B is positive definite and $\mathbf{d}^* \neq \bar{\mathbf{d}}$, we know $(\mathbf{d}^*-\bar{\mathbf{d}})^\top B(\mathbf{d}^*-\bar{\mathbf{d}}) > 0$. This implies that $t_0 = 0$, i.e., $\mathbf{d}^*$ is on $\partial\mathbb{S}$. □

How to identify $\bar{\mathbf{d}} \in \mathbb{S}$ or not? We say that $\mathbf{d} \in \mathbb{S}$ if and only if $\lambda_{\min}(\mathbf{d})$, the smallest Z-eigenvalue of $\mathbf{d}$, is nonnegative.

Theorem 6.6 *Suppose all tensors* $\mathbf{d}$ *are regular. Then, the smallest Z-eigenvalue* $\lambda_{\min}(\mathbf{d})$ *is a continuous concave function.*

Proof From Theorem 1.2, $\lambda_{\min}(\mathbf{d})$ is a root of the E-characteristic polynomial, whose coefficients are continuous polynomials in elements of $\mathbf{d}$. By the fact that the roots of any polynomial are continuous functions of the coefficients, we get the continuity of $\lambda_{\min}(\mathbf{d})$.

Let $\mathbf{d} = t\mathbf{d}^{(1)} + (1-t)\mathbf{d}^{(2)}$, where $\mathbf{d}^{(1)}, \mathbf{d}^{(2)} \in \mathbf{R}^N$ and $0 \le t \le 1$. Suppose $\mathbf{x}^*$ is the Z-eigenvector corresponding to the smallest Z-eigenvalue of $\mathbf{d}$. Then ${\mathbf{x}^*}^\top \mathbf{x}^* = 1$ and

$$\begin{aligned}\lambda_{\min}(\mathbf{d}) &= \mathbf{u}(\mathbf{x}^*)^\top \mathbf{d} \\ &= t\mathbf{u}(\mathbf{x}^*)^\top \mathbf{d}^{(1)} + (1-t)\mathbf{u}(\mathbf{x}^*)^\top \mathbf{d}^{(2)} \\ &\ge t\lambda_{\min}(\mathbf{d}^{(1)}) + (1-t)\lambda_{\min}(\mathbf{d}^{(2)}).\end{aligned}$$

Hence, $\lambda_{\min}(\mathbf{d})$ is a concave function. □

Since $\lambda_{\min}(\cdot)$ is a concave function, we have $-\lambda_{\min}(\cdot)$ is a convex function. Then, we could discuss its subgradient [235]. $\widehat{\mathbf{g}^*}$ is a subgradient of $-\lambda_{\min}(\cdot)$ at $\mathbf{d} \in \mathbf{R}^N$ if

$$-\lambda_{\min}(\widetilde{\mathbf{d}}) \ge -\lambda_{\min}(\mathbf{d}) + \widehat{\mathbf{g}^*}^\top (\widetilde{\mathbf{d}} - \mathbf{d})$$

for all $\widetilde{\mathbf{d}}$.

We consider the following positive semidefinite tensor (PSDT) model:

$$L(\mathbf{d}^*) = \min\{L(\mathbf{d}) : -\lambda_{\min}(\mathbf{d}) \le 0\}. \tag{6.15}$$

Since $L(\mathbf{d})$ is a strictly convex quadratic function and $-\lambda_{\min}(\cdot)$ is continuous and convex, PSDT is a convex optimization problem. We have the following theorem.

Theorem 6.7 *Suppose* $\bar{\mathbf{d}} \notin \mathbb{S}$. *Then* $\mathbf{d}^*$ *is the unique global minimizer of PSDT (6.15) if and only if there is a positive number* μ *such that*

$$\begin{cases} B(\mathbf{d}^* - \bar{\mathbf{d}}) + \mu\widehat{\mathbf{g}^*} = 0, \\ \lambda_{\min}(\mathbf{d}^*) \qquad\qquad = 0, \end{cases} \tag{6.16}$$

where $\widehat{\mathbf{g}^*}$ *is a subgradient of the convex function* $\lambda_{\min}$ *at* $\mathbf{d}^*$. *By (6.16), we have*

$$\begin{cases} (\mathbf{d}^*)^\top B(\mathbf{d}^* - \bar{\mathbf{d}}) = 0, \\ \widehat{\mathbf{g}^*}^\top \mathbf{d}^* \qquad\qquad = 0. \end{cases} \tag{6.17}$$

Proof By Theorem 6.5, PSDT (6.15) has a unique global minimizer $\mathbf{d}^*$, which is on the boundary of $\mathbb{S}$. By the continuity of $\lambda_{\min}(\cdot)$, $\lambda_{\min}(\mathbf{d}^*) = 0$. Since $\bar{\mathbf{d}} \neq \mathbb{S}$, we get that $\mathbf{d}^* \neq \bar{\mathbf{d}}$ and hence $\nabla L(\mathbf{d}^*) \neq 0$. Now, from the optimality condition of the convex optimization problem PSDT (6.15), we know that there exists a positive scaler μ such that (6.16) holds.

From $\lambda_{min}(\mathbf{d}^*) = \mathbf{u}(\mathbf{x}^*)^\top \mathbf{d}^*$, where $\mathbf{x}^*$ is the Z-eigenvector corresponding to the smallest Z-eigenvalue of $\mathbf{d}^*$, we see $\widehat{\mathbf{g}^*} = \mathbf{u}(\mathbf{x}^*)$. By the second equation of (6.16), we get the second equation of (6.17). By taking inner products on both sides of the first equation of (6.16) with $\mathbf{d}^*$, we have the first equation of (6.17). □

We may solve PSDT (6.15) by a standard convex programming method [125]. First, we calculate $\bar{\mathbf{d}} = B^{-1}A\mathbf{f}$. If $\lambda_{\min}(\bar{\mathbf{d}}) \geq 0$, then $\mathbf{d}^* = \bar{\mathbf{d}}$ and the task is completed. If $\lambda_{\min}(\bar{\mathbf{d}}) < 0$, by Theorem 6.7, $\lambda_{\min}(\mathbf{d}^*) = 0$. Hence, in this case, we only need to solve the following model:

$$L(\mathbf{d}^*) = \min\{L(\mathbf{d}) : \lambda_{\min}(\mathbf{d}) = 0\}, \tag{6.18}$$

which has only an equality constraint. However, it is not a convex optimization problem. On the other hand, (6.16) is still its optimality condition. If we use the subgradient of $\lambda_{\min}(\mathbf{d})$ as a substitute of its gradient, according to numerical optimization [211], we may use a gradient descent method to solve (6.18).

According to Theorem 1.2, the Z-eigenvalues are also rotationally invariant. Hence, we may use them and their functions as characteristic quantities of PSDT. After find the global minimizer $\mathbf{d}^*$ of PSDT, we calculate $\lambda_{\min} = \lambda_{\min}(\mathbf{d}^*)$ and the other Z-eigenvalues of $\mathbf{d}^*$ as $\lambda_1 \geq \lambda_2 \geq \cdots \geq \lambda_\nu \geq 0$. Then $\lambda_1 = \lambda_{\max}$ and $\lambda_\nu = \lambda_{\min}$.

As we discussed before, $\lambda_{\min}$ is a measure of the extent of positive definiteness of $\mathbf{d}^*$. On the other hand, if $(\lambda_{\max}, \mathbf{x}^{\max})$ is a Z-eigenpair of $\mathbf{d}^*$, then $\mathbf{x}^{\max}$ is the principal ADC direction. Along this principal direction $\mathbf{x}^{\max}$, the ADC value of $\mathbf{d}^*$ attains its maximum.

We define the PSDT mean value as

$$M_{PSDT} = \frac{1}{\nu}\sum_{i=1}^{\nu} \lambda_i,$$

and the PSDT fractional anisotropy as

$$FA_{PSDT} = \sqrt{\frac{\nu}{\nu - 1}}\sqrt{\frac{\sum_{i=1}^{\nu}(\lambda_i - M_{PSDT})^2}{\sum_{i=1}^{\nu}\lambda_i^2}}.$$

Then we have $0 \leq FA_{PSDT} \leq 1$. If $FA_{PSDT} = 0$, the diffusion is isotropic. If $FA_{PSDT} = 1$, the diffusion is anisotropic.

Chen et al. [50] employed a sum-of-squares (SOS) tensor to approximate the positive definite generalized diffusion tensor.

6.4 Nonnegative Diffusion Orientation Distribution Function

To address multiple fiber crossing of biological tissues, Tuch [266] introduced Q-ball imaging (QBI) to reconstruct the *diffusion orientation distribution function* (ODF) of these multiple fiber crossing, based on HARDI. As the water molecules in normal tissues tend to diffuse along fibers when contained in fiber bundles [9], the principal directions (maxima) of the ODF agree with the true synthetic fiber directions. Tuch [266] showed that the ODF $\Psi(\mathbf{x})$ can be estimated directly from the raw HARDI signal on a single sphere of q-space by the Funk-Radon transformation, which could be regarded as a smoothed version of the true ODF. In fact, the **Funk–Radon transformation** at a given direction $\mathbf{x}$ is the great circle integral of the signal $S(\mathbf{w})$ on the sphere defined by the plane through the origin with normal vector $\mathbf{x}$, i.e.,

$$\Psi(\mathbf{x}) = \int_{\|\mathbf{w}\|=1} \delta(\mathbf{x}^\top \mathbf{w}) S(\mathbf{w}) d\mathbf{w}, \tag{6.19}$$

where $\delta(\cdot)$ is the Dirac delta function that is zero everywhere except at zero with value one.

Suppose that the order m is even and the HARDI signal at the direction $\mathbf{x}$ has the form

$$S(\mathbf{x}) = \mathscr{A}\mathbf{x}^m = \mathbf{u}(\mathbf{x})^\top \mathbf{s}, \tag{6.20}$$

where $\mathscr{A}$ is an mth order symmetric tensor, $\mathbf{u}(\mathbf{x})$ is a basis of mth degree homogeneous polynomials in $\mathbf{x}$, and $\mathbf{s}$ is a vector collecting all independent elements of the symmetric tensor $\mathscr{A}$. Similarly, we denote

$$\Psi(\mathbf{x}) = \mathscr{B}\mathbf{x}^m = \mathbf{u}(\mathbf{x})^\top \mathbf{r}, \tag{6.21}$$

where $\mathscr{B}$ is an mth order symmetric tensor and and $\mathbf{r}$ is a vector collecting independent elements of $\mathscr{B}$.

Now, we establish the relationship between $\mathbf{s}$ and $\mathbf{r}$ using the Funk–Radon transformation (6.19) and the spherical harmonics. For a unit vector $\mathbf{x} \in \mathbb{S}$, its spherical coordinate (θ, ϕ) is helpful:

$$\mathbf{x} = \begin{pmatrix} x_1 \\ x_2 \\ x_3 \end{pmatrix} = \begin{pmatrix} \sin\theta\cos\phi \\ \sin\theta\sin\phi \\ \cos\theta \end{pmatrix},$$

where $0 \le \theta \le \pi$ and $0 \le \phi \le 2\pi$.

The **spherical harmonics** (SH) is a basis for functions on the unit sphere. Let Y_ℓ^q denote SH of order ℓ and phase factor q. Explicitly, it is given as follows

$$Y_\ell^q(\theta, \phi) = \sqrt{\frac{2\ell+1}{4\pi} \cdot \frac{(\ell-|q|)!}{(\ell+|q|)!}} P_\ell^{|q|}(\cos\theta) e^{iq\phi},$$

where P_ℓ^q is associated Legendre polynomials. Let $\ell = 0, 2, \ldots, m$ and $q = -\ell, \ldots, 0, \ldots, \ell$. A single index p in terms of ℓ and q is used such that $p \equiv p(\ell, q) = (\ell^2 + \ell + 2)/2 + q$. Then $p = 1, \ldots, N$. As in [77, 78], the real spherical harmonics of order less than or equal to m, are

$$R_p = \begin{cases} \sqrt{2}\mathrm{Re}(Y_\ell^q) & \text{if } -\ell \le q < 0, \\ Y_\ell^0 & \text{if } q = 0, \\ \sqrt{2}\mathrm{Im}(Y_\ell^q) & \text{if } 0 < q \le \ell, \end{cases}$$

for $p = 1, \ldots, N$, where $\mathrm{Re}(Y_\ell^q)$ and $\mathrm{Im}(Y_\ell^q)$ represent the real and imaginary parts of Y_ℓ^q, respectively. This basis (R_p) is real, symmetric, and orthonormal. Thus, the HARDI signal S can be described as

$$S(\theta, \phi) = \sum_{p=1}^{N} c_p R_p(\theta, \phi).$$

Moreover, Descoteaux et al. [78] showed that the ODF can be expressed as

$$\Psi(\theta, \phi) = \sum_{p=1}^{N} 2\pi P_{\ell(p)}(0) c_p R_p(\theta, \phi), \tag{6.22}$$

where $P_\ell(0)$ is a Legendre polynomial with simple expression

$$P_\ell(0) = (-1)^{\ell/2} \frac{1 \cdot 3 \cdots (\ell - 1)}{2 \cdot 4 \cdots \ell},$$

when ℓ is even. This demonstrated that the ODF can be estimated by scaling of the HARDI signal's spherical harmonic coefficients.

Theorem 6.8 *For $p, k = 1, \ldots, N$, let*

$$t_{pk} = \int_{\mathbb{S}} (\mathbf{u}(\mathbf{x}))_k R_p(\mathbf{x}) d\mathbb{S}.$$

Then $T = (t_{pk})$ is an $n \times n$ invertible matrix. Let D be an $n \times n$ diagonal matrix with its diagonal elements as $P_{l(1)}(0), \ldots, P_{l(N)}(0)$. Let $M = 2\pi T^{-1} D T$. Then we have $\mathbf{r} = M\mathbf{s}$.

Proof We note that both the mth order tensor polynomials restricted to the sphere and the even order spherical harmonics up to order m, are bases for the same space of polynomials. Hence, for the vector version $\mathbf{s}$ of a HARDI signal S, there exists a vector $\mathbf{c}$ of spherical harmonic coefficients such that $\mathbf{c} = T\mathbf{s}$. This fact means that T is invertible. By (6.22), we know

$$\Psi(\mathbf{x}) = \sum_{p=1}^{N} 2\pi P_{\ell(p)}(0) c_p R_p(\mathbf{x}),$$

where $\mathbf{c} = (c_p)$ is the spherical harmonics series coefficients of $S(\mathbf{x})$. Hence, the spherical harmonics series coefficients of $\Psi(\mathbf{x})$ are $\psi_p = 2\pi P_{\ell(p)}(0) c_p$. Let $\psi =$

(ψ_p). Then we obtain $\mathbf{c} = T\mathbf{s}$ and $\psi = T\mathbf{r}$. Thus, $\mathbf{r} = 2\pi T^{-1}DT\mathbf{s} = M\mathbf{s}$. The proof is complete. □

Suppose that $r(\geq N)$ HARDI signals $\mathbf{f} \in \mathbf{R}^r$ are measured along gradient directions $\{\mathbf{x}^{(j)} : j = 1, \ldots, r\}$. Let $A = (\mathbf{u}(\mathbf{x}^{(j)}), \ldots, \mathbf{u}(\mathbf{x}^{(j)})) \in \mathbf{R}^{N \times r}$ and $B = AA^\top$. Then, A has a full row rank and B is positive definite. To fit the HARDI signals, we solve a linear least squares problem

$$\min \quad F(\mathbf{s}) = \|A^\top \mathbf{s} - \mathbf{f}\|^2.$$

As a probability distribution function, ODF should be nonnegative, i.e., $\mathbf{r}$ is positive semidefinite. By Theorem 6.8, $\mathbf{r}$ is positive semidefinite if and only if $A\mathbf{s}$ is positive semidefinite, i.e., its smallest Z-eigenvalue $\lambda_{\min}(As) \geq 0$. Then, we formulate the convex optimization model as

$$F(\mathbf{s}^*) = \min\{F(\mathbf{s}) : \lambda_{\min}(A\mathbf{s}) \geq 0\}. \tag{6.23}$$

We can solve this convex optimization problems using the approach discussed in Sect. 6.3.

Next, we give an example on the case that two fibers cross in one voxel as shown in the right image of Fig. 6.1. Using the approach introduced in this section, we obtain a diffusion ODF illustrated in Fig. 6.2. Clearly, two crossing fibers with horizontal and vertical directions are observed.

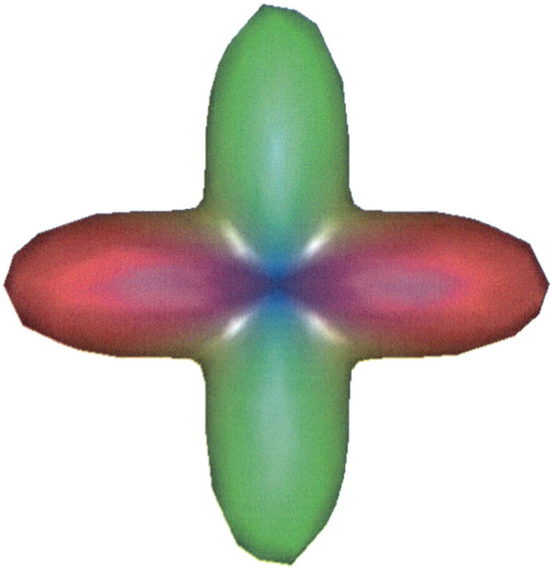

Fig. 6.2 A diffusion ODF for two crossing fibers

6.5 Nonnegative Fiber Orientation Distribution Function

The spherical deconvolution (SD) model is another approach to capture crossing fibers in a voxel of biological tissues. Compared with Q-ball imaging, SD methods impose a different assumption: all the nerve fibers share a common response function. Hence, the ODF generated by SD is usually called the *fiber orientation distribution function* or FOD. Using HARDI signals, the **spherical deconvolution** model [265] expresses the signal attenuation as a convolution of the FOD and the common response function

$$S(\mathbf{x})/S_0 = \int_{\mathbb{S}} f(\mathbf{v}) R(\mathbf{x}, \mathbf{v})\, \mathrm{d}\mathbf{v}, \tag{6.24}$$

where $\mathbf{v}$ is an orientation of a nerve fiber, $f(\mathbf{v})$ is the FOD to be estimated, and $R(\mathbf{x}, (0,0,1)^\top)$ is the common response function corresponding to HARDI signals of a typical fiber with the orientation $(0,0,1)^\top$. To the sense of statistics, the FOD should be nonnegative

$$f(\mathbf{v}) \geq 0, \qquad \forall \mathbf{v} \in \mathbb{S} \tag{6.25}$$

and has a unit-mass

$$\int_{\mathbb{S}} f(\mathbf{v})\, \mathrm{d}\mathbf{v} = 1. \tag{6.26}$$

Let the FOD function be a homogeneous polynomial $f(\mathbf{v}) = \mathscr{A}\mathbf{v}^m$, where $\mathscr{A}$ is a symmetric tensor and m is even for symmetry. To deal with the nonnegative constraint (6.25), our basic idea is to restrict $\mathscr{A}$ as a sum-of-squares (SOS) tensor. We collect independent elements of the SOS tensor $\mathscr{A}$ in a vector $\mathbf{w} \in \mathbf{R}^N$, and rewrite the FOD as

$$f(\mathbf{v}) = \mathbf{u}(\mathbf{x})^\top \mathbf{w}.$$

Suppose that $\bar{\mathbf{u}}(\mathbf{v})$ is a basis of $m/2$th order homogeneous polynomials in $\mathbf{v}$. Then, there exists a positive semidefinite matrix X such that

$$f(\mathbf{v}) = \bar{\mathbf{u}}(\mathbf{v})^\top X \bar{\mathbf{u}}(\mathbf{v}).$$

Furthermore, there is a linear map A such that $\mathbf{w} = A(X)$. In this way, we relax (6.25) as

$$\mathbf{w} = A(X) \quad \text{and} \quad X \succ 0. \tag{6.27}$$

We consider the unit-mass constraint (6.26). It can be rewritten as

$$\int_{\mathbb{S}} f(\mathbf{v})\, \mathrm{d}\mathbf{v} = \int_{\mathbb{S}} \sum_{j=1}^{N} w_j \mathbf{u}_j(\mathbf{v})\, \mathrm{d}\mathbf{v} = \sum_{j=1}^{N} w_j \underbrace{\int_{\mathbb{S}} \mathbf{u}_j(\mathbf{v})\, \mathrm{d}\mathbf{v}}_{:=\mathbf{s}_j} = 1.$$

Hence, we rewrite the unit-mass constraint as

$$\mathbf{w}^\top \mathbf{s} = 1. \tag{6.28}$$

The involved integrals in the vector $\mathbf{s} = (s_j)$ could be calculated exactly using the MATLAB symbolic integration.

Suppose that HARDI measures r gradient directions $\mathbf{x}^{(i)}$ and generates corresponding signals $S^{(i)}$ for $i = 1, \ldots, r$. We represent the SD equation (6.24) similarly

$$S^{(i)}/S_0 = \int_{\mathbb{S}} \sum_{j=1}^{N} w_j \phi_j(\mathbf{v})\, R(\mathbf{x}, \mathbf{v})\, \mathrm{d}\mathbf{v} = \sum_{j=1}^{N} w_j \underbrace{\int_{\mathbb{S}} \phi_j(\mathbf{v}) R(\mathbf{x}, \mathbf{v})\, \mathrm{d}\mathbf{v}}_{\gamma_{ij}}.$$

Denote $\mathbf{f} = (S^{(i)}/S_0) \in \mathbf{R}^r$. Due to the full rank assumption, the deconvolution matrix $\Gamma = (\gamma_{ij})$ has a full column rank. Then, the SD fitting could be rewritten as

$$\mathbf{f} \approx \Gamma \mathbf{w}. \tag{6.29}$$

For the response function corresponding to a single fiber, we employ the bipolar Watson function [286]

$$R(\mathbf{x}, \mathbf{v}) = \lim_{\alpha \to +\infty} \exp[-\alpha(\mathbf{x}^\top \mathbf{v})^2].$$

It is a refined Gaussian response corresponding to MRI signal attenuation of a signal fiber, as long as α is large enough.

Taking all the components (6.27)–(6.29) together, we obtain a convex semidefinite programming model

$$\begin{cases} \min & \frac{1}{2}\|\mathbf{f} - \Gamma \mathbf{w}\|^2 \\ \text{s.t.} & \mathbf{w} = A(X), \quad X \succeq 0, \\ & \mathbf{w}^\top \mathbf{s} = 1. \end{cases} \tag{6.30}$$

This convex semidefinite programming model could be solved efficient by alternating direction methods of multipliers [49].

We now consider the case of two fibers crossing as shown in the right image in Fig. 6.1. Using the model (6.30), we obtain a fiber ODF illustrated in Fig. 6.3. Clearly, there are two crossing fibers with horizontal and vertical directions.

Finally, we study a real-world human brain HARDI data. Each voxel is of size $1.875 \times 1.875 \times 2\,\text{mm}^3$, the diffusion-weighted b-value is $b = 3000\,\text{s/mm}^2$, and 200 gradient directions are equipped.

For instance, we select a coronal slice of a healthy human brain, whose fractional anisotropy (FA) map is reported in Fig. 6.4. The interesting region is marked by a yellow box. We illustrate detailed contour profiles of estimated FOD for this area in Fig. 6.5. From this figure, we find that there are three nervous fibers detected in this

Fig. 6.3 A fiber ODF for two crossing fibers

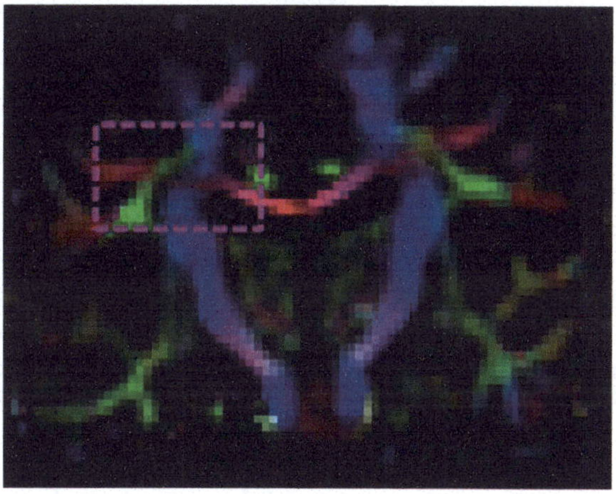

Fig. 6.4 FA map

area. The left-to-right directional fiber is the corpus callosum, which connects the left and right cerebral hemispheres and facilitates interhemispheric communication. The up-to-down directional fiber is the corticospinal tracts, which control voluntary movement of skeletal muscles. The front-to-back directional fiber is the superior longitudinal fasciculus, which connects the front and the back of the cerebrum. The reconstruction of nerve fibers are consistent with neuroanatomy.

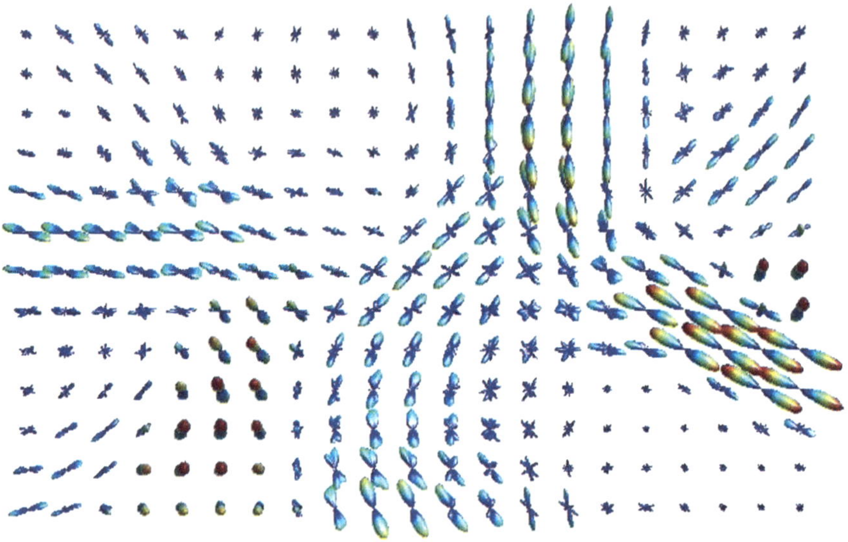

Fig. 6.5 The reconstruction of corpus callosum crossing corticospinal tracts for the interesting region

6.6 Image Authenticity Verification

In imaging science [302, 304], the directional characteristic of illumination in an image could be measured by the skewness of oriented gradients. Therefore, we introduce the gradient skewness tensor (GST) to describe the third order statistics of gradients. The estimating local directional characteristic of illumination can be modeled as solving the largest D-eigenvalue of the corresponding GST.

Now, we give the definition of GST, which involves a third order symmetric tensor and a second order symmetric tensor from image gradients. By applying two 1-D filters in both directions such as

$$\begin{pmatrix} -1 & 0 & 1 \end{pmatrix} \quad \text{and} \quad \begin{pmatrix} -1 \\ 0 \\ 1 \end{pmatrix},$$

we obtain the horizontal and vertical gradient matrices of an image I denoted by G_x and G_y, respectively. The oriented gradient matrix G_θ for a fixed angle $\theta \in [0, 2\pi)$ is defined as

$$G_\theta = G_x \cdot \cos\theta + G_y \cdot \sin\theta.$$

For a rectangle region L in the image I, we denote $G_{\theta,L} = (g_1, g_2, \ldots, g_n)$ as all gradient values of G_θ within the region L written in vector form, and $G_{x,L} = (x_1, x_2, \ldots, x_n)$ and $G_{y,L} = (y_1, y_2, \ldots, y_n)$ similarly for horizontal and vertical gradients G_x and G_y respectively. Let $\bar{g}$ be the mean of $G_{\theta,L}$, the **skewness** of $G_{\theta,L}$ is defined as

$$s(\theta) = \frac{\frac{1}{n}\Sigma_{i=1}^{n}(g_i - \bar{g})^3}{\left(\frac{1}{n}\Sigma_{i=1}^{n}(g_i - \bar{g})^2\right)^{3/2}}, \tag{6.31}$$

and it is a function of θ. Because of

$$g_i = x_i \cos\theta + y_i \sin\theta$$

for a fixed θ, we have

$$\bar{g} = \bar{x}\cos\theta + \bar{y}\sin\theta,$$

where $\bar{x}$ and $\bar{y}$ is the mean of $G_{x,L}$, and $G_{y,L}$. Therefore, the skewness function is represented as

$$\begin{aligned} s(\theta) &= \frac{\frac{1}{n}\sum_{i=1}^{n}((x_i\cos\theta + y_i\sin\theta) - (\bar{x}\cos\theta + \bar{y}\sin\theta))^3}{\left(\frac{1}{n}\sum_{i=1}^{n}((x_i\cos\theta + y_i\sin\theta) - (\bar{x}\cos\theta + \bar{y}\sin\theta))^2\right)^{3/2}} \\ &= \frac{\frac{1}{n}\sum_{i=1}^{n}((x_i - \bar{x})\cos\theta + (y_i - \bar{y})\sin\theta)^3}{(\frac{1}{n}\sum_{i=1}^{n}((x_i - \bar{x})\cos\theta + (y_i - \bar{y})\sin\theta)^2)^{3/2}}. \end{aligned} \tag{6.32}$$

Furthermore, we define a series of discrete moments

$$\mu_{p,q} = \frac{\sum_{i=1}^{n}(x_i - \bar{x})(y_i - \bar{y})}{n}$$

for $p, q = 0, 1, 2, 3,$. We can get another form of (6.32),

$$s(\theta) = \frac{\mu_{3,0}\cos^3\theta + 3\mu_{2,1}\cos^2\theta\sin\theta + 3\mu_{1,2}\cos\theta\sin^2\theta + \mu_{0,3}\sin^3\theta}{(\mu_{2,0}\cos^2\theta + 2\mu_{1,1}\cos\theta\sin\theta + \mu_{0,2}\sin^2\theta)^{3/2}}. \tag{6.33}$$

Consequently, we get the skewness function $s(\theta)$ which represents the skewness value of image gradients for angle θ in the region L.

We use $\mathbf{x} = (\cos\theta, \sin\theta)^\top$ to denote the direction vector. By (6.33), we get

$$s(x) = \frac{\mu_{3,0}x_1^3 + 3\mu_{2,1}x_1^2x_2 + 3\mu_{1,2}x_1x_2^2 + \mu_{0,3}x_2^3}{(\mu_{2,0}x_1^2 + 2\mu_{1,1}x_1x_2 + \mu_{0,2}x_2^2)^{3/2}}. \tag{6.34}$$

Then, the gradient skewness tensor $\mathscr{A}$ is a third order two dimensional symmetric tensor with elements

$$\begin{aligned} a_{111} &= \mu_{3,0}, \\ a_{112} = a_{121} = a_{211} &= \mu_{2,1}, \\ a_{122} = a_{212} = a_{221} &= \mu_{1,2}, \\ a_{222} &= \mu_{0,3}. \end{aligned} \tag{6.35}$$

A related second order symmetric tensor is

$$D = \begin{pmatrix} \mu_{2,0} & \mu_{1,1} \\ \mu_{1,1} & \mu_{0,2} \end{pmatrix}.$$

From (6.31), D is positive definite when all gradients within the region are not the same everywhere, and this situation is easily satisfied in natural images. As a result, the skewness function $s(x)$ can be written as

$$s(x) = \frac{\mathscr{A}\mathbf{x}^3}{(D\mathbf{x}^2)^{3/2}}.$$

By optimization theory, the problem

$$\max s(\mathbf{x}) = \frac{\mathscr{A}\mathbf{x}^3}{(D\mathbf{x}^2)^{3/2}}$$

is just equivalent to solving the largest D-eigenvalue of M, i.e., the solution of the system of equations:

$$\begin{cases} \mathscr{A}\mathbf{x}^2 = \lambda D\mathbf{x}, \\ D\mathbf{x}^2 = 1. \end{cases}$$

6.7 Notes

Starting from the basic model of DTI [9, 10], there are mainly two directions involving tensors in our opinion. The first one is to develop rotation invariant indices of non-Gaussian diffusion of water molecules, such as D-eigenvalues of diffusion kurtosis tensor, mean kurtosis, fractional kurtosis anisotropy [229], Z-eigenvalues of generalized diffusion tensor, PSDT mean value, PSDT fractional anisotropy [231], and so on. The second one is trying to obtain the inherent fiber orientations of nerve fibers in biological tissues. Typical examples are the diffusion ODF estimated by Q-ball imaging [232] and the fiber ODF estimated by spherical deconvolution model [49], where the involved higher order tensors are positive semidefinite. To extract fiber orientations, tensor-based methods were established [2, 18]. Additionally, Schultz et al. [239] wrote a review paper on higher order tensors in MRI.

Section 6.1 In DKI, Qi, Wang and Wu [229] considered the problem of maximizing the apparent kurtosis coefficient K_{app} and proposed the definition of D-eigenvalues. They proved that D-eigenvalues are invariant under orthogonal transformations. Hence, D-eigenvalues of the diffusion kurtosis tensor, as well as the mean kurtosis and the fractional kurtosis anisotropy, are valuable indexes for medical imaging. Cui, Dai and Nie [66] gave an efficient numerical method for all D-eigenvalues of a symmetric real tensor.

Han, Qi, and Wu [116] studied the largest diffusion value and the associated diffusion direction of water molecules in biological tissues by the following optimization model

$$\begin{cases} \max & D\mathbf{x}^2 - \dfrac{1}{6}b\mathrm{MD}^2\mathcal{W}\mathbf{x}^4 \\ \text{s.t.} & \mathbf{x}^\top\mathbf{x} = 1, \mathbf{x} \in \mathbf{R}^3. \end{cases}$$

They pointed out that the extreme diffusion values are also invariant under orthogonal transformations.

For real-world diffusion tensors and diffusion kurtosis tensors derived from MRI experiments on rat spinal cord specimen fixed in formalin, plenty of results on D-eigenvalues of the diffusion kurtosis tensors and associated D-eigenvectors are reported in [227, 229, 288, 308].

Section 6.2 Owing to the signal attenuation in the MRI process, the estimated diffusion tensor and kurtosis tensor should satisfy

$$bD\mathbf{x}^2 - \tfrac{1}{6}b^2\mathrm{MD}^2\mathcal{W}\mathbf{x}^4 > 0$$

for all directions $\mathbf{x}$ and b-values $b \in (0, \bar{b}]$. This is a semi-infinite programming problem. Hu, Huang, Ni and Qi [128] established an exact conic linear optimization model for the positive definiteness of diffusion kurtosis imaging.

Section 6.3 The diffusion kurtosis imaging used a second order diffusion tensor and a fourth order kurtosis tensor. Can we combine these two tensors into one tensor? That is the generalized diffusion tensor [215], whose order is even and may great than four. Qi, Yu and Wu [231] studied the positive definite generalized diffusion tensor imaging. Chen, Dai, Han and Sun [50] approximated the positive definite generalized diffusion tensor by a sum-of-squares tensor. If the order of the generalized diffusion tensor is four, this approximation is exact.

Section 6.4 To estimate the inherent fiber orientations of nerve fibers in biological tissues, Tuch [266] proposed the Q-ball imaging to estimate diffusion orientation distribution function. Qi, Yu and Xu [232] applied nonnegative tensors for enforcing nonnegative diffusion orientation distribution function.

Section 6.5 The spherical deconvolution [265] is another method to estimate the inherent fiber orientations of nerve fibers in biological tissues. Chen, Dai and Han [49] employed a sum-of-squares polynomial to approximate the fiber orientation distribution function.

Section 6.6 Another important application of D-eigenvalues is for image authenticity verification. Zhang, Zhou and Peng [302] explored skewness of oriented gradients to detect local illumination. By using the D-eigenvector corresponding to the largest D-eigenvalue of the skewness tensor of oriented gradients, Zhang, Zhou and Peng [303, 304] proposed a novel approach for local illumination detection.

6.8 Exercises

1 Suppose the elements of a diffusion tensor D are

$$d_{11} = 0.1755 \times 10^{-3},\ d_{22} = 0.1390 \times 10^{-3},\ d_{33} = 0.4006 \times 10^{-3},$$
$$d_{23} = 0.0017 \times 10^{-3},\ d_{13} = 0.0132 \times 10^{-3},\ d_{12} = 0.0035 \times 10^{-3}.$$

Elements of a diffusion kurtosis tensor $\mathscr{W}$ are

$$\begin{array}{llll} w_{1111} = 0.4982, & w_{2222} = 0, & w_{3333} = 2.6311, & w_{1112} = -0.0582, \\ w_{1113} = -1.1719, & w_{1222} = 0.4880, & w_{2223} = -0.6162, & w_{1333} = 0.7639, \\ w_{2333} = 0.7631, & w_{1122} = 0.2336, & w_{1133} = 0.4597, & w_{2233} = 0.1519, \\ w_{1123} = -0.0171, & w_{1223} = 0.1582, & w_{1233} = -0.4087. & \end{array}$$

Use "NSolve" command in Mathematica to compute all D-eigenvalues of $\mathscr{W}$, and then calculate mean kurtosis, and fractional kurtosis anisotropy.
(Hint: See [229].)
2 Let $\mathbf{d}$ be the representation vector of generalized diffusion tensor. Its elements are

$$\begin{array}{llll} d_{0,0,4} = 0.1287, & d_{0,1,3} = 0.7023, & d_{0,2,2} = 0.6931, & d_{0,3,1} = 0.0, \\ d_{0,4,0} = 0.0409, & d_{1,0,3} = 0.0101, & d_{1,1,2} = 0.0363, & d_{1,2,1} = -0.0246, \\ d_{1,3,0} = -0.0140, & d_{2,0,2} = -0.5627, & d_{2,1,1} = -0.5331, & d_{2,2,0} = 1.5083, \\ d_{3,0,1} = -0.0739, & d_{3,1,0} = -0.1141, & d_{4,0,0} = 0.0049. & \end{array}$$

Use "NSolve" command in Mathematica to compute all Z-eigenvalues of $\mathscr{D}$, and then calculate PSDT mean value, PSDT fractional anisotropy.
(Hint: See [231].)

Chapter 7
Third Order Tensors in Physics and Mechanics

Third order tensors have wide applications in physics and mechanics. Examples include piezoelectric tensors in crystal study, third order symmetric traceless tensors in liquid crystal study and third order susceptibility tensors in nonlinear optics study. On the other hand, the Levi-Civita tensor is famous in tensor calculus.

In 2017, Qi [225] studied third order tensors and hypermatrices systematically, by regarding a third order tensor as a linear operator which transforms a second order tensor into a first order tensor, or a first order tensor into a second order tensor. For a third order tensor, its transpose, kernel tensor, L-inverse and nonsingularity are defined in [225]. Especially, the transpose of a third order tensor is uniquely defined. Note that the transpose of the piezoelectric tensor is the inverse piezoelectric tensor (the electrostriction tensor). Furthermore, the kernel tensor of a third order tensor is a second order positive semi-definite symmetric tensor, which is the product of that third order tensor and its transpose. A third order tensor has an L-inverse if and only if it is nonsingular. Here, "L" is named after Levi-Civita. Qi [225] also defined L-eigenvalues, singular values, C-eigenvalues and Z-eigenvalues for a third order tensor. They are all invariants of the third order tensor. A third order tensor is nonsingular if and only if all of its L-eigenvalues are positive. Physical meanings of these new concepts were discussed. Qi showed that the Levi-Civita tensor is nonsingular, its L-inverse is a half of itself, and its three L-eigenvalues are all the square root of two. Qi [225] also introduced third order orthogonal tensors. Third order orthogonal tensors are nonsingular as well, and their L-inverses are their transposes.

Piezoelectricity was discovered by Jacques Curie and Pierre Curie in 1880 [70]. In the next year, the converse piezoelectric effect was predicted by Lippmann [182] and confirmed by Curies [71] immediately. Now they have wide applications in the production and detection of sounds, generation of high voltages, electronic frequency generation, microbalances, driving an ultrasonic nuzzle, and ultrafine focusing of optical assemblies.

In the piezoelectric effect and the converse piezoelectric effect, the piezoelectric tensor, a third order three dimensional partially symmetric tensor, plays a key role. A

L. Qi et al., *Tensor Eigenvalues and Their Applications*, Advances in Mechanics and Mathematics 39, https://doi.org/10.1007/978-981-10-8058-6_7

third order real tensor is called a piezoelectric-type tensor if it is partially symmetric with respect to its last two indices. The piezoelectric tensor is a piezoelectric-type tensor of dimension three.

In 2017, Chen, Jákli and Qi [51] addressed C-eigenvalues and C-eigenvectors for piezoelectric-type tensors. Here, "C" names after Curie brothers (See Fig. 7.2). They showed that C-eigenvalues, which are invariant under orthonormal transformations, always exist. For a piezoelectric-type tensor, the largest C-eigenvalue and its C-eigenvectors form the best rank-one piezoelectric-type approximation of that tensor. This means that for the piezoelectric tensor, its largest C-eigenvalue determines the highest piezoelectric coupling constant. They further showed that for the piezoelectric tensor, the largest C-eigenvalue corresponds to the electric displacement vector with the largest 2-norm in the piezoelectric effect under unit uniaxial stress and the strain tensor with the largest 2-norm in the converse piezoelectric effect under unit electric field vector. Thus, C-eigenvalues and C-eigenvectors have concrete physical meanings in the piezoelectric effect and the converse piezoelectric effect. Chen, Jákli and Qi [51] also studied computational methods for C-eigenvalues and C-eigenvectors of the piezoelectric tensor for various crystal classes.

Liquid crystals are well-known for visualization applications of flat panel electronic displays. Beyond that, various optical and electronic devices, such as laser printers, light-emitting diodes, field-effect transistors, and holographic data storage, were invented with the development of bent-core (banana-shaped) liquid crystals. To characterize condensed phases exhibited by these bent-core molecules, a third order three dimensional symmetric traceless tensor was employed. Based on such a tensor, the orientationally ordered octupolar (tetrahedratic) phase has been predicted theoretically and confirmed experimentally. Whereafter, the octupolar order parameters of liquid crystals are widely studied.

In 2015 and 2016, Virga [270] and Gaeta and Virga [99] studied third order octupolar tensors in two and three dimensions, respectively. The octupolar potential, a scalar-valued function of the octupolar tensor, was also introduced. In particular, Gaeta and Virga [99] declared that the admissible region of the octupolar potential is within the form of a dome and there are two generic octupolar states, divided by the separatrix surface, in the space of these three parameters. Such a separatrix surface may physically represent an intra-octupolar transition.

In 2017, utilizing the resultant theory in algebraic geometry and the E-characteristic polynomial in spectral theory of tensors [228], Chen, Qi and Virga [52] gave the algebraic expressions of the dome and the separatrix surface explicitly. This may turn such an intra-octupolar transition an observable prediction.

7.1 Third Order Tensors and Hypermatrices

In this section, we regard a tensor as a linear operator with dimension 3. Vectors in $\mathbf{R}^3$ are denoted as $x_i, y_i, z_i, \ldots$, and matrices are denoted as $u_{ij}, v_{ij}, \ldots$, where indices i and j range from 1 to 3. For a product of vectors and matrices, if one index

Table 7.1 Operators for second order tensors

No.	Tensor operators	Matrix operators
1	$\mathbf{xU} = \mathbf{y}$	$x_i u_{ij} = y_j$
2	$\mathbf{x} = \mathbf{Uy}$	$x_i = u_{ij} y_j$
3	$\mathbf{U} = \mathbf{x} \otimes \mathbf{y}$	$u_{ij} = x_i y_j$
4	$\mathbf{xUy}$	$x_i u_{ij} y_j$
5	$\mathbf{U} \bullet \mathbf{V}$	$u_{ij} v_{ij}$
6	$\mathbf{W} = \mathbf{UV}$	$w_{ij} = u_{ik} v_{kj}$

is repeated twice, we calculate the sum on this index. This is the common usage in the literature of physics and mechanics. If we single out one vector or one matrix, their indices can be substituted with any other ones. However, if we address a product or a relation, or an equation of vectors and matrices, indices therein are related, and cannot be changed arbitrarily. In this hypermatrix notation, the Kronecker symbol δ_{ij} denotes the identity matrix. A square matrix p_{ij} is called orthogonal if it satisfies $p_{ik} p_{jk} = \delta_{ij}$.

In a three dimensional physical space $\mathbb{H}$, small bold letters $\mathbf{x}, \mathbf{y}, \mathbf{z}, \ldots$ denote first order tensors. First order tensors have linear operators, i.e., additions among themselves and multiplications with scalars. Hence, there must be a first order zero tensor $\mathbf{0}$. For any $\mathbf{x}, \mathbf{y} \in \mathbb{H}$, there exists a scalar $\mathbf{x} \bullet \mathbf{y}$ that satisfies the inner product rules, and $\mathbb{H}$ is complete with regard to this inner product. Then, $\mathbb{H}$ is a Hilbert space mathematically. Thus, $\mathbf{x} \bullet \mathbf{y}$ is called the inner product of $\mathbf{x}$ and $\mathbf{y}$. If $\mathbf{x} \bullet \mathbf{y} = 0$, $\mathbf{x}$ and $\mathbf{y}$ are orthogonal. If $\mathbf{x} \bullet \mathbf{x} = 1$, $\mathbf{x}$ is a first order unit tensor. Let $\{\mathbf{e}_1, \mathbf{e}_2, \mathbf{e}_3\}$ be an orthonormal basis. Under this basis, two first order tensors $\mathbf{x}$ and $\mathbf{y}$ in $\mathbb{H}$ are represented by two vectors x_i and y_i in $\mathbf{R}^3$, respectively. Then we get

$$\mathbf{x} \bullet \mathbf{y} = x_i y_i.$$

The summation value $x_i y_i$ is independent from the choice of the orthonormal basis. The first order tensor $\mathbf{x}$, which satisfies $\mathbf{x} \bullet \mathbf{x} = 0$ is called a zero tensor $\mathbf{0}$. It is always represented by the zero vector, and vice versa.

Second order tensors on $\mathbb{H}$ are denoted as capital bold letters $\mathbf{U}, \mathbf{V}, \ldots$. We denote $B(\mathbb{H})$ the set of all second order tensors on $\mathbb{H}$. Under an orthonormal basis $\{\mathbf{e}_1, \mathbf{e}_2, \mathbf{e}_3\}$, a second order tensor $\mathbf{U}$ is represented by a matrix u_{ij} in $\mathbf{R}^{3\times3}$. Assume that a first order tensor $\mathbf{x}$ is represented by x_i under this basis. Suppose that we have another orthonormal basis $\{\mathbf{g}_1, \mathbf{g}_2, \mathbf{g}_3\}$. Under $\{\mathbf{g}_1, \mathbf{g}_2, \mathbf{g}_3\}$, $\mathbf{x}$ and $\mathbf{U}$ are represented by a vector z_q in $\mathbf{R}^3$ and a matrix v_{qr} in $\mathbf{R}^{3\times3}$, respectively. Then there is an orthogonal matrix p_{iq}, which is determined by these two bases, such that $x_i = p_{iq} z_q$ and $u_{ij} = p_{iq} p_{jr} v_{qr}$.

Second order tensors are physical quantities. They also have linear operators. There exists a second order zero tensor $\mathcal{O}$. Suppose that $\mathbf{U}, \mathbf{V}, \mathbf{W} \in B(\mathbb{H})$, $\mathbf{x}, \mathbf{y} \in \mathbb{H}$ are represented by $u_{ij}, v_{ij}, w_{ij}, x_i$ and y_j under an orthonormal basis, respectively. These operations can be described in Table 7.1.

We note that the tensor $\mathbf{U} = \mathbf{x} \otimes \mathbf{y}$ is called a second order rank-one tensor. The value of $\mathbf{U} \bullet \mathbf{V}$ is also independent from the choice of the basis, and called the inner product of $\mathbf{U}$ and $\mathbf{V}$. The second order tensor $\mathbf{W} = \mathbf{UV}$ is called the product of $\mathbf{U}$ and $\mathbf{V}$.

For $\mathbf{U} \in B(\mathbb{H})$, if there is $\mathbf{y} \in \mathbb{H}$, $\mathbf{y} \neq \mathbf{0}$ such that $\mathbf{Uy} = \mathbf{0}$, $\mathbf{U}$ is called a second order singular tensor. Otherwise, $\mathbf{U}$ is called nonsingular.

For any $\mathbf{U} \in B(\mathbb{H})$, there exists a unique second order tensor $\mathbf{U}^\top \in B(\mathbb{H})$ such that for any $\mathbf{x}, \mathbf{y} \in \mathbb{H}$, $\mathbf{xUy} = \mathbf{yU}^\top\mathbf{x}$. This second order tensor $\mathbf{U}^\top$ is coined the transpose of $\mathbf{U}$. Clearly, under an orthonormal basis, if $\mathbf{U}$ is represented by u_{ij}, its transpose $\mathbf{U}^\top$ must be represented by u_{ji}. Moreover, for any $\mathbf{U} \in B(\mathbb{H})$, we have $(\mathbf{U}^\top)^\top = \mathbf{U}$. A second order tensor $\mathbf{U}$ is nonsingular if and only if $\mathbf{U}^\top$ is nonsingular. If $\mathbf{U} = \mathbf{U}^\top$, then $\mathbf{U}$ is called a symmetric tensor. If $\mathbf{U} = -\mathbf{U}^\top$, then $\mathbf{U}$ is called an anti-symmetric tensor. Under an orthonormal basis, symmetric tensors and anti-symmetric tensors are always represented by symmetric matrices and anti-symmetric matrices respectively.

If $\mathbf{U} = \mathbf{UV}$ for any $\mathbf{U} \in B(\mathbb{H})$, then $\mathbf{V}$ is called the second order identity tensor, and denoted by $\mathbf{I}$. For any $\mathbf{U} \in B(\mathbb{H})$, we also have $\mathbf{U} = \mathbf{IU}$. The identity tensor $\mathbf{I}$ is represented by δ_{ij}.

If $\mathbf{UV} = \mathbf{I}$, we say $\mathbf{V}$ is the inverse tensor of $\mathbf{U}$, and denote it as $\mathbf{U}^{-1}$. A second order tensor $\mathbf{U}$ has an inverse tensor if and only if it is nonsingular. The inverse tensor is unique if it exists, and we always have $(\mathbf{U}^{-1})^{-1} = \mathbf{U}$.

If $\mathbf{UU}^\top = \mathbf{I}$, $\mathbf{U}$ is named an orthogonal tensor. Hence, an orthogonal tensor $\mathbf{U}$ is nonsingular, and $\mathbf{U}^{-1} = \mathbf{U}^\top$. A second order tensor is orthogonal if and only if its representative matrix is orthogonal.

Because the value of $\mathbf{U} \bullet \mathbf{V}$ is independent from the choice of the basis, for any $\mathbf{U} \in B(\mathbb{H})$, $\mathbf{U} \bullet \mathbf{I}$ is independent from the choice of the basis, and is called the **trace** of $\mathbf{U}$. The representation of $\mathbf{U} \bullet \mathbf{I}$ is u_{ii} under an orthonormal basis. A second order tensor is called a traceless tensor if its trace is zero.

We say that $\mathbf{U}$ and $\mathbf{V}$ are orthogonal if $\mathbf{U} \bullet \mathbf{V} = 0$. If $\mathbf{U} \bullet \mathbf{U} = 1$, then $\mathbf{U}$ is called a second order unit tensor.

If there exist a scalar λ and a first order nonzero tensor $\mathbf{y}$ such that $\mathbf{Uy} = \lambda\mathbf{y}$, λ is called an eigenvalue of $\mathbf{U}$ and $\mathbf{y}$ is called the associated eigenvector. The eigenvalue λ of $\mathbf{U}$ is independent from the choice of the basis. A sufficient and necessary condition for a second order tensor being nonsingular is that this tensor has no zero eigenvalue. There are three real eigenvalues of a second order symmetric tensor. The eigenvectors associated with different eigenvalues of a second order symmetric tensor are mutually orthogonal.

For a given $\mathbf{U} \in B(\mathbb{H})$, $\{\mathbf{y} \in \mathbb{H} : \mathbf{Uy} = \mathbf{0}\}$ forms a linear subspace of $\mathbb{H}$, which is called the **null space** of $\mathbf{U}$. We define the rank of $\mathbf{U}$ as 3 minus the dimension of the null space of $\mathbf{U}$.

Eigenvalues of a second order tensor are invariants of the tensor. For a second order symmetric tensor $\mathbf{U}$, its trace is the sum of its eigenvalues, and the determinant of any of its representative matrix is equal to the product of its eigenvalues. Hence, this determinant is also an invariant of that tensor, and can be called the determinant of the tensor. This is also true for second order nonsymmetric tensors. The discussion of this may involve complex eigenvalues.

Let $\mathbf{U}$ be a second order symmetric tensor. We say that $\mathbf{U}$ is positive semi-definite if for any first order tensor $\mathbf{x} \in \mathbb{H}$, $\mathbf{xUx} \geq 0$. Moreover, we say that $\mathbf{U}$ is positive definite if for any nonzero $\mathbf{x} \in \mathbb{H}$, we have $\mathbf{xUx} > 0$. The tensor $\mathbf{U}$ is positive semi-definite if and only if all of its eigenvalues are nonnegative. The tensor $\mathbf{U}$ is positive definite if and only if all of its eigenvalues are positive.

Now, we turn to third order tensors. We use $a_{ijk}, b_{ijk}, \dots$ with triple indices to denote $3 \times 3 \times 3$ hypermatrices, where the indices i, j and k range from 1 to 3. A hypermatrix a_{ijk} is called orthogonal if

$$a_{ijk}a_{ljk} = \delta_{il}.$$

Third order tensors on $\mathbb{H}$ are denoted as calligraphic letters $\mathscr{A}, \mathscr{B}, \dots$, and $T(\mathbb{H})$ stands for the set of third order tensors on $\mathbb{H}$. When we discuss products of two third order tensors, we may also refer to a fourth order tensor that is also denoted by a calligraphic letter $\mathscr{T}$.

Under an orthonormal basis $\{\mathbf{e}_1, \mathbf{e}_2, \mathbf{e}_3\}$, a first order tensor $\mathbf{x}$ in $\mathbb{H}$ is represented by a three dimensional vector x_i, a second order tensor $\mathbf{U}$ on $\mathbb{H}$ is represented by a 3-by-3 matrix u_{ij}, a third order tensor $\mathscr{A}$ is represented by a 3-by-3-by-3 hypermatrix a_{ijk}. Let $\{\mathbf{g}_1, \mathbf{g}_2, \mathbf{g}_3\}$ be another orthonormal basis. Under that basis, assume that $\mathbf{x}$, $\mathbf{U}$, and $\mathscr{A}$ are represented by y_q, v_{qr}, and b_{qrs} in $\mathbf{R}^3$, respectively. In fact, there exists an orthogonal matrix p_{iq}, which is only dependent on these two bases, such that $x_i = p_{iq}y_q$, $u_{ij} = p_{iq}p_{jr}v_{qr}$, and $a_{ijk} = p_{iq}p_{jr}p_{ks}b_{qrs}$.

A third order tensor is also a physical quantity. They also have linear operations. Thus, there is a third order zero tensor $\mathscr{O}$. Let $\mathscr{A}, \mathscr{B} \in T(\mathbb{H})$, $\mathbf{U}, \mathbf{V} \in B(\mathbb{H})$, $\mathbf{x}, \mathbf{y}, \mathbf{z} \in \mathbb{H}$, and a fourth order tensor $\mathscr{T}$ be represented by a_{ijk}, b_{ijk}, u_{ij}, v_{jk}, x_i, y_j, z_k and t_{ijkl} under an orthonormal basis, respectively. Some of these operations can be described in Table 7.2.

The third order tensor $\mathbf{x} \otimes \mathbf{y} \otimes \mathbf{z}$ is called a rank-one tensor. The value of $\mathscr{A} \bullet \mathscr{B}$ is called the inner product of $\mathscr{A}$ and $\mathscr{B}$. The tensor $\mathscr{A}\mathscr{B}$ is called the second order tensor product of $\mathscr{A}$ and $\mathscr{B}$, and $\mathscr{A} \oplus \mathscr{B}$ is called the fourth order tensor product of $\mathscr{A}$ and $\mathscr{B}$.

Under an orthonormal basis $\{\mathbf{e}_1, \mathbf{e}_2, \mathbf{e}_3\}$ of $\mathbb{H}$, $\mathscr{A} \in T(\mathbb{H})$ is represented by a hypermatrix a_{ijk}. That is to say,

$$a_{ijk} = \mathbf{e}_i \mathscr{A} \mathbf{e}_j \mathbf{e}_k.$$

Note that $\{-\mathbf{e}_1, -\mathbf{e}_2, -\mathbf{e}_3\}$ is also an orthonormal basis of $\mathbb{H}$. Under the same orthonormal basis, $\mathscr{A}$ could also be represented by hypermatrix

$$-\mathbf{e}_i \mathscr{A}(-\mathbf{e}_j)(-\mathbf{e}_k) = -a_{ijk}.$$

Table 7.2 Operations for third order tensors

No.	Tensor operators	Matrix operators
1	$\mathbf{x}\mathscr{A} = \mathbf{V}$	$x_i a_{ijk} = v_{jk}$
2	$\mathbf{U} = \mathscr{A}\mathbf{z}$	$u_{ij} = a_{ijk} z_k$
3	$\mathbf{U}\mathscr{A} = \mathbf{z}$	$u_{ij} a_{ijk} = z_k$
4	$\mathbf{x} = \mathscr{A}\mathbf{V}$	$x_i = a_{ijk} v_{jk}$
5	$\mathbf{x}\mathbf{y}\mathscr{A} = \mathbf{z}$	$x_i y_j a_{ijk} = z_k$
6	$\mathbf{x}\mathscr{A}\mathbf{z} = \mathbf{y}$	$x_i a_{ijk} z_k = y_j$
7	$\mathbf{x} = \mathscr{A}\mathbf{y}\mathbf{z}$	$x_i = a_{ijk} y_j z_k$
8	$\mathscr{A} = \mathbf{U} \otimes \mathbf{z}$	$a_{ijk} = u_{ij} z_k$
9	$\mathscr{A} = \mathbf{x} \otimes \mathbf{V}$	$a_{ijk} = x_i v_{jk}$
10	$\mathscr{A} = \mathbf{x} \otimes \mathbf{y} \otimes \mathbf{z}$	$a_{ijk} = x_i y_j z_k$
11	$\mathbf{x}\mathscr{A}\mathbf{y}\mathbf{z}$	$x_i a_{ijk} y_j z_k$
12	$\mathscr{A} \bullet \mathscr{B}$	$a_{ijk} b_{ijk}$
13	$\mathbf{U} = \mathscr{A}\mathscr{B}$	$u_{il} = a_{ijk} b_{jkl}$
14	$\mathscr{T} = \mathscr{A} \oplus \mathscr{B}$	$t_{ijkl} = a_{ijm} b_{mkl}$

A third order tensor $\mathscr{A} \in T(\mathbb{H})$ is named a singular tensor if there exists a first order nonzero tensor $\mathbf{x} \in \mathbb{H}$ such that $\mathbf{x}\mathscr{A} = \mathscr{O}$. Otherwise, $\mathscr{A}$ is called a **nonsingular tensor**.

For any $\mathbf{y}, \mathbf{z} \in \mathbb{H}$, if $\mathscr{A}\mathbf{y}\mathbf{z} = \mathscr{A}\mathbf{z}\mathbf{y}$, then the third order tensor $\mathscr{A} \in T(\mathbb{H})$ is called **right-side symmetric**. Under an orthonormal basis, if the tensor $\mathscr{A}$ is represented by a_{ijk}, then we have $a_{ijk} = a_{ikj}$ for all i, j, and k. A typical example of a third order right-side symmetric tensor is the piezoelectric tensor in solid crystals; See Sect. 7.3. However, in liquid crystals, the piezoelectric tensor is not right-side symmetric generically.

A third order tensor $\mathscr{A} \in T(\mathbb{H})$ is called **left-side symmetric** if for any $\mathbf{x}, \mathbf{y} \in \mathbb{H}$, we have $\mathbf{x}\mathbf{y}\mathscr{A} = \mathbf{y}\mathbf{x}\mathscr{A}$. Under an orthonormal basis, if $\mathscr{A}$ is represented by a_{ijk}, we have $a_{ijk} = a_{jik}$. Inverse piezoelectric tensor is a typical example of a third order left-side symmetric tensor.

A third order tensor $\mathscr{A} \in T(\mathbb{H})$ is called **left-side anti-symmetric** if for any $\mathbf{x}, \mathbf{y} \in \mathbb{H}$, $\mathbf{x}\mathbf{y}\mathscr{A} = -\mathbf{y}\mathbf{x}\mathscr{A}$. Under an orthonormal basis, if $\mathscr{A}$ is represented by a_{ijk}, then we have $a_{ijk} = -a_{jik}$ for all i, j and k. A typical example of a third order left-side anti-symmetric tensor is the Hall tensor.

A third order tensor $\mathscr{A} \in T(\mathbb{H})$ is called **right-side anti-symmetric** if for any $\mathbf{y}, \mathbf{z} \in \mathbb{H}$, $\mathscr{A}\mathbf{y}\mathbf{z} = -\mathscr{A}\mathbf{z}\mathbf{y}$. A third order tensor $\mathscr{A} \in T(\mathbb{H})$ is called **centrally anti-symmetric** if for any $\mathbf{x}, \mathbf{z} \in \mathbb{H}$, $\mathbf{x}\mathscr{A}\mathbf{z} = -\mathbf{z}\mathscr{A}\mathbf{x}$. If a third order tensor $\mathscr{A} \in T(\mathbb{H})$ is right-side, left-side and centrally anti-symmetric, the tensor is called **totally anti-symmetric**.

The **Levi-Civita tensor** $\mathscr{E}$, which can be called the **permutation tensor**, is a famous example of third order three dimensional totally anti-symmetric tensors. Under a certain orthonormal basis, $\mathscr{E}$ is represented by the Levi-Civita hypermatrix

ϵ_{ijk}, whose elements are $\epsilon_{123} = \epsilon_{312} = \epsilon_{231} = 1$, $\epsilon_{213} = \epsilon_{321} = \epsilon_{132} = -1$ and $\epsilon_{ijk} = 0$ otherwise. Other third order three dimensional totally anti-symmetric tensors are multiples of the Levi-Civita tensor. Then, the representative hypermatrix of the the Levi-Civita tensor $\mathscr{E}$ is either ϵ_{ijk} or $-\epsilon_{ijk}$.

For any $\mathbf{x}, \mathbf{y}, \mathbf{z} \in \mathbb{H}$, we assume that $\mathscr{A}, \mathscr{B} \in T(\mathbb{H})$ satisfy $\mathbf{x}\mathscr{A}\mathbf{yz} = \mathbf{y}\mathscr{B}\mathbf{zx}$. Then we call $\mathscr{B}$ the **transpose tensor** of $\mathscr{A}$ and denote $\mathscr{B} = \mathscr{A}^\top$.

Proposition 7.1 *Let $\mathscr{A} \in T(\mathbb{H})$. Then its transpose exists and is unique. Furthermore, we always have $[(\mathscr{A}^\top)^\top]^\top = \mathscr{A}$.*

We say that $\mathscr{A}$ is **cyclically symmetric**, if $\mathscr{A}^\top = \mathscr{A}$. Then the Levi-Civita tensor is a cyclically symmetric tensor. If for any $\mathbf{x}, \mathbf{y}, \mathbf{z} \in \mathbb{H}$, $\mathbf{x}\mathscr{A}\mathbf{yz}$ is invariant under any permutation of the positions of $\mathbf{x}$, $\mathbf{y}$ and $\mathbf{z}$, then we say that $\mathscr{A}$ is a third order symmetric tensor. A hypermatrix a_{ijk} is called symmetric if a_{ijk} is invariant under any change of its indices. Clearly, a third order tensor is symmetric if and only if its represented hypermatrix is symmetric under an orthonormal basis. It is also easy to know that a third order tensor is symmetric if and only if it is both right-side and cyclically symmetric.

If $\mathscr{A}\mathscr{A}^\top = \mathbf{I}$, then we call $\mathscr{A}$ a third order three dimensional **orthogonal tensor**. Since the representative matrix of $\mathbf{I}$ is δ_{ij}, we have a proposition.

Proposition 7.2 *Let $\mathscr{A} \in T(\mathbb{H})$. Then $\mathscr{A}$ is an orthogonal tensor if and only if its representative hypermatrix is orthogonal, under an orthonormal basis.*

Let $\mathscr{A} \in T(\mathbb{H})$ and $\mathbf{U} = \mathscr{A}\mathscr{A}^\top$. Then $\mathbf{U}$ is called the **kernel tensor** of $\mathscr{A}$.

Theorem 7.1 *Let $\mathbf{U} \in B(\mathbb{H})$ be a kernel tensor of $\mathscr{A} \in T(\mathbb{H})$. Then $\mathbf{U}$ is symmetric and positive semi-definite. Moreover, $\mathbf{U}$ is positive definite if and only if $\mathscr{A}$ is nonsingular. Specially, a third order orthogonal tensor is nonsingular.*

The kernel tensor $\mathbf{U} = \mathscr{A}\mathscr{A}^\top$ is uniquely determined by $\mathscr{A}$. Thus, some invariants of $\mathbf{U}$, such as the trace and the determinant of $\mathbf{U}$, are also invariants of $\mathscr{A}$. We call the rank of $\mathbf{U}$ as the **rank** of $\mathscr{A}$.

Let $\mathscr{A}$ and $\mathscr{B}$ be two third order three dimensional tensors on $\mathbb{H}$. We say that $\mathscr{B}$ is an **L-inverse** of $\mathscr{A}$, if $\mathscr{A}$ and $\mathscr{B}$ satisfy

$$\mathscr{A}\mathscr{B} = \mathbf{I} \tag{7.1}$$

and

$$\mathscr{B} \oplus \mathscr{A} = \mathscr{A}^\top \oplus (\mathscr{B}^\top)^\top. \tag{7.2}$$

Moreover, we denote that $\mathscr{B} = \mathscr{A}^{-1}$.

Theorem 7.2 *(1) For a third order tensor $\mathscr{A}$, we assume that its L-inverse exists. Then such an L-inverse is unique, denoted as $\mathscr{A}^{-1}$. Moreover, we have $(\mathscr{A}^{-1})^{-1} = \mathscr{A}$.*

(2) The tensor $\mathscr{A}$ has an L-inverse if and only if $\mathscr{A}$ is nonsingular.

(3) If $\mathscr{A}\mathscr{A}^\top = t\mathbf{I}$ for some $t > 0$, then $\mathscr{A}$ is nonsingular, and $\mathscr{A}^{-1} = \frac{1}{t}\mathscr{A}^\top$. In particular, $\mathscr{A}^{-1} = \mathscr{A}^\top$ if $\mathscr{A}$ is an orthogonal tensor.

Fig. 7.1 Tullio Levi-Civita (1873–1941)

Here, the letter "L" is named after Levi-Civita (see Fig. 7.1). We argue that the concept of L-inverse is different from the Moore–Penrose inverse of a rectangular matrix. The Moore–Penrose inverse of a rectangular always exists. However, the L-inverse of a third order tensor exists if and only if the tensor is nonsingular. Second, the L-inverse is a tensor concept. It is corresponding to the transformation from a first order tensor to a second order tensor. The Moore–Penrose inverse of a rectangular matrix does not have such a property. An application of L-inverse confirms this.

Proposition 7.3 *Suppose that $\mathscr{A}^{-1}$ of a third order tensor $\mathscr{A}$ exists and the following tensor relation holds:*

$$\mathscr{A}\mathbf{z} = \mathbf{U}.$$

Then we obtain

$$\mathbf{z} = \mathscr{A}^{-1}\mathbf{U}.$$

A typical example satisfying $\mathscr{A}\mathbf{z} = \mathbf{U}$ is the inverse piezoelectric effect, where $\mathscr{A}$ is the inverse piezoelectric tensor, $\mathbf{z}$ is the electric field strength and $\mathbf{U}$ is the deformation. Thus, if $\mathscr{A}^{-1}$ exists, we may calculate $\mathbf{z}$ from $\mathbf{U}$.

Let $\mathscr{A} \in T(\mathbb{H})$, $\mathbf{V} \in B(\mathbb{H})$, $\mathbf{x} \in \mathbb{H}$ and $\sigma \in \mathbf{R}$. If $\sigma \geq 0$ satisfies

$$\mathscr{A}\mathbf{V} = \sigma\mathbf{x}, \quad \mathscr{A}^\top\mathbf{x} = \sigma\mathbf{V}, \quad \mathbf{V} \bullet \mathbf{V} = 1, \quad \mathbf{x} \bullet \mathbf{x} = 1, \tag{7.3}$$

σ is an **L-eigenvalue** of $\mathscr{A}$, $\mathbf{V}$ and $\mathbf{x}$ are associated L-eigentensor and L-eigenvector respectively.

Theorem 7.3 *Let $\mathscr{A} \in T(\mathbb{H})$. Then $\mathscr{A}$ has three L-eigenvalues $\sigma_1 \geq \sigma_2 \geq \sigma_3 \geq 0$, with associated L-eigentensors $\mathbf{V}_1, \mathbf{V}_2, \mathbf{V}_3$, and L-eigenvectors $\mathbf{x}_1, \mathbf{x}_2, \mathbf{x}_3$, respectively. The following assertions hold.*

(1) $\mathbf{V}_i \bullet \mathbf{V}_j = \delta_{ij}$ and $\mathbf{x}_i \bullet \mathbf{x}_j = \delta_{ij}$.
(2) Let $\mathbf{U} = \mathscr{A}\mathscr{A}^\top$ be the kernel tensor of $\mathscr{A}$, which has eigenvalues $\lambda_1 \geq \lambda_2 \geq \lambda_3$. Then $\sigma_j^2 = \lambda_j$ for $j = 1, 2, 3$, and $\mathbf{x}_j$ for $j = 1, 2, 3$ are associated eigenvectors of $\mathbf{U}$. Hence, L-eigenvalues of $\mathscr{A}$ are invariants of the tensor $\mathscr{A}$.
(3) We obtain

$$\mathscr{A} = \sum_{j=1}^{3} \sigma_j \mathbf{x}_j \otimes \mathbf{V}_j. \tag{7.4}$$

(4) $\mathscr{A}^{-1}$ exists if and only if $\sigma_1 \geq \sigma_2 \geq \sigma_3 > 0$. In this case, we know

$$\mathscr{A}^{-1} = \sum_{j=1}^{3} \frac{1}{\sigma_j} \mathbf{V}_j \otimes \mathbf{x}_j. \tag{7.5}$$

(5) We get

$$\sigma_1 = \max\{\sqrt{(\mathscr{A}\mathbf{V}) \bullet (\mathscr{A}\mathbf{V})} : \mathbf{V} \bullet \mathbf{V} = 1\}. \tag{7.6}$$

(6) If $\mathscr{A}$ is right-side symmetric, its L-eigentensors associated with a positive L-eigenvalue are symmetric.
(7) If $\mathscr{A}\mathscr{A}^\top = \alpha \mathbf{I}$ for some $\alpha \geq 0$, then the L-eigenvalues of $\mathscr{A}$ are $\sigma_1 = \sigma_2 = \sigma_3 = \sqrt{\alpha}$. In particular, if $\mathscr{A}$ is a third order orthogonal tensor, then its L-eigenvalues are all ones, i.e., $\sigma_1 = \sigma_2 = \sigma_3 = 1$.

If $\mathscr{A}\mathscr{A}^\top = \alpha \mathbf{I}$ for some $\alpha \geq 0$, then the three L-eigenvalues of $\mathscr{A}$ are the same. On the other hand, if the three L-eigenvalues of a third order tensor $\mathscr{A}$ are the same, do we always have $\mathscr{A}\mathscr{A}^\top = \alpha \mathbf{I}$ for some $\alpha \geq 0$?

We call formulas (7.4) and (7.5) the L-eigenvalue decomposition of $\mathscr{A}$ and its L-inverse. We do not say that $\frac{1}{\sigma_j}$ for $j = 1, 2, 3$ are L-eigenvalues of $\mathscr{A}^{-1}$, as $\mathscr{A}^{-1}$ may have different L-eigenvalues.

We do not call σ_j ($j = 1, 2, 3$) singular values, although they are associated with the singular value decomposition theory of rectangular matrices. One reason is that they are third order tensors, but not rectangular matrices. They are only associated with rectangular matrices unfolded from the representative hypermatrices with respect to the last two indices. If we unfold the related hypermatrices with respect to the other two indices, results may be different. The second reason is that singular values for third order tensors will be defined later.

Given a third order tensor $\mathscr{A} \in T(\mathbb{H})$, a linear subspace $\{\mathbf{V} \in B(\mathbb{H}) : \mathscr{A}\mathbf{V} = \mathbf{0}\}$ is called the null space of $\mathscr{A}$. We have the following proposition.

Proposition 7.4 *Let $\mathscr{A} \in T(\mathbb{H})$. The dimension of the null space of $\mathscr{A}$ is at least* 6. *The sum of the rank and the dimension of this null space is* 9.

This proposition means that the null space of a third order tensor is quite "large", because its dimension is at least 6.

Suppose that $\mathscr{A} \in T(\mathbb{H})$, $\mathbf{x}, \mathbf{y}, \mathbf{z} \in \mathbb{H}$ and $\eta \in \mathbf{R}$. We say that η is a **singular value** of the tensor $\mathscr{A}$, $\mathbf{x}$, $\mathbf{y}$ and $\mathbf{z}$ are associated left, central and right singular vectors respectively if $\eta \geq 0$ and the following system of equations hold.

$$\mathscr{A}\mathbf{yz} = \eta\mathbf{x}, \quad \mathbf{x}\mathscr{A}\mathbf{z} = \eta\mathbf{y}, \quad \mathbf{xy}\mathscr{A} = \eta\mathbf{z}, \quad \mathbf{x} \bullet \mathbf{x} = 1, \quad \mathbf{y} \bullet \mathbf{y} = 1, \quad \mathbf{z} \bullet \mathbf{z} = 1. \tag{7.7}$$

Theorem 7.4 *Let $\mathscr{A} \in T(\mathbb{H})$. Then singular values of $\mathscr{A}$ always exist and are invariants of $\mathscr{A}$. Let η be a singular value of $\mathscr{A}$, with associated left, central, and right singular vectors* $\mathbf{x}$, $\mathbf{y}$, *and* $\mathbf{z}$. *Then $\eta = \mathbf{x}\mathscr{A}\mathbf{yz}$. For the maximum singular value η_1 of $\mathscr{A}$, we have*

$$\eta_1 = \max\{\mathbf{x}\mathscr{A}\mathbf{yz} : \mathbf{x} \bullet \mathbf{x} = 1, \mathbf{y} \bullet \mathbf{y} = 1, \mathbf{z} \bullet \mathbf{z} = 1\}. \tag{7.8}$$

Assume that $\mathscr{A} \in T(\mathbb{H})$, $\mathbf{x}, \mathbf{y}, \in \mathbb{H}$ and $\mu \in \mathbf{R}$. We say that μ is a **C-eigenvalue** of $\mathscr{A}$, $\mathbf{x}$ and $\mathbf{y}$ are associated left and right C-eigenvectors respectively if $\mu \geq 0$ and the following system of equations hold.

$$\mathscr{A}\mathbf{yy} = \mu\mathbf{x}, \quad \mathbf{x}\mathscr{A}\mathbf{y} = \mu\mathbf{y}, \quad \mathbf{x} \bullet \mathbf{x} = 1, \quad \mathbf{y} \bullet \mathbf{y} = 1. \tag{7.9}$$

C-eigenvalues were introduced in [51] for third order right-side symmetric tensors. We may extend them to general third order tensors; See Sect. 7.2. Here, "C" names after Jacques Curie and Pierre Curie (see Fig. 7.2).

Theorem 7.5 *Let $\mathscr{A} \in T(\mathbb{H})$ be right-side symmetric. Then $\mathscr{A}$ always has C-eigenvalues that are invariants of $\mathscr{A}$. Let μ be a C-eigenvalue of $\mathscr{A}$ with associated left and right C-eigenvectors* $\mathbf{x}$ *and* $\mathbf{y}$. *Then $\mu = \mathbf{x}\mathscr{A}\mathbf{yy}$. Let μ_1 be the maximum C-eigenvalue of $\mathscr{A}$. Then we get*

$$\mu_1 = \max\{\mathbf{x}\mathscr{A}\mathbf{yy} : \mathbf{x} \bullet \mathbf{x} = 1, \mathbf{y} \bullet \mathbf{y} = 1\}. \tag{7.10}$$

By comparing (7.10) with (7.8), we have $\mu_1 \leq \eta_1$ for a third order right-side symmetric tensor $\mathscr{A}$.

Next, we recall the definition of Z-eigenvalues of a tensor $\mathscr{A}$. Suppose that $\mathscr{A} \in T(\mathbb{H})$, $\mathbf{x}, \in \mathbb{H}$ and $\nu \in \mathbf{R}$. We say that ν is a Z-eigenvalue of $\mathscr{A}$ with an associated Z-eigenvector $\mathbf{x}$, if $\nu \geq 0$ and the following system of equations hold.

$$\mathscr{A}\mathbf{xx} = \nu\mathbf{x}, \quad \mathbf{x} \bullet \mathbf{x} = 1. \tag{7.11}$$

Theorem 7.6 *Let $\mathscr{A} \in T(\mathbb{H})$ be symmetric. Then Z-eigenvalues of $\mathscr{A}$ always exist and are invariants of $\mathscr{A}$. Let ν be a Z-eigenvalue of $\mathscr{A}$ with an associated*

Fig. 7.2 Pierre Curie (1859–1906)

Z-eigenvector $\mathbf{x}$. *Then* $\nu = \mathbf{x}\mathscr{A}\mathbf{x}\mathbf{x}$. *Let* ν_1 *be the maximum Z-eigenvalue of* $\mathscr{A}$. *Then*

$$\nu_1 = \max\{\mathbf{x}\mathscr{A}\mathbf{x}\mathbf{x} : \mathbf{x} \bullet \mathbf{x} = 1\}. \quad (7.12)$$

Comparing (7.12) with (7.8) and (7.10), we have $\nu_1 \le \mu_1 \le \eta_1$ *for a third order symmetric tensor* $\mathscr{A}$. *If* $\mathscr{A}$ *is symmetric, then* $\nu_1 = \mu_1$.

We address more about the Levi-Civita tensor $\mathscr{E}$.

Theorem 7.7 *The Levi-Civita tensor* $\mathscr{E}$ *is a nonsingular tensor. The kernel tensor of* $\mathscr{E}$ *is* $2\mathbf{I}$, *the L-inverse is* $\mathscr{E}^{-1} = \frac{1}{2}\mathscr{E}$, *and its three L-eigenvalues are* $\sigma_1 = \sigma_2 = \sigma_3 = \sqrt{2}$. *For any* $\mathbf{z} \in \mathbb{H}$, *if* $\mathbf{U} = \mathscr{E}\mathbf{z}$, *then* $\mathbf{z} = \frac{1}{2}\mathscr{E}\mathbf{U}$. *Furthermore,* $\frac{1}{\sqrt{2}}\mathscr{E}$ *is a third order orthogonal tensor.*

Now, the question is what the largest singular value of $\mathscr{E}$ is?

We have known that right-side symmetry and left-side symmetry are tensor properties. Some tensor symmetric properties by using the Levi-Civita tensor could be defined. Let a_{ijk} be a hypermatrix. The hypermatrix a_{ijk} is selectively right-side symmetric if $a_{ijk} = a_{ikj}$ for $i \neq k$ and $i \neq j$. Similarly, the hypermatrix a_{ijk} is selectively left-side symmetric if $a_{ijk} = a_{jik}$ for $k \neq i$ and $i \neq k \neq j$.

Proposition 7.5 *Suppose $\mathscr{A} \in T(\mathbb{H})$. Under an orthonormal basis, the representative hypermatrix of $\mathscr{A}$ is selectively right-side symmetric if and only if $\mathscr{A}\mathscr{E} = \mathscr{O}$, and the representative hypermatrix of $\mathscr{A}$ is selectively left-side symmetric if and only if $\mathscr{E}\mathscr{A} = \mathscr{O}$. Thus, both the selectively right-side symmetric property and the selectively left-side symmetric property are tensor properties.*

7.2 C-Eigenvalues of the Piezoelectric Tensors

We consider a special kind of tensors, named piezoelectric-type tensors, which is the generalization of right-side symmetric tensor.

Definition 7.1 Suppose that $\mathscr{A} = (a_{ijk}) \in T_{3,n}$. If $\mathscr{A}$ is symmetric with respect to its last two indices, i.e., $a_{ijk} = a_{ikj}$ for all i, j, and k, then $\mathscr{A}$ is called a **piezoelectric-type tensor**. In particular, if $n = 3$, $\mathscr{A}$ is a right-side symmetric tensor.

The total number of independent elements of a piezoelectric-type tensor $\mathscr{A}$ is $\frac{1}{2}n^2(n+1)$.

Suppose that λ is a real number and $\mathbf{x}, \mathbf{y} \in \mathbf{R}^n$ are unit vectors. We say a piezoelectric-type tensor $\mathscr{A}$ is a rank-one tensor if its elements are $a_{ijk} = \lambda x_i y_j y_k$ for $i, j, k = 1, 2, \ldots, n$. Shortly, we denote this rank-one piezoelectric-type tensor as $\lambda\mathbf{x} \otimes \mathbf{y} \otimes \mathbf{y} \in T_{m,n}$, where "$\otimes$" stands for the outer product. If there exist a scalar $\lambda \in \mathbf{R}$ and vectors $\mathbf{x}, \mathbf{y} \in \mathbf{R}^n$ minimizing the optimization problem

$$\min\left\{\|\mathscr{A} - \lambda\mathbf{x} \otimes \mathbf{y} \otimes \mathbf{y}\|_F^2 : \lambda \in \mathbf{R}, \mathbf{x}^\top\mathbf{x} = 1, \mathbf{y}^\top\mathbf{y} = 1\right\}, \tag{7.13}$$

then the tensor $\lambda\mathbf{x}\otimes\mathbf{y}\otimes\mathbf{y}$ is called the best rank-one piezoelectric-type approximation of $\mathscr{A}$.

By these notations, we represent the definition of C-eigenvalues and C-eigenvectors of a piezoelectric tensor formally.

Definition 7.2 Suppose that $\mathscr{A} \in T_{3,n}$ is a piezoelectric-type tensor. If there exist a scalar $\lambda \in \mathbf{R}$ and vectors $\mathbf{x} \in \mathbf{R}^n$ and $\mathbf{y} \in \mathbf{R}^n$ satisfying

$$\mathscr{A}\mathbf{y}\mathbf{y} = \lambda\mathbf{x}, \quad \mathbf{x}\mathscr{A}\mathbf{y} = \lambda\mathbf{y}, \quad \mathbf{x}^\top\mathbf{x} = 1, \quad \text{and} \quad \mathbf{y}^\top\mathbf{y} = 1, \tag{7.14}$$

then λ is called a C-eigenvalue of $\mathscr{A}$, $\mathbf{x}$ and $\mathbf{y}$ are called its associated left and right C-eigenvectors, respectively. $(\lambda, \mathbf{x}, \mathbf{y})$ is called a C-eigentriple of $\mathscr{A}$.

Immediately, we get the following theorem.

Theorem 7.8 *Suppose that $\mathscr{A}$ is a piezoelectric-type tensor. The following three assertions hold.*

(1) The piezoelectric-type tensor $\mathscr{A}$ has at least one C-eigentriple.
(2) Suppose that $(\lambda, \mathbf{x}, \mathbf{y})$ is a C-eigentriple of $\mathscr{A}$. Then

$$\lambda = \mathbf{x}\mathscr{A}\mathbf{y}\mathbf{y}.$$

Moreover, $(\lambda, \mathbf{x}, -\mathbf{y})$, $(-\lambda, -\mathbf{x}, \mathbf{y})$, and $(-\lambda, -\mathbf{x}, -\mathbf{y})$ are also C-eigentriples of $\mathscr{A}$.
(3) Let $(\lambda^*, \mathbf{x}^*, \mathbf{y}^*)$ be a C-eigentriple of $\mathscr{A}$, where λ^* is the largest C-eigenvalue of $\mathscr{A}$. Then

$$\lambda^* = \max \ \{\mathbf{x}\mathscr{A}\mathbf{y}\mathbf{y} : \mathbf{x}^\top \mathbf{x} = 1, \mathbf{y}^\top \mathbf{y} = 1\}. \tag{7.15}$$

Furthermore, the tensor $\lambda^*\mathbf{x}^* \otimes \mathbf{y}^* \otimes \mathbf{y}^*$ is the best rank-one piezoelectric-type approximation of $\mathscr{A}$.

Theorem 7.8 (c) implies that the largest C-eigenvalue λ^* determines the highest piezoelectric coupling constant, and $\mathbf{y}^*$ is the direction of the stress where this appears. In the sense of this point, the largest C-eigenvalue of a piezoelectric tensor has concrete physical meaning.

The invariant property under orthogonal transformations is important for many physical problems. Next, we are going to show that C-eigenvalues of a piezoelectric-type tensor $\mathscr{A} = (a_{ijk})$ are invariant under orthogonal transformations. Let $\mathscr{A}$ be a piezoelectric-type tensor and $Q = (q_{ir}) \in \mathbf{R}^{n\times n}$ be an orthogonal matrix. Then, the rotated tensor $\mathscr{A}Q^3$ is also a piezoelectric-type tensor.

Theorem 7.9 *Suppose that $Q \in \mathbf{R}^{n\times n}$ is an orthogonal matrix and $\mathscr{A} \in T_{3,n}$ is a piezoelectric-type tensor. Let $(\lambda, \mathbf{x}, \mathbf{y})$ be a C-eigentriple of $\mathscr{A}$. Then, $(\lambda, Q^\top\mathbf{x}, Q^\top\mathbf{y})$ is a C-eigentriple of $\mathscr{A}Q^3$.*

Next, we address two intervals for locating all C-eigenvalues of piezoelectric-type tensors, which provide upper bounds for the largest C-eigenvalue [166].

Theorem 7.10 *Let $\mathscr{A} \in T_{3,n}$ be a piezoelectric-type tensor, and λ be one of its C-eigenvalues. Then, we have*

$$|\lambda| \le \kappa_1(\mathscr{A}), \tag{7.16}$$

where

$$r_i^{(1)}(\mathscr{A}) := \sum_{j,k\in[n]} |a_{ijk}|, \qquad r_j^{(2)}(\mathscr{A}) := \sum_{i,k\in[n]} |a_{ijk}|,$$

and

$$\kappa_1(\mathscr{A}) := \max_{i,j\in[n]} \sqrt{r_i^{(1)}(\mathscr{A}) r_j^{(2)}(\mathscr{A})}.$$

Proof Since λ is a C-eigenvalue of $\mathscr{A}$, we assume that $\mathbf{x} = (x_i) \in \mathbf{R}^n$ and $\mathbf{y} = (y_i) \in \mathbf{R}^n$ are associated left and right C-eigenvectors. Denote that

$$|x_p| = \max_{i\in[n]} |x_i| \in (0, 1] \qquad \text{and} \qquad |y_q| = \max_{i\in[n]} |y_i| \in (0, 1].$$

Considering the pth equation of $\mathscr{A}\mathbf{yy} = \lambda\mathbf{x}$, we obtain

$$\lambda x_p = \sum_{j,k\in[n]} a_{pjk} y_j y_k,$$

and hence

$$\begin{aligned} |\lambda||x_p| &\le \sum_{j,k\in[n]} |a_{pjk}||y_j||y_k|, \\ &\le \sum_{j,k\in[n]} |a_{pjk}||y_q||y_q| \\ &\le |y_q| r_p^{(1)}(\mathscr{A}). \end{aligned} \tag{7.17}$$

In a similar way, we have

$$|\lambda||y_q| \le |x_p| r_q^{(2)}(\mathscr{A}). \tag{7.18}$$

Multiplying (7.17) and (7.18) and eliminating a factor $|x_p||y_q| > 0$, we get

$$|\lambda|^2 \le r_p^{(1)}(\mathscr{A}) r_q^{(2)}(\mathscr{A}) \le \max_{i,j\in[n]} r_i^{(1)}(\mathscr{A}) r_j^{(2)}(\mathscr{A}).$$

Let $\kappa_1(\mathscr{A}) = \max_{i,j\in[n]} \sqrt{r_i^{(1)}(\mathscr{A}) r_j^{(2)}(\mathscr{A})}$. Then the proof is complete. □

From this theorem, $\kappa_1(\mathscr{A})$ is an upper bound for the largest C-eigenvalue of a piezoelectric-type tensor.

Let S be a subset of $[n]$. Denote that

$$\begin{aligned} \Omega_S &:= \{(i,j) : i \in S \text{ or } j \in S\}, \\ \Omega_S^c &:= \{(i,j) : i \notin S \text{ and } j \notin S\}. \end{aligned}$$

For a given piezoelectric-type tensor $\mathscr{A} \in T_{3,n}$, we define

$$\begin{aligned} r_k^{\Omega_S}(\mathscr{A}) &:= \sum_{(i,j)\in\Omega_S} |a_{ijk}|, \\ r_k^{\Omega_S^c}(\mathscr{A}) &:= \sum_{(i,j)\in\Omega_S^c} |a_{ijk}|. \end{aligned}$$

Here, $r_k^{\Omega_S}(\mathscr{A}) = 0$ if $S = \emptyset$, and $r_k^{\Omega_S^c}(\mathscr{A}) = 0$ if $S = [n]$. Moreover, we have $r_k^{(2)}(\mathscr{A}) = r_k^{\Omega_S}(\mathscr{A}) + r_k^{\Omega_S^c}(\mathscr{A})$. Define

$$r_S(\mathscr{A}) := \max_{i,j\in[n]} \frac{1}{2}\left(r_j^{\Omega_S}(\mathscr{A}) + \sqrt{(r_j^{\Omega_S}(\mathscr{A}))^2 + 4 r_i^{(1)}(\mathscr{A}) r_j^{\Omega_S^c}(\mathscr{A})} \right),$$

for any subset S of $[n]$, and

$$\kappa_2(\mathscr{A}) := \min_{S\subseteq[n]} r_S(\mathscr{A}).$$

Then, we have another bound for λ.

Theorem 7.11 *Suppose that λ is a C-eigenvalue of a piezoelectric-type tensor $\mathscr{A} \in T_{3,n}$. Then, we get*

$$|\lambda| \le \kappa_2(\mathscr{A}).$$

Proof Let S be a subset of $[n]$. We assume that $\mathbf{x} = (x_i) \in \mathbf{R}^n$ and $\mathbf{y} = (y_i) \in \mathbf{R}^n$ are left and right C-eigenvectors of $\mathscr{A}$ corresponding to the C-eigenvalue λ. We denote

$$|x_p| = \max_{i\in[n]} |x_i| \in (0, 1] \qquad \text{and} \qquad |y_q| = \max_{i\in[n]} |y_i| \in (0, 1].$$

Considering the qth equation of $\mathbf{x}\mathscr{A}\mathbf{y} = \lambda\mathbf{y}$, we have

$$\begin{aligned}
|\lambda||y_q| &= \left|\sum_{i,j\in[n]} a_{ijq}x_i y_j\right| \\
&\le \sum_{i,j\in[n]} |a_{ijq}||x_i||y_j| \\
&\le |x_p||y_q| \sum_{i,j\in[n]} |a_{ijq}| \\
&\le |x_p||y_q|(r_q^{\Omega_S}(\mathscr{A}) + r_q^{\Omega_S^c}(\mathscr{A})) \\
&\le r_q^{\Omega_S}(\mathscr{A})|y_q| + r_q^{\Omega_S^c}(\mathscr{A})|x_p|.
\end{aligned}$$

Thus,

$$(|\lambda| - r_q^{\Omega_S}(\mathscr{A}))|y_q| \le r_q^{\Omega_S^c}(\mathscr{A})|x_p|. \tag{7.19}$$

By multiplying (7.17) and (7.19) and eliminating $|x_p||y_q|$, it yields that

$$|\lambda|(|\lambda| - r_q^{\Omega_S}(\mathscr{A})) \le r_p^{(1)}(\mathscr{A})r_q^{\Omega_S^c}(\mathscr{A}).$$

Then, we solve the above inequality and get

$$|\lambda| \le \frac{1}{2}\left(r_q^{\Omega_S}(\mathscr{A}) + \sqrt{(r_q^{\Omega_S}(\mathscr{A}))^2 + 4r_p^{(1)}(\mathscr{A})r_q^{\Omega_S^c}(\mathscr{A})}\right) \le r_S(\mathscr{A}).$$

Because $|\lambda| \le r_S(\mathscr{A})$ is valid for all $S \subseteq [n]$, we obtain this theorem. $\square$

Note that $\kappa_2(\mathscr{A})$ is also an upper bound of the largest C-eigenvalue of a piezoelectric-type tensor $\mathscr{A}$. By setting $S = \emptyset$, we find that

$$\kappa_2(\mathscr{A}) \le r_\emptyset = \max_{i,j\in[n]} \sqrt{r_i^{(1)}(\mathscr{A}) r_j^{(2)}(\mathscr{A})} = \kappa_1(\mathscr{A}).$$

Hence, the upper bound of the largest C-eigenvalue of a piezoelectric-type tensor obtain by Theorem 7.11 is sharper than the upper bound established in Theorem 7.10.

In physics, for non-centrosymmetric materials, the linear piezoelectric equation is expressed as

$$p_i = \sum_{j,k=1}^{3} a_{ijk} T_{jk},$$

where $\mathscr{A} = (a_{ijk})$ is a third order **piezoelectric tensor**, $T = (T_{jk}) \in \mathbf{R}^{3\times 3}$ is the second order stress tensor, and $\mathbf{p} = (p_i) \in \mathbf{R}^3$ is the electric change density displacement (polarization), which is a first order tensor. Since $\mathscr{A}$ is a piezoelectric-type tensor, there are 18 independent elements in $\mathscr{A}$.

Under a unit uniaxial stress, what situations will trigger the maximal piezoelectricity? An idea example of uniaxial stress is the stress in a long, vertical rod loaded by hanging a weight on the end [212, p. 90]. In this case, the second order stress tensor could be rewritten as $T = \mathbf{y}\mathbf{y}^\top$ with $\mathbf{y}^\top\mathbf{y} = 1$. Then, we formulate this maximal piezoelectricity problem into an optimization model

$$\begin{cases} \max & \|\mathbf{p}\|_2 \\ \text{s.t.} & \mathbf{p} = \mathscr{A}\mathbf{y}\mathbf{y}, \\ & \mathbf{y}^\top\mathbf{y} = 1. \end{cases} \tag{7.20}$$

By a dual norm, we know $\|\mathbf{p}\|_2 = \max_{\mathbf{x}^\top\mathbf{x}=1} \mathbf{x}^\top\mathbf{p} = \max_{\mathbf{x}^\top\mathbf{x}=1} \mathbf{x}\mathscr{A}\mathbf{y}\mathbf{y}$. Hence, it suffices to consider the following optimization problem

$$\max\ \mathbf{x}\mathscr{A}\mathbf{y}\mathbf{y} \qquad \text{s.t. } \mathbf{x}^\top\mathbf{x} = 1,\ \mathbf{y}^\top\mathbf{y} = 1. \tag{7.21}$$

Assume that $(\mathbf{x}^*, \mathbf{y}^*)$ is an optimal solution of the above optimization problem. Then, the optimal value of objective $\lambda^* = \mathbf{x}^*\mathscr{A}\mathbf{y}^*\mathbf{y}^*$ is the largest C-eigenvalue of the piezoelectric tensor $\mathscr{A}$, and $\mathbf{y}^*$ is the unit uniaxial direction that the maximal piezoelectric effect take place along. These results imply the following theorem.

Theorem 7.12 *Suppose that $(\lambda^*, \mathbf{x}^*, \mathbf{y}^*)$ is a C-eigentriple of the piezoelectric tensor $\mathscr{A}$, and λ^* is the largest C-eigenvalue of $\mathscr{A}$. Then, under a unit uniaxial stress, the maximum value of the 2-norm of the electric polarization is λ^* and the optimal axial direction is $\mathbf{y}^*$.*

We turn to the converse piezoelectric effect and consider the following linear equation

$$S_{jk} = \sum_i a_{ijk} e_i,$$

where $S = (S_{jk}) \in \mathbf{R}^{3\times 3}$ is the strain tensor and $\mathbf{e} = (e_i) \in \mathbf{R}^3$ is the electric field strength. Denote $\|\cdot\|_2$ as the matrix spectral norm, i.e., $\|S\|_2 = \max_{\mathbf{y}^\top\mathbf{y}=1} \mathbf{y}^\top S\mathbf{y}$. Now, we study the maximization problem on the spectral norm of S:

$$\begin{cases} \max & \|S\|_2 \\ \text{s.t.} & S_{jk} = \sum_{i=1}^{3} e_i a_{ijk} \qquad \forall j,k \in \{1,2,3\}, \\ & \mathbf{e}^\top\mathbf{e} = 1. \end{cases} \tag{7.22}$$

Since $\|S\|_2 = \max_{\mathbf{y}^\top\mathbf{y}=1} \mathbf{y}^\top S\mathbf{y} = \max_{\mathbf{y}^\top\mathbf{y}=1} \mathbf{e}\mathscr{A}\mathbf{y}\mathbf{y}$, we could rewrite (7.22) as follows:

$$\max \ \{\mathbf{e}\mathscr{A}\mathbf{y}\mathbf{y} : \mathbf{e}^\top\mathbf{e} = 1, \mathbf{y}^\top\mathbf{y} = 1\}.$$

Let $(\mathbf{e}^*, \mathbf{y}^*)$ be an optimal solution of the above optimization problem. Then, $\lambda^* = \mathbf{e}^*\mathscr{A}\mathbf{y}^*\mathbf{y}^*$ is the largest C-eigenvalue of the piezoelectric tensor $\mathscr{A}$, $\mathbf{e}^*$ and $\mathbf{y}^*$ are its associated left and right C-eigenvectors.

Theorem 7.13 *Suppose that* $(\lambda^*, \mathbf{x}^*, \mathbf{y}^*)$ *is a C-eigentriple of the piezoelectric tensor* $\mathscr{A}$ *and* λ^* *is the largest C-eigenvalue. Then, under unit electric field strength* $\|\mathbf{x}^*\| = 1$, λ^* *is the largest spectral norm of a strain tensor generated by the converse piezoelectric effect.*

Due to the crystallographic symmetry of piezoelectric materials, there are 32 classes in crystals [119]. However, 11 classes of crystals enjoy the center of symmetry, hence there are no piezoelectricity. In addition, for a special class numbered 432, piezoelectric changes counteract each other. Thus, there remains 20 crystallographic classes, where piezoelectricity may exist. We now examine some typical crystals in these crystallographic classes.

By Theorem 7.8 (b), we know that $(-\lambda, -\mathbf{x}, \mathbf{y})$, $(\lambda, \mathbf{x}, -\mathbf{y})$, and $(-\lambda, -\mathbf{x}, -\mathbf{y})$ are all C-eigentriples of a piezoelectric tensor if $(\lambda, \mathbf{x}, \mathbf{y})$ is a C-eigentriple of this tensor. For compactness, we use $(\lambda, \mathbf{x}, \mathbf{y})$ to present the group of these four C-eigentriples of the piezoelectric tensor.

Piezoelectric tensors of crystals have special symmetric structures microscopically. First, we examine crystals in 23 and $\bar{4}3m$ crystallographic point groups. The corresponding piezoelectric tensor $\mathscr{A}(\alpha)$ has only one independent parameter $\alpha > 0$:

$$a_{123} = a_{213} = a_{312} = -\alpha.$$

Other elements of $\mathscr{A}(\alpha)$ are zeros.

Proposition 7.6 *There are* 13 *groups of C-eigentriples of the piezoelectric tensor* $\mathscr{A}(\alpha)$.

Next, we present some C-eigentriples of piezoelectric tensors arising from known piezoelectric materials with different symmetries. All piezoelectric tensors measured

Table 7.3 C-eigenvalues and associated C-eigenvectors of the piezoelectric tensor of $\mathscr{A}_{Na_2O}$

No.	λ	$\mathbf{x}^\top$			$\mathbf{y}^\top$		
1	9.61902	0	0	−1	0	0	1
2	2.47286	0.2726	−0.17527	−0.946028	0.951413	0.307917	0
3	2.47286	0.0154875	0.323714	−0.946028	0.74237	−0.66999	0
4	2.47286	−0.288088	−0.148444	−0.946028	0.209043	0.977906	0
5	2.4721	0.286464	0.147607	−0.946652	0.977851	−0.209031	0.0106522
6	2.4721	−0.271064	0.174281	−0.946652	0.307899	−0.951359	−0.0106521
7	2.4721	−0.0153997	−0.321888	−0.946652	0.669952	0.742327	−0.0106521

here are reported in [74] with unit (pC/N) that is omitted for convenience. Here, "p" represents pico (10^{-12}), "C" means Coulomb (electric), and "N" stands for Newton (force).

Example 1 A compound VFeSb belongs to the $\bar{4}3m$ crystallographic point group [74]. The nonzero parameter $\alpha = 3.68180667$ for the piezoelectric tensor $\mathscr{A}_{VFeSb}$. By Proposition 7.6 and Problem 3 in Exercises, we know that the largest C-eigenvalue of $\mathscr{A}_{VFeSb}$ is about 4.25138.

Example 2 The compound Na_2O belongs to the 3 crystallographic point group. There are six independent parameters in the piezoelectric tensor $\mathscr{A}_{Na_2O}$

$$a_{111} = -a_{122} = -a_{212} = 0.80022, \quad a_{123} = -a_{213} = 0.138975,$$
$$a_{113} = a_{223} = -0.16963, \quad a_{222} = -a_{112} = -a_{211} = -0.043735,$$
$$a_{311} = a_{322} = -2.339395, \quad \text{and } a_{333} = -9.61902.$$

Other elements of the piezoelectric tensor is zero. Positive C-eigenvalues of $\mathscr{A}_{Na_2O}$ and associated C-eigenvectors are reported in Table 7.3.

Example 3 The compound $LiMnO_2$ belongs to the m crystallographic point group. There are ten independent parameters in the piezoelectric tensor $\mathscr{A}_{LiMnO_2}$

$$a_{111} = 2.34136, \quad a_{122} = 0.06249, \quad a_{133} = 0.50554, \quad a_{113} = -0.70406,$$
$$a_{223} = -0.40829, \quad a_{212} = 0.76378, \quad a_{311} = 0.33075, \quad a_{322} = -0.23931,$$
$$a_{333} = -0.02956, \quad \text{and} \quad a_{313} = 0.21625.$$

Other elements of the piezoelectric tensor is zero. Positive C-eigenvalues of $\mathscr{A}_{LiMnO_2}$ and associated C-eigenvectors are reported in Table 7.4.

Finally, we consider the difference between C-eigenvalues of the piezoelectric tensor and singular values of a matrix. There are $\frac{n(n+1)}{2}$ independent elements in an n-by-n symmetric matrix $S = (s_{ij})$. Hence we may record S as a vector

Table 7.4 C-eigenvalues and associated C-eigenvectors of the piezoelectric tensor of $\mathscr{A}_{LiMnO_2}$

No.	λ	$\mathbf{x}^\top$			$\mathbf{y}^\top$		
1	2.5855	0.997928	0	0.0643332	0.949448	0	−0.313925
2	0.294235	0.925346	0	0.379124	0.274158	0	0.961685
3	0.247334	0.252654	0	−0.967557	0	1	0
4	0.170592	0.870132	0.121659	−0.477566	0.264222	−0.808591	0.525706
5	0.170592	0.870132	−0.121659	−0.477566	0.264222	0.808591	0.525706

$$\mathrm{vec}(S) = (s_{11}, s_{22}, \ldots, s_{nn}, \sqrt{2}s_{(n-1)n}, \ldots, \sqrt{2}s_{12})^\top \in \mathbf{R}^{\frac{n(n+1)}{2}}.$$

We equip off-diagonal elements of S with a multiple $\sqrt{2}$, while diagonal elements of S are with coefficient 1. Hence, we get $\|S\|_F = \|\mathrm{vec}(S)\|_2$.

Suppose that $\mathscr{A} \in T_{3,n}$ is a piezoelectric-type tensor which contains $\frac{n^2(n+1)}{2}$ independent elements. Owing to the partly symmetry of piezoelectric-type tensors $\mathscr{A}$, we represent each symmetric slice-matrix as a vector. By collecting these vectors, we obtain an n-by-$\frac{n(n+1)}{2}$ matrix

$$M(\mathscr{A}) = \begin{pmatrix} a_{111} & a_{122} & \cdots & a_{1nn} & \sqrt{2}a_{1(n-1)n} & \cdots & \sqrt{2}a_{112} \\ \vdots & \vdots & \vdots & \vdots & \vdots & \vdots & \vdots \\ a_{n11} & a_{n22} & \cdots & a_{nnn} & \sqrt{2}a_{n(n-1)n} & \cdots & \sqrt{2}a_{n12} \end{pmatrix}. \tag{7.23}$$

Each row of the above matrix records a symmetric slice-matrix of the piezoelectric-type tensor. Let $\mathbf{y} = (y_1, \ldots, y_n)^\top \in \mathbf{R}^n$. By direct calculations, we obtain the following useful equality

$$\mathscr{A}\mathbf{y}\mathbf{y} = M(\mathscr{A})vec(\mathbf{y}\mathbf{y}^\top).$$

Theorem 7.14 *Suppose that λ^* and μ^* are the largest C-eigenvalue of a piezoelectric-type tensor $\mathscr{A}$ and the largest singular value of the matrix $M(\mathscr{A})$, respectively. Then,*

$$\lambda^* \leq \mu^*. \tag{7.24}$$

Let $\mathscr{A} \in T_{3,2}$ be a piezoelectric-type tensor, where $a_{112} = a_{222} = 1$ and other elements are zero. Then, by some calculations, we get that

$$\lambda^* = \frac{2}{\sqrt{3}} < \sqrt{2} = \mu_*.$$

The strict inequality holds in this case.

7.3 Third Order Three Dimensional Symmetric Traceless Tensors and Liquid Crystals

In 1888, an Austrian botanist Friedrich Reinitzer (1857–1927) discovered liquid crystals which were further studied and named later by German physicist Otto Lehmann (1855–1922) in 1904. Liquid crystals are special matter in a state which has properties between those of conventional liquids and solid crystals. A liquid crystal phase of matter is described as one where constituent molecules are sufficiently disordered to confer the flow properties of a liquid, yet still preserve some degree of ordering such that the phase is anisotropic. In recent advance of liquid crystals, the octupolar (tetrahedratic) phase for bent-core (banana-shaped) liquid crystals molecules is a novel phase that was predicted theoretically [24, 186] and confirmed experimentally [287].

Each liquid crystal molecule has a microscopic polar axis $\mathbf{p}$. In a macroscopic molecular assembly, we could consider the orientational distribution ρ of these polar axes. By Buckingham's formula [28], the probability density of an orientational distribution $\rho : S^2 \mapsto R_+$ over the unit sphere $S^2 = \{\mathbf{p} \in \mathbf{R}^3 : \mathbf{p}^\top \mathbf{p} = 1\}$ could be represented as

$$\rho(\mathbf{p}) = \frac{1}{4\pi}\left(1 + \sum_{k=1}^{\infty} \frac{(2k+1)!!}{k!} \langle \overline{\mathbf{p}^{\otimes k}} \rangle_\rho \cdot \mathbf{p}^{\otimes k}\right),$$

where $\mathbf{p}^{\otimes k}$ is a kth order rank-one tensor, $\langle \overline{\mathbf{p}^{\otimes k}} \rangle_\rho$ is the corresponding multipole average, and the overline $\overline{\cdots}$ denotes the irreducible, symmetric, and traceless part of the tensor it surmounted. As a probability density function, the spherical integral

$$\int_{S^2} \rho(\mathbf{p}) \mathrm{d}\mathbf{p} = 1,$$

which is equal to a spherical integral of the constant term $\frac{1}{4\pi}$. Hence, spherical integrals of other terms should be zero.

Particularly, the first three multipole averages are called the dipolar, quadrupolar, and octupolar order tensors

$$\mathbf{d} = \langle \mathbf{p} \rangle_\rho, \qquad Q = \langle \overline{\mathbf{p} \otimes \mathbf{p}} \rangle_\rho, \qquad \mathscr{A} = \langle \overline{\mathbf{p} \otimes \mathbf{p} \otimes \mathbf{p}} \rangle_\rho.$$

They are meaningful for resolving the characteristic features of ρ. Since their spherical integrals are zero, we know that Q is a symmetric and traceless matrix, and the **octupolar tensor** $\mathscr{A} = (a_{ijk})$ is a third order three dimensional symmetric and traceless tensor.

We now give the definition of traceless tensors formally.

Definition 7.3 Let $\mathscr{T} = (t_{i_1 i_2 \ldots i_m}) \in S_{m,n}$. If

$$\sum_{i=1}^{n} t_{iii_3 \ldots i_m} = 0 \qquad \text{for all } i_3, \ldots, i_m = 1, 2, \ldots, n,$$

then $\mathscr{T}$ is called a **traceless tensor**.

The following theorem shows that under orthogonal transformations, the traceless property of a symmetric tensor is invariant [99].

Theorem 7.15 *Let $\mathscr{T} = (t_{i_1 i_2 \ldots i_m}) \in S_{m,n}$ be a traceless tensor and $Q = (q_{ij}) \in \mathbf{R}^{n \times n}$ be an orthogonal matrix. Then, $\mathscr{T} Q^m$ is also a traceless tensor.*

Proof It is straightforward to see that the new tensor $\mathscr{T} Q^m$ is real-valued and symmetric. Now, we consider its slice matrices. As the sum of all matrix eigenvalues, the trace of a symmetric matrix is invariant under an orthogonal transformation. Hence, we get

$$\sum_{i=1}^{n} \sum_{j_1=1}^{m} \sum_{j_2=1}^{m} t_{j_1 j_2 j_3 \cdots j_m} q_{ij_1} q_{ij_2} = \sum_{i=1}^{n} t_{iij_3 \cdots j_m} = 0 \tag{7.25}$$

for all $j_3, \ldots, j_m = 1, 2, \ldots, n$. By some calculations,

$$\begin{aligned}
\sum_{i=1}^{n} [\mathscr{T} Q^m]_{iii_3 \cdots i_m} &= \sum_{i=1}^{n} \sum_{j_1=1}^{n} \sum_{j_2=1}^{n} \sum_{j_3=1}^{n} \cdots \sum_{j_m=1}^{n} t_{j_1 j_2 j_3 \cdots j_m} q_{ij_1} q_{ij_2} q_{i_3 j_3} \cdots q_{i_m j_m} \\
&= \sum_{j_3=1}^{n} \cdots \sum_{j_m=1}^{n} \left(\sum_{i=1}^{n} \sum_{j_1=1}^{n} \sum_{j_2=1}^{n} t_{j_1 j_2 j_3 \cdots j_m} q_{ij_1} q_{ij_2} \right) q_{i_3 j_3} \cdots q_{i_m j_m} \\
&= 0,
\end{aligned}$$

where the last equality is valid because of (7.25). Hence, the new tensor $\mathscr{T} Q^m$ is also traceless. □

In the remainder of this chapter, we focus on the octupolar tensor $\mathscr{A}$. The symmetry means that ten independent parameters are needed for presenting a generic octupolar tensor, i.e.,

$$\mathscr{A} = \left(\begin{array}{ccc|ccc|ccc} a_{111} & a_{112} & a_{113} & a_{112} & a_{122} & a_{123} & a_{113} & a_{123} & a_{133} \\ a_{112} & a_{122} & a_{123} & a_{122} & a_{222} & a_{223} & a_{123} & a_{223} & a_{233} \\ a_{113} & a_{123} & a_{133} & a_{123} & a_{223} & a_{233} & a_{133} & a_{233} & a_{333} \end{array} \right) \in S_{3,3}. \tag{7.26}$$

The word "traceless" means that

$$\begin{cases} a_{111} + a_{122} + a_{133} = 0, \\ a_{112} + a_{222} + a_{233} = 0, \\ a_{113} + a_{223} + a_{333} = 0. \end{cases} \tag{7.27}$$

Hence, seven independent parameters are sufficient for describing a generic octupolar tensor $\mathscr{A}$.

For convenience, we denote seven independent parameters of the octupolar tensor $\mathscr{A}$ as

$$\alpha_0 = a_{123},$$
$$\alpha_1 = a_{111},\ \alpha_2 = a_{222},\ \alpha_3 = a_{333},$$
$$\beta_1 = a_{122},\ \beta_2 - a_{233},\ \beta_3 - a_{113}.$$

By the traceless property (7.27), we could rewrite (7.26) as

$$\mathscr{A} = \left(\begin{array}{ccc|ccc|ccc} \alpha_1 & -\alpha_2-\beta_2 & \beta_3 & -\alpha_2-\beta_2 & \beta_1 & \alpha_0 & \beta_3 & \alpha_0 & -\alpha_1-\beta_1 \\ -\alpha_2-\beta_2 & \beta_1 & \alpha_0 & \beta_1 & \alpha_2 & -\alpha_3-\beta_3 & \alpha_0 & -\alpha_3-\beta_3 & \beta_2 \\ \beta_3 & \alpha_0 & -\alpha_1-\beta_1 & \alpha_0 & -\alpha_3-\beta_3 & \beta_2 & -\alpha_1-\beta_1 & \beta_2 & \alpha_3 \end{array}\right).$$

Then, the associated **octupolar potential** defined on the unit sphere S^2 [99] is

$$\begin{aligned} \Phi(\mathbf{x}) \equiv \mathscr{A}\mathbf{x}^3 &= \sum_{i=1}^{3}\sum_{j=1}^{3}\sum_{k=1}^{3} a_{ijk}x_ix_jx_k \\ &= \alpha_1x_1^3 + \alpha_2x_2^3 + \alpha_3x_3^3 + 6\alpha_0x_1x_2x_3 + 3\beta_1x_1x_2^2 + 3\beta_2x_2x_3^2 + 3\beta_3x_1^2x_3 \\ &\quad - 3(\alpha_1+\beta_1)x_1x_3^2 - 3(\alpha_2+\beta_2)x_1^2x_2 - 3(\alpha_3+\beta_3)x_2^2x_3. \end{aligned} \tag{7.28}$$

We say that the potential $\Phi(\mathbf{x})$ has at least one maximum point on the unit sphere S^2. Without loss of generality, we rotate the Cartesian coordinate system such that the maximum point being the North pole $(0, 0, 1)^\top$, i.e.,

$$\alpha_3 = \Phi(0, 0, 1) \geq 0.$$

From the spectral theory of tensors [221], α_3 is a Z-eigenvalue of $\mathscr{A}$ with an associated Z-eigenvector $(0, 0, 1)^\top$. Hence, all Z-eigenvectors λ and associated Z-eigenvalues $\mathbf{x}$ of $\mathscr{A}$ must satisfy the following system of Z-eigenvalues:

$$\begin{cases} \alpha_1x_1^2 + 2\alpha_0x_2x_3 + \beta_1x_2^2 + 2\beta_3x_1x_3 - (\alpha_1+\beta_1)x_3^2 - 2(\alpha_2+\beta_2)x_1x_2 = \lambda x_1, \\ \alpha_2x_2^2 + 2\alpha_0x_1x_3 + 2\beta_1x_1x_2 + \beta_2x_3^2 - (\alpha_2+\beta_2)x_1^2 - 2(\alpha_3+\beta_3)x_2x_3 = \lambda x_2, \\ \alpha_3x_3^2 + 2\alpha_0x_1x_2 + 2\beta_2x_2x_3 + \beta_3x_1^2 - 2(\alpha_1+\beta_1)x_1x_3 - (\alpha_3+\beta_3)x_2^2 = \lambda x_3, \\ x_1^2 + x_2^2 + x_3^2 = 1. \end{cases}$$

Hence, by substituting $\lambda = \alpha_3$ and $\mathbf{x} = (0, 0, 1)^\top$ to the above system, we obtain

$$\alpha_1 + \beta_1 = 0 \qquad \text{and} \qquad \beta_2 = 0.$$

Moreover, because $\Phi(\mathbf{x})$ is a continuous function and $\Phi(-x_1, 0, 0) = -\Phi(x_1, 0, 0)$, we can rotate the Cartesian coordinate system so that $\Phi(1, 0, 0) = 0$ and we get

$$\alpha_1 = 0.$$

In this way, we determined three parameters in $\mathscr{A}$ that are $\alpha_1 = \beta_1 = \beta_2 = 0$.

Then, the octupolar tensor in (7.26) is indeed

$$\mathscr{A} = \left(\begin{array}{ccc|ccc|ccc} 0 & -\alpha_2 & \beta_3 & -\alpha_2 & 0 & \alpha_0 & \beta_3 & \alpha_0 & 0 \\ -\alpha_2 & 0 & \alpha_0 & 0 & \alpha_2 & -\alpha_3 - \beta_3 & \alpha_0 & -\alpha_3 - \beta_3 & 0 \\ \beta_3 & \alpha_0 & 0 & \alpha_0 & -\alpha_3 - \beta_3 & 0 & 0 & 0 & \alpha_3 \end{array}\right),$$

which features four independent parameters: α_0, α_2, α_3, and β_3. Correspondingly, the octupolar potential (7.28) reduces to

$$\begin{aligned} \Phi(\mathbf{x}; \alpha_0, \alpha_2, \alpha_3, \beta_3) = {} & \alpha_2 x_2^3 + \alpha_3 x_3^3 + 6\alpha_0 x_1 x_2 x_3 \\ & + 3\beta_3 x_1^2 x_3 - 3\alpha_2 x_1^2 x_2 - 3(\alpha_3 + \beta_3) x_2^2 x_3 \end{aligned} \qquad (7.29)$$

for all $\mathbf{x} \in S^2$. Without loss of generality, we can assume

$$\alpha_2 \geq 0$$

because of the symmetry revealed by the following proposition.

Proposition 7.7 *For the octupolar potential (7.29), we get*

$$\Phi(x_1, x_2, x_3; \alpha_0, \alpha_2, \alpha_3, \beta_3) = \Phi(x_1, -x_2, x_3; -\alpha_0, -\alpha_2, \alpha_3, \beta_3).$$

We exploit the assumption that the North pole $(0, 0, 1)^\top$ is a maximum point of the octupolar potential $\Phi(\mathbf{x})$ on the unit sphere S^2 with a value α_3.

Theorem 7.16 *Suppose that the North pole* $(0, 0, 1)^\top$ *is a local maximum point of* $\Phi(\mathbf{x})$ *on* S^2. *Then, the following inequality holds*

$$3\alpha_3^2 - 4\alpha_3\beta_3 - 4\beta_3^2 - 4\alpha_0^2 \geq 0. \qquad (7.30)$$

If the strict inequality holds in (7.30), $(0, 0, 1)^\top$ *is a strict local maximum point of* $\Phi(\mathbf{x})$ *on* S^2.

Proof We consider the spherical optimization problem:

$$\begin{cases} \max \ \Phi(\mathbf{x}) = \mathscr{A}\mathbf{x}^3 \\ \text{s.t.} \ \ \mathbf{x}^\top \mathbf{x} = 1. \end{cases} \qquad (7.31)$$

Its Lagrangian is

$$L(\mathbf{x}, \lambda) = -\frac{1}{3}\mathscr{A}\mathbf{x}^3 + \frac{\lambda}{2}(\mathbf{x}^\top \mathbf{x} - 1).$$

The Hessian of the Lagrangian is

$$\nabla^2_{\mathbf{xx}} L(\mathbf{x}, \lambda) = \lambda I - 2\mathscr{A}\mathbf{x},$$

which is positive semidefinite on the tangent space $\mathbf{x}^\perp \equiv \{\mathbf{y} \in \mathbf{R}^3 : \mathbf{x}^\top \mathbf{y} = 0\}$ if $\mathbf{x}$ is a (local) maximum point of $\Phi(\mathbf{x})$ on S^2 [211]. Let $P \equiv I - \mathbf{x}\mathbf{x}^\top \in \mathbf{R}^{3\times 3}$ be the projection matrix onto $\mathbf{x}^\perp$. Then, the matrix $P^\top \nabla^2_{\mathbf{xx}} L(\mathbf{x}, \lambda) P$ is positive semidefinite. By use of the first-order necessary condition,

$$\mathscr{A}\mathbf{x}^2 = \lambda \mathbf{x} \qquad \text{and} \qquad \mathbf{x}^\top \mathbf{x} = 1,$$

we have that

$$\begin{aligned} P^\top \nabla^2_{\mathbf{xx}} L(\mathbf{x}, \lambda) P &= (I - \mathbf{x}\mathbf{x}^\top)(\lambda I - 2\mathscr{A}\mathbf{x})(I - \mathbf{x}\mathbf{x}^\top) \\ &= \lambda(I - \mathbf{x}\mathbf{x}^\top) - 2(I - \mathbf{x}\mathbf{x}^\top)(\mathscr{A}\mathbf{x})(I - \mathbf{x}\mathbf{x}^\top) \\ &= \lambda(I - \mathbf{x}\mathbf{x}^\top) - 2(\mathscr{A}\mathbf{x} - \mathbf{x}(\mathscr{A}\mathbf{x}^2)^\top - (\mathscr{A}\mathbf{x}^2)\mathbf{x}^\top + (\mathscr{A}\mathbf{x}^3)\mathbf{x}\mathbf{x}^\top) \\ &= \lambda(I - \mathbf{x}\mathbf{x}^\top) - 2(\mathscr{A}\mathbf{x} - \lambda\mathbf{x}\mathbf{x}^\top - \lambda\mathbf{x}\mathbf{x}^\top + \lambda\mathbf{x}\mathbf{x}^\top) \\ &= \lambda(I + \mathbf{x}\mathbf{x}^\top) - 2\mathscr{A}\mathbf{x}. \end{aligned} \tag{7.32}$$

Because the North pole $(0, 0, 1)^\top$ is a local maximum point with associated multiplier $\lambda = \alpha_3$, we arrive at

$$[P^\top \nabla^2_{\mathbf{xx}} L(\mathbf{x}, \lambda) P]_{\lambda=\alpha_3, \mathbf{x}=(0,0,1)^\top} = \begin{bmatrix} \alpha_3 - 2\beta_3 & -2\alpha_0 & 0 \\ -2\alpha_0 & 3\alpha_3 + 2\beta_3 & 0 \\ 0 & 0 & 0 \end{bmatrix}.$$

As easily seen, this projected Hessian has eigenvalues

$$\mu_1 = 0, \qquad \mu_{2,3} = 2\alpha_3 \pm \sqrt{(\alpha_3 + 2\beta_3)^2 + 4\alpha_0^2},$$

which are all required to be non-negative [211]. Hence, we obtain the following inequality

$$3\alpha_3^2 - 4\alpha_3\beta_3 - 4\beta_3^2 - 4\alpha_0^2 \geq 0.$$

If the strict inequality holds, i.e., if $3\alpha_3^2 - 4\alpha_3\beta_3 - 4\beta_3^2 - 4\alpha_0^2 > 0$, then the North pole $(0, 0, 1)^\top$ is a strict local maximum point of $\Phi(\mathbf{x})$ on S^2 [211]. □

In the case $\alpha_3 = 0$, we know that $\alpha_0 = \beta_3 = 0$ by Theorem 7.16. If $\alpha_2 > 0$ holds, $\Phi(0, 1, 0) = \alpha_2 > \alpha_3 = \Phi(0, 0, 1)$. This contradicts that the North pole $(0, 0, 1)^\top$ is the maximum point. Hence we also have $\alpha_2 = 0$. The octupolar tensor $\mathscr{A}$ is a trivial zero tensor.

In the remainder, we shall consider the case that α_3 is positive. Without loss of generality, by Proposition 7.7 and Theorem 7.16, we can restrict our discussion on the following admissive region

$$\alpha_3 = 1, \quad \alpha_2 \geq 0, \quad \text{and} \quad \alpha_0^2 + (\beta_3 + \tfrac{1}{2})^2 \leq 1. \tag{7.33}$$

Then, there are three parameters in the octupolar tensor

$$\mathscr{A}(\alpha_0, \beta_3, \alpha_2) = \left(\begin{array}{ccc|ccc|ccc} 0 & -\alpha_2 & \beta_3 & -\alpha_2 & 0 & \alpha_0 & \beta_3 & \alpha_0 & 0 \\ -\alpha_2 & 0 & \alpha_0 & 0 & \alpha_2 & -1-\beta_3 & \alpha_0 & -1-\beta_3 & 0 \\ \beta_3 & \alpha_0 & 0 & \alpha_0 & -1-\beta_3 & 0 & 0 & 0 & 1 \end{array} \right) \tag{7.34}$$

and the associated octupolar potential

$$\Phi(\mathbf{x}) = \alpha_2 x_2^3 + x_3^3 + 6\alpha_0 x_1 x_2 x_3 + 3\beta_3 x_1^2 x_3 - 3\alpha_2 x_1^2 x_2 - 3(1+\beta_3) x_2^2 x_3 \tag{7.35}$$

for all $\mathbf{x} \in S^2$. In the following sections, we will give algebraic expressions of the dome surface and the separatrix surface of the octupolar tensor (7.34).

7.4 Algebraic Expression of the Dome Surface

On the unit sphere S^2, we have assumed that the global maximum point of the octupolar potential $\Phi(\mathbf{x})$ is the North pole $(0, 0, 1)^\top$; See Sect. 7.3. That is to say, $\lambda = 1$ is the largest Z-eigenvalue of the octupolar tensor $\mathscr{A}$, and $(0, 0, 1)^\top$ is an associated Z-eigenvector. In the admissible cylinder (7.33), there is a reduced region such that the maximal Z-eigenvalue of $\mathscr{A}(\alpha_0, \beta_3, \alpha_2)$ is 1. The boundary of this reduced admissible region is called the dome [99]: its apex is at $(\alpha_0, \beta_3, \alpha_2) = (0, -\frac{1}{2}, \frac{\sqrt{2}}{2})$, and it meets the plane $\alpha_2 = 0$ along the circle $\alpha_0^2 + \beta_3^2 + \beta_3 = 0$. Next, we try to find the algebraic expression of this dome.

We introduce our main tool: the **resultant** from algebraic geometry [64]. We consider a system of homogeneous polynomials $f(\mathbf{x}) = 0$, where $f(\mathbf{x}) = (f_1(\mathbf{x}), \dots, f_n(\mathbf{x}))^\top : \mathbf{R}^n \mapsto \mathbf{R}^n$ and each component $f_i(\mathbf{x})$ has positive degree d_i for $i = 1, \dots, n$. By algebraic geometry, the resultant of f is an irreducible polynomial $\mathrm{RES}(f)$ in coefficients of $f(\mathbf{x})$, which vanishes if and only if that polynomial system $f(\mathbf{x}) = 0$ has nonzero solutions; and if $f_i(\mathbf{x}) = x_i^{d_i}$, the value of that resultant should be 1.

For example, we consider a system of linear equations $A\mathbf{x} = 0$. Each component function is a first order homogeneous polynomial. Then, the resultant is the determinant $\det(A)$. By linear algebra, we know that $A\mathbf{x} = 0$ has a nonzero solution if and only if $\det(A)$ vanishes; and if $A = I$, then $\det(A) = 1$.

For a system with only two variables

$$\begin{cases} f_1(x, y) = a_m x^m + a_{m-1} x^{m-1} y + \cdots + a_0 y^m & = 0, \\ f_2(x, y) = b_0 x^n + b_1 x^{n-1} y + \cdots + b_n y^n & = 0, \end{cases}$$

the resultant is the following determinant

$$\mathrm{RES}(f_1, f_2) = \begin{vmatrix} a_m & a_{m-1} & \cdots & a_0 & & & \\ & \ddots & \ddots & \ddots & \ddots & & \\ & & & a_m & a_{m-1} & \cdots & a_0 \\ b_0 & b_1 & \cdots & b_n & & & \\ & \ddots & \ddots & \ddots & \ddots & & \\ & & & b_0 & b_1 & \cdots & b_n \end{vmatrix}.$$

By direct observation, we find that $\mathrm{RES}(f_1, f_2)$ is the determinant of the following system

$$\begin{cases} x^{n-1} f_1(x, y) = & 0, \\ & \vdots \\ y^{n-1} f_1(x, y) = & 0, \\ x^{m-1} f_2(x, y) = & 0, \\ & \vdots \\ y^{m-1} f_2(x, y) = & 0. \end{cases}$$

If the system $f_1(x, y) = f_2(x, y) = 0$ has non-trivial solutions, it goes the same for the system above. Hence, the determinant $\mathrm{RES}(f_1, f_2)$ vanishes.

Next, we turn to the general cases. For simplicity, we take the resultant $\mathrm{RES}(\mathscr{A}\mathbf{x}^2)$ for example, where

$$\mathscr{A}\mathbf{x}^2 = \begin{pmatrix} -2\alpha_2 x_1 x_2 + 2\beta_3 x_1 x_3 + 2\alpha_0 x_2 x_3 \\ -\alpha_2 x_1^2 + \alpha_2 x_2^2 + 2\alpha_0 x_1 x_3 - 2(1+\beta_3) x_2 x_3 \\ \beta_3 x_1^2 - (1+\beta_3) x_2^2 + x_3^2 + 2\alpha_0 x_1 x_2 \end{pmatrix} \equiv \begin{pmatrix} f_3(x_1, x_2, x_3) \\ f_1(x_1, x_2, x_3) \\ f_2(x_1, x_2, x_3) \end{pmatrix}.$$

By the spectral theory of tensors, $\mathrm{RES}(\mathscr{A}\mathbf{x}^2)$ is called the hyperdeterminant of the octupolar tensor $\mathscr{A}$ [228].

The following process is a standard approach introduced in Chap. 3, Sect. 4 of [64]. Since the degree of each component homogeneous polynomial f_i in variables x_1, x_2, x_3 is $d_i = 2$, we may set the total degree

$$d = \sum_{i=1}^{3} (d_i - 1) + 1 = 4$$

and divide monomials $\mathbf{x}^{\boldsymbol{\upsilon}} = x_1^{\upsilon_1} x_2^{\upsilon_2} x_3^{\upsilon_3}$ of total degree $|\boldsymbol{\upsilon}| \equiv \upsilon_1 + \upsilon_2 + \upsilon_3 = 4$ into three sets:

$$\begin{aligned} S_1 &= \{\mathbf{x}^{\boldsymbol{\upsilon}} : |\boldsymbol{\upsilon}| = d, x_1^2 \text{ divides } \mathbf{x}^{\boldsymbol{\upsilon}}\} = \{x_1^4, x_1^3 x_2, x_1^3 x_3, x_1^2 x_2^2, x_1^2 x_2 x_3, x_1^2 x_3^2\}, \\ S_2 &= \{\mathbf{x}^{\boldsymbol{\upsilon}} : |\boldsymbol{\upsilon}| = d, x_1^2 \text{ doesn't divide } \mathbf{x}^{\boldsymbol{\upsilon}} \text{ but } x_2^2 \text{ does}\} = \{x_1 x_2^3, x_1 x_2^2 x_3, x_2^4, x_2^3 x_3, x_2^2 x_3^2\}, \\ S_3 &= \{\mathbf{x}^{\boldsymbol{\upsilon}} : |\boldsymbol{\upsilon}| = d, x_1^2, x_2^2 \text{ don't divide } \mathbf{x}^{\boldsymbol{\upsilon}} \text{ but } x_3^2 \text{ does}\} = \{x_1 x_2 x_3^2, x_1 x_3^3, x_2 x_3^3, x_3^4\}. \end{aligned}$$

Clearly, there exist $\binom{d+2}{2} = 15$ monomials $\mathbf{x}^{\upsilon}$ with total degree 4 and each belongs to one of sets S_1, S_2, and S_3, which are mutually disjoint. Then, the system of 15 equations are listed here

$$\begin{cases} \mathbf{x}^{\upsilon}/x_1^2 \cdot f_1 = 0 & \text{for all } \mathbf{x}^{\upsilon} \in S_1, \\ \mathbf{x}^{\upsilon}/x_2^2 \cdot f_2 = 0 & \text{for all } \mathbf{x}^{\upsilon} \in S_2, \\ \mathbf{x}^{\upsilon}/x_3^2 \cdot f_3 = 0 & \text{for all } \mathbf{x}^{\upsilon} \in S_3. \end{cases}$$

Its coefficient matrix D in the unknowns $\mathbf{x}^{\upsilon}$ with total degree 4 is

$$D = \begin{pmatrix}
-\alpha_2 & & 2\alpha_0 & \alpha_2 & -2(1+\beta_3) & & & & & & & & & & \\
 & -\alpha_2 & & & 2\alpha_0 & & \alpha_2 & -2(1+\beta_3) & & & & & & & \\
 & & -\alpha_2 & & & 2\alpha_0 & & \alpha_2 & & & & -2(1+\beta_3) & & & \\
 & & & -\alpha_2 & & & & 2\alpha_0 & \alpha_2 & -2(1+\beta_3) & & & & & \\
 & & & & -\alpha_2 & & & & & \alpha_2 & -2(1+\beta_3) & 2\alpha_0 & & & \\
 & & & & & -\alpha_2 & & & & & \alpha_2 & & 2\alpha_0 & -2(1+\beta_3) & \\
 & \beta_3 & & 2\alpha_0 & & & -1-\beta_3 & & & & & 1 & & & \\
 & & \beta_3 & & 2\alpha_0 & & & -1-\beta_3 & & & & & 1 & & \\
 & & & \beta_3 & & & 2\alpha_0 & & -1-\beta_3 & & 1 & & & & \\
 & & & & \beta_3 & & & 2\alpha_0 & & -1-\beta_3 & & & & 1 & \\
 & & & & & \beta_3 & & & & & -1-\beta_3 & 2\alpha_0 & & & 1 \\
 & & & -2\alpha_2 & 2\beta_3 & & & 2\alpha_0 & & & & & & & \\
 & & & & -2\alpha_2 & 2\beta_3 & & & & & & 2\alpha_0 & & & \\
 & & & & & & & -2\alpha_2 & & & 2\alpha_0 & 2\beta_3 & & & \\
 & & & & & & & & & & & -2\alpha_2 & 2\beta_3 & 2\alpha_0 &
\end{pmatrix}.$$

It has an important property that

$$\det(D) = \mathrm{RES}(\mathscr{A}\mathbf{x}^2) \cdot \text{extraneous factor}.$$

A monomial $\mathbf{x}^v$ of total degree $d = 4$ is called reduced if x_i^2 divides $\mathbf{x}^v$ for exactly one i. Let D' be the determinant of the submatrix of D obtained by deleting all rows and columns corresponding to reduced monomials, i.e., in our context,

$$D' = \begin{pmatrix} -\alpha_2 & & \\ & -\alpha_2 & \alpha_2 \\ & \beta_3 & -1-\beta_3 \end{pmatrix}.$$

According to Theorem 4.9 in Chap. 3 of [64], to within a sign, the resultant $\mathrm{RES}(\mathscr{A}\mathbf{x}^2)$ reads as

$$\begin{aligned} \mathrm{RES}(\mathscr{A}\mathbf{x}^2) &= \frac{\det D}{\det D'} \qquad (7.36)\\ &= 16\alpha_2^2(48\alpha_0^8\beta_3 + 4\alpha_0^6(\alpha_2^2 + \beta_3(32\beta_3^2 + 24\beta_3 - 9)) + 3\alpha_0^4(\alpha_2^2(52\beta_3^2 + 28\beta_3 - 1)\\ &\quad + 4\beta_3^2(8\beta_3^3 + 8\beta_3^2 - 9\beta_3 - 9)) + 6\alpha_0^2(\alpha_2^4(4\beta_3 + 1) - \alpha_2^2\beta_3(14\beta_3^3 + 36\beta_3^2 + 35\beta_3\\ &\quad + 10) - 2\beta_3^3(\beta_3 + 1)^2(8\beta_3 + 9)) + (\alpha_2^2 - 4(\beta_3 + 1)^3)(\alpha_2^2 - \beta_3^2(2\beta_3 + 3))^2). \end{aligned}$$

Theorem 7.17 *The hyperdeterminant of the octupolar tensor $\mathscr{A}$ is presented in (7.36). The system $\mathscr{A}\mathbf{x}^2 = 0$ has nonzero solutions if and only if the hyperdeterminant vanishes, i.e.,* $\mathrm{RES}(\mathscr{A}\mathbf{x}^2) = 0$.

Next, we compute the E-characteristic polynomial $\phi_{\mathscr{A}}(\lambda)$ of the octupolar tensor (7.34), using the approach of resultants. By introducing an auxiliary x_0, the E-characteristic polynomial $\phi_{\mathscr{A}}(\lambda)$ is indeed a resultant of the following system of homogeneous polynomial equations:

$$\begin{cases} x_1^2 + x_2^2 + x_3^2 - x_0^2 = 0, \\ -\alpha_2 x_1^2 + \alpha_2 x_2^2 + 2\alpha_0 x_1 x_3 - 2(1+\beta_3)x_2 x_3 - \lambda x_0 x_2 = 0, \\ \beta_3 x_1^2 - (1+\beta_3)x_2^2 + x_3^2 + 2\alpha_0 x_1 x_2 - \lambda x_0 x_3 = 0, \\ -2\alpha_2 x_1 x_2 + 2\beta_3 x_1 x_3 + 2\alpha_0 x_2 x_3 - \lambda x_0 x_1 = 0. \end{cases} \qquad (7.37)$$

Using Mathematica, we get the explicit expression of the E-characteristic polynomial $\phi_{\mathscr{A}}(\lambda)$.

Theorem 7.18 *The E-characteristic polynomial of the octupolar tensor $\mathscr{A}$ in (7.34) is*

$$\phi_{\mathscr{A}}(\lambda) = (\lambda^2 - 1)\sum_{i=0}^{6} c_{2i}\lambda^{2i},$$

where, in particular,

$$\begin{aligned}c_0 = & \, 256\alpha_2^4(48\alpha_0^8\beta_3 + 4\alpha_0^6(\alpha_2^2 + \beta_3(32\beta_3^2 + 24\beta_3 - 9)) + 3\alpha_0^4(\alpha_2^2(52\beta_3^2 \\ & + 28\beta_3 - 1) + 4\beta_3^2(8\beta_3^3 + 8\beta_3^2 - 9\beta_3 - 9)) + 6\alpha_0^2(\alpha_2^4(4\beta_3 + 1) \\ & - \alpha_2^2\beta_3(14\beta_3^3 + 36\beta_3^2 + 35\beta_3 + 10) - 2\beta_3^3(\beta_3 + 1)^2(8\beta_3 + 9)) \\ & + (\alpha_2^2 - 4(\beta_3 + 1)^3)(\alpha_2^2 - \beta_3^2(2\beta_3 + 3))^2)^2.\end{aligned}$$

Cartwright and Sturmfels [29] pointed out that if the tensor $\mathscr{A} \in T_{m,n}$ has finitely many equivalence classes of eigenpairs over complex field, then their number, counted with multiplicity, is equal to

$$\frac{(m-1)^n - 1}{m-2}.$$

For a tensor $\mathscr{A} \in T_{m,n}$, if $(\lambda, \mathbf{x})$ satisfies $\mathscr{A}\mathbf{x}^{m-1} = \lambda\mathbf{x}$, then all $(t^{m-2}\lambda, tx)$ with free t is called a equivalence class of $\mathscr{A}$. In the context of the octupolar tensor, (λ, x) and $(-\lambda, -x)$ belong to the same equivalence class. Hence, the E-characteristic polynomial of the octupolar tensor $\mathscr{A}$ should be a polynomial in λ of degree 14, which coincides with Theorem 7.18. Moreover, we find that the E-characteristic polynomial $\phi_{\mathscr{A}}(\lambda)$ has no odd-degree terms. By comparing the expression of c_0 and (7.36), we get

$$c_0 = \left[\mathrm{RES}(\mathscr{A}\mathbf{x}^2)\right]^2.$$

Theorem 7.19 *We assume that $\mathscr{A}\mathbf{x}^2 = \mathbf{0}$ has only the trivial solution $\mathbf{x} = \mathbf{0}$. Then, the system (7.37) has a non-trivial solution $(x_0, \ldots, x_3)$ over $\mathbf{C}$ if and only if the E-characteristic polynomial $\phi_{\mathscr{A}}(\lambda) = 0$.*

Corollary 7.1 *If $\mathrm{RES}(\mathscr{A}\mathbf{x}^2) \neq 0$, all E-eigenvalues of the octupolar tensor $\mathscr{A}$ are non-zero.*

Now, we are ready to give an explicit formula for the dome. Recalling the E-characteristic polynomial $\phi_{\mathscr{A}}(\lambda)$ in Theorem 7.18, we can obtain that $\lambda = 1$ is its root. Since the dome is the locus where the maximal Z-eigenvalue is $\lambda = 1$, $\lambda = 1$ is at least a double root of $\phi_{\mathscr{A}}(\lambda)$. By substituting $\lambda = 1$ into $\phi_{\mathscr{A}}(\lambda)/(\lambda^2 - 1) = 0$, we obtain the following equation

$$c_1(\alpha_0, \beta_3, \alpha_2)^3 \cdot c_2(\alpha_0, \beta_3, \alpha_2) \cdot c_3(\alpha_0, \beta_3, \alpha_2) = 0, \tag{7.38}$$

where

$$c_1(\alpha_0, \beta_3, \alpha_2) = 3 - 4\alpha_0^2 - 4\beta_3^2 - 4\beta_3,$$

$$c_2(\alpha_0, \beta_3, \alpha_2) = 64\alpha_2^4 - 16\alpha_2^2(1 + 2\beta_3)(-12\alpha_0^2 + (1 + 2\beta_3)^2) + (4\alpha_0^2 + (1 + 2\beta_3)^2)^3,$$

and

$$\begin{aligned}c_3(\alpha_0,\beta_3,\alpha_2) &= \alpha_2^6(2\beta_3-1)((2\beta_3+5)^2-12\alpha_0^2)+\alpha_2^4(-48\alpha_0^4(3\beta_3^2-1)\\&+12\alpha_0^2(8\beta_3^4+24\beta_3^3+26\beta_3^2-4\beta_3-11)-16\beta_3^6-96\beta_3^5-168\beta_3^4-72\beta_3^3\\&-21\beta_3^2-24\beta_3+40)+8\alpha_2^2(8\alpha_0^6+6\alpha_0^4(4\beta_3^2-2\beta_3-5)+3\alpha_0^2(8\beta_3^4+8\beta_3^3\\&-12\beta_3^2-3\beta_3+6)+8\beta_3^6+36\beta_3^5+42\beta_3^4+3\beta_3^3-9\beta_3^2-2)\\&-16(\alpha_0^2+\beta_3^2+\beta_3)^2(4\alpha_0^2+4\beta_3^2+4\beta_3-3).\end{aligned} \tag{7.39}$$

In the admissive region (7.33), $c_1(\alpha_0,\beta_3,\alpha_2)\geq 0$ and the equality holds only on the boundary of the admissible region. Thus, the first part $c_1(\alpha_0,\beta_3,\alpha_2)=0$ is a trivial solution of (7.38).

As for the second part, $c_2(\alpha_0,\beta_3,\alpha_2)$ is a quadratic function in α_2^2 that attains its minimum value $4\alpha_0^2(4\alpha_0^2-3(1+2\beta_3)^2)^2\geq 0$. If $\alpha_0=0$, we have $c_2(\alpha_0,\beta_3,\alpha_2)=(-8\alpha_2^2+(1+2\beta_3)^3)^2$. Hence, when

$$\alpha_0=0 \qquad \text{and} \qquad 8\alpha_2^2-(1+2\beta_3)^3=0, \tag{7.40}$$

we have $c_2(\alpha_0,\beta_3,\alpha_2)=0$. If $4\alpha_0^2-3(1+2\beta_3)^2=0$, we get $c_2=64(\alpha_2^2+(1+2\beta_3)^3)^2$. Hence, when

$$4\alpha_0^2-3(1+2\beta_3)^2=0 \qquad \text{and} \qquad \alpha_2^2+(1+2\beta_3)^3=0, \tag{7.41}$$

we also have $c_2(\alpha_0,\beta_3,\alpha_2)=0$. However, under either (7.40) or (7.41), there are two E-eigenvectors corresponding to the E-eigenvalue 1. The first one is the North pole $(0,0,1)^\top$ and the other one is always a complex vector by direct explorations. Hence, we omit both $c_1=0$ and $c_2=0$.

Let us turn attention to the third part $c_3(\alpha_0,\beta_3,\alpha_2)=0$, which has multiple roots in α_2 for fixed α_0 and β_3. For instance, when $\alpha_0=0$ and $\beta_3=-0.8$, $\alpha_2^{(1)}=\alpha_2^{(2)}=\frac{2}{\sqrt{17}}\approx 0.4851$ and $\alpha_2^{(3)}=\frac{4\sqrt{7}}{5\sqrt{5}}\approx 0.9466$ are roots of the equation. When $\alpha_0=0.1$ and $\beta_3=-0.8$, roots of $c_3=0$ are $\alpha_2^{(1)}=0.3765$, $\alpha_2^{(2)}=0.5862$, and $\alpha_2^{(3)}=9459$. Which value of α_2 lies on the dome? Clearly, if the largest Z-eigenvalue of $\mathscr{A}(\alpha_0,\beta_3,\alpha_2)$ is larger than 1, then the triple $(\alpha_0,\beta_3,\alpha_2)$ is above the dome. By direct numerical explorations, we see that, for fixed α_0 and β_3, the dome is marked by the smallest non-negative value of α_2 such that $c_3(\alpha_0,\beta_3,\alpha_2)=0$, i.e.,

$$\alpha_2^{(\text{dome})}(\alpha_0,\beta_3)=\min\{\alpha_2\geq 0: c_3(\alpha_0,\beta_3,\alpha_2)=0\} \qquad \text{for } \alpha_0^2+\beta_3^2+\beta_3\leq 0. \tag{7.42}$$

The contour profile of this dome is illustrated in Fig. 7.3.

We go to details for the apex and the base of the dome. At the apex $(\alpha_0,\beta_3,\alpha_2)=(0,-\frac{1}{2},\frac{\sqrt{2}}{2})$, the E-characteristic polynomial of $\mathscr{A}(0,-\frac{1}{2},\frac{\sqrt{2}}{2})$ is

$$\phi_{\mathscr{A}}(\lambda)=19683\lambda^6(\lambda^2-1)^4.$$

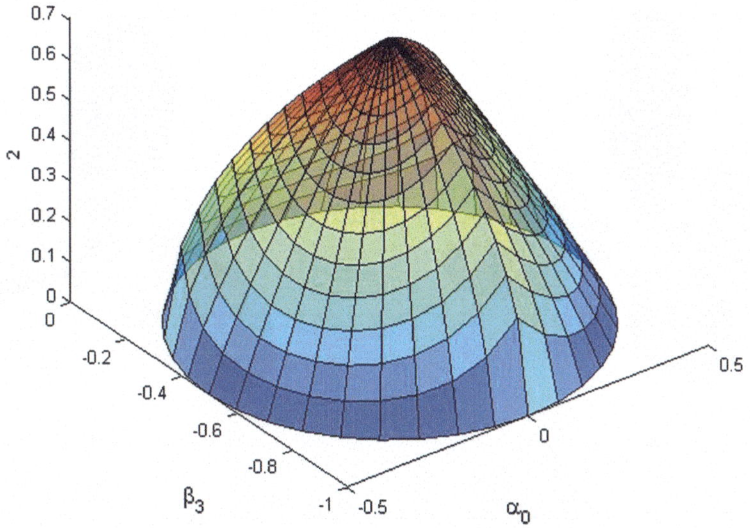

Fig. 7.3 The dome that bounds the reduced admissible region as represented by (7.42)

Clearly, $\lambda^2 = 1$ is a quadruple root of $\phi_{\mathscr{A}}(\lambda)$, and hence it is a Z-eigenvalue of $\mathscr{A}$. Four associated Z-eigenvectors are listed as follows:

$$\mathbf{x}^{(1)} = (0, 0, 1)^\top, \qquad \mathbf{x}^{(2)} = \left(0, \tfrac{2\sqrt{2}}{3}, -\tfrac{1}{3}\right)^\top,$$
$$\mathbf{x}^{(3)} = \left(\tfrac{\sqrt{6}}{3}, -\tfrac{\sqrt{2}}{3}, -\tfrac{1}{3}\right)^\top, \quad \mathbf{x}^{(4)} = \left(-\tfrac{\sqrt{6}}{3}, -\tfrac{\sqrt{2}}{3}, -\tfrac{1}{3}\right)^\top.$$

The corresponding contour profile of the octupolar potential $\Phi(\mathbf{x})$ is illustrated in Fig. 7.4a.

At the base of the dome, we know $\alpha_2 = 0$ and $\alpha_0^2 + \beta_3^2 + \beta_3 = 0$, which represents a circle of center in $(\alpha_0, \beta_3) = (0, -\frac{1}{2})$ and radius $\frac{1}{2}$. Then, the E-characteristic polynomial reduces to

$$\phi_{\mathscr{A}}(\lambda) = -64\lambda^8(\lambda^2 - 1)^3.$$

Clearly, $\lambda^2 = 1$ is a triple root of $\phi_{\mathscr{A}}(\lambda)$. Specifically, the largest Z-eigenvalue $\lambda = 1$ of $\mathscr{A}(0, -1, 0)$ has three Z-eigenvectors, namely,

$$\mathbf{x}^{(1)} = (0, 0, 1)^\top, \ \mathbf{x}^{(2)} = \left(-\tfrac{\sqrt{3}}{2}, 0, -\tfrac{1}{2}\right)^\top, \ \mathbf{x}^{(3)} = \left(\tfrac{\sqrt{3}}{2}, 0, -\tfrac{1}{2}\right)^\top.$$

In this case, the contour profile of the octupolar potential $\Phi(\mathbf{x})$ is shown in Fig. 7.4b. The two profiles in Fig. 7.4 illustrate the typical appearance of the octupolar potential in two highly symmetric states.

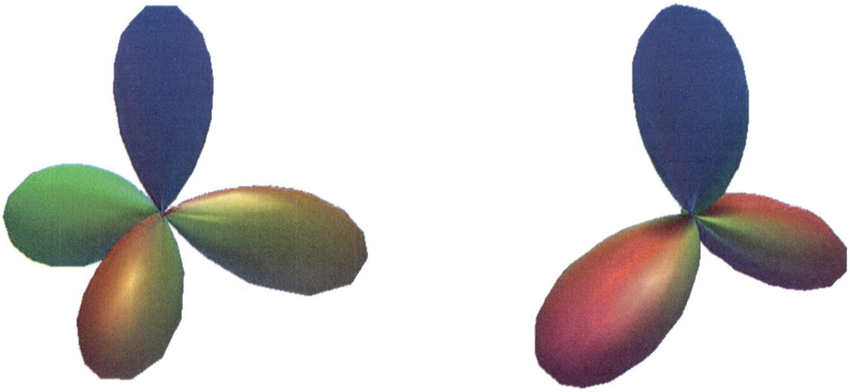

(a) $\mathscr{A}(0, -\frac{1}{2}, \frac{\sqrt{2}}{2})$ (b) $\mathscr{A}(0, -1, 0)$

Fig. 7.4 Two typical (symmetric) polar plots of the octupolar potential

7.5 Algebraic Expression of the Separatrix Surface

Gaeta and Virga [99] pointed out that there exists a separatrix surface connecting two different generic states of the octupolar potential Φ: Φ has four maxima and three (positive) saddles in one generic state; in the other generic state, Φ has three maxima and two (positive) saddles; See Fig. 7.4. Whereafter, we try to determine explicitly the separatrix surface.

We recall the spherical optimization problem (7.31) in Sect. 7.3:

$$\begin{cases} \max \ \Phi(\mathbf{x}) = \mathscr{A}\mathbf{x}^3 \\ \text{s.t. } \ \mathbf{x}^\top \mathbf{x} = 1. \end{cases}$$

Its projected Hessian (7.32) reads as

$$P^\top \nabla^2_{\mathbf{xx}} L(\mathbf{x}, \lambda) P = \lambda(I + \mathbf{x}\mathbf{x}^\top) - 2\mathscr{A}\mathbf{x}.$$

Clearly, this projected Hessian $P^\top \nabla^2_{\mathbf{xx}} L(\mathbf{x}, \lambda) P$ has two zero eigenvalues. $\mu_1 = 0$ is an eigenvalue of the projected Hessian with the associated eigenvector $\mathbf{x}$. Suppose that μ_2 and μ_3 are the other two eigenvalues of the projected Hessian. Let $\sigma \equiv \mu_2\mu_3 = \mu_1\mu_2 + \mu_1\mu_3 + \mu_2\mu_3$. Then, by linear algebra, σ is equal to the sum of all 2-by-2 principal minors of the projected Hessian. By calculations, we get

$$\sigma = \lambda^2(3 + 2x_1^2 + 2x_2^2 + 2x_3^2) + 2\lambda \left(\alpha_2 x_2^3 + x_3^3 + 6\alpha_0 x_1 x_2 x_3 + 3\beta_3 x_1^2 x_3 - 3\alpha_2 x_1^2 x_2 \right.$$
$$\left. -3(1 + \beta_3) x_2^2 x_3 \right) - 4\left((\alpha_0^2 + \alpha_2^2 + \beta_3^2) x_1^2 + (\alpha_0^2 + \alpha_2^2 + (\beta_3 + 1)^2) x_2^2 \right.$$

$$
\begin{aligned}
&+(\alpha_0^2+\beta_3^2+\beta_3+1)x_3^2-2\alpha_0x_1x_2-2\alpha_0\alpha_2x_1x_3-\alpha_2(2\beta_3+1)x_2x_3\Big)\\
&=7\lambda^2-4\Big((\alpha_0^2+\alpha_2^2+\beta_3^2)x_1^2+(\alpha_0^2+\alpha_2^2+(\beta_3+1)^2)x_2^2+(\alpha_0^2+\beta_3^2+\beta_3+1)x_3^2\\
&\quad-2\alpha_0x_1x_2-2\alpha_0\alpha_2x_1x_3-\alpha_2(2\beta_3+1)x_2x_3) \qquad (7.43)\\
&=0.
\end{aligned}
$$

Furthermore, if $\lambda \neq 0$ and $\mathbf{x} \neq 0$ satisfy $\mathscr{A}\mathbf{x}^2 = \lambda\mathbf{x}$, $\frac{\lambda}{\|\mathbf{x}\|}$ and $\frac{1}{\|\mathbf{x}\|}\mathbf{x}$ satisfy the system of Z-eigenvalues. Hence, we omit the spherical constraint $\mathbf{x}^\top\mathbf{x} = 1$ temporarily and just focus on the system of homogeneous polynomial equations (7.43) and

$$
\begin{cases}
-2\alpha_2x_1x_2+2\beta_3x_1x_3+2\alpha_0x_2x_3-\lambda x_1=0,\\
-\alpha_2x_1^2+\alpha_2x_2^2+2\alpha_0x_1x_3-2(1+\beta_3)x_2x_3-\lambda x_2=0,\\
\beta_3x_1^2-(1+\beta_3)x_2^2+x_3^2+2\alpha_0x_1x_2-\lambda x_3=0.
\end{cases} \qquad (7.44)
$$

Using the resultant theory, we obtain the resultant of (7.43) and (7.44), which is the separatrix:

$$
1792(4\alpha_0^2+4\beta_3^2+4\beta_3-3)^2\sum_{i=0}^{8}d_{2i}(\alpha_0,\beta_3)\alpha_2^{2i}=0, \qquad (7.45)
$$

where

$$
\begin{aligned}
d_{16}&=27(-16\alpha_0^4-8\alpha_0^2(4\beta_3^2-44\beta_3+13)-(2\beta_3-1)(2\beta_3+7)^3),\\
d_{14}&=-54(128\alpha_0^6-16\alpha_0^4(48\beta_3^2+78\beta_3-29)+16\alpha_0^2(72\beta_3^4+124\beta_3^3+190\beta_3^2\\
&\quad-101\beta_3-69)+(2\beta_3+7)^2(40\beta_3^3+44\beta_3^2+62\beta_3-47)),\\
d_{12}&=-9(4096\alpha_0^8-128\alpha_0^6(277\beta_3^2-92\beta_3-55)+48\alpha_0^4(152\beta_3^4+1944\beta_3^3\\
&\quad-7094\beta_3^2-1548\beta_3+53)+8\alpha_0^2(5648\beta_3^6+18912\beta_3^5+60408\beta_3^4+115368\beta_3^3\\
&\quad+86625\beta_3^2-44964\beta_3-18410)-1664\beta_3^8-22400\beta_3^7-124064\beta_3^6\\
&\quad-377088\beta_3^5-624840\beta_3^4-383256\beta_3^3+109994\beta_3^2+181940\beta_3-17605),
\end{aligned}
$$

$$
\begin{aligned}
d_{10}&=-2(22528\alpha_0^{10}+256\alpha_0^8(800\beta_3^2+3620\beta_3+599)+64\alpha_0^6(5440\beta_3^4-195290\beta_3^3\\
&\quad-97221\beta_3^2-44476\beta_3+8375)+16\alpha_0^4(12800\beta_3^6+1073640\beta_3^5+2832444\beta_3^4\\
&\quad+2369838\beta_3^3-242151\beta_3^2-492540\beta_3-270455)+4\alpha_0^2(17920\beta_3^8\\
&\quad-1188320\beta_3^7-6499376\beta_3^6-13648368\beta_3^5-10198728\beta_3^4+1289514\beta_3^3\\
&\quad+3579185\beta_3^2+123260\beta_3+206555)+32768\beta_3^{10}+483200\beta_3^9+3111744\beta_3^8\\
&\quad+10647136\beta_3^7+19890064\beta_3^6+19640424\beta_3^5+5479324\beta_3^4-6109790\beta_3^3\\
&\quad-3422445\beta_3^2+504920\beta_3+3560),\\
d_8&=5(40960\alpha_0^{12}-12288\alpha_0^{10}(97\beta_3^2+88\beta_3+44)+256\alpha_0^8(39921\beta_3^4+34176\beta_3^3\\
&\quad+42870\beta_3^2+12132\beta_3+1667)-128\alpha_0^6(141080\beta_3^6+419208\beta_3^5+389430\beta_3^4
\end{aligned}
$$

$$
\begin{aligned}
&+ 82228\beta_3^3 - 15613\beta_3^2 - 100402\beta_3 - 37063) + 48\alpha_0^4(212768\beta_3^8 + 1023680\beta_3^7 \\
&+ 1963504\beta_3^6 + 1378192\beta_3^5 - 304390\beta_3^4 - 508976\beta_3^3 + 63582\beta_3^2 - 57076\beta_3 \\
&- 52349) - 8\alpha_0^2(149376\beta_3^{10} + 1199488\beta_3^9 + 4718496\beta_3^8 + 9599232\beta_3^7 \\
&+ 9822584\beta_3^6 + 3227448\beta_3^5 - 2548818\beta_3^4 - 2029036\beta_3^3 - 53961\beta_3^2 \\
&+ 37902\beta_3 - 40196) + (2\beta_3 + 1)^2(10304\beta_3^{10} + 154304\beta_3^9 + 911472\beta_3^8 \\
&+ 2786464\beta_3^7 + 4828732\beta_3^6 + 3895212\beta_3^5 + 22345\beta_3^4 - 1558688\beta_3^3 \\
&- 352512\beta_3^2 + 133184\beta_3 - 7840)),
\end{aligned}
$$

$$
\begin{aligned}
d_6 = {} & 16(28672\alpha_0^{14} - 512\alpha_0^{12}(688\beta_3^2 + 1102\beta_3 + 941) - 128\alpha_0^{10}(9696\beta_3^4 \\
&- 40380\beta_3^3 - 33951\beta_3^2 - 20148\beta_3 - 11743) - 160\alpha_0^8(5632\beta_3^6 - 26064\beta_3^5 \\
&- 7644\beta_3^4 + 35134\beta_3^3 - 57181\beta_3^2 - 51958\beta_3 - 5454) + 40\alpha_0^6(18944\beta_3^8 \\
&- 103552\beta_3^7 - 737312\beta_3^6 - 1217152\beta_3^5 - 320576\beta_3^4 + 504962\beta_3^3 + 120149\beta_3^2 \\
&- 112824\beta_3 - 29175) + 2\alpha_0^4(577536\beta_3^{10} + 1111040\beta_3^9 - 2474880\beta_3^8 \\
&- 6705600\beta_3^7 - 341600\beta_3^6 + 9137976\beta_3^5 + 5840100\beta_3^4 - 884330\beta_3^3 \\
&- 684765\beta_3^2 + 374580\beta_3 + 132449) + \alpha_0^2(2\beta_3 + 1)^2(80896\beta_3^{10} + 999808\beta_3^9 \\
&+ 3452640\beta_3^8 + 5398208\beta_3^7 + 3717992\beta_3^6 - 367068\beta_3^5 - 2064016\beta_3^4 \\
&- 746875\beta_3^3 + 150774\beta_3^2 + 30796\beta_3 - 25928) - (2\beta_3 + 1)^4(2048\beta_3^{10} \\
&+ 25824\beta_3^9 + 135752\beta_3^8 + 385692\beta_3^7 + 535154\beta_3^6 + 253167\beta_3^5 \\
&- 114083\beta_3^4 - 118464\beta_3^3 - 4364\beta_3^2 + 7632\beta_3 - 656)),
\end{aligned}
$$

$$
\begin{aligned}
d_4 = {} & 16(-32768\alpha_0^{16} + 2048\alpha_0^{14}(241\beta_3^2 - 284\beta_3 - 83) + 256\alpha_0^{12}(9208\beta_3^4 \\
&- 5384\beta_3^3 + 6390\beta_3^2 + 25496\beta_3 + 8589) + 128\alpha_0^{10}(25680\beta_3^6 - 32160\beta_3^5 \\
&+ 7368\beta_3^4 + 257000\beta_3^3 + 212292\beta_3^2 + 23286\beta_3 - 10209) + 80\alpha_0^8(8064\beta_3^8 \\
&- 169344\beta_3^7 - 304224\beta_3^6 + 311872\beta_3^5 + 774736\beta_3^4 + 306928\beta_3^3 - 61620\beta_3^2 \\
&- 34824\beta_3 + 1209) - 8\alpha_0^6(281856\beta_3^{10} + 2529280\beta_3^9 + 6835200\beta_3^8 \\
&+ 6572800\beta_3^7 + 316800\beta_3^6 - 2303424\beta_3^5 - 174400\beta_3^4 + 593920\beta_3^3 \\
&+ 124725\beta_3^2 - 2310\beta_3 + 6019) - 2\alpha_0^4(2\beta_3 + 1)^2(230144\beta_3^{10} + 1285120\beta_3^9 \\
&+ 3244032\beta_3^8 + 4304128\beta_3^7 + 2583584\beta_3^6 + 28128\beta_3^5 - 669240\beta_3^4 \\
&- 261024\beta_3^3 + 369\beta_3^2 + 30952\beta_3 - 3232) - 4\alpha_0^2(2\beta_3 + 1)^4(5408\beta_3^{10} \\
&+ 10240\beta_3^9 - 22272\beta_3^8 - 72224\beta_3^7 - 83578\beta_3^6 - 75384\beta_3^5 - 40635\beta_3^4 \\
&+ 8889\beta_3^3 + 10338\beta_3^2 - 2444\beta_3 - 184) + 2(\beta_3 + 1)^2(2\beta_3 + 1)^6(400\beta_3^8 + 4000\beta_3^7 \\
&+ 15408\beta_3^6 + 16240\beta_3^5 - 2449\beta_3^4 - 6128\beta_3^3 + 104\beta_3^2 + 272\beta_3 - 24)),
\end{aligned}
$$

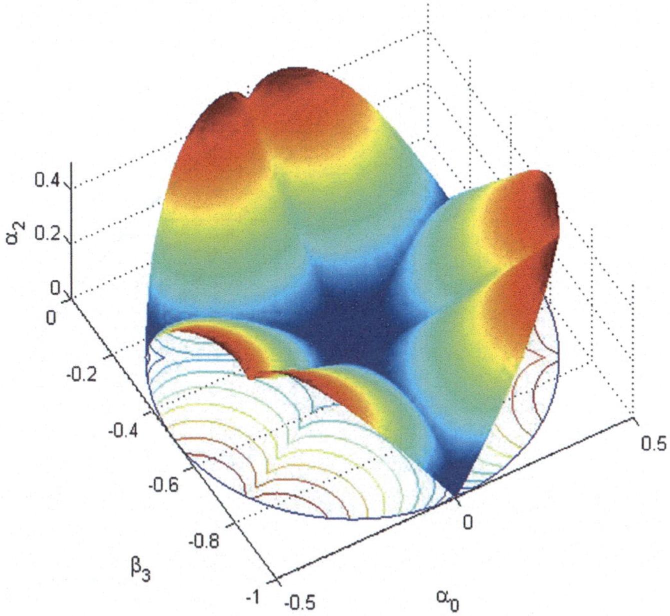

Fig. 7.5 The separatrix below the dome as represented by (7.45)

$$
\begin{aligned}
d_2 = & -256(\alpha_0^2 + \beta_3^2 + \beta_3)^2(4\alpha_0^2 + (2\beta_3 + 1)^2)^3(64\alpha_0^8 + 8\alpha_0^6(32\beta_3^2 - 112\beta_3 - 81) \\
& + 2\alpha_0^4(192\beta_3^4 - 576\beta_3^3 - 1356\beta_3^2 - 78\beta_3 + 245) + \alpha_0^2(256\beta_3^6 + 384\beta_3^5 \\
& - 408\beta_3^4 - 136\beta_3^3 + 764\beta_3^2 + 215\beta_3 - 62) + (2\beta_3 + 1)^2(16\beta_3^6 + 144\beta_3^5 \\
& + 266\beta_3^4 + 87\beta_3^3 - 89\beta_3^2 - 32\beta_3 + 6)), \\
d_0 = & \ 256(\alpha_0^2 + \beta_3^2 + \beta_3)^4(4\alpha_0^2 + 4\beta_3^2 + 4\beta_3 - 3)(4\alpha_0^2 + (2\beta_3 + 1)^2)^5.
\end{aligned}
$$

Below the dome, the contour profile of the separatrix surface is illustrated in Fig. 7.5, which shows a 6-fold symmetry. We now compare the separatrix and the dome given by (7.45) and (7.42). Using the polar coordinates, we introduce two parameters ρ and χ to define

$$\alpha_0 = \rho \cos \chi \qquad \text{and} \qquad \beta_3 = -\tfrac{1}{2} + \rho \sin \chi, \tag{7.46}$$

where $\rho \in \left[0, \frac{1}{2}\right]$ and $\chi \in (-\pi, \pi]$. In Fig. 7.6, when $\chi = -\frac{\pi}{2}, -\frac{5\pi}{12}, -\frac{\pi}{3}, -\frac{\pi}{4}$, and $-\frac{\pi}{6}$, the cross-sections of the dome and the separatrix are illustrated in dash-dot lines and solid lines, respectively.

Finally, we study the special case $\chi = -\frac{\pi}{2}$ in (7.46), so as to describe a cross-section of the separatrix that reaches the base of the dome. The dome equation $c_3(\alpha_0, \beta_3, \alpha_2) = 0$ reduces to

$$-4(\rho + 1)(2\alpha_2^2 - (1 - \rho)(1 + 2\rho)^2)(\alpha_2^2(\rho - 2) - 2\rho + 1)^2 = 0.$$

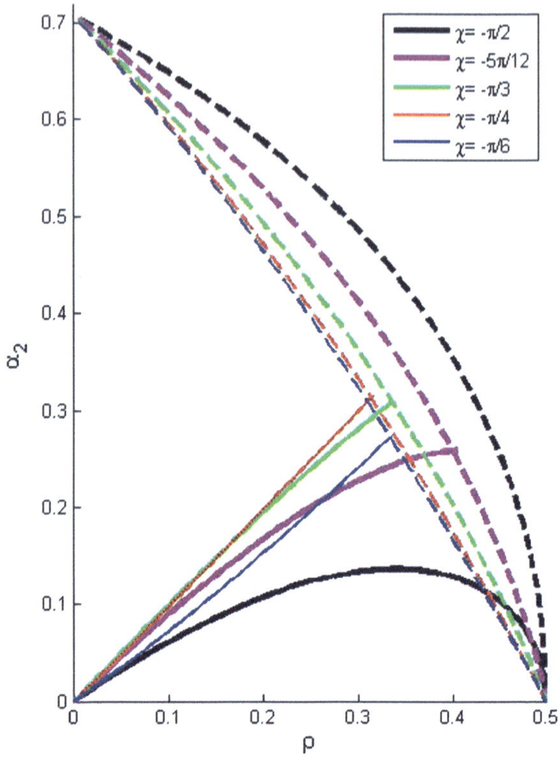

Fig. 7.6 The cross-sections of dome and separatrix

Clearly, $\rho + 1 > 0$. Because $(1-\rho)(1+2\rho)^2$ is monotonically increasing in $\rho \in [0, \frac{1}{2}]$, we have $(1-\rho)(1+2\rho)^2 \geq 1$. Hence, by $2\alpha_2^2 - (1-\rho)(1+2\rho)^2 = 0$ and $\alpha_2 \geq 0$, we obtain the region in parameter space where $\alpha_2 \geq \frac{1}{\sqrt{2}}$ lies above the dome. Moreover, according to $\alpha_2^2(\rho-2) - 2\rho + 1 = 0$, the cross-section of the dome is the curve

$$\alpha_2^{(\text{dome})}(\rho) = \sqrt{\frac{1-2\rho}{2-\rho}} \qquad \text{for } \chi = -\frac{\pi}{2}. \tag{7.47}$$

When $\chi = -\frac{\pi}{2}$, the separatrix is rewritten as

$$458752(1-\rho)^2(1+\rho)^3(\alpha_2^2 + 4(1-\rho)(1-2\rho))\left(3(3-\rho)\alpha_2^2 - 4\rho^2(1-2\rho)\right)^3 \cdot \left(\alpha_2^4 + \alpha_2^2(6\rho+4) + \rho^2(2\rho+1)^2\right) = 0.$$

For $\rho \in [0, \frac{1}{2}]$, $(1-\rho)^2(1+\rho)^3 > 0$. Moreover, $\alpha_2^2 + 4(1-\rho)(1-2\rho) \geq 0$ and the equality holds if and only if $(\rho, \alpha_2) = (\frac{1}{2}, 0)$. Also, $\alpha_2^4 + \alpha_2^2(6\rho+4) + \rho^2(2\rho+1)^2 \geq 0$ and the equality holds if and only if $(\rho, \alpha_2) = (0, 0)$. From $3(3-\rho)\alpha_2^2 - 4\rho^2(1-$

$2\rho) = 0$, we obtain that the cross-section of the separatrix is the curve

$$\alpha_2^{(\text{sepa})}(\rho) = \frac{2\rho}{\sqrt{3}}\sqrt{\frac{1-2\rho}{3-\rho}} \qquad \text{for } \chi = -\frac{\pi}{2}. \tag{7.48}$$

These two curves in (7.47) and (7.48) have a common vertical tangent at $\rho = \frac{1}{2}$. Furthermore, since

$$[\alpha_2^{(\text{dome})}(\rho)]^2 - [\alpha_2^{(\text{sepa})}(\rho)]^2 = \frac{(3-2\rho)^2(1-\rho-2\rho^2)}{3(2-\rho)(3-\rho)} \geq 0$$

in $\rho \in [0, \frac{1}{2}]$ and the equality holds if and only if $\rho = \frac{1}{2}$, we confirm analytically that the separatrix lies below the dome when $\chi = -\frac{\pi}{2}$.

Applying a similar discussion for the case $\chi = -\frac{\pi}{6}$, the following curves are obtained as representations for cross-sections of the dome and separatrix, respectively,

$$\alpha_2^{(\text{dome})}(\rho) = \frac{1-2\rho}{\sqrt{2}}\sqrt{1+\rho}, \qquad \alpha_2^{(\text{sepa})}(\rho) = \frac{2\rho}{\sqrt{3}}\sqrt{\frac{1+2\rho}{3+\rho}}. \tag{7.49}$$

They intersect at $(\rho, \alpha_2) = \left(\frac{1}{3}, \frac{\sqrt{2}}{3\sqrt{3}}\right)$.

7.6 Eigendiscriminant from Algebraic Geometry

At the beginning, we review the traceless property of the symmetric octupolar tensor from the viewpoint of harmonic functions.

Definition 7.4 Let $f : U \mapsto \mathbf{R}$ (where U is an open subset of $\mathbf{R}^n$) be a twice continuously differentiable function. We say f is a harmonic function if it satisfies Laplace's equation everywhere on U:

$$\Delta f = 0, \qquad \text{i.e.,} \qquad \frac{\partial^2 f}{\partial x_1^2} + \frac{\partial^2 f}{\partial x_2^2} + \cdots + \frac{\partial^2 f}{\partial x_n^2} = 0.$$

Suppose that $\mathscr{A} \in S_{3,3}$ is an octupolar tensor, and $\Phi(\mathbf{x}) = \mathscr{A}\mathbf{x}^3$ is the octupolar potential. Using the Laplace operator, we get

$$\begin{aligned}\Delta\Phi(\mathbf{x}) &= \frac{\partial^2}{\partial x_1^2}\Phi(\mathbf{x}) + \frac{\partial^2}{\partial x_2^2}\Phi(\mathbf{x}) + \frac{\partial^2}{\partial x_3^2}\Phi(\mathbf{x}) \\ &= 6\left((a_{111}+a_{122}+a_{133})x_1 + (a_{112}+a_{112}+a_{233})x_2 + (a_{113}+a_{223}+a_{333})x_3\right) \\ &= 0.\end{aligned}$$

Hence, the traceless property (7.27) holds. In this sense, we say that the traceless of the octupolar tensor $\mathscr{A}$ is annihilated by the Laplace operator.

Second, we claim that the separatrix (7.45) is indeed the eigendiscriminant of the octupolar tensor (7.34) [1, 256]. Before we start, let the given tensor $\mathscr{A} \in T_{m,n}$ be generic, which means that $\mathscr{A}$ lies in a certain dense open subset in the tensor space $T_{m,n}$. This subset will be characterized later as the nonvanishing locus of the eigendiscriminant. In algebraic geometry, the solution set of a system of polynomial equations is called a **variety** [19]. We now consider the system $\mathscr{A}\mathbf{x}^{m-1} = \lambda\mathbf{x}$, which is equivalent to

$$\operatorname{rank}\left((\mathscr{A}\mathbf{x}^{m-1} \;\; \mathbf{x})\right) \leq 1. \tag{7.50}$$

Clearly, the E-eigenvectors of $\mathscr{A}$ are the solutions of the above system. The eigenconfiguration of $\mathscr{A}$ is the variety defined by the 2-by-2 minors of the matrix $(\mathscr{A}\mathbf{x}^{m-1} \;\; \mathbf{x})$. Then, we have the following theorem.

Theorem 7.20 (Cartwright, Sturmfels 2013) *Let $\mathscr{A} \in S_{m,n}$ be generic. The number of solutions of the system* (7.50) *is*

$$\rho(n, m) = \frac{(m-1)^n - 1}{m-2} = \sum_{i=0}^{n-1} (m-1)^i.$$

The same count holds for eigenconfigurations of symmetric tensors.

If the tensor is not generic, then its eigenconfiguration consists of fewer than $\rho(n, m)$ points or is not zero-dimensional. The elements of all these non-generic tensors satisfy an irreducible homogeneous polynomial, called the eigendiscriminant $\Delta_{n,m}$. The degree of the eigendiscriminant is presented in the following theorem.

Theorem 7.21 (Abo, Seigal, Sturmfels 2017) *The eigendiscriminant is an irreducible homogeneous polynomial of degree*

$$\operatorname{degree}(\Delta_{n,m}) = n(n-1)(m-1)^{n-1}.$$

From this theorem, we know that the eigendiscriminant of the octupolar tensors $\mathscr{A}(\alpha_0, \beta_3, \alpha_2)$ in (7.34) is a homogeneous polynomial in variables $\alpha_0, \beta_3, \alpha_2$ with degree 24. This fact is coincide with the separatrix in (7.45).

Motivated by this result, we apply the eigendiscriminant for the octupolar tensors $\mathscr{A} \in S_{3,3}$. The corresponding octupolar potentials are ternary cubics that are harmonic. We recall that the octupolar phase of a molecule assembly of bent-core liquid crystals has two generic states: the octupolar potential has four/three maximum points on the unit sphere. An ideal octupolar phase has four maximum points. When passing through the separatrix surface, two maximum points coincide. That is to say, two E-/Z-eigenvalues of the octupolar tensor meet and integrate into one. This corresponds to vanishing of the eigendiscriminant.

7.7 Notes

The conception "tensor" was originally arising from physics and mechanics [134]. Hence, it is necessary to address tensors in physics and mechanics. However, tensors in physics and mechanics are different from hypermatrices. Specifically, let $\mathscr{A}$ be a third order tensor. Under a Cartesian basis $\{\mathbf{e}_1, \mathbf{e}_2, \mathbf{e}_3\}$, the components of $\mathscr{A}$ in $\{\mathbf{e}_1, \mathbf{e}_2, \mathbf{e}_3\}$ could be represented as a hypermatrix

$$\left(\begin{array}{ccc|ccc|ccc} a_{111} & a_{112} & a_{113} & a_{211} & a_{212} & a_{213} & a_{311} & a_{312} & a_{313} \\ a_{121} & a_{122} & a_{123} & a_{221} & a_{222} & a_{223} & a_{321} & a_{322} & a_{323} \\ a_{131} & a_{132} & a_{133} & a_{231} & a_{232} & a_{233} & a_{331} & a_{332} & a_{333} \end{array}\right).$$

The representation of the tensor $\mathscr{A}$ in terms of its components can also be expressed as

$$\mathscr{A} = \sum_{i,j,k=1}^{3} a_{ijk} \mathbf{e}_i \otimes \mathbf{e}_j \otimes \mathbf{e}_k.$$

This representation is particularly convenient when using a general coordinate system.

Section 7.1 The difference between tensors and hypermatrices was studied by Qi [225]. Some valuable viewpoints and new notions, such as tensor transpose, kernel tensor, nonsingularity, and L-inverse of tensors, were introduced. The L-inverse of third order tensors is a nontrivial generalization of inverse of second order tensors.

Section 7.2 The piezoelectric effects was discovered by Curie brothers. Nowadays, piezoelectric material as well as its piezoelectric tensor are active research area. Chen, Jákli and Qi [51] proposed the C-eigenvalues for piezoelectric tensors. Furthermore, Li and Li [166] gave bounds for the C-eigenvalues of piezoelectric tensors.

Section 7.3 Virga [270] studied two dimensional symmetric traceless tensors arising from liquid crystals. Gaeta and Virga [99] addressed three dimensional symmetric traceless tensors. Here, traces of each slice matrix of a traceless tensor are all zeros. This reflects an important property in physics. Some other papers on liquid crystal tensors include [24, 184, 186].

Gaeta and Virga [99] demonstrated an interesting approach for reducing independent parameters of a symmetric traceless tensor from seven to three. The expressions of the dome and the separatrix get benefit from this approach.

Section 7.4 The existence of the dome was pointed by Gaeta and Virga [99]. Chen, Qi and Virga [52] used the E-characteristic polynomial from spectral tensor theory [221, 228] to obtain the algebraic expression of the dome.

Section 7.5 Gaeta and Virga [99] pointed out that there are two generic octupolar states. Moreover, they gave a numerical simulation for the separatrix. Chen, Qi and Virga [52] employed resultant theory from algebraic geometry to gave a close-form solution for the separatrix.

Section 7.6 The traceless property of a tensor is due to the Laplace operator. In simple terms, the algebraic expression of the dome was motivated by two meeting Z-eigenvalues, and the closed-form solution of the separatrix was motivated by two meeting Z-eigenvectors. Such a viewpoint from algebraic geometry may give us some insights to this problem.

Finally, we discuss octupolar tensors in two dimensions. We consider third order two dimensional symmetric traceless tensors

$$\mathscr{A} = \left(\begin{array}{cc|cc} a_{111} & a_{112} & a_{112} & a_{122} \\ a_{112} & a_{122} & a_{122} & a_{222} \end{array}\right) \in S_{3,2}. \tag{7.51}$$

By using traceless property

$$a_{111} + a_{122} = 0 \quad \text{and} \quad a_{112} + a_{222} = 0,$$

and introducing parameters $\alpha = a_{111}$ and $\beta = a_{222}$, we obtain

$$\mathscr{A} = \left(\begin{array}{cc|cc} \alpha & -\beta & -\beta & -\alpha \\ -\beta & -\alpha & -\alpha & \beta \end{array}\right).$$

On the compact circle $x_1^2 + x_2^2 = 1$, the potential function $\Phi(\mathbf{x}) = \mathscr{A}\mathbf{x}^3$ has at least one maximum point. Without lose of generality, we assume the maximum point being $(1, 0)^\top$. By writing down the KKT system $\mathscr{A}\mathbf{x}^2 = \lambda\mathbf{x}$ explicitly as

$$\begin{cases} \alpha x_1^2 - 2\beta x_1 x_2 - \alpha x_2^2 = \lambda x_1, \\ -\beta x_1^2 - 2\alpha x_1 x_2 + \beta x_2^2 = \lambda x_2, \end{cases}$$

we get $(1, 0)^\top$ satisfying the above system and obtain

$$\beta = 0.$$

Hence, the octupolar tensor

$$\mathscr{A} = \left(\begin{array}{cc|cc} \alpha & 0 & 0 & -\alpha \\ 0 & -\alpha & -\alpha & 0 \end{array}\right)$$

has only one parameter α.

Without loss of generality, we only need to consider the constant tensor

$$\mathscr{A} = \left(\begin{array}{cc|cc} 1 & 0 & 0 & -1 \\ 0 & -1 & -1 & 0 \end{array}\right).$$

Its E-characteristic polynomial is

Fig. 7.7 The potential of an octupolar tensor in two dimensions

$$\phi_{\mathscr{A}}(\lambda) = \mathrm{RES}\begin{pmatrix} x_1^2 - x_2^2 - \lambda x_0 x_1 \\ -2x_1x_2 - \lambda x_0 x_2 \\ x_1^2 + x_2^2 - x_0^2 \end{pmatrix} = 16(\lambda - 1)^3(\lambda + 1)^3.$$

Hence, there are three Z-eigenvectors corresponding to the largest Z-eigenvalue $\lambda = 1$:

$$\mathbf{x}_1 = (1, 0)^\top, \quad \mathbf{x}_2 = (-\tfrac{1}{2}, \tfrac{\sqrt{3}}{2})^\top, \quad \mathbf{x}_2 = (-\tfrac{1}{2}, -\tfrac{\sqrt{3}}{2})^\top.$$

The contour profile of the octupolar potential $\Phi(\mathbf{x}) = x_1^3 - 3x_1x_2^2$ is illustrated in Fig. 7.7. For more information on octupolar tensors in two dimensions, we refer to [270].

7.8 Exercises

1. For a given tensor

$$\left(\begin{array}{ccc|ccc|ccc} 1 & 0 & 0 & 0 & 0 & 0 & 0 & 0 & 0 \\ 0 & 0 & 0 & 0 & 2 & 0 & 0 & 0 & -1 \\ 0 & 0 & 0 & 0 & 0 & -1 & 0 & -1 & 3 \end{array}\right),$$

is $\mathscr{A}$ orthogonal and nonsingular? Compute its kernel tensor and L-inverse tensor if it is nonsingular.

2. A piezoelectric tensor $\mathscr{A} \in S_{3,2}$ contains only two nonzero elements $a_{112} = a_{222} = 1$. Thus,

$$M(\mathscr{A}) = \begin{pmatrix} 0 & 0 & \sqrt{2} \\ 0 & 1 & 0 \end{pmatrix}.$$

Compute the largest singular value of $M(\mathscr{A})$ and the largest C-eigenvalue of $\mathscr{A}$, and compare them.

3. For the tensor $\mathscr{A}(\alpha)$ with $\alpha > 0$ in Proposition 7.6: $a_{123} = a_{213} = a_{312} = -\alpha$ and other elements of $\mathscr{A}(\alpha)$ are all zeros, prove that the largest C-eigenvalue of $\mathscr{A}(\alpha)$ is $\frac{2}{\sqrt{3}}\alpha$.

4. Compute positive Z-eigenvalues and associated Z-eigenvectors of $\mathscr{A}(0, -\frac{1}{2}, \frac{\sqrt{2}}{2})$ and $\mathscr{A}(0, -1, 0)$, where $\mathscr{A}(\alpha_0, \beta_3, \alpha_2)$ is defined in (7.34).

5. When $\chi = -\frac{\pi}{6}$ in the polar coordinates (7.46), prove that the algebraic expressions of the dome and separatrix curves are (7.49).

Chapter 8
Fourth Order Tensors in Physics and Mechanics

Fourth order tensors have also wide applications in physics and mechanics. Examples include the piezo-optical tensor, the elasto-optical tensor and the flexoelectric tensor. The most well-known fourth order tensor is the elasticity tensor [134, 212, 318]. It is closely related to the strong ellipticity condition in nonlinear mechanics.

The strong ellipticity condition plays an important role in nonlinear elasticity and in materials. In this chapter, we will discuss the application of the spectral theory of tensors to the studying of strong ellipticity of elasticity.

First of all, we will define M-eigenvalues for an elasticity tensor. It will be shown that the strong ellipticity condition holds if and only if the smallest M-eigenvalue of the elasticity tensor is positive. We will say that the elasticity tensor is M-positive definite (M-PD) if the strong ellipticity condition holds. We will also define S-positive definiteness for the elasticity tensor. The elasticity tensor is S-positive definite (S-PD) if its unfolded symmetric matrix is positive definite. If the elasticity tensor is S-PD, then the strong ellipticity condition holds. The converse conclusion is not right. It is easy to check the elasticity tensor is S-PD or not, as this only needs to check if the smallest eigenvalue of the unfolding matrix of the elasticity tensor is positive or not. Thus, S-positive definiteness provides a checkable sufficient condition for strong ellipticity.

Then, we will derive necessary and sufficient conditions for the strong ellipticity condition of anisotropic elastic materials. It can be observed that the strong ellipticity condition holds if and only if a second order tensor function is positive definite for any unit vectors. Then one can further link this condition to positive definiteness of three second-order symmetric tensors, a fourth order symmetric tensor and a sixth-order symmetric tensor. In particular, we will consider conditions of strong ellipticity of the rhombic classes, for which one needs to check the co-positivity of three second-order symmetric tensors and the positive definiteness of a sixth-order symmetric tensor. One may check this necessary and sufficient condition by computing the smallest Z-eigenvalues of these symmetric tensors via the method of Cui, Dai and Nie [66].

L. Qi et al., *Tensor Eigenvalues and Their Applications*, Advances in Mechanics and Mathematics 39, https://doi.org/10.1007/978-981-10-8058-6_8

Thus, numerically, this provides a practical method for checking the strong ellipticity condition of a given elasticity tensor.

Since the range of S-positive (semi)definiteness is too narrow, to extend the coverage of S-PSD tensors, then we will study several sufficient conditions for strong ellipticity, other than S-positive definiteness. An alternating algorithm for checking this sufficient condition is described. Some other sufficient conditions are also discussed. Furthermore, we define elasticity M-tensors and nonsingular elasticity M-tensors. If the elasticity M-tensor is a nonsingular elasticity M-tensor, then the strong ellipticity condition holds. We give an example to show that a nonsingular elasticity M-tensor may not be S-PD. We then present a sufficient condition which says that the strong ellipticity holds if unfolding matrix can be modified to a positive definite one by preserving the summation of some entries.

We also study strong ellipticity for higher order elasticity tensors. Paired symmetric tensors and strongly paired symmetric tensors are defined. It is easy to know that all fourth order three dimensional elasticity tensors are paired symmetric tensors. The notion of M-eigenvalue for elasticity tensors is generalized for sixth order paired symmetric tensors as well as the strong ellipticity. In addition, we study conditions under which the paired symmetric tensor is M-positive semidefiniteness or M-positive definiteness. At last, a sequential semidefinite programming method is proposed by using semidefinite relaxation technique.

8.1 The Elasticity Tensor, Strong Ellipticity and M-Eigenvalues

Hooke's law, named after 17th-century British physicist Robert Hooke (1635–1703), is a principle of physics that the force needed to extend or compress a spring by some distance is proportional to that distance. The modern theory of elasticity generalizes Hooke's law to say that the strain (deformation) tensor $S = (S_{ij})$ (a second order three dimensional symmetric tensor) of an elastic object or material is the tensor product of the elasticity tensor (a fourth order tensor) $\mathscr{C} = (C_{ijkl})$ and the stress tensor $E = (E_{kl})$ (another second order three dimensional symmetric tensor) applied to that elastic object or material, i.e.,

$$S_{ij} = \sum_{k,l=1}^{3} C_{ijkl} E_{kl}.$$

Since S and E are symmetric, $\mathscr{C}$ is symmetric with respect to the first two indices i and j, and the last two indices k and l, respectively, i.e.,

$$C_{ijkl} = C_{jikl} = C_{ijlk},$$

for $i, j, k, l = 1, 2, 3$. This is called minor symmetry in the mechanics literature. Furthermore, the elasticity tensor $\mathscr{C}$ has a major symmetry, i.e.,

$$C_{ijkl} = C_{klij},$$

for $i, j, k, l = 1, 2, 3$. Thus, the elasticity tensor has 21 independent entries, which are called elasticities.

Elasticities are related to elastic moduli. Depending upon how stress and strain are to be measured, including directions, there are various types of elastic moduli. The three primary ones are Young's modulus, the shear modulus or modulus of rigidity, and the bulk modulus. Three other elastic moduli are Poisson's ratio, Lamé's first parameter, and P-wave modulus.

An important topic in elasticity theory is the strong ellipticity condition which guarantees the existence of solutions of basic boundary-value problems of elastostatics and thus ensures an elastic material to satisfy some mechanical properties. Research works on this topic include Knowles and Sternberg [158, 159], Simpson and Spector [244], Rosakis [236], Wang and Aron [276], Walton and Wilber [271], Chiriţă, Danescu, and Ciarletta [59], Zubov and Rudev [319], and Gourgiotis and Bigoni [106], etc.

We now use $\mathscr{A} = (a_{ijkl})$ to denote the elasticity tensor of some elastic material. By the discussion at the beginning of this section, it satisfies the following symmetry:

$$a_{ijkl} = a_{jikl} = a_{ijlk} = a_{klij}. \tag{8.1}$$

And the strong (ordinary) ellipticity condition holds if

$$f(\mathbf{x}, \mathbf{y}) = \mathscr{A}\mathbf{xyxy} = \sum_{i,j,k,l=1}^{n} a_{ijkl} x_i y_j x_k y_l > 0 (\geq 0)$$

for all unit vectors $\mathbf{x}, \mathbf{y} \in \mathbf{R}^n$. According to Knowles and Sternberg [158, 159], the term "strong ellipticity condition" is a common usage in mechanics.

Now, for a given elasticity tensor $\mathscr{A} = (a_{ijkl})$, we define another fourth order tensor $\bar{\mathscr{A}} = (\bar{a}_{ijkl})$ with entries satisfying

$$\bar{a}_{ijkl} = \frac{a_{ijkl} + a_{kjil} + a_{ilkj} + a_{klij}}{4}, \quad \text{for } i, j, k, l \in [n].$$

So, the new tensor $\bar{\mathscr{A}}$ satisfies

$$\bar{a}_{ijkl} = \bar{a}_{kjil} = \bar{a}_{klij}, \tag{8.2}$$

and it is obvious that

$$\mathscr{A}\mathbf{xyxy} = \bar{\mathscr{A}}\mathbf{xyxy}, \quad \text{for any } \mathbf{x}, \mathbf{y} \in \mathbf{R}^n.$$

In the following analysis, we always consider the fourth order tensors with symmetry (8.2). Denote $\mathbb{E}_{4,n}$ the set containing all fourth order n dimensional tensors satisfying (8.2). We mainly consider the cases $n = 2$ or $n = 3$ in this section.

By optimization theory, we may see that the strong ellipticity condition holds if and only if the optimal value of the following optimization problem is positive, and the ordinary ellipticity condition holds if and only if the optimal value of this optimization problem is nonnegative.

$$\begin{array}{ll} \min & \mathscr{A}\mathbf{xyxy} \\ \text{s.t.} & \mathbf{x}^\top \mathbf{x} = 1, \ \ \mathbf{y}^\top \mathbf{y} = 1. \end{array} \tag{8.3}$$

Based this observation, we introduce the definition of M-eigenvalues for fourth order n-dimensional tensors which satisfy (8.2).

Definition 8.1 Let $\mathscr{A} = (a_{ijkl}) \in \mathbb{E}_{4,n}$, $i, j, k, l \in [n]$, where $n = 3$. If there are $\lambda \in \mathbf{R}$, $\mathbf{x}, \mathbf{y} \in \mathbf{R}^n \backslash \{\mathbf{0}\}$ satisfying that

$$\begin{cases} \mathscr{A} \cdot \mathbf{yxy} = \lambda \mathbf{x}, \\ \mathscr{A}\mathbf{xyx} \cdot = \lambda \mathbf{y}, \\ \mathbf{x}^\top \mathbf{x} = 1, \\ \mathbf{y}^\top \mathbf{y} = 1, \end{cases} \tag{8.4}$$

where $\mathscr{A} \cdot \mathbf{yxy}$, $\mathscr{A}\mathbf{xyx}\cdot$ are vectors whose ith components are

$$\sum_{j,k,l=1}^{n} a_{ijkl} y_j x_k y_l \quad \text{and} \quad \sum_{i,j,k=1}^{n} a_{ijkl} x_i y_j x_k,$$

then we call λ an M-eigenvalue of $\mathscr{A}$, and call $\mathbf{x}, \mathbf{y}$ left and right M-eigenvectors of $\mathscr{A}$, associated with the M-eigenvalue λ.

Here, "M" means mechanics. We now have the following result.

Theorem 8.1 *Let $\mathscr{A} \in \mathbb{E}_{4,n}$. Then its M-eigenvalue always exists. Let $\mathbf{x}, \mathbf{y}$ be left and right M-eigenvectors of $\mathscr{A}$, associated with an M-eigenvalue λ. Then $\lambda = \mathscr{A}\mathbf{xyxy}$. Furthermore, the strong ellipticity condition holds if and only if the smallest M-eigenvalue of $\mathscr{A}$ is positive, and the ordinary ellipticity condition holds if and only if the smallest M-eigenvalue of $\mathscr{A}$ is nonnegative.*

Proof For the optimization problem (8.3), it is obvious that the problem has at least one maximizer and one minimizer since the objective function is continuous and its feasible region is compact. By the optimality conditions of problem (8.3), there are $\lambda \in \mathbf{R}$, $\mathbf{x}, \mathbf{y} \in \mathbf{R}^n \backslash \{\mathbf{0}\}$ such that (8.4) holds, which implies that M-eigenvalues always exist, and it is not difficult to see that

$$\lambda = \mathscr{A}\mathbf{xyxy}.$$

Then the other results follow this directly. □

Thus, for any elasticity tensor $\mathscr{A} \in \mathbb{E}_{4,n}$ satisfying (8.2), if the strong ellipticity condition holds, we also say that $\mathscr{A}$ is M-positive definite (M-PD), and if the ordinary ellipticity condition holds, we also say that $\mathscr{A}$ is M-positive semi-definite (MPSD).

In the following analysis, denote $\lambda_{\mathscr{A}}$ as the M-eigenvalue of tensor $\mathscr{A}$ with the largest absolute value. For $\lambda \in \mathbf{R}$, we say $\lambda\mathbf{xyxy} = \lambda\mathbf{x} \otimes \mathbf{y} \otimes \mathbf{x} \otimes \mathbf{y}$ is the best rank-one approximation of $\mathscr{A}$, if it can minimize $\|\mathscr{A} - \lambda\mathbf{xyxy}\|_F$, where the norm $\|\cdot\|_F$ is the Frobenius norm, and $\mathbf{x}^\top\mathbf{x} = 1, \mathbf{y}^\top\mathbf{y} = 1$. The next theorem provides an application of the M-eigenvalue in the best rank-one approximation for fourth order three dimensional tensors satisfying (8.2).

Theorem 8.2 *Let $\mathscr{A} \in \mathbb{E}_{4,n}$. Let $\mathbf{x}, \mathbf{y}$ be left and right M-eigenvectors of $\mathscr{A}$, associated with $\lambda_{\mathscr{A}}$. Then, $\lambda_{\mathscr{A}}\mathbf{xyxy}$ is the best rank-one approximation of $\mathscr{A}$.*

Proof Let $\lambda \in \mathbf{R}$ be any scalar. Suppose $\mathbf{x} \in \mathbf{R}^m, \mathbf{y} \in \mathbf{R}^n$ such that $\mathbf{x}^\top\mathbf{x} = 1, \mathbf{y}^\top\mathbf{y} = 1$. Then, we know that

$$\begin{aligned}\|\mathscr{A} - \lambda\mathbf{xyxy}\|_F^2 &= \|\mathscr{A}\|_F^2 - 2\lambda\mathscr{A}\mathbf{xyxy} + \lambda^2(\mathbf{x}^\top\mathbf{x})(\mathbf{y}^\top\mathbf{y}) \\ &= \|\mathscr{A}\|_F^2 - 2\lambda\mathscr{A}\mathbf{xyxy} + \lambda^2.\end{aligned}$$

It is obvious that the minimum of $\|\mathscr{A} - \lambda\mathbf{xyxy}\|$ must be attained at $\lambda = \mathscr{A}\mathbf{xyxy}$. Thus, it holds that

$$\begin{aligned}&\min\{\|\mathscr{A} - \lambda\mathbf{xyxy}\|_F^2 \mid \lambda \in \mathbf{R},\ \mathbf{x}^\top\mathbf{x} = 1, \mathbf{y}^\top\mathbf{y} = 1\} \\ &= \min\{\|\mathscr{A}\|_F^2 - (\mathscr{A}\mathbf{xyxy})^2 \mid \mathbf{x}^\top\mathbf{x} = 1, \mathbf{y}^\top\mathbf{y} = 1\} \\ &= \|\mathscr{A}\|_F^2 - \max\{(\mathscr{A}\mathbf{xyxy})^2 \mid \mathbf{x}^\top\mathbf{x} = 1, \mathbf{y}^\top\mathbf{y} = 1\},\end{aligned}$$

and the desired result holds. □

Let $P = (p_{ii'}) \in \mathbf{R}^{m\times m}$ and $Q = (q_{jj'}) \in \mathbf{R}^{n\times n}$ be two orthogonal matrices. Suppose tensor $\mathscr{B} = (b_{ijkl})$ is defined such that

$$b_{ijkl} = \sum_{i',k'=1}^{m} \sum_{j',l'=1}^{n} p_{ii'}q_{jj'}p_{kk'}q_{ll'}a_{i'j'k'l'}.$$

Then we say $\mathscr{A} \in \mathbb{E}_{4,n}$ and $\mathscr{B}$ are orthogonally similar. It is not difficult to see that if $\mathscr{A}$ satisfies the partially symmetric condition (8.2), then $\mathscr{B}$ also satisfies (8.2). The following conclusion holds, and the proof is omitted since it can be proved by definition directly.

Theorem 8.3 *Let $\mathscr{A} \in \mathbb{E}_{4,3}$. Assume $\mathscr{B}$ is another fourth order three dimensional tensor orthogonal similar to $\mathscr{A}$ via orthogonal matrices P and Q. Then the following results hold:*

(1) $\mathscr{A}$ and $\mathscr{B}$ have the same M-eigenvalues;

(2) *if λ is an M-eigenvalue of $\mathscr{A}$ with left and right M-eigenvectors $\mathbf{x}$ and $\mathbf{y}$, then λ is also an M-eigenvalue of $\mathscr{B}$ with left and right M-eigenvectors $P\mathbf{x}$ and $Q\mathbf{y}$.*

Theorems 8.1 and 8.3 reveal that the M-eigenvalues are some intrinsic parameters of the elasticity tensor. Recently, Xiang, Qi and Wei [289] showed that there are two M-eigenvalues in the isotopic case. The first M-eigenvalue is the shear modulus G, and must be positive. The second M-eigenvalue is the P-wave modulus. Hence, in the isotropic case, strong ellipticity holds if and only if the P-wave modulus is positive. Xiang, Qi and Wei [289] further showed that the positiveness condition of the M-eigenvalues coincides with the existing conditions for strong ellipticity in the cubic case and the polar anisotropic case. This further reveals the physical meanings of M-eigenvalues and indicates that the positiveness condition of M-eigenvalues is more general for strong ellipticity.

We now define S-positive definiteness for elasticity tensors $\mathscr{A} \in \mathbb{E}_{4,n}$. Such property has been considered by Lord Kelvin 160 years ago [261, 262]. Let $\mathscr{A} = (a_{ijkl}) \in \mathbb{E}_{4,n}$ be an elasticity tensor satisfying (8.2). We say that $\mathscr{A}$ is S-positive definite (S-PD) if

$$\mathscr{A} D^2 = \sum_{i,j,k,l=1}^{n} a_{ijkl} d_{ij} d_{kl} > 0$$

for all $D = (d_{ij}) \in \mathbf{R}^{n \times n}$. Similarly, we say that $\mathscr{A}$ is S-positive semi-definite (S-PSD) if

$$\mathscr{A} D^2 = \sum_{i,j,k,l=1}^{n} a_{ijkl} d_{ij} d_{kl} \geq 0$$

for all $D = (d_{ij}) \in \mathbf{R}^{n \times n}$. For any $\mathbf{x}, \mathbf{y} \in \mathbf{R}^n$, we can define $d_{ij} = x_i y_j$. This implies that the S-positive definiteness is a sufficient condition for the M-positive definiteness. The following example shows that the converse is not true.

Let $\mathscr{A} = (a_{ijkl}) \in \mathbb{E}_{4,2}$ be an elasticity tensor satisfying (8.2). Its independent entries are defined as

$$a_{1111} = 1,\ a_{1112} = 2,\ a_{1122} = 4,\ a_{1212} = 12,\ a_{2121} = 12,$$

$$a_{1222} = 1,\ a_{1121} = 2,\ a_{2122} = 1,\ a_{2222} = 2.$$

Then, the corresponding homogeneous polynomial can be written as

$$\begin{aligned} f(\mathbf{x}, \mathbf{y}) &= \mathscr{A}\mathbf{xyxy} \\ &= x_1^2 y_1^2 + 4x_1^2 y_1 y_2 + 12 x_1^2 y_2^2 + 4x_1 x_2 y_1^2 + 16 x_1 x_2 y_1 y_2 \\ &\quad + 2x_1 x_2 y_2^2 + 12 x_2^2 y_1^2 + 2x_2^2 y_1 y_2 + 2x_2^2 y_2^2. \end{aligned}$$

By a direct computation, we can find its smallest M-eigenvalue $\lambda = 0.5837$. Thus, the tensor $\mathscr{A}$ is M-PD. On the other hand, to discuss its S-positive definiteness, we can rewrite the matrix $D \in \mathbf{R}^{n\times n}$ to an n^2-dimensional vector, and $\mathscr{A}$ can be rewritten as a symmetric matrix in $\mathbf{R}^{n^2\times n^2}$. Then, it holds that $\mathscr{A}$ is S-PD if and only if the corresponding symmetric matrix is positive definite. For this example, $n = 2$, the 4×4 symmetric matrix is

$$\begin{pmatrix} a_{1111} & a_{1112} & a_{1121} & a_{1122} \\ a_{1112} & a_{1212} & a_{1122} & a_{1222} \\ a_{1121} & a_{1222} & a_{2121} & a_{2122} \\ a_{1122} & a_{1122} & a_{2122} & a_{2222} \end{pmatrix} = \begin{pmatrix} 1 & 2 & 2 & 4 \\ 2 & 12 & 4 & 1 \\ 2 & 4 & 12 & 1 \\ 4 & 1 & 1 & 2 \end{pmatrix}.$$

Since the smallest eigenvalue of this symmetric matrix is $-2.6110 < 0$, we conclude that $\mathscr{A}$ is not S-PD. Thus, S-positive definiteness is a sufficient but not necessary condition for M-positive definiteness.

We use the name "S-positive definiteness" as this is related with S-eigenvalues of fourth order tensors defined in [228], where the eigenvalues of the $n \times n$ unfolded symmetric matrix described above are called the S-eigenvalues of the tensor $\mathscr{A}$.

To end this section, we describe a direct method for computing M-eigenvalues of fourth order two dimensional tensors satisfying (8.2) i.e. tensors such that $\mathscr{A} \in \mathbb{E}_{4,n}$ and $n = 2$. The key technique is the Sylvester formula of the resultant, which is used to solve a system involving only two variables. In fact, by Definition 8.1, we can always homogenize the first two equations of (8.4) by the following form

$$\begin{cases} \mathscr{A} \cdot \mathbf{yxy} = \lambda(\mathbf{y}^\top \mathbf{y})\mathbf{x}, \\ \mathscr{A}\mathbf{xyx}\cdot = \lambda(\mathbf{x}^\top \mathbf{x})\mathbf{y} \end{cases} \tag{8.5}$$

According to the resultant theory [64], we know that (8.5) has a nonzero complex solution $(\mathbf{x}, \mathbf{y})$ if and only if λ is a root of the resultant. Particularly, for the case $n = 2$, the following results hold.

Theorem 8.4 *Let $\mathscr{A} = (a_{ijkl}) \in \mathbb{E}_{4,n}$ with $n = 2$. Then the following results hold for its M-eigenvalues and the corresponding M-eigenvectors:*

(1) if $a_{1112} = a_{1121} = 0$, then $\lambda = a_{1111}$ is an M-eigenvalue of $\mathscr{A}$ and the corresponding left and right M-eigenvector is $\mathbf{x} = \mathbf{y} = (1, 0)^\top$;
(2) for any real roots $(u, v)^\top$ of the following equations

$$\begin{cases} a_{1121}u^2 + (a_{2121} - a_{1111})uv - a_{1121}v^2 = 0, \\ a_{1112}u^2 + 2a_{1122}uv + a_{2122}v^2 = 0, \end{cases} \tag{8.6}$$

it holds that $\lambda = a_{1111}u^2 + 2a_{1121}uv + a_{2121}v^2$ is an M-eigenvalue with the left and right M-eigenvector such that

$$\mathbf{x} = \frac{(u, v)^\top}{\sqrt{u^2 + v^2}}, \quad \mathbf{y} = (\pm 1, 0);$$

(3) for any real roots $(u, v)^\top$ *of the following equations*

$$\begin{cases} a_{1121}u^2 + 2a_{1122}uv + a_{1222}v^2 = 0, \\ a_{1112}u^2 + (a_{1212} - a_{1111})uv - a_{1112}v^2 = 0, \end{cases} \tag{8.7}$$

it holds that $\lambda = a_{1111}u^2 + 2a_{1112}uv + a_{1212}v^2$ *is an M-eigenvalue with the left and right M-eigenvectors such that*

$$\mathbf{x} = (\pm 1, 0)^\top, \quad \mathbf{y} = \frac{(u, v)^\top}{\sqrt{u^2 + v^2}};$$

(4) $\lambda = \mathscr{A}\mathbf{xyxy}$ *is an M-eigenvalue with left and right M-eigenvectors such that*

$$\mathbf{x} = \pm\frac{(u, 1)^\top}{\sqrt{u^2 + 1}}, \quad \mathbf{y} = \pm\frac{(v, 1)^\top}{\sqrt{v^2 + 1}}.$$

Furthermore, u, v are real solutions of the following system of polynomial equations

$$\begin{cases} a_{1121}u^2v^2 + 2a_{1122}u^2v + a_{1222}u^2 + (a_{2121} - a_{1111})uv^2 - a_{1121}v^2 \\ \quad + 2(a_{2122} - a_{1112})uv + (a_{2222} - a_{1212})u - 2a_{1122}v - a_{1222} = 0, \\ a_{1112}u^2v^2 + (a_{1212} - a_{1111})u^2v - a_{1112}u^2 + 2(a_{1222} - a_{1121})uv \\ \quad + 2a_{1122}uv^2 - 2a_{1122}u + a_{2122}v^2 + (a_{2222} - a_{2121})v - a_{2122} = 0. \end{cases}$$

Moreover, if $a_{1112} = a_{1121} = 0$*, all M-eigenvalues and related M-eigenvectors of* $\mathscr{A}$ *are given by (1)–(4); otherwise, they are given by (2)–(4).*

Proof (1) If $a_{1112} = a_{1121} = 0$, according to Definition 8.1, by a direct computation that (1) holds.

(2) If $y_2 = 0$, then $y_1 = 1$ or $y_1 = -1$, by Definition 8.1, we have

$$\begin{cases} a_{1111}x_1 + a_{1121}x_2 = \lambda x_1, \\ a_{2111}x_1 + a_{2121}x_2 = \lambda x_2, \\ a_{1111}x_1^2 + 2a_{1121}x_1x_2 + a_{2121}x_2^2 = \lambda, \\ a_{1112}x_1^2 + 2a_{1122}x_1x_2 + a_{2122}x_2^2 = 0, \\ x_1^2 + x_2^2 = 1. \end{cases}$$

By the first two equations, we have

$$\begin{cases} a_{1121}x_1^2 + (a_{2121} - a_{1111})x_1x_2 - a_{1121}x_2^2 = 0, \\ a_{1112}x_1^2 + 2a_{1122}x_1x_2 + a_{2122}x_2^2 = 0, \\ x_1^2 + x_2^2 = 1. \end{cases}$$

Suppose $u = \frac{x_1}{\sqrt{x_1^2+x_2^2}}$, $v = \frac{x_2}{\sqrt{x_1^2+x_2^2}}$, the desired result (2) holds.

(3) We can prove this by a similar way as (2).

(4) If $x_2 \neq 0$ and $y_2 \neq 0$, by (8.4), it follows that

$$\begin{cases} a_{1111}x_1y_1^2 + 2a_{1112}x_1y_1y_2 + a_{1212}x_1y_2^2 + a_{1121}x_2y_1^2 \\ \quad + 2a_{1122}x_2y_1y_2 + a_{1222}x_2y_2^2 = \lambda x_1, \\ a_{1121}x_1y_1^2 + 2a_{1122}x_1y_1y_2 + a_{1222}x_1y_2^2 + a_{2121}x_2y_1^2 \\ \quad + 2a_{2122}x_2y_1y_2 + a_{2222}x_2y_2^2 = \lambda x_2, \\ a_{1111}x_1^2y_1 + a_{1112}x_1^2y_2 + 2a_{1121}x_1x_2y_1 + 2a_{1122}x_1x_2y_2 \\ \quad + a_{2121}x_2^2y_1 + a_{2122}x_2^2y_2 = \lambda y_1, \\ a_{1112}x_1^2y_1 + a_{1212}x_1^2y_2 + 2a_{1122}x_1x_2y_1 + 2a_{1222}x_1x_2y_2 \\ \quad + a_{2122}x_2^2y_1 + a_{2222}x_2^2y_2 = \lambda y_2, \\ x_1^2 + x_2^2 = 1, \\ y_1^2 + y_2^2 = 1. \end{cases} \tag{8.8}$$

Denote $u = \frac{x_1}{x_2}$ and $v = \frac{y_1}{y_2}$. By the first two equations of (8.8), it holds that

$$\begin{aligned} &a_{1121}u^2v^2 + 2a_{1122}u^2v + a_{1222}u^2 + (a_{2121} - a_{1111})uv^2 - a_{1121}v^2 \\ &+ 2(a_{2122} - a_{1112})uv + (a_{2222} - a_{1212})u - 2a_{1122}v - a_{1222} = 0. \end{aligned} \tag{8.9}$$

From the third equation and the fourth equation of (8.8), it follows that

$$\begin{aligned} &a_{1112}u^2v^2 + (a_{1212} - a_{1111})u^2v - a_{1112}u^2 + 2(a_{1222} - a_{1121})uv \\ &+ 2a_{1122}uv^2 - 2a_{1122}u + a_{2122}v^2 + (a_{2222} - a_{2121})v - a_{2122} = 0. \end{aligned} \tag{8.10}$$

Combining (8.9) and (8.10) and the fact that $x_1^2 + x_2^2 = 1$, $y_1^2 + y_2^2 = 1$, the desired results hold automatically. □

8.2 Strong Ellipticity via Z-Eigenvalues of Symmetric Tensors

In the last section, we studied the strong ellipticity condition for the elasticity tensor for isotropic materials. Several sufficient conditions for preserving the strong ellipticity condition were given. Now, we study some necessary and sufficient conditions for the strong ellipticity condition. The main tool for this study is Z-eigenvalues of symmetric tensors. Particularly, the strong ellipticity condition of the elasticity tensor can be connected with the smallest Z-eigenvalue of some symmetric tensors.

For both planar and three dimensional cases, checkable conditions are given to find all Z-eigenvalues of any tensor with even order.

The other tool that will be used to establish the necessary and sufficient condition for the strong ellipticity condition of the elasticity tensor is the strict copositivity of a second-order symmetric tensor. To move on, we first cite a useful lemma.

Lemma 8.1 (Simpson, Spector 1983) *Suppose $M = (m_{ij}) \in \mathbf{R}^{3\times 3}$ is a symmetric matrix. Then, M is strictly copositive if and only if the following conditions hold:*

$$\begin{aligned}
&m_{11} > 0, \quad m_{22} > 0, \quad m_{33} > 0,\\
&R \equiv m_{12} + (m_{11}m_{22})^{\frac{1}{2}} > 0, \qquad S \equiv m_{13} + (m_{11}m_{33})^{\frac{1}{2}} > 0,\\
&T \equiv m_{23} + (m_{22}m_{33})^{\frac{1}{2}} > 0,\\
&m_{33}^{\frac{1}{2}}m_{12} + m_{22}^{\frac{1}{2}}m_{13} + m_{11}^{\frac{1}{2}}m_{23} + (2RST)^{\frac{1}{2}} + (m_{11}m_{22}m_{33})^{\frac{1}{2}} > 0.
\end{aligned}$$

In Theorem 8.4, we studied a method to compute all the M-eigenvalues of the elasticity tensor for $n = 2$. Now, we study the case that $n = 3$.

Suppose $\mathscr{A} = (a_{ijkl})$ is the elasticity tensor with $i, j, k, l \in [3]$. Here, [3] stands for a set $\{1, 2, 3\}$. Let $Q(\mathbf{x}) = (q_{jl}) \in \mathbf{R}^{3\times 3}$ denote the acoustic tensor, which is defined by

$$q_{jl} = \sum_{i,k=1}^{3} a_{ijkl}x_i x_k. \tag{8.11}$$

Then $Q(\mathbf{x})$ is a second-order symmetric tensor. By the analysis of the last section, the tensor $\mathscr{A}$ is M-PD or the strong ellipticity condition holds if and only if the acoustic tensor $Q(\mathbf{x})$ is positive definite for all unit vector $\mathbf{x}$. As we all know that a symmetric matrix is positive definite if and only if all of its leading principal minors are positive definite. For the sake of simplicity, we define the following matrices:

$$M_1 = \begin{pmatrix} a_{1111} & a_{1121} & a_{1131}\\ a_{1121} & a_{2121} & a_{2131}\\ a_{1131} & a_{2131} & a_{3131}\end{pmatrix}, \quad M_2 = \begin{pmatrix} a_{1212} & a_{1222} & a_{1232}\\ a_{1222} & a_{2222} & a_{2232}\\ a_{1232} & a_{2232} & a_{3232}\end{pmatrix}, \tag{8.12}$$

and

$$M_3 = \begin{pmatrix} a_{1313} & a_{1323} & a_{1333}\\ a_{1323} & a_{2323} & a_{2333}\\ a_{1333} & a_{2333} & a_{3333}\end{pmatrix}. \tag{8.13}$$

We now consider the diagonal entries of $Q(\mathbf{x})$. By (8.11), it holds that

$$\begin{aligned}
q_{11} &= a_{1111}x_1^2 + 2a_{1121}x_1x_2 + 2a_{1131}x_1x_3 + a_{2121}x_2^2 + 2a_{2131}x_2x_3 + a_{3131}x_3^2,\\
q_{22} &= a_{1212}x_1^2 + 2a_{1222}x_1x_2 + 2a_{1232}x_1x_3 + a_{2222}x_2^2 + 2a_{2232}x_2x_3 + a_{3232}x_3^2,\\
q_{33} &= a_{1313}x_1^2 + 2a_{1323}x_1x_2 + 2a_{1333}x_1x_3 + a_{2323}x_2^2 + 2a_{2333}x_2x_3 + a_{3333}x_3^2.
\end{aligned} \tag{8.14}$$

From (8.12)–(8.14), its obvious that

$$\begin{aligned} q_{11} > 0,\ \forall \mathbf{x} \neq \mathbf{0} &\Leftrightarrow M_1 \text{ is positive definite },\\ q_{22} > 0,\ \forall \mathbf{x} \neq \mathbf{0} &\Leftrightarrow M_2 \text{ is positive definite },\\ q_{33} > 0,\ \forall \mathbf{x} \neq \mathbf{0} &\Leftrightarrow M_3 \text{ is positive definite }. \end{aligned}$$

For all principal second-order minors of $Q(\mathbf{x})$, it is not difficult to see that

$$q_{11}q_{22} - q_{12}^2, \quad q_{11}q_{33} - q_{13}^2, \quad q_{22}q_{33} - q_{23}^2$$

are all fourth order homogeneous polynomials corresponding to $\mathbf{x}$. Suppose $\mathscr{T}$, is a fourth order three dimensional symmetric tensors such that

$$\mathscr{T}\mathbf{x}^4 = q_{11}q_{22} - q_{12}^2,$$

By Theorem 1.2 in Chap. 1, we know that the second-order leading principal minors is positive if and only if the smallest Z-eigenvalue of the fourth order tensor $\mathscr{T}$ is positive.

Similar to the discussion above, let $\mathscr{W}$ be a sixth-order three dimensional tensor such that

$$\mathscr{W}\mathbf{x}^4 = \sum_{i_1,\ldots,i_6=1}^{3} w_{i_1\ldots i_6} x_{i_1} x_{i_2} \ldots x_{i_6} = det(Q(\mathbf{x})),$$

which implies that $det(Q(\mathbf{x})) > 0$ if and only if the smallest Z-eigenvalue of $\mathscr{W}$ is positive. In summary, we have the following conclusion. Here, the positive definiteness of a symmetric tensor is defined in Chap. 1.

Theorem 8.5 *The strong ellipticity condition holds for the elasticity tensor $\mathscr{A}$ if and only if the following three conditions hold:*

(1) the symmetric matrices $M_i, i \in [3]$ are positive definite;
(2) the fourth order symmetric tensor $\mathscr{T}$ is positive definite;
(3) the sixth order symmetric tensor $\mathscr{W}$ is positive definite.

Based on Theorem 8.5, we now discuss a kind of practical elasticity tensors from the rhombic system, where $\mathscr{A} = (a_{ijkl})$ satisfies the symmetries

$$a_{ijkl} = a_{klij} = a_{ijlk}, \quad \forall\, i, j, k, l \in [3]. \tag{8.15}$$

As stated in [59], the elasticity tensor $\mathscr{A}$ has entries such that

$$\begin{aligned} a_{1123} = a_{1131} = a_{1112} = a_{2223} = a_{2231} = a_{2212} = 0,\\ a_{3323} = a_{3331} = a_{3312} = a_{2331} = a_{2312} = a_{3112} = 0. \end{aligned} \tag{8.16}$$

And the nonzero components are denoted shortly by

$$a_{11} = a_{1111},\ \ a_{22} = a_{2222},\ \ a_{33} = a_{3333},\ \ a_{12} = a_{1122},\ \ a_{23} = a_{2233},$$

$$a_{31} = a_{3311},\ \ a_{44} = a_{2323},\ \ a_{55} = a_{1313},\ \ a_{66} = a_{1212}.$$

To discuss its strong ellipticity condition, the following matrices are needed:

$$P = \begin{pmatrix} a_{11}a_{66} & \frac{a_{11}a_{22}+a_{12}^2-2a_{12}a_{66}}{2} & \frac{a_{11}a_{44}+a_{55}a_{66}}{2} \\ \frac{a_{11}a_{22}+a_{12}^2-2a_{12}a_{66}}{2} & a_{22}a_{66} & \frac{a_{44}a_{66}+a_{22}a_{55}}{2} \\ \frac{a_{11}a_{44}+a_{55}a_{66}}{2} & \frac{a_{44}a_{66}+a_{22}a_{55}}{2} & a_{44}a_{55} \end{pmatrix}, \tag{8.17}$$

Furthermore, suppose the second order symmetric tensor $Q(\mathbf{x})$ and the corresponding sixth order tensor $\mathscr{W}$ are defined as in Theorem 8.5. The following theorem shows the strong ellipticity for the rhombic system.

Theorem 8.6 *Suppose that $\mathscr{A}$ is the elasticity tensor with entries satisfying (8.16). Then the strong ellipticity condition holds for $\mathscr{A}$ if and only if the following three conditions hold:*

(1) $a_{ii} > 0$ for all $i \in [6]$;
(2) the symmetric matrix P defined in (8.17) is copositive;
(3) the sixth order symmetric tensor $\mathscr{W}$ is positive definite.

Proof Combining (8.14) with (8.15), we know that

$$q_{11} = a_{11}x_1^2 + a_{66}x_2^2 + a_{55}x_3^2,\ \ q_{22} = a_{66}x_1^2 + a_{22}x_2^2 + a_{44}x_3^2,$$
$$q_{33} = a_{55}x_1^2 + a_{44}x_2^2 + a_{33}x_3^2.$$

Together with (8.12) and (8.13), we obtain

$$M_1 = \begin{pmatrix} a_{11} & 0 & 0 \\ 0 & a_{66} & 0 \\ 0 & 0 & a_{55} \end{pmatrix},\ M_2 = \begin{pmatrix} a_{66} & 0 & 0 \\ 0 & a_{22} & 0 \\ 0 & 0 & a_{44} \end{pmatrix},\ M_3 = \begin{pmatrix} a_{55} & 0 & 0 \\ 0 & a_{44} & 0 \\ 0 & 0 & a_{33} \end{pmatrix}.$$

Thus, from (1) of Theorem 8.5 with (8.14), we know that $a_{ii} > 0$ for all $i \in [6]$.

On the other hand, by (8.11) and (8.15)–(8.16), it is easy to know that

$$\begin{aligned} q_{11}q_{22} - q_{12}^2 &= a_{11}a_{66}x_1^4 + a_{11}a_{22}x_1^2x_2^2 + a_{11}a_{44}x_1^2x_3^2 \\ &+ a_{66}^2x_1^2x_2^2 + a_{66}a_{22}x_2^4 + a_{66}a_{44}x_2^2x_3^2 \\ &+ a_{55}a_{66}x_3^2x_1^2 + a_{55}a_{22}x_3^2x_2^2 + a_{55}a_{44}x_3^4 \\ &- a_{12}^2x_1^2x_2^2 - a_{66}^2\mathbf{x}_1^2x_2^2 - 2a_{12}a_{66}x_1^2x_2^2 \\ &= a_{11}a_{66}x_1^4 + a_{66}a_{22}x_2^4 + a_{55}a_{44}x_3^4 \\ &+ (a_{11}a_{22} - a_{12}^2 - 2a_{12}a_{66})x_1^2x_2^2 \\ &+ (a_{11}a_{44} + a_{55}a_{66})x_1^2x_3^2 + (a_{66}a_{44} + a_{55}a_{22})x_2^2x_3^2 \\ &= (x_1^2, x_2^2, x_3^2)P(x_1^2, x_2^2, x_3^2)^\top,\ \ \forall\, \mathbf{x} \in \mathbf{R}^3\backslash\{\mathbf{0}\}, \end{aligned}$$

By (2) and (3) of Theorem 8.5, we know that the desired results hold. □

It should be noted that the key part of the necessary and sufficient condition in Theorem 8.5 is to find the smallest Z-eigenvalues of the symmetric tensors $\mathscr{T}$ and $\mathscr{W}$. Cui, Dai and Nie [66] has proposed an efficient method for computing all the Z-eigenvalues of symmetric tensors of low orders and dimensions. A package of their method is available via the following link:

AReigSTensors: http://www.math.ucsd.edu/njw/CODES/reigsymtensor/areigs tsrweb.html

Note that the order of $\mathscr{T}$ is 4, the order of $\mathscr{W}$ is 6, and their dimensions are 3, one may use the package of Cui, Dai and Nie [66] to find the smallest Z-eigenvalues of the symmetric tensors $\mathscr{T}$ and $\mathscr{W}$. Thus, numerically, Theorem 8.5 provides an efficient way to determine strong ellipticity of the elasticity tensor of a material.

8.3 Other Sufficient Condition for Strong Ellipticity

In Sect. 8.1, we learned that a sufficient condition for the strong ellipticity condition is that the elasticity tensor is S-PD. The elasticity tensor is S-PD if the unfolded symmetric matrix is positive definite. It is easy to check a symmetric matrix is positive definite or not. Thus, this provides an easily checkable condition for the strong ellipticity condition. Unfortunately, the range of the tensors satisfying S-positive definiteness are too narrow. Therefore, it is important to find some kinds of tensors that satisfy M-positive definiteness but not S-positive definiteness. In this section, we first study another sufficient condition for strong ellipticity, and present an alternating projection method to verify whether the given elasticity tensor satisfies this sufficient condition or not. Then we study properties of elasticity M-tensors, which is a big class of structured tensors satisfying the strong ellipticity condition, but do not satisfy S-positive definiteness. All the results here were originally studied by Ding, Qi and Yan [85], and Ding, Liu, Qi and Yan [79] very recently.

Since the strong ellipticity condition holds for $\mathscr{A} \in \mathbb{E}_{4,3}$ if and only if

$$f(\mathbf{x}, \mathbf{y}) = \mathscr{A}\mathbf{xyxy} := \sum_{i,j,k,l=1}^{3} a_{ijkl} x_i y_j x_k y_l > 0, \quad \text{for } \mathbf{x},\ \mathbf{y} \in \mathbf{R}^3 \backslash \{\mathbf{0}\}.$$

We recall that $\mathscr{A}$ is M-positive semidefinite if $\mathscr{A}\mathbf{xyxy} \geq 0$ holds in the above equation, and $\mathscr{A}$ is S-positive definite (S-positive semi-definite) if

$$\mathscr{A} D^2 = \sum_{i,j,k,l=1}^{3} a_{ijkl} d_{ij} d_{kl} > 0 (\geq 0), \quad \forall\, D \in \mathbf{R}^{3\times 3},\ D \neq 0\ (D \in \mathbf{R}^{3\times 3}).$$

On the other hand, let matrix $A_{jl} = (a_{ijkl}) \in \mathbf{R}^{3\times 3}$ and $\mathbf{d}_j = (d_{ij}) \in \mathbf{R}^3$ for each $i, k \in [3]$. Then the tensor $\mathscr{A}$ and matrix D can be unfolded such that

$$A = \begin{pmatrix} A_{11} & A_{12} & A_{13} \\ A_{21} & A_{22} & A_{23} \\ A_{31} & A_{32} & A_{33} \end{pmatrix} \in \mathbf{R}^{9\times 9}, \quad d = \begin{pmatrix} \mathbf{d}_1 \\ \mathbf{d}_2 \\ \mathbf{d}_3 \end{pmatrix} \in \mathbf{R}^9. \tag{8.18}$$

It is not difficult to verify that $\mathscr{A}$ is S-positive definite (or semidefinite respectively) if and only if the matrix A is positive definite (or semidefinite respectively), which is a sufficient condition for the strong ellipticity of elasticity tensors as discussed in Sect. 8.1.

On the other hand, for any $\mathscr{A} \in \mathbb{E}_{4,3}$ and $\mathbf{y} \in \mathbf{R}^3$, similar to the discussion in Sect. 8.1, define

$$\mathscr{A}\mathbf{y}^2 \in \mathbf{R}^{3\times 3} \quad \text{with} \quad (\mathscr{A}\mathbf{y}^2)_{ik} = \sum_{j,l=1}^{3} a_{ijkl} y_j y_l,$$

$$\mathscr{A}\mathbf{x}^2 \in \mathbf{R}^{3\times 3} \quad \text{with} \quad (\mathscr{A}\mathbf{x}^2)_{jl} = \sum_{i,k=1}^{3} a_{ijkl} x_i x_k.$$

Then it holds that $\mathscr{A}\mathbf{xyxy} = \mathbf{x}^\top(\mathscr{A}\mathbf{y}^2)\mathbf{x} = \mathbf{y}^\top(\mathscr{A}\mathbf{x}^2)\mathbf{y}$. By matrix theory, we know that every positive definite matrix can be factorized into the sum of several rank-one matrices and the minimal number of the rank-one matrices is exactly its rank. Hence the elasticity tensor $\mathscr{A} \in \mathbb{E}_{4,3}$ is M-PSD if

$$\mathscr{A}\mathbf{y}^2 = \sum_{s=1}^{r} \alpha_s (U_s\mathbf{y})(U_s\mathbf{y})^\top, \quad \alpha_s > 0, \tag{8.19}$$

where $U_s \in \mathbf{R}^{3\times 3}$, $s \in [r]$. Apparently, we know that $r \geq 3$ if $\mathscr{A}$ is M-PD. Without loss of generality, suppose $U_s = (u_{ik}^{(s)}) \in \mathbf{R}^3$, $s \in [r]$. By (8.19), it follows that

$$\begin{aligned}
(\mathscr{A}\mathbf{y}^2)_{ik} &= \sum_{j,l=1}^{3} a_{ijkl} y_j y_l \\
&= \sum_{s=1}^{r} \alpha_s \left(\sum_{j=1}^{3} u_{ij}^{(s)} y_j\right)\left(\sum_{l=1}^{(3)} u_{kl}^{s} y_l\right) \\
&= \sum_{j,l=1}^{3} \left(\sum_{s=1}^{r} \alpha_s u_{ij}^{(s)} u_{kl}^{(s)}\right) y_j y_l,
\end{aligned}$$

which implies that given U_s, the entries of $\mathscr{A}$ are uniquely determined by

$$a_{ijkl} = \frac{1}{2} \sum_{s=1}^{r} \alpha_s (u_{ij}^{(s)} u_{kl}^{(s)} + u_{kj}^{(s)} u_{il}^{(s)}). \tag{8.20}$$

Then we need to consider when the tensor $\mathscr{A}$ can be written as (8.20). By (8.19), let $\mathscr{B} = (b_{ijkl}), i, j, k, l \in [3]$ be defined by

$$b_{ijkl} = \sum_{s=1}^{r} \alpha_s u_{ij}^{(s)} u_{kl}^{(s)}.$$

It is easy to check that $b_{ijkl} = b_{klij}$ but $\mathscr{B}$ may not obey the partial symmetry condition (8.2). Moreover, similar to (8.18), we know that the unfolding matrix B of $\mathscr{B}$ is positive semidefinite such that

$$B = \sum_{s=1}^{r} \alpha_s \mathbf{u}_s \mathbf{u}_s^\top,$$

where $\mathbf{u}_s \in \mathbf{R}^9$ is the unfolding vector of U_s, $s \in [r]$. Therefore, we know that $\mathscr{B}$ is S-PSD since $\alpha_s > 0$, $s \in [r]$. By a direct computation, we obtain that

$$a_{ijkl} = a_{kjil} = \frac{1}{2}(b_{ijkl} + b_{kjil}), \quad \forall\, i, j, k, l \in [3],$$

which implies that

$$\mathscr{A}\mathbf{xyxy} = \mathscr{B}\mathbf{xyxy} = \mathscr{B}(\mathbf{xy}^\top)^2.$$

Hence, $\mathscr{A}$ is M-PD or M-PSD if $\mathscr{B}$ is S-PD or S-PSD. We now define two new tensor sets such that

$$\mathbb{T}_{\mathscr{A}} := \{\mathscr{T} \;:\; t_{ijkl} = t_{klij},\; t_{ijkl} + t_{klij} = 2a_{ijkl}\},$$

$$\mathbb{S} := \{\mathscr{T} \;:\; t_{ijkl} = t_{klij},\; \mathscr{T} \text{ is S-PSD}\}.$$

Then we have the following result.

Theorem 8.7 *Suppose $\mathscr{A} \in \mathbb{E}_{4,3}$, the following results hold:*

(1) if $\mathbb{T}_{\mathscr{A}} \cap \mathbb{S} \neq \emptyset$, then $\mathscr{A}$ is M-positive semi-definite;
(2) if $\mathbb{T}_{\mathscr{A}} \cap (\mathbb{S} \backslash \partial\mathbb{S}) \neq \emptyset$, then $\mathscr{A}$ is M-positive definite.

In addition, an alternating projection method is proposed in [85] to check whether the intersection of two closed convex sets is empty or not. Assume P_1 and P_2 are projection operators onto $\mathbb{T}_{\mathscr{A}}$ and $\mathbb{S}$ respectively such that

$$\begin{cases} \mathscr{B}^{t+1} = P_2(\mathscr{A}^t), \\ \mathscr{A}^{t+1} = P_1(\mathscr{B}^{t+1}), \end{cases} \quad t = 0, 1, 2, \cdots.$$

Here, for any given tensor $\mathscr{A} \in \mathbb{E}_{4,3}$, we can take $\mathscr{A}^0 = \mathscr{A}$. So the mainly iterative process can be defined such that

Algorithm 8.1 For given tensor $\mathscr{A} \in \mathbb{E}_{4,3}$, set $\mathscr{A}^0 = \mathscr{A}$. Let $A^{(t)}$ and $B^{(t)}$ be the unfolding matrices of $\mathscr{A}^t$ and $\mathscr{B}^t$, respectively.

Step 0. Give the eigenvalue decomposition of matrix $A^{(t)}$ such that

$$A^t = V^{(t)} D^{(t)} (V^{(t)})^\top;$$

Step 1. let $B^{t+1} = V^{(t)} D_+^{(t)} (V^{(t)})^\top$, where $D_+^{(t)} = \text{diag}\,(\max\{d_{ii}^{(t)}, 0\})$;
Step 2. set

$$a_{ijil}^{(t+1)} = a_{ijil} \ \text{ for } i, j, l = 1, 2, 3, \ \ a_{ijkj}^{(t+1)} = a_{ijkj} \ \text{ for } i, j, k = 1, 2, 3;$$

$$a_{ijkl}^{(t+1)} = a_{ijkl} + \frac{1}{2}(b_{ijkl}^{(t+1)} - b_{kjil}^{(t+1)}) \ \text{ for } i \neq k, j \neq l;$$

Step 3. take $t = t + 1$ and return to **Step 0**.
For the convergence of the algorithm, we have the following results.

Theorem 8.8 *Suppose $\mathscr{A} \in \mathbb{E}_{4,3}$ is a given tensor. If $\mathbb{T}_{\mathscr{A}} \cap \mathbb{S} \neq \emptyset$. Then the sequences $\{\mathscr{A}^t\}$ and $\{\mathscr{B}^t\}$ produced by Algorithm 8.1 both converge to a point $\mathscr{A}^* \in \mathbb{T}_{\mathscr{A}} \cap \mathbb{S}$.*

Next, we will study some further sufficient conditions for the strong ellipticity condition. For the tensor $\mathscr{A} \in \mathbb{E}_{4,3}$, we can always compute the eigenvalue factorization of its unfolding matrix $A \in \mathbf{R}^{9\times 9}$ such that $A = \sum_{s=1}^r \alpha_s \mathbf{u}_s \mathbf{u}_s^\top$, which implies that

$$\mathscr{A}\mathbf{y}^2 = \sum_{s=1}^r \alpha_s (U_s \mathbf{y})(U_s \mathbf{y})^\top, \tag{8.21}$$

with $\mathbf{u}_s$ is the unfolding vector of U_s. Note that the coefficients α_s in (8.21) are not necessarily positive, otherwise that reduces the case discussed above. In fact, the orthogonality of $\mathbf{u}_s$ is not required in the following discussion, and the number of terms r may also be larger than nine. In the following conclusions, without loss of generality, we assume that $\alpha_1, \ldots, \alpha_q > 0$ and $\alpha_{q+1}, \ldots, \alpha_r < 0$.

Theorem 8.9 *Let $\mathscr{A} \in \mathbb{E}_{4,3}$ be a given tensor defined in (8.21). Then we have the following results.*

(1) Suppose $q = 3$ and $\alpha_1, \alpha_2, \alpha_3 > 0$ and $\alpha_4, \ldots, \alpha_r < 0$. Let $U_s = \mathbf{v}_s \mathbf{w}_s^\top$ for $s \in [3]$. If $V = (\mathbf{v}_1, \mathbf{v}_2, \mathbf{v}_3)$ and $W = (\mathbf{w}_1, \mathbf{w}_2, \mathbf{w}_3)$ are nonsingular, then $\mathscr{A}$ is MPSD if and only if (i) $U_s = V \Sigma_s W^\top$ with $\Sigma_s = \text{diag}\,(\sigma_{s1}, \sigma_{s2}, \sigma_{s3})$, $s = 4, \ldots, r$ and (ii) the matrix $\text{diag}(\alpha_1, \alpha_2, \alpha_3) + \sum_{s=4}^r \alpha_s \sigma_s \sigma_s^\top$ is positive semidefinite, where $\sigma_s = (\sigma_{s1}, \sigma_{s2}, \sigma_{s3})^\top$.
(2) If $r = 7$, $q = 6$, suppose $\alpha_s > 0$, $s \in [6]$ and $\alpha_7 < 0$. Assume that $U_s = \mathbf{v}_s \mathbf{w}_s^\top$ for $s \in [6]$ and $\mathbf{v}_s = \mathbf{v}_{s+3}$ for $s = 1, 2, 3$. Suppose $V = (\mathbf{v}_1, \mathbf{v}_2, \mathbf{v}_3)$, $W = (\mathbf{w}_1, \mathbf{w}_2, \mathbf{w}_3)$ and $\widetilde{W} = (\mathbf{w}_4, \mathbf{w}_5, \mathbf{w}_6)$ are nonsingular matrices, and suppose $\mathbf{w}_s$ and $\mathbf{w}_{s+3}$, $s \in [3]$ are linearly independent. Then $\mathscr{A}$ is MPSD if and only if (i) $U_7 = V\Sigma_7 W^\top + V\widetilde{\Sigma}_7 \widetilde{W}^\top$

with $\Sigma_7 = diag(\sigma_1, \sigma_2, \sigma_3)$ *and* $\widetilde{\Sigma}_7 = diag(\sigma_4, \sigma_5, \sigma_6)$, *and (ii)* $\sup\{\eta(\mathbf{y}) : \mathbf{y} \notin \bigcup_{s\in[3]}(\mathbf{w}_s^{\perp} \cap \mathbf{w}_{s+3}^{\perp})\} \leq \frac{1}{-\alpha_7}$, *where* $\eta(\mathbf{y})$ *is defined by*

$$\eta(\mathbf{y}) := \sum_{s=1}^{3} \frac{(\sigma_s \mathbf{w}_s^\top \mathbf{y} + \sigma_{s+3}\mathbf{w}_{s+3}^\top \mathbf{y})^2}{\alpha_s(\mathbf{w}_s^\top \mathbf{y})^2 + \alpha_{s+3}(\mathbf{w}_{s+3}^\top \mathbf{y})^2}.$$

(3) If $r = 10, q = 9$ *such that* $\alpha_s > 0, s \in [9]$ *and* $\alpha_{10} < 0$, *assume that* $U_s = \mathbf{v}_s\mathbf{w}_s^\top$ *for* $s \in [9]$ *and* $\mathbf{v}_{s+6} = \mathbf{v}_{s+3}$ *for* $s = 1, 2, 3$. *Suppose that* $V = (\mathbf{v}_1, \mathbf{v}_2, \mathbf{v}_3)$, $W = (\mathbf{w}_1, \mathbf{w}_2, \mathbf{w}_3)$, $\widetilde{W} = (\mathbf{w}_4, \mathbf{w}_5, \mathbf{w}_6)$ *and* $\widehat{W} = (\mathbf{w}_7, \mathbf{w}_8, \mathbf{w}_9)$ *are nonsingular matrices, and suppose that* $\mathbf{w}_s, \mathbf{w}_{s+3}, \mathbf{w}_{s+6}, s \in [3]$ *are linearly independent. Then* $\mathscr{A}$ *is MPSD if and only if (i)* $U_{10} = V\Sigma_{10}\mathscr{W}^\top + V\widetilde{\Sigma}_{10}\widetilde{\mathscr{W}}^\top + V\widehat{\Sigma}_{10}\widehat{\mathscr{W}}^\top$ *with* $\Sigma_{10} = diag(\sigma_1, \sigma_2, \sigma_3)$, $\widetilde{\Sigma}_{10} = diag(\sigma_4, \sigma_5, \sigma_6)$, $\widehat{\Sigma}_{10} = diag(\sigma_7, \sigma_8, \sigma_9)$, *and (ii)* $\max\{\eta(\mathbf{y}) : \mathbf{y}^\top\mathbf{y} = 1\} \leq \frac{1}{-\alpha_{10}}$, *where* $\eta(\mathbf{y})$ *is defined by*

$$\eta(\mathbf{y}) := \sum_{s=1}^{3} \frac{(\sigma_s \mathbf{w}_s^\top \mathbf{y} + \sigma_{s+3}\mathbf{w}_{s+3}^\top \mathbf{y} + \sigma_{s+6}\mathbf{w}_{s+6}^\top \mathbf{y})^2}{\alpha_s(\mathbf{w}_s^\top \mathbf{y})^2 + \alpha_{s+3}(\mathbf{w}_{s+3}^\top \mathbf{y})^2 + \alpha_{s+6}(\mathbf{w}_{s+6}^\top \mathbf{y})^2}.$$

It should be noted that $\mathscr{A} \in \mathbb{E}_{4,3}$ is M-PD when the strict inequality holds in the last condition.

Next, we will introduce a big class of structured tensors satisfying strong ellipticity. Recent years, all kinds of structured tensors were studied based on the notion of tensor eigenvalues, which was defined by Qi [221] and Lim [177] independently. In the structure point of view, fourth order three dimensional elasticity tensors is another kinds of structured tensors. Similarly, some kinds of structured tensors can be defined based on the definition of M-eigenvalue for elasticity tensors. In the following analysis, we mainly study a new kind of tensor named elasticity M-tensors. Furthermore, several equivalent definitions of nonsingular elasticity M-tensors are given.

Let $\mathscr{E} = (e_{ijkl}) \in \mathbb{E}_{4,n}$ be defined by

$$e_{ijkl} = \begin{cases} 1, \text{ if } i = j \text{ and } k = l, \\ 0, \text{ otherwise.} \end{cases} \tag{8.22}$$

For the sake of simplicity, we call $\mathscr{E}$ above the identity tensor since it serves as an identity element in $\mathbb{E}_{4,n}$.

By a direct computation, we obtain that

$$\mathscr{E} \cdot \mathbf{yxy} = \mathbf{x}(\mathbf{y}^\top\mathbf{y}), \quad \mathscr{E}\mathbf{xyx}\cdot = (\mathbf{x}^\top\mathbf{x})\mathbf{y}.$$

By Definition 8.1, for any $\mathscr{A} \in \mathbb{E}_{4,n}$, we can define the M-eigenvalue for any tensors $\mathscr{A} \in \mathbb{E}_{4,n}$ such that

$$\begin{cases} \mathscr{A} \cdot \mathbf{yxy} = \lambda\mathscr{E} \cdot \mathbf{yxy}, \\ \mathscr{A}\mathbf{xyx}\cdot = \lambda\mathscr{E}\mathbf{xyx}\cdot, \end{cases} \tag{8.23}$$

where $\lambda \in \mathbf{R}$, $\mathbf{x} \in \mathbf{R}^n\backslash\{\mathbf{0}\}$, and $\mathbf{y} \in \mathbf{R}^n\backslash\{\mathbf{0}\}$. Noted that (8.23) is exactly the KKT condition of the optimization problem (8.4). In addition, we can get the following result from the definition of identity tensor $\mathscr{E}$.

Proposition 8.1 *Assume $\mathscr{A} \in \mathbb{E}_{4,n}$ is a given tensor. Let $\mathscr{B} = \alpha(\mathscr{A} + \beta\mathscr{E})$, where $\alpha, \beta \in \mathbf{R}$. It holds that μ is an M-eigenvalue of $\mathscr{B}$ if and only if $\mu = \alpha(\lambda + \beta)$ and λ is an M-eigenvalue of $\mathscr{A}$. Furthermore, λ and μ correspond to the same M-eigenvectors.*

For any tensor $\mathscr{A} \in \mathbb{E}_{4,n}$, let $\sigma_M(\mathscr{A})$ be the set containing all M-eigenvalues of $\mathscr{A}$. Define $\rho_M(\mathscr{A})$ as the M-spectral radius of $\mathscr{A}$:

$$\rho_M(\mathscr{A}) := \max\{|\lambda| \, : \, \lambda \in \sigma_M(\mathscr{A})\}.$$

On the other hand, let $\lambda_{\max}(\mathscr{A})$ and $\lambda_{\min}(\mathscr{A})$ be the maximal and the minimal M-eigenvalues of $\mathscr{A}$, respectively. Then by the optimization theory, one can easily get that

$$\begin{aligned} \lambda_{\max}(\mathscr{A}) &= \max\left\{\mathscr{A}\mathbf{xyxy} \, : \, \mathbf{x}, \ \mathbf{y} \in \mathbf{R}^n, \ \mathbf{x}^\top\mathbf{x} = \mathbf{y}^\top\mathbf{y} = 1\right\}, \\ \lambda_{\min}(\mathscr{A}) &= \min\left\{\mathscr{A}\mathbf{xyxy} \, : \, \mathbf{x}, \ \mathbf{y} \in \mathbf{R}^n, \ \mathbf{x}^\top\mathbf{x} = \mathbf{y}^\top\mathbf{y} = 1\right\}. \end{aligned} \tag{8.24}$$

Similar to the Perron–Frobenius theorem for nonnegative tensors (details see Chap. 3 of [228]), we have the following result.

Theorem 8.10 *Let $\mathscr{A} = (a_{ijkl}) \in \mathbb{E}_{4,n}$ be a nonnegative tensor i.e. $a_{ijkl} \geq 0$ for all $i, j, k, l \in [n]$. Then the following results hold.*

(1) The M-spectral radius $\rho_M(\mathscr{A})$ is exactly the gretest M-eigenvalue of $\mathscr{A}$.
(2) There is a pair of nonnegative M-eigenvectors corresponding to the M-spectral radius $\rho_M(\mathscr{A})$.

Combining (8.24) with Theorem 8.10, the following corollary holds automatically.

Corollary 8.1 *Suppose $\mathscr{A} \in \mathbb{E}_{4,n}$ is nonnegative, then it follows that*

$$\rho_M(\mathscr{A}) := \max\left\{\mathscr{A}\mathbf{xyxy} \, : \, \mathbf{x}, \mathbf{y} \in \mathbf{R}^n_+, \ \mathbf{x}^\top\mathbf{x} = \mathbf{y}^\top\mathbf{y} = 1\right\}.$$

For the sake of description, for any tensor $\mathscr{A} = (a_{ijkl}) \in \mathbb{E}_{4,n}$, we call the entries $a_{ijij}, i, j \in [n]$ diagonal entries of $\mathscr{A}$, and other entries are called off-diagonal entries. From a direct computation such that, for any $i, j \in [n]$

$$\mathscr{A}\mathbf{1}_i\mathbf{1}_j\mathbf{1}_i\mathbf{1}_j = a_{ijij},$$

we know that all diagonal entries are positive (nonnegative) if $\mathscr{A}$ is M-PD (M-PSD). Now, we give the following definition, which is similar to the usual notion of Z-tensors [228].

Definition 8.2 Let $\mathscr{A} \in \mathbb{E}_{4,n}$. Then $\mathscr{A}$ is called an **elasticity Z-tensor** if all its off-diagonal entries are nonpositive.

By this definition, we obtain that $\mathscr{A} \in \mathbb{E}_{4,n}$ is an elasticity Z-tensor if and only if it can be written as

$$\mathscr{A} = s\mathscr{E} - \mathscr{B},$$

where $s \in \mathbf{R}$ and $\mathscr{B} \in \mathbb{E}_{4,n}$ is a nonnegative tensor. If the elasticity Z-tensor $\mathscr{A} = s\mathscr{E} - \mathscr{B} \in \mathbb{E}_{4,n}$ satisfies that $s \geq \rho_M(\mathscr{A})$ ($s > \rho_M(\mathscr{A})$), then $\mathscr{A}$ is called an elasticity M-tensor (nonsingular elasticity M-tensor).

Theorem 8.11 *Suppose* $\mathscr{A} = (a_{ijkl}) \in \mathbb{E}_{4,n}$ *is an elasticity Z-tensor. Let*

$$\alpha = \max\left\{a_{ijij} : i, j = 1, 2, \ldots, n\right\}.$$

Then the following results hold.

(1) $\mathscr{A}$ *is a nonsingular elasticity M-tensor if and only if* $\alpha > \rho_M(\alpha\mathscr{E} - \mathscr{A})$.
(2) $\mathscr{A}$ *is an elasticity M-tensor if and only if* $\mathscr{A} + t\mathscr{E}$ *is a nonsingular elasticity M-tensor for any* $t > 0$.
(3) $\mathscr{A}$ *is a nonsingular elasticity M-tensor if and only if* $\mathscr{A}$ *is M-positive definite.*
(4) $\mathscr{A}$ *is an elasticity M-tensor if and only if* $\mathscr{A}$ *is M-positive semidefinite.*

Recall the unfolded matrix (8.18) for tensors $\mathscr{A} \in \mathbb{E}_{4,3}$, we can extend the unfolded matrix to any tensors $\mathscr{A} = (a_{ijkl}) \in \mathbb{E}_{4,n}$ in the following two ways such that

$$\mathbf{A}_\mathbf{x} = \begin{bmatrix} \mathscr{A}(:,1,:,1) & \mathscr{A}(:,1,:,2) & \cdots & \mathscr{A}(:,1,:,n) \\ \mathscr{A}(:,2,:,1) & \mathscr{A}(:,2,:,2) & \cdots & \mathscr{A}(:,2,:,n) \\ \vdots & \vdots & \ddots & \vdots \\ \mathscr{A}(:,n,:,1) & \mathscr{A}(:,n,:,2) & \cdots & \mathscr{A}(:,n,:,n) \end{bmatrix} \in \mathbf{R}^{n^2 \times n^2},$$

$$\mathbf{A}_\mathbf{y} = \begin{bmatrix} \mathscr{A}(1,:,1,:) & \mathscr{A}(1,:,2,:) & \cdots & \mathscr{A}(1,:,n,:) \\ \mathscr{A}(2,:,1,:) & \mathscr{A}(2,:,2,:) & \cdots & \mathscr{A}(2,:,n,:) \\ \vdots & \vdots & \ddots & \vdots \\ \mathscr{A}(n,:,1,:) & \mathscr{A}(n,:,2,:) & \cdots & \mathscr{A}(n,:,n,:) \end{bmatrix} \in \mathbf{R}^{n^2 \times n^2}.$$

Thus, for any $\mathbf{x}, \mathbf{y} \in \mathbf{R}^n$, it holds that

$$\mathscr{A}\mathbf{xyxy} = (\mathbf{y} \odot \mathbf{x})^\top \mathbf{A}_x (\mathbf{y} \odot \mathbf{x}) = (\mathbf{x} \odot \mathbf{y})^\top \mathbf{A}_y (\mathbf{x} \odot \mathbf{y}),$$

where $\odot$ denotes the Kronecker product. Therefore, an elasticity M-tensor is M-PD (M-PSD) if the corresponding matrix $A_\mathbf{x}$ is positive definite (positive semidefinite) and $A_\mathbf{y}$ is positive definite (positive semidefinite). The following example shows that the converse may not be true.

Example 8.1 Let $\mathscr{A} \in \mathbb{E}_{4,2}$ be a given elasticity M-tensor with entries such that

$$\begin{aligned} a_{1111} &= 13, \ a_{1212} = 2, \quad a_{2121} = 2, \\ a_{2222} &= 12, \ a_{1112} = -2, \ a_{1121} = -2, \\ a_{1122} &= -4, \ a_{1222} = -1, \ a_{2122} = -1. \end{aligned}$$

With the help of Mathematica, we obtain that $\mathscr{A}$ has six M-eigenvalues: 13.4163, 12.1118, 11.2036, 6.1778, 0.2442, and 0.1964. Thus $\mathscr{A}$ is a nonsingular elasticity M-tensor since the minimal M-eigenvalue of $\mathscr{A}$ is positive. Nonetheless, the unfolding matrices of A are

$$\mathbf{A}_{\mathrm{x}} = \mathbf{A}_{\mathrm{y}} = \begin{bmatrix} 13 & -2 & -2 & -4 \\ -2 & 2 & -4 & -1 \\ -2 & -4 & 2 & -1 \\ -4 & -1 & -1 & 12 \end{bmatrix},$$

with four eigenvalues: −2.8331, 6.0000, 9.2221, and 16.6110, which implies that A_{x} and A_{y} are not positive semidefinite and thus not M-matrices. □

By the two unfolded matrices A_{x} and A_{y} above, we define the following two matrices such that

$$(\mathscr{A}\mathbf{x}^2\cdot) = \begin{bmatrix} \mathbf{x}^\top\mathscr{A}(:,1,:,1)\mathbf{x} & \mathbf{x}^\top\mathscr{A}(:,1,:,2)\mathbf{x} & \cdots & \mathbf{x}^\top\mathscr{A}(:,1,:,n)\mathbf{x} \\ \mathbf{x}^\top\mathscr{A}(:,2,:,1)\mathbf{x} & \mathbf{x}^\top\mathscr{A}(:,2,:,2)\mathbf{x} & \cdots & \mathbf{x}^\top\mathscr{A}(:,2,:,n)\mathbf{x} \\ \vdots & \vdots & \ddots & \vdots \\ \mathbf{x}^\top\mathscr{A}(:,n,:,1)\mathbf{x} & \mathbf{x}^\top\mathscr{A}(:,n,:,2)\mathbf{x} & \cdots & \mathbf{x}^\top\mathscr{A}(:,n,:,n)\mathbf{x} \end{bmatrix},$$

$$(\mathscr{A}\cdot\mathbf{y}^2) = \begin{bmatrix} \mathbf{y}^\top\mathscr{A}(1,:,1,:)\mathbf{y} & \mathbf{y}^\top\mathscr{A}(1,:,2,:)\mathbf{y} & \cdots & \mathbf{y}^\top\mathscr{A}(1,:,n,:)\mathbf{y} \\ \mathbf{y}^\top\mathscr{A}(2,:,1,:)\mathbf{y} & \mathbf{y}^\top\mathscr{A}(2,:,2,:)\mathbf{y} & \cdots & \mathbf{y}^\top\mathscr{A}(2,:,n,:)\mathbf{y} \\ \vdots & \vdots & \ddots & \vdots \\ \mathbf{y}^\top\mathscr{A}(n,:,1,:)\mathbf{y} & \mathbf{y}^\top\mathscr{A}(n,:,2,:)\mathbf{y} & \cdots & \mathbf{y}^\top\mathscr{A}(n,:,n,:)\mathbf{y} \end{bmatrix}.$$

Then, we have the following conclusions.

Theorem 8.12 *Suppose $\mathscr{A} \in \mathbb{E}_{4,n}$ is an elasticity Z-tensor. The following results hold.*

(1) $\mathscr{A}$ is a nonsingular elasticity M-tensor if and only if $(\mathscr{A}\mathbf{x}^2\cdot)$ is a nonsingular M-matrix for each $\mathbf{x} \geq \mathbf{0}, \mathbf{x} \neq \mathbf{0}$.
(2) $\mathscr{A}$ is an elasticity M-tensor if and only if $(\mathscr{A}\mathbf{x}^2\cdot)$ is an M-matrix for each $\mathbf{x} \geq 0$.
(3) $\mathscr{A}$ is a nonsingular elasticity M-tensor if and only if $(\mathscr{A}\cdot\mathbf{y}^2)$ is a nonsingular M-matrix for each $\mathbf{y} \geq \mathbf{0}, \mathbf{y} \neq \mathbf{0}$.
(4) $\mathscr{A}$ is an elasticity M-tensor if and only if $(\mathscr{A}\cdot\mathbf{y}^2)$ is an M-matrix for each $\mathbf{y} \geq \mathbf{0}$.

In [82], it proved that a Z-tensor is a nonsingular M-tensor if and only if there exists a positive vector $\mathbf{x}$ such that $\mathscr{A}\mathbf{x}^{m-1}$ is positive. Now, we extend the results to nonsingular elasticity M-tensors.

Theorem 8.13 *Suppose $\mathscr{A} \in \mathbb{E}_{4,n}$ is an elasticity Z-tensor. The following conditions are equivalent:*

(1) $\mathscr{A}$ is a nonsingular elasticity M-tensor;
(2) For each $\mathbf{x} \geq \mathbf{0}$, $\mathbf{x} \neq \mathbf{0}$, there exists $\mathbf{y} > \mathbf{0}$ such that $\mathscr{A}\mathbf{xyx}\cdot > \mathbf{0}$;
(3) For each $\mathbf{x} \geq \mathbf{0}$, $\mathbf{x} \neq \mathbf{0}$, there exists $\mathbf{y} \geq \mathbf{0}$, $\mathbf{y} \neq \mathbf{0}$, such that $\mathscr{A}\mathbf{xyx}\cdot > \mathbf{0}$;
(4) For each $\mathbf{y} \geq \mathbf{0}$, $\mathbf{y} \neq \mathbf{0}$, there exists $\mathbf{x} > \mathbf{0}$ such that $\mathscr{A} \cdot \mathbf{yxy} > \mathbf{0}$;
(5) For each $\mathbf{y} \geq \mathbf{0}$, $\mathbf{y} \neq \mathbf{0}$, there exists $\mathbf{x} \geq \mathbf{0}$, $\mathbf{x} \neq \mathbf{0}$, such that $\mathscr{A} \cdot \mathbf{yxy} > \mathbf{0}$.

By a direct computation, the condition (2) in Theorem 8.13 implies that $(\mathscr{A}\mathbf{x}^2\cdot)\mathbf{y} = \mathscr{A}\mathbf{xyx}\cdot > \mathbf{0}$. Let $\mathbf{y} = (y_1, y_2, \ldots, y_n)^\top > \mathbf{0}$ and $D = \text{diag}(y_1, \ldots, y_n) \in \mathbf{R}^{n\times n}$. Denote $\widetilde{A} := (\mathscr{A}\mathbf{x}^2\cdot)D$. If $\mathscr{A}$ is an elasticity Z-tensor, we know that $\widetilde{A}$ is also a Z-matrix, and

$$|\widetilde{a}_{ii}| - \sum_{j\neq i} |\widetilde{a}_{ij}| = \widetilde{a}_{ii} + \sum_{j\neq i} \widetilde{a}_{ij} = (\mathscr{A}\mathbf{xyx}\cdot)_i > 0, \quad \text{for } i \in [n].$$

Therefore $\widetilde{A}$ is a strictly diagonally dominant matrix. Based on this, we have the following result.

Corollary 8.2 *Suppose $\mathscr{A} \in \mathbb{E}_{4,n}$ is an elasticity Z-tensor. The following conditions hold.*

(1) $\mathscr{A}$ is a nonsingular elasticity M-tensor if and only if or each $\mathbf{x} \geq \mathbf{0}$, $\mathbf{x} \neq \mathbf{0}$, there exists a positive diagonal matrix D such that $D(\mathscr{A}\mathbf{x}^2\cdot)D$ is strictly diagonally dominant;
(2) $\mathscr{A}$ is a nonsingular elasticity M-tensor if and only if for each $\mathbf{y} \geq 0$, $\mathbf{y} \neq \mathbf{0}$, there exists a positive diagonal matrix D such that $D(\mathscr{A} \cdot \mathbf{y}^2)D$ is strictly diagonally dominant.

8.4 Computational Methods for M-Eigenvalues

From the previous discussion, we know that the M-eigenvalue plays an important role in checking the strong ellipticity of tensors arising in nonlinear elastic materials. Hence, any numerical method for computing the minimum M-eigenvalue of a given elasticity tensor is very applicable in the study of nonlinear elasticity and materials. In this section, we will study a practical algorithm for computing the largest M-eigenvalue of fourth order tensors, which can also be used to compute the smallest M-eigenvalue of the proposed tensors. Hence it can be used to check whether the given tensor satisfies the strong ellipticity or not.

For finding the largest M-eigenvalue of a tensor satisfying (8.2), we mainly aims to consider the following optimization problem:

$$\begin{array}{ll} \max & f(\mathbf{x},\mathbf{y}) = \sum_{i,k=1}^{m}\sum_{j,l=1}^{n} a_{ijkl}x_i y_j x_k y_l \\ s.t. & \mathbf{x}^\top\mathbf{x} = 1,\ \mathbf{y}^\top\mathbf{y} = 1 \\ & \mathbf{x}\in\mathbf{R}^m,\ \mathbf{y}\in\mathbf{R}^n. \end{array} \tag{8.25}$$

And the main frame of the algorithm is as follows.

Algorithm 8.2 Initialization Step. Take initial points $\mathbf{x}_0 \in \mathbf{R}^m$, $\mathbf{y}_0 \in \mathbf{R}^n$, and let $k = 0$;
Iterative Step. Execute the following procedure alternatively until convergence:

$$\bar{\mathbf{x}}^{(k+1)} = \mathscr{A}\cdot\mathbf{y}^{(k)}\mathbf{x}^{(k)}\mathbf{y}^{(k)}, \quad \mathbf{x}^{(k+1)} = \frac{\bar{\mathbf{x}}^{(k+1)}}{\|\bar{\mathbf{x}}^{(k+1)}\|};$$

$$\bar{\mathbf{y}}^{(k+1)} = \mathscr{A}\mathbf{x}^{(k+1)}\mathbf{y}^{(k)}\mathbf{x}^{(k+1)}\cdot, \quad \mathbf{y}^{(k+1)} = \frac{\bar{\mathbf{y}}^{(k+1)}}{\|\bar{\mathbf{y}}^{(k+1)}\|};$$

$$k = k+1.$$

For any given fourth order tensor $\mathscr{A} = (a_{ijkl})$ satisfying (8.2), similar to discussions in the previous section, let $B(\mathbf{y}) \in \mathbf{R}^{m\times m}$ and $C(\mathbf{x}) \in \mathbf{R}^{n\times n}$ are two symmetric matrices such that

$$(B(\mathbf{y}))_{ik} = \sum_{j,l=1}^{n} a_{ijkl}y_j y_l, \quad (C(\mathbf{x}))_{jl} = \sum_{i,k=1}^{m} a_{ijkl}x_i x_k.$$

So, based on these two matrices, the following conclusion holds.

Theorem 8.14 *For any given tensor $\mathscr{A}$ satisfying (8.2) and $\mathbf{x} \in \mathbf{R}^m, \mathbf{y} \in \mathbf{R}^n$, assume that $B(\mathbf{y})$ and $C(\mathbf{x})$ are both positive definite. Then the generated sequence $\{f(\mathbf{x}^{(k)},\mathbf{y}^{(k)}) = \mathscr{A}\mathbf{x}^{(k)}\mathbf{y}^{(k)}\mathbf{x}^{(k)}\mathbf{y}^{(k)}\}$ in Algorithm 8.2 is nondecreasing.*

Proof By conditions that $B(\mathbf{y})$ and $C(\mathbf{x})$ are both positive definite, it follows that the function $f(\mathbf{x},\mathbf{y})$ is strictly convex corresponding to $\mathbf{x} \in \mathbf{R}^m$ and $\mathbf{y} \in \mathbf{R}^n$ respectively. Hence, for any $k \geq 0$, we obtain

$$f(\mathbf{x}^{(k+1)},\mathbf{y}^{(k)}) - f(\mathbf{x}^{(k)},\mathbf{y}^{(k)}) \geq \langle \mathbf{x}^{(k+1)} - \mathbf{x}^{(k)}, \nabla_{\mathbf{x}} f(\mathbf{x}^{(k)},\mathbf{y}^{(k)})\rangle. \tag{8.26}$$

Since $\nabla_{\mathbf{x}} f(\mathbf{x}^{(k)},\mathbf{y}^{(k)}) = 2\mathscr{A}\cdot\mathbf{y}^{(k)}\mathbf{x}^{(k)}\mathbf{y}^{(k)}$, by Algorithm 8.2, one has

$$\mathbf{x}^{(k+1)} = \frac{\nabla_{\mathbf{x}} f(\mathbf{x}^{(k)},\mathbf{y}^{(k)})}{\|\nabla_{\mathbf{x}} f(\mathbf{x}^{(k)},\mathbf{y}^{(k)})\|}.$$

Combining this with (8.27), recalling the Cauchy–Schwartz inequality, it holds that $f(\mathbf{x}^{(k+1)},\mathbf{y}^{(k)}) \geq f(\mathbf{x}^{(k)},\mathbf{y}^{(k)})$ and the strict inequality holds when $\mathbf{x}^{(k+1)} \neq \mathbf{x}^{(k)}$.

On the other hand, it is not difficult to prove that $f(\mathbf{x}^{(k+1)},\mathbf{y}^{(k+1)}) \geq f(\mathbf{x}^{(k+1)},\mathbf{y}^{(k)})$ and the strict inequality holds when $\mathbf{y}^{(k+1)} \neq \mathbf{y}^{(k)}$, and the desired results hold. □

From Theorem 8.14, we know that the generated sequence converges to a stationary point of problem (8.25) in the case that $B(\mathbf{y})$ and $C(\mathbf{x})$ are both positive definite, which is equivalent with the condition that tensor $\mathscr{A}$ is M-PD. Now, a question can be raised naturally: how can we do when either $B(\mathbf{y})$ or $C(\mathbf{x})$ is not positive definite. Actually, one can take a real scalar $\tau \in \mathbf{R}$ such that

$$\tau > \max\{|\lambda| \ : \ \lambda \text{ is an M-eigenvalue of tensor } \mathscr{A}\}.$$

By the identity tensor $\mathscr{E}$ defined as in (8.22), let $\bar{f}(\mathbf{x}, \mathbf{y}) = \tau\mathscr{E}\mathbf{xyxy} + \mathscr{A}\mathbf{xyxy} = \bar{\mathscr{A}}\mathbf{xyxy}$. It is obvious that if $\mathbf{x}$ and $\mathbf{y}$ constitute a pair of M-eigenvectors of tensor $\mathscr{A}$ associated with M-eigenvalue λ, then they are also a pair of M-eigenvectors of tensor $\bar{\mathscr{A}}$ associated with M-eigenvalue $\tau+\lambda$. So, we can always compute the largest M-eigenvalue of $\mathscr{A}$ since $\bar{f}(\mathbf{x}, \mathbf{y})$ satisfies the assumption of Theorem 8.14.

Next, we will show how to choose a suitable scalar τ. In fact, this can be done according to the estimation of the largest eigenvalue of the unfolded matrix of tensor $\mathscr{A}$ defined below, which is different with the unfolding matrix discussed in Sect. 8.3. For any elasticity tensor satisfying (8.2), define the unfolding matrix $A = (A_{st}) \in \mathbf{R}^{mn\times mn}$ such that

$$A_{st} = a_{ijkl}, \quad \text{when } s = n(i-1)+j, \ t = n(k-1)+l.$$

With the help of the Kronecker product $\mathbf{x} \odot \mathbf{y} \in \mathbf{R}^{mn}$, by a direct computation, it follows that

$$f(\mathbf{x}, \mathbf{y}) = \mathscr{A}\mathbf{xyxy} = (\mathbf{x} \odot \mathbf{y})^\top A(\mathbf{x} \odot \mathbf{y}).$$

Then, the following conclusion holds. We omit the proof here since it is simple.

Proposition 8.2 *Assume $\mathscr{A}$ is a given tensor satisfying (8.2). Suppose the unfolded matrix of $\mathscr{A}$ is A. If A is positive definite, then $\mathscr{A}$ is M-PD. Moreover, all the M-eigenvalues of tensor $\mathscr{A}$ lie in the interval composed by the smallest eigenvalue and the largest eigenvalue of A.*

By Theorem 1.1 in Chap. 1, we know that the magnitude of any eigenvalue of the matrix $A = (A_{st})$ must be less than or equal to $\max_{1\le s\le mn}\sum_{t=1}^{mn}|A_{st}|$, which can be easily computed. Actually, it is sufficient to guarantee $\tau\mathscr{E} + \mathscr{A}$ satisfies the assumption of Theorem 8.14 if we choose $\tau = \max_{1\le s\le mn}\sum_{t=1}^{mn}|A_{st}|$.

On the other hand, the following example shows that a fourth order partially symmetric tensor $\mathscr{A}$ may be M-PD but the corresponding unfolded matrix A is not positive definite on $\mathbf{R}^{mn}$.

Example 8.2 Consider the following fourth order two-dimensional $\mathscr{A}$ with all independent entries such that

$$a_{1111} = 12, \ a_{1112} = 1, \ a_{1121} = 1, \ a_{1122} = 5, \ a_{1212} = 2,$$

$$a_{1222} = 1, \ a_{2121} = 2, \ a_{2122} = 1, \ a_{2222} = 12.$$

For this symmetric tensor, the unfolded matrix is

$$A = \begin{pmatrix} 12 & 1 & 1 & 5 \\ 1 & 2 & 5 & 1 \\ 1 & 5 & 2 & 1 \\ 5 & 1 & 1 & 12 \end{pmatrix}.$$

It is not difficult to see that, for any $\mathbf{x}, \mathbf{y} \in \mathbf{R}^2$, the corresponding polynomial holds that

$$\begin{aligned} f(\mathbf{x}, \mathbf{y}) =& \mathscr{A}\mathbf{xyxy} = \sum_{i,j,k,l=1}^{2} a_{ijkl} x_i y_j x_k y_l \\ =& (x_1 y_1 + x_1 y_2)^2 + (x_1 y_2 + x_2 y_2)^2 + (x_2 y_1 + x_1 y_1)^2 + (x_2 y_1 + x_2 y_2)^2 \\ &+ 10(x_1 y_1 + x_2 y_2)^2. \end{aligned}$$

Apparently, we know that $f(\mathbf{x}, \mathbf{y}) > 0$ for any nonzero vectors $\mathbf{x}, \mathbf{y} \in \mathbf{R}^2$. However, for $\omega = (0, 1, -1, 0)^\top$, it holds that $\omega^\top A \omega = -6 < 0$, which means that the unfolded matrix A is not positive definite on $\mathbf{R}^4$. $\square$

As discussed above, the Algorithm 8.2 can generate a stationary point of problem (8.25) generally. However, one cannot confirm whether the accumulation point is the best maximizer or not. Now, we introduce an initialization technique to make Algorithm 8.2 accumulate to a good maximizer of problem (8.25), which was first presented by Wang et al. [280].

By the algebra theory, we know that one of the good maximizer for the function $\mathbf{x}^\top G \mathbf{x}$ over the unit sphere is the unit eigenvector corresponding to the largest eigenvalue of positive definite and symmetric matrix G. For a given tensor $\mathscr{A}$ satisfying the symmetry (8.2) with the unfolded matrix $A \in \mathbf{R}^{mn \times mn}$, assume $\omega \in \mathbf{R}^{mn}$ is a unit eigenvector of A associated with the largest eigenvalue μ. Let $(\mathbf{x}^*, \mathbf{y}^*)$ be a solution of problem (8.25). Then it follows that $f(\mathbf{x}^*, \mathbf{y}^*) \le \mu$ and the equality holds only when $\mathbf{x}^* \odot \mathbf{y}^* = \pm\omega$. Stimulated by this, we may take an initial point in Algorithm 8.2 that maximizes the inner product $\omega^\top(\mathbf{x} \odot \mathbf{y})$ over unit spheres.

In order to solve the subproblem successfully, we have to fold the vector $\omega \in \mathbf{R}^{mn}$ to a matrix $W \in \mathbf{R}^{m \times n}$. For any $k \in [mn]$, denote $i = \lceil \frac{k}{n} \rceil$, $j = (k-1) \bmod n + 1$, and let

$$W_{ij} = \omega_k.$$

Then, from a direct computation, we obtain

$$\langle \mathbf{x} \odot \mathbf{y}, \omega \rangle = \mathbf{x}^\top W \mathbf{y},$$

and the involved subproblem is as follows:

$$\begin{array}{ll} \max & \mathbf{x}^\top W \mathbf{y} \\ s.t. & \mathbf{x}^\top \mathbf{x} = 1, \ \mathbf{y}^\top \mathbf{y} = 1, \end{array}$$

which is equivalent with

$$\begin{array}{ll} \min & \|W - \mu \mathbf{x}\mathbf{y}^\top\|_F^2 \\ s.t. & \mathbf{x}^\top \mathbf{x} = 1, \ \mathbf{y}^\top \mathbf{y} = 1, \\ & \mu \in \mathbf{R}. \end{array}$$

Furthermore, the involved subproblem can be solved successfully by the singular eigenvalue decomposition of the matrix W [105], which means that if it holds that

$$W = U^\top \Sigma V = \sum_{i=1}^{r} \sigma_i \mathbf{u}^{(i)} \mathbf{v}^{(i)\top},$$

where $\sigma_1 > \sigma_2 > \cdots > \sigma_r > 0$, and r is the rank of the matrix, then $\mathbf{u}^{(1)}$ and $\mathbf{v}^{(1)}$ constitute the solution of the subproblem. Hence, the initialization subproblem can be solved by letting $\mathbf{x}^{(0)} = \mathbf{u}^{(1)}$, $\mathbf{y}^{(0)} = \mathbf{v}^{(1)}$.

By the discussion above, we have the following complete version for computing the largest M-eigenvalue of a fourth order partially symmetric tensor.

Algorithm 8.3 **Step 1**: Given $\mathscr{A} = (a_{ijkl})$. Compute its unfolded matrix $A = (A_{ij})$.
1.1: Choose $\tau = \sum_{1 \le i \le j \le mn} |A_{ij}|$ and let $\bar{\mathscr{A}} = \tau \mathscr{E} + \mathscr{A}$. Then unfold $\bar{\mathscr{A}}$ to matrix $\bar{A}$.
1.2: Compute the eigenvector ω of matrix $\bar{A}$ associated with the largest eigenvalue and fold it into the matrix W.
1.3: Compute the singular vectors $\mathbf{u}^{(1)}$ and $\mathbf{v}^{(1)}$ corresponding to the largest singular value of the matrix W.
1.4: Take $\mathbf{x}^0 = \mathbf{u}^{(1)}$, $\mathbf{y}^0 = \mathbf{v}^{(1)}$, and let $k = 0$.
Step 2: Execute the following procedure alternatively until certain convergence criterion is satisfied and output $\mathbf{x}^*$, $\mathbf{y}^*$:

$$\begin{array}{ll} \bar{\mathbf{x}}^{(k+1)} = \mathscr{A} \cdot \mathbf{y}^{(k)} \mathbf{x}^{(k)} \mathbf{y}^{(k)}, & \mathbf{x}^{(k+1)} = \dfrac{\bar{\mathbf{x}}^{(k+1)}}{\|\bar{\mathbf{x}}^{(k+1)}\|}; \\ \bar{\mathbf{y}}^{(k+1)} = \mathscr{A} \mathbf{x}^{(k+1)} \mathbf{y}^{(k)} \mathbf{x}^{(k+1)} \cdot, & \mathbf{y}^{(k+1)} = \dfrac{\bar{\mathbf{y}}^{(k+1)}}{\|\bar{\mathbf{y}}^{(k+1)}\|}; \\ k = k + 1. & \end{array}$$

Step 3: Output the largest M-eigenvalue of tensor $\mathscr{A} : \lambda = f(\mathbf{x}^*, \mathbf{y}^*) - \tau$, and the associated M-eigenvectors: $\mathbf{x}^*$, $\mathbf{y}^*$.

For Algorithm 8.3, the computation complexity at each iterative step is of order $O(m^2 n + mn^2)$. Thus the algorithm can be said to be practical if the largest M-eigenvalue of tensor $\mathscr{A}$ can be generated within a few steps.

To end this section, we show that how to compute the smallest M-eigenvalue with the proposed algorithms in this section for tensors satisfying (8.2). It is obvious that $\alpha\mathcal{E} - A$ also satisfies the partially symmetry condition (8.2) for all $\alpha \in \mathbf{R}$. Hence, for any given tensor $\mathcal{A} \in \mathcal{E}_{4,n}$, we can always choose a sufficient large $\alpha \in \mathbf{R}$ such that $\alpha\mathcal{E} - A$ is M-PD, which implies that the smallest M-eigenvalue of $\mathcal{A}$ equals α minus the largest M-eigenvalue of $\alpha\mathcal{E} - A$. Therefore, the algorithm proposed in this section for computing the largest M-eigenvalue can be used to check the strong ellipticity condition of a given elasticity tensor.

8.5 Higher Order Elasticity Tensors

In this section, we mainly study the M-positive definiteness (semidefiniteness) for higher order elasticity tensors. Furthermore, for a given elasticity tensor, we consider the sum-of-squares (SOS) decomposition of its corresponding homogeneous polynomials. It is obvious that an elasticity tensor is M-positive semidefinite if its corresponding polynomial has an SOS decomposition. Moreover, necessary and sufficient conditions are proposed to guarantee the SOS decomposition.

The notions such that paired symmetric tensor and strongly paired symmetric tensor will be defined, and it is easy to verify that the fourth order three dimensional elasticity tensors studied in the previous sections are all paired symmetric tensors. In addition, some higher order elasticity tensors from solid mechanics and elasticity mechanics are strongly paired symmetric tensors [123, 155, 263]. For simplicity of symbols, we only consider properties of sixth order three dimensional (strongly) paired symmetric tensors, and the obtained conclusions can be naturally extended to the case of more higher order (strongly) paired symmetric tensors.

First of all, we introduce the definition of (strongly) paired symmetric tensor.

Definition 8.3 Let $\mathcal{A} = (a_{i_1 j_1 i_2 j_2 \dots i_m j_m})$ be a $2m$th order n dimensional tensor. The indices for the entries of $\mathcal{A}$ can be divided into m adjacent blocks $\{i_1 j_1\}, \dots, \{i_m j_m\}$. If the entries are invariant under any permutation of indices in every block $\{i_l j_l\}$ for $l \in [m]$, i.e.,

$$a_{i_1 j_1 i_2 j_2 \dots i_m j_m} = a_{j_1 i_1 i_2 j_2 \dots i_m j_m} = a_{i_1 j_1 j_2 i_2 \dots i_m j_m} = \cdots = a_{i_1 j_1 i_2 j_2 \dots j_m i_m},$$

then $\mathcal{A}$ is called a $2m$th order n dimensional paired symmetric tensor. If the tensor further satisfies that the entries are invariant under any permutation of every block, then $\mathcal{A}$ is called a $2m$th order n dimensional strongly paired symmetric tensor.

For the sake of simplicity, let $PS_{2m,n}$ ($SPS_{2m,n}$) denote the set of all $2m$th order n dimensional (strongly) paired symmetric tensors. It should be noted that, the higher order elasticity tensors in [123, 155, 263] are strongly paired symmetric tensors and every entry $a_{i_1 j_1 i_2 j_2 \dots i_m j_m}$ is called an mth order elastic constant, which is an important quantity in studies of elasticity theory.

As discussed in the previous sections, we say a $2m$th order n dimensional paired symmetric tensor $\mathscr{A} = (a_{i_1 j_1 i_2 j_2 \ldots i_m j_m})$ is M-PSD(M-PD) if and only if

$$\begin{aligned}\mathscr{A}\mathbf{xxyy}\cdots\mathbf{zz} &= \mathscr{A}\mathbf{x}^2\mathbf{y}^2\cdots\mathbf{z}^2 \\ &= \sum_{i_1,j_1,\ldots,i_m,j_m\in[n]} a_{i_1 j_1 i_2 j_2 \ldots i_m j_m} x_{i_1} x_{j_1} y_{i_2} y_{j_2} \cdots z_{i_m} z_{j_m} \geq (>)0,\end{aligned}$$

for all (nonzero) $\mathbf{x}, \mathbf{y}, \ldots, \mathbf{z} \in \mathbf{R}^n$.

Particularly, a sixth order three dimensional paired symmetric tensor $\mathscr{A}$ satisfies

$$a_{ijklpq} = a_{jiklpq} = a_{ijlkpq} = a_{ijklqp}, \quad \forall\, i, j, k, l, p, q \in [3]. \tag{8.27}$$

If $\mathscr{A}$ is strongly paired symmetric, it further holds that

$$a_{ijklpq} = a_{klijpq} = a_{ijpqkl}, \quad \forall\, i, j, k, l, p, q \in [3]. \tag{8.28}$$

The smallest M-eigenvalue for any fourth order three dimensional elasticity tensor plays an important role in verifying whether or not the proposed elasticity tensor holds the strong ellipticity. Similar to Definition 8.1, the definition of M-eigenvalue can be extended to the sixth order three dimensional cases. Before that, we introduce the following symbols. For any paired symmetric tensor $\mathscr{A} \in PS_{6,3}$, let $\mathscr{A}\cdot\mathbf{xy}^2\mathbf{z}^2$, $\mathscr{A}\mathbf{x}^2\cdot\mathbf{yz}^2$, $\mathscr{A}\mathbf{x}^2\mathbf{y}^2\cdot\mathbf{z} \in \mathbf{R}^3$ be defined by

$$\begin{aligned}(\mathscr{A}\cdot\mathbf{xy}^2\mathbf{z}^2)_i &:= \sum_{j,k,l,p,q=1}^{3} a_{ijklpq} x_j y_k y_l z_p z_q, \quad \forall\, i \in \{1,2,3\},\\ (\mathscr{A}\mathbf{x}^2\cdot\mathbf{yz}^2)_k &:= \sum_{i,j,l,p,q=1}^{3} a_{ijklpq} x_i x_j y_l z_p z_q, \quad \forall\, k \in \{1,2,3\},\\ (\mathscr{A}\mathbf{x}^2\mathbf{y}^2\cdot\mathbf{z})_p &:= \sum_{i,j,k,l,q=1}^{3} a_{ijklpq} x_i x_j y_k y_l z_q, \quad \forall\, p \in \{1,2,3\}.\end{aligned}$$

Then it can be verified that

$$\langle \mathbf{x}, \mathscr{A}\cdot\mathbf{xy}^2\mathbf{z}^2\rangle = \langle \mathbf{y}, \mathscr{A}\mathbf{x}^2\cdot\mathbf{yz}^2\rangle = \langle \mathbf{z}, \mathscr{A}\mathbf{x}^2\mathbf{y}^2\cdot\mathbf{z}\rangle = \mathscr{A}\mathbf{x}^2\mathbf{y}^2\mathbf{z}^2.$$

Definition 8.4 For any paired symmetric tensor $\mathscr{A} = (a_{ijklpq}) \in PS_{6,3}$, if there exist $\lambda \in \mathbf{R}$ and $\mathbf{x}, \mathbf{y}, \mathbf{z} \in \mathbf{R}^3$ such that

$$\begin{cases}\mathscr{A}\cdot\mathbf{xy}^2\mathbf{z}^2 = \lambda\mathbf{x}, \quad \mathscr{A}\mathbf{x}^2\cdot\mathbf{yz}^2 = \lambda\mathbf{y}, \quad \mathscr{A}\mathbf{x}^2\mathbf{y}^2\cdot\mathbf{z} = \lambda\mathbf{z},\\ \mathbf{x}^\top\mathbf{x} = 1, \quad \mathbf{y}^\top\mathbf{y} = 1, \quad \mathbf{z}^\top\mathbf{z} = 1,\end{cases}$$

then λ is called an M-eigenvalue of $\mathscr{A}$ and $\mathbf{x}$, $\mathbf{y}$, $\mathbf{z}$ are the eigenvectors of $\mathscr{A}$ associated with the M-eigenvalue λ.

From the definition, similar to Theorem 8.1, we have the following results.

Theorem 8.15 *For any (strongly) paired symmetric tensor* $\mathscr{A} = (a_{ijklpq}) \in PS_{6,3}$, *the following results hold:*

(1) *its M-eigenvalues always exist. Moreover, if* $\mathbf{x}, \mathbf{y}, \mathbf{z}$ *are the eigenvectors of* $\mathscr{A}$ *associated with the M-eigenvalue* λ, *then* $\lambda = \mathscr{A}\mathbf{x}^2\mathbf{y}^2\mathbf{z}^2$;
(2) $\mathscr{A}$ *is M-PD if and only if its smallest M-eigenvalue is positive.*

Combining Theorem 8.15 with Definition 8.4, we obtain the following result.

Theorem 8.16 *For any (strongly) paired symmetric tensor* $\mathscr{A} = (a_{ijklpq}) \in PS_{6,3}$,

(1) λ *is an M-eigenvalue of* $\mathscr{A}$ *if and only if* $-\lambda$ *is an M-eigenvalue of* $-\mathscr{A}$.
(2) $\mathscr{A}$ *is positive definite if and only if the largest M-eigenvalue of* $-\mathscr{A}$ *is negative.*

We say that

$$\lambda_*(\mathbf{x}^*)^2(\mathbf{y}^*)^2(\mathbf{z}^*)^2 = \lambda_*\mathbf{x}^* \otimes \mathbf{x}^* \otimes \mathbf{y}^* \otimes \mathbf{y}^* \otimes \mathbf{z}^* \otimes \mathbf{z}^*$$

is the best rank-one approximation of a given tensor $\mathscr{A} \in T_{6,3}$ if $(\lambda_*, x^*, y^*, z^*) \in \mathbf{R} \times \mathbf{R}^3 \times \mathbf{R}^3 \times \mathbf{R}^3$ solves the optimization problem:

$$\begin{aligned} &\min \ \|\mathscr{A} - \lambda\mathbf{x}^2\mathbf{y}^2\mathbf{z}^2\|_F^2 \\ &\text{s.t. } \lambda \in \mathbf{R} \quad \text{and} \quad \mathbf{x}^\top\mathbf{x} = 1, \ \mathbf{y}^\top\mathbf{y} = 1, \ \mathbf{z}^\top\mathbf{z} = 1, \ \forall\, \mathbf{x}, \mathbf{y}, \mathbf{z} \in \mathbf{R}^3. \end{aligned}$$

Based on this, we have the following conclusion for the best rank-one approximation of sixth order three dimensional paired symmetric tensors.

Theorem 8.17 *For any (strongly) paired symmetric tensor* $\mathscr{A} = (a_{ijklpq}) \in PS_{6,3}$, *if* λ_* *is an M-eigenvalue of* $\mathscr{A}$ *with the largest absolute value among all M-eigenvalues of* $\mathscr{A}$, *and* $\mathbf{x}^*, \mathbf{y}^*, \mathbf{z}^* \in \mathbf{R}^3$ *are the eigenvectors of* $\mathscr{A}$ *associated with the M-eigenvalue* λ_*, *then* $\lambda_*(\mathbf{x}^*)^2(\mathbf{y}^*)^2(\mathbf{z}^*)^2$ *is the best rank-one approximation of* $\mathscr{A}$.

Recall the discussion in Sect. 8.2, the M-positive definiteness of a fourth order three dimensional elasticity tensor can be equivalently transferred to check the positive definiteness of three second order tensors, positive definiteness of a fourth order symmetric tensor and a sixth order symmetric tensor. Similarly, the conclusions above can be extended to sixth order (strongly) paired symmetric tensors. To move on, we present three block sub-tensors for a sixth order three dimensional paired symmetric tensor $\mathscr{A} \in PS_{6,3}$. Suppose $\mathscr{A} = (a_{ijklpq}) \in PS_{6,3}$, $i, j, k, l, p, q \in [3]$. Denote block sub-tensors such that

$$\mathscr{A}_{ij} := (a_{ijklpq})_{k,l,p,q\in[3]}, \quad \mathscr{B}_{kl} := (a_{ijklpq})_{i,j,p,q\in[3]} \quad \text{and} \quad \mathscr{C}_{pq} := (a_{ijklpq})_{i,j,k,l\in[3]}. \tag{8.29}$$

Then the following results are obvious.

Proposition 8.3 *For any $\mathscr{A} \in T_{6,3}$, let sub-tensors $\mathscr{A}_{ij}$, $\mathscr{B}_{kl}$ and $\mathscr{C}_{pq}$ be defined by (8.29). Then it holds that*

(1) if $\mathscr{A}$ is a paired symmetric tensor, then all sub-tensors $\mathscr{A}_{ij}$, $\mathscr{B}_{kl}$ and $\mathscr{C}_{pq}$ are paired symmetric tensors.
(2) if $\mathscr{A}$ is a strongly paired symmetric tensor, then $\mathscr{A}_{st} = \mathscr{B}_{st} = \mathscr{C}_{st}$ for all $s, t \in [3]$.

By sub-tensors defined in (8.29), another three matrices can be defined as below

$$
\begin{aligned}
A(\mathbf{y},\mathbf{z}) &:= \begin{pmatrix} \mathscr{A}_{11}\mathbf{y}^2\mathbf{z}^2 & \mathscr{A}_{12}\mathbf{y}^2\mathbf{z}^2 & \mathscr{A}_{13}\mathbf{y}^2\mathbf{z}^2 \\ \mathscr{A}_{21}\mathbf{y}^2\mathbf{z}^2 & \mathscr{A}_{22}\mathbf{y}^2\mathbf{z}^2 & \mathscr{A}_{23}\mathbf{y}^2\mathbf{z}^2 \\ \mathscr{A}_{31}\mathbf{y}^2\mathbf{z}^2 & \mathscr{A}_{32}\mathbf{y}^2\mathbf{z}^2 & \mathscr{A}_{33}\mathbf{y}^2\mathbf{z}^2 \end{pmatrix}, \\
B(\mathbf{x},\mathbf{z}) &:= \begin{pmatrix} \mathscr{B}_{11}\mathbf{x}^2\mathbf{z}^2 & \mathscr{B}_{12}\mathbf{x}^2\mathbf{z}^2 & \mathscr{B}_{13}\mathbf{x}^2\mathbf{z}^2 \\ \mathscr{B}_{21}\mathbf{x}^2\mathbf{z}^2 & \mathscr{B}_{22}\mathbf{x}^2\mathbf{z}^2 & \mathscr{B}_{23}\mathbf{x}^2\mathbf{z}^2 \\ \mathscr{B}_{31}\mathbf{x}^2\mathbf{z}^2 & \mathscr{B}_{32}\mathbf{x}^2\mathbf{z}^2 & \mathscr{B}_{33}\mathbf{x}^2\mathbf{z}^2 \end{pmatrix}, \qquad (8.30) \\
C(\mathbf{x},\mathbf{y}) &:= \begin{pmatrix} \mathscr{C}_{11}\mathbf{x}^2\mathbf{y}^2 & \mathscr{C}_{12}\mathbf{x}^2\mathbf{y}^2 & \mathscr{C}_{13}\mathbf{x}^2\mathbf{y}^2 \\ \mathscr{C}_{21}\mathbf{x}^2\mathbf{y}^2 & \mathscr{C}_{22}\mathbf{x}^2\mathbf{y}^2 & \mathscr{C}_{23}\mathbf{x}^2\mathbf{y}^2 \\ \mathscr{C}_{31}\mathbf{x}^2\mathbf{y}^2 & \mathscr{C}_{32}\mathbf{x}^2\mathbf{y}^2 & \mathscr{C}_{33}\mathbf{x}^2\mathbf{y}^2 \end{pmatrix}.
\end{aligned}
$$

Then we know that for all $\mathbf{x}, \mathbf{y}, \mathbf{z} \in \mathbf{R}^3$

$$\mathscr{A}\mathbf{x}^2\mathbf{y}^2\mathbf{z}^2 = \mathbf{x}^\top A(\mathbf{y},\mathbf{z})\mathbf{x} = \mathbf{y}^\top B(\mathbf{x},\mathbf{z})\mathbf{y} = \mathbf{z}^\top C(\mathbf{x},\mathbf{y})\mathbf{z}.$$

Theorem 8.18 *Let $\mathscr{A} \in PS_{6,3}$. Suppose that matrices $A(\mathbf{y},\mathbf{z})$, $B(\mathbf{x},\mathbf{z})$ and $C(\mathbf{x},\mathbf{y})$ are defined by (8.30). Then the following conditions are equivalent:*

(1) The tensor $\mathscr{A}$ is M-PD;
(2) The matrix $A(\mathbf{y},\mathbf{z})$ is PD for all $\mathbf{y}, \mathbf{z} \in \mathbf{R}^3\backslash\{\mathbf{0}\}$;
(3) The matrix $B(\mathbf{x},\mathbf{z})$ is PD for all $\mathbf{x}, \mathbf{z} \in \mathbf{R}^3\backslash\{\mathbf{0}\}$;
(4) The matrix $C(\mathbf{x},\mathbf{y})$ is PD for all $\mathbf{x}, \mathbf{y} \in \mathbf{R}^3\backslash\{\mathbf{0}\}$.

Furthermore, if $\mathscr{A}$ is a strongly paired symmetric tensor, then the above (2), (3) and (4) are equivalent.

By Sylvester's criterion that a matrix is positive definite if and only if all its leading principal submatrices are positive definite, we have the following results.

Theorem 8.19 *Let $\mathscr{A} \in PS_{6,3}$, and $A(\mathbf{y},\mathbf{z})$, $B(\mathbf{x},\mathbf{z})$ and $C(\mathbf{x},\mathbf{y})$ are defined by (8.30). Then, tensor $\mathscr{A}$ is M-PD if and only if one of the following results holds.*

(1) $\mathscr{A}_{11}\mathbf{y}^2\mathbf{z}^2 > 0$, $(\mathscr{A}_{11}\mathbf{y}^2\mathbf{z}^2)(\mathscr{A}_{22}\mathbf{y}^2\mathbf{z}^2) - (\mathscr{A}_{12}\mathbf{y}^2\mathbf{z}^2)(\mathscr{A}_{21}\mathbf{y}^2\mathbf{z}^2) > 0$ and $\det(A(\mathbf{y},\mathbf{z})) > 0$ for all $\mathbf{y}, \mathbf{z} \in \mathbf{R}^3\backslash\{\mathbf{0}\}$.
(2) $\mathscr{B}_{11}\mathbf{x}^2\mathbf{z}^2 > 0$, $(\mathscr{B}_{11}\mathbf{x}^2\mathbf{z}^2)(\mathscr{B}_{22}\mathbf{x}^2\mathbf{z}^2) - (\mathscr{B}_{12}\mathbf{x}^2\mathbf{z}^2)(\mathscr{B}_{21}\mathbf{x}^2\mathbf{z}^2) > 0$ and $\det(B(\mathbf{x},\mathbf{z})) > 0$ for all $\mathbf{x}, \mathbf{z} \in \mathbf{R}^3\backslash\{\mathbf{0}\}$.
(3) $\mathscr{C}_{11}\mathbf{x}^2\mathbf{y}^2 > 0$, $(\mathscr{C}_{11}\mathbf{x}^2\mathbf{y}^2)(\mathscr{C}_{22}\mathbf{x}^2\mathbf{y}^2) - (\mathscr{C}_{12}\mathbf{x}^2\mathbf{y}^2)(\mathscr{C}_{21}\mathbf{x}^2\mathbf{y}^2) > 0$ and $\det(C(\mathbf{x},\mathbf{y})) > 0$ for all $\mathbf{x}, \mathbf{y} \in \mathbf{R}^3\backslash\{\mathbf{0}\}$.

Furthermore, if $\mathscr{A} \in PS_{6,3}$ is a strongly paired symmetric tensor, then the above (1), (2) and (3) are the same.

Recall the M-positive definiteness of fourth order three dimensional elasticity tensors in Sect. 8.2, we know that it can be checked equivalently by the positive semidefiniteness of three second order symmetric tensors, a fourth order symmetric tensor and a sixth order symmetric tensor. While an even order symmetric tensor is positive definite if and only if its smallest Z-eigenvalue is positive. Therefore, with the help of algorithm proposed by Cui, Dai and Nie [66], we can check the M-positive semidefiniteness of a fourth order elasticity tensor by computing the smallest Z-eigenvalues of the corresponding symmetric tensors. Unfortunately, the leading principal sub-matrices of $A(\mathbf{y}, \mathbf{z})$ or $B(\mathbf{x}, \mathbf{z})$ or $C(\mathbf{x}, \mathbf{y})$ do not correspond to any symmetric tensors. Hence, we can not check the M-positive definiteness of a sixth order strongly paired symmetric tensor by computing Z-eigenvalues of some symmetric tensors.

To give the next result, the following definition is needed.

Definition 8.5 Let $\mathscr{A} = (a_{i_1 i_2 \ldots i_{2m}}) \in T_{2m,3}$ be a given tensor. Suppose $t \in [1, 2m]$ is an even number. Then $\mathscr{A}$ is called a bi-block symmetric tensor if its indices $\{i_1, i_2, \ldots, i_{2m}\}$ can be divided into two adjacent blocks

$$\{i_1, i_2, \ldots, i_t\} \text{ and } \{i_{t+1}, i_{t+2}, \ldots, i_{2m}\}$$

such that the entries of $\mathscr{A}$ being invariant under any permutation of indices in every block of $\{i_1, i_2, \ldots, i_t\}$ and $\{i_{t+1}, i_{t+2}, \ldots, i_{2m}\}$, i.e.,

$$a_{i_1 i_2 \ldots i_t i_{t+1} i_{t+2} \ldots i_{2m}} = a_{\sigma(i_1 i_2 \ldots i_t)\sigma(i_{t+1} i_{t+2} \ldots i_{2m})} \tag{8.31}$$

for all $i_1, i_2, \ldots, i_{2m} \in \{1, 2, 3\}$, where $\sigma(i_1 i_2 \ldots i_t)$ denotes an arbitrary permutation of $i_1 i_2 \ldots i_t$.

To extend the definition of M-eigenvalue to bi-block symmetric tensors, the following symbols are needed. Suppose $\mathscr{A} \in T_{2m,3}$ and $t \in [1, 2m]$. Then we denote

$$\mathscr{A}\mathbf{x}^t\mathbf{y}^{2m-t} := \sum_{i_1,\ldots,i_t,i_{t+1},\ldots,i_{2m}=1}^{3} a_{i_1\ldots i_t i_{t+1}\ldots i_{2m}} x_{i_1} \ldots x_t y_{t+1} \ldots y_{2m};$$

$\mathscr{A}\mathbf{x}^{t-1}\mathbf{y}^{2m-t} \in \mathbf{R}^3$ with entries such that

$$(\mathscr{A}\mathbf{x}^{t-1}\mathbf{y}^{2m-t})_i := \sum_{i_2,\ldots,i_t,i_{t+1},\ldots,i_{2m}=1}^{3} a_{i i_2 \ldots i_t i_{t+1}\ldots i_{2m}} x_{i_2} \ldots x_t y_{t+1} \ldots y_{2m}, \quad \forall\, i \in [3];$$

$\mathscr{A}\mathbf{x}^t\mathbf{y}^{2m-t-1} \in \mathbf{R}^3$ with entries such that

$$(\mathscr{A}\mathbf{x}^t\mathbf{y}^{2m-t-1})_i := \sum_{i_1,\ldots,i_t,i_{t+2},\ldots,i_{2m}=1}^{3} a_{i_1\ldots i_t i i_{t+2}\ldots i_{2m}} x_{i_1} \ldots x_t y_{t+2} \ldots y_{2m}, \quad \forall\, i \in [3].$$

Definition 8.6 Let $\mathscr{A} = (a_{i_1 \ldots i_t i_{t+1} \ldots i_{2m}}) \in T_{2m,3}$ with entries satisfying bi-block symmetry given by (8.31). Assume $t = m$, and if there exist $\lambda \in \mathbf{R}$ and $\mathbf{x}, \mathbf{y} \in \mathbf{R}^3$ such that

$$\begin{cases} \mathscr{A}\mathbf{x}^{m-1}\mathbf{y}^m = \lambda\mathbf{x}, & \mathscr{A}\mathbf{x}^m\mathbf{y}^{m-1} = \lambda\mathbf{y}, \\ \mathbf{x}^\top\mathbf{x} = 1, & \mathbf{y}^\top\mathbf{y} = 1, \end{cases}$$

then λ is called an M-eigenvalue of $\mathscr{A}$ and $\mathbf{x}, \mathbf{y}$ are called eigenvectors associated with the M-eigenvalue λ.

Similar to Theorems 8.15–8.17, we obtain the following conclusions for the bi-block symmetric tensors.

Theorem 8.20 *Let $\mathscr{A} = (a_{i_1 \ldots i_t i_{t+1} \ldots i_{2m}}) \in T_{2m,3}$ with entries satisfying bi-block symmetry given by (8.31). Assume $t = m$, then the following results hold.*

(1) The M-eigenvalues of $\mathscr{A}$ always exist. Moreover, if $\mathbf{x}, \mathbf{y} \in \mathbf{R}^3$ are the eigenvectors of $\mathscr{A}$ associated with the M-eigenvalue λ, then $\lambda = \mathscr{A}\mathbf{x}^m\mathbf{y}^m$.
(2) The tensor $\mathscr{A}$ is M-positive definite if and only if the smallest M-eigenvalue of $\mathscr{A}$ is positive.
(3) λ is an M-eigenvalue of $\mathscr{A}$ if and only if $-\lambda$ is an M-eigenvalue of $-\mathscr{A}$. Furthermore, $\mathscr{A}$ is M-positive definite if and only if the largest M-eigenvalue of $-\mathscr{A}$ is negative.
(4) If λ_* is an M-eigenvalues of $\mathscr{A}$ with the largest absolute value among all M-eigenvalues of $\mathscr{A}$, and $\mathbf{x}^*, \mathbf{y}^* \in \mathbf{R}^3$ are the eigenvectors of $\mathscr{A}$ associated with the M-eigenvalue λ_*, then $\lambda_*(\mathbf{x}^*)^m(\mathbf{y}^*)^m$ is the best rank-one approximation of $\mathscr{A}$.

From Theorem 8.19, one can define bi-block symmetric tensors $\mathscr{T}_A^1, \mathscr{T}_B^1, \mathscr{T}_C^1 \in T_{8,3}$ and $\mathscr{T}_A^2, \mathscr{T}_B^2, \mathscr{T}_C^2 \in T_{12,3}$ such that

$$\mathscr{T}_A^1\mathbf{y}^4\mathbf{z}^4 = (\mathscr{A}_{11}\mathbf{y}^2\mathbf{z}^2)(\mathscr{A}_{22}\mathbf{y}^2\mathbf{z}^2) - (\mathscr{A}_{12}\mathbf{y}^2\mathbf{z}^2)(\mathscr{A}_{21}\mathbf{y}^2\mathbf{z}^2), \quad \mathscr{T}_A^2\mathbf{y}^6\mathbf{z}^6 = \det(A(\mathbf{y},\mathbf{z}));$$

$$\mathscr{T}_B^1\mathbf{x}^4\mathbf{z}^4 = (\mathscr{B}_{11}\mathbf{x}^2\mathbf{z}^2)(\mathscr{B}_{22}\mathbf{x}^2\mathbf{z}^2) - (\mathscr{B}_{12}\mathbf{x}^2\mathbf{z}^2)(\mathscr{B}_{21}\mathbf{x}^2\mathbf{z}^2), \quad \mathscr{T}_B^2\mathbf{x}^6\mathbf{z}^6 = \det(B(\mathbf{x},\mathbf{z}));$$

$$\mathscr{T}_C^1\mathbf{x}^4\mathbf{y}^4 = (\mathscr{C}_{11}\mathbf{x}^2\mathbf{y}^2)(\mathscr{C}_{22}\mathbf{x}^2\mathbf{y}^2) - (\mathscr{C}_{12}\mathbf{x}^2\mathbf{y}^2)(\mathscr{C}_{21}\mathbf{x}^2\mathbf{y}^2), \quad \mathscr{T}_C^2\mathbf{x}^6\mathbf{y}^6 = \det(C(\mathbf{x},\mathbf{y})).$$

Combining this with Definition 8.6, we obtain the following results.

Theorem 8.21 *Let $\mathscr{A} \in T_{6,3}$ be a paired symmetric tensor. Then $\mathscr{A}$ is M-positive definite if and only if one of the following results holds.*

(1) The smallest M-eigenvalues of tensors $\mathscr{A}_{11}$, $\mathscr{T}_A^1$ and $\mathscr{T}_A^2$ are positive.
(2) The smallest M-eigenvalues of tensors $\mathscr{B}_{11}$, $\mathscr{T}_B^1$ and $\mathscr{T}_B^2$ are positive.
(3) The smallest M-eigenvalue of tensors $\mathscr{C}_{11}$, $\mathscr{T}_C^1$ and $\mathscr{T}_C^2$ are positive.

Furthermore, if $\mathscr{A}$ is a strongly paired symmetric tensor, then the above (1), (2) and (3) are the same.

In the following, we will study the sum-of-squares (SOS) decomposition for the homogeneous polynomial corresponding to the paired symmetric tensors. In this case, we say that the related paired symmetric tensor is SOS if the corresponding polynomial has an SOS decomposition. It is clear that a paired symmetric tensor is M-PSD if it is SOS. Particularly, we study the case $m = 4$ and $m = 6$, and obtain the conclusions for any even number $m > 6$.

The unfolded matrix for a paired symmetric tensor is defined as below.

Definition 8.7 Suppose $\mathscr{A} = (a_{ijkl}) \in T_{4,3}$ and $\mathscr{B} = (b_{ijklpq}) \in T_{6,3}$ are two given tensors.
(1) Define a matrix by

$$M = (m_{st}) \quad \text{with } m_{st} = a_{i_s i_t j_s j_t} \; \forall\, s, t \in [9], \tag{8.32}$$

where $i_1 i_2 \ldots i_9$ and $j_1 j_2 \ldots j_9$ are two arbitrary permutations of 123123123 which satisfy that, for any $r \in \{1, 2, 3\}$, at least one of $i_{3(r-1)+1} i_{3(r-1)+2} i_{3(r-1)+3}$ and $j_{3(r-1)+1} j_{3(r-1)+2} j_{3(r-1)+3}$ is an arbitrary permutation of 123.
(2) Define a matrix by

$$N = (n_{st}) \quad \text{with } n_{st} = b_{i_s i_t j_s j_t k_s k_t} \; \forall\, s, t \in [27], \tag{8.33}$$

where $i_1 i_2 \ldots i_{27}$, $j_1 j_2 \ldots j_{27}$ and $k_1 k_2 \ldots k_{27}$ are three arbitrary permutations of $\underbrace{123123 \cdot 123}_{27}$ which satisfy that, for any $r \in \{1, 2, \ldots, 9\}$, at least one of $i_{3(r-1)+1} i_{3(r-1)+2} i_{3(r-1)+3}$, $j_{3(r-1)+1} j_{3(r-1)+2} j_{3(r-1)+3}$ and $k_{3(r-1)+1} k_{3(r-1)+2} k_{3(r-1)+3}$ is an arbitrary permutation of 123.

Then we say that the matrices M is an unfolded matrix of tensor $\mathscr{A}$ with respect to indices $i_1 i_2 \ldots i_9$ and $j_1 j_2 \ldots j_9$; N is an unfolded matrix of tensor $\mathscr{B}$ with respect to indices $i_1 i_2 \ldots i_{27}$, $j_1 j_2 \ldots j_{27}$ and $k_1 k_2 \ldots k_{27}$

It should be noted that, in the previous sections, the unfolded matrices for an fourth order elasticity tensor are concrete cases of this definition. For example, suppose $\mathscr{A} \in PS_{4,3}$, and M^1 and M^2 are defined by

$$M^1 = \begin{bmatrix} A_{11} & A_{12} & A_{13} \\ A_{21} & A_{22} & A_{23} \\ A_{31} & A_{32} & A_{33} \end{bmatrix} \quad \text{and} \quad M^2 = \begin{bmatrix} B_{11} & B_{12} & B_{13} \\ B_{21} & B_{22} & B_{23} \\ B_{31} & B_{32} & B_{33} \end{bmatrix}, \tag{8.34}$$

where A_{ij} are block sub-matrices tensor $\mathscr{A}$ for any $i, j, k, l \in [3]$ such that

$$A_{ij} := (a_{ijkl})_{kl} \quad \text{and} \quad B_{kl} := (a_{ijkl})_{ij}.$$

We can easily get that M^1 is an unfolded matrix of $\mathscr{A}$ with respect to indices 111222333 and 123123123, and M^2 is an unfolded matrix with respect to indices 123123123 and 111222333. In addition, we have that M^1 and M^2 are symmetric matrices, and they further satisfy that $M^1 = M^2$ when $\mathscr{A}$ is a strongly paired symmetric tensor.

By Definition 8.7, we first give several sufficient conditions or checkable necessary conditions for the SOS decomposition of fourth order three dimensional symmetric tensors.

Theorem 8.22 *Suppose $i_1 i_2 \ldots i_9$ and $j_1 j_2 \ldots j_9$ are two arbitrary permutations of 123123123. Let $\mathscr{A} = (a_{ijkl}) \in PS_{4,3}$ be a paired symmetric tensor. Assume M is an unfolded matrix of $\mathscr{A}$ with respect to indices $i_1 i_2 \ldots i_9$ and $j_1 j_2 \ldots j_9$. Then, the following results hold.*

(1) If M is positive semidefinite, then the polynomial $\mathscr{A}\mathbf{x}^2\mathbf{y}^2$ is an SOS of bilinear forms.
(2) If M is positive definite, then the tensor $\mathscr{A}$ is M-positive definite.
(3) If $\mathscr{A}\mathbf{x}^2\mathbf{y}^2$ is an SOS of bilinear forms, then it holds that $a_{iikk} \geq 0, \ \forall\, i, k \in [3]$.

To investigate the necessary and sufficient conditions for the SOS decomposition of tensor $\mathscr{A} \in PS_{4,3}$, we define tensor $\mathscr{B} \in T_{4,3}$ with entries such that

$$b_{ijkl} = b_{jilk}, \quad b_{ijlk} = b_{jikl}, \quad \text{and } b_{ijkl} + b_{jilk} + b_{ijlk} + b_{jikl} = 4a_{ijkl}. \quad (8.35)$$

Here, we call that $\mathscr{B}$ is a **semi-paired symmetric tensor** of $\mathscr{A}$. Therefore, it follows that $\mathscr{A}\mathbf{x}^2\mathbf{y}^2 = \mathscr{B}\mathbf{x}^2\mathbf{y}^2$, and the conclusion below holds.

Theorem 8.23 *Let $\mathscr{A} \in PS_{4,3}$. Suppose $\mathscr{B}$ is defined by (8.35) is a semi-paired symmetric tensor of $\mathscr{A}$. Then, it holds that*

(1) $\mathscr{A}\mathbf{x}^2\mathbf{y}^2$ is an SOS of bilinear forms if and only if the quadratic form $\mathscr{B}\mathbf{x}^2\mathbf{y}^2$ is an SOS of bilinear forms;
(2) tensor $\mathscr{A}$ is M-PD (M-PSD) if and only if the tensor $\mathscr{B}$ is M-PD (M-PSD).

The following lemma is useful in the following discussion.

Lemma 8.2 *Let $\mathscr{A} = (a_{ijkl}) \in PS_{4,3}$. Assume $M = (m_{st}) \in \mathbf{R}^{9\times 9}$ is a symmetric matrix such that*

$$\mathscr{A}\mathbf{x}^2\mathbf{y}^2 = \sum_{i,j,k,l=1}^{3} a_{ijkl} x_i x_j y_k y_l = \mathbf{u}^\top M \mathbf{u}, \quad \forall\, \mathbf{x}, \mathbf{y} \in \mathbf{R}^3,$$

where the vector $\mathbf{u} = (x_1 y_1, x_1 y_2, x_1 y_3, x_2 y_1, x_2 y_2, x_2 y_3, x_3 y_1, x_3 y_2, x_3 y_3)^\top$. Then the matrix M is an unfolded matrix of some semi-paired symmetric tensor of $\mathscr{A}$.

According to Lemma 8.2 and the notion of semi-paired symmetric tensor, we now give a necessary and sufficient condition for the bilinear SOS decomposition of the polynomial corresponding to a fourth order three dimensional paired symmetric tensor.

Theorem 8.24 *Let $\mathscr{A} = (a_{ijkl}) \in PS_{4,3}$. Then $\mathscr{A}\mathbf{x}^2\mathbf{y}^2$ is an SOS of bilinear forms if and only if an unfolding matrix of some semi-paired symmetric tensor of $\mathscr{A}$ is positive semidefinite.*

Now, we consider the SOS decomposition for the homogeneous polynomial defined by a sixth order three dimensional paired symmetric tensor. Similar to Theorem 8.22, the following results holds.

Theorem 8.25 *Let $\mathscr{A} = (a_{ijklpq}) \in PS_{6,3}$. Suppose that N defined by Definition 8.7 is an unfold matrix of $\mathscr{A}$ with respect to indices $i_1 i_2 \ldots i_{27}$, $j_1 j_2 \ldots j_{27}$ and $k_1 k_2 \ldots k_{27}$. Then, we have the following conclusions.*

(1) If N is positive semidefinite, then $\mathscr{A}\mathbf{x}^2\mathbf{y}^2\mathbf{z}^2$ is an SOS of trilinear forms.
(2) If N is positive definite, then the tensor $\mathscr{A}$ is M-positive definite.
(3) If $\mathscr{A}\mathbf{x}^2\mathbf{y}^2\mathbf{z}^2$ is an SOS of trilinear forms, then $a_{iikkpp} \geq 0,\ \forall\, i, k, p \in [3]$.

Similar to (8.35), we can define the semi-paired symmetric tensor for a sixth order three dimensional paired symmetric tensor. Suppose $\mathscr{A} = (a_{ijklpq}) \in PS_{6,3}$. Define tensor $\mathscr{B} = (b_{ijklpq}) \in T_{6,3}$ with entries satisfying

$$b_{ijklpq} = b_{jilkqp},\quad b_{jiklpq} = b_{ijlkqp},\quad b_{ijlkpq} = b_{jiklqp},\quad b_{ijklqp} = b_{jilkpq}, \tag{8.36}$$

and

$$b_{ijklpq} + b_{jiklpq} + b_{ijlkpq} + b_{ijklqp} = 4a_{ijklpq}. \tag{8.37}$$

We say that $\mathscr{B}$ is a **semi-paired symmetric tensor** of $\mathscr{A}$.

Similar to Theorems 8.23 and 8.24, we have the following results.

Theorem 8.26 *Let $\mathscr{A} = (a_{ijklpq}) \in PS_{6,3}$. Suppose that $\mathscr{B}$ is defined by (8.36) and (8.37) is an semi-paired symmetric tensor of $\mathscr{A}$. Then, we have the following results.*

(1) The polynomial $\mathscr{A}\mathbf{x}^2\mathbf{y}^2\mathbf{z}^2$ is an SOS of trilinear forms if and only if $\mathscr{B}\mathbf{x}^2\mathbf{y}^2\mathbf{z}^2$ is an SOS of trilinear forms.
(2) Tensor $\mathscr{A}$ is M-positive definite (M-positive semidefinite) if and only if tensor $\mathscr{B}$ is M-PD (M-PSD).
(3) $\mathscr{A}\mathbf{x}^2\mathbf{y}^2\mathbf{z}^2$ an SOS of trilinear forms if and only if an unfolded matrix of some semi-paired symmetric tensor of $\mathscr{A}$ is positive semidefinite.

To end this section, we note that a sequential semidefinite programming method is established for computing the smallest M-eigenvalue of any fourth order three dimensional paired symmetric tensor [138]. Generally speaking, for a given tensor $\mathscr{A} \in T_{4,3}$, a new polynomial $F_{r,s} : \mathbf{R}^3 \times \mathbf{R}^3 \to \mathbf{R}$ is defined by

$$F_{r,s}(\mathbf{x}, \mathbf{y}) := \left(\sum_{i=1}^{3} x_i^2\right)^r \left(\sum_{i=1}^{3} y_i^2\right)^s \mathscr{A}\mathbf{x}^2\mathbf{y}^2,$$

which is a homogeneous polynomial with $\deg(F_r) = 2(r+s)+4$. If $\mathscr{A}$ is M-PD, it is proved that $F_{r,s}(x, y)$ is an SOS for some sufficiently large integers $r, s \geq 0$. Denote

$$\mathbb{K} := \{\mathscr{A} \in T_{4,3} : F_{r,s}(\mathbf{x}, \mathbf{y}) \text{ is an SOS for some } r, s \geq 0\}.$$

For any paired symmetric tensor $\mathscr{A} \in PS_{4,3}$, its smallest M-eigenvalue can be computed by the optimization problem

$$\begin{aligned} &\min \mathscr{A}\mathbf{x}^2\mathbf{y}^2 \\ &\text{s.t. } \mathbf{x}^\top\mathbf{x} = 1, \ \mathbf{y}^\top\mathbf{y} = 1, \end{aligned}$$

which can be written equivalently as

$$\begin{aligned} &\max \gamma \\ &\text{s.t. } \mathscr{A}\mathbf{x}^2\mathbf{y}^2 \geq \gamma, \ \forall\, (\mathbf{x}, \mathbf{y}) \in \{(\mathbf{x}, \mathbf{y}) \in \mathbf{R}^3 \times \mathbf{R}^3 : \mathbf{x}^\top\mathbf{x} = 1, \ \mathbf{y}^\top\mathbf{y} = 1\}. \end{aligned} \tag{8.38}$$

Suppose $\mathscr{E} = (e_{ijkl}) \in \mathbb{T}_{4,3}$ is defined by

$$e_{ijkl} := \begin{cases} 1 & \text{if } i = j, k = l, \\ 0 & \text{otherwise,} \end{cases} \quad \forall i, j, k, l \in [3].$$

Combining this with the set $\mathbb{K}$, the optimization problem (8.38) is equivalent to

$$\begin{aligned} &\max \gamma \\ &\text{s.t. } \mathscr{A} + \gamma\mathscr{E} \in \mathbb{K}, \end{aligned}$$

which can also be relaxed to another form by the polynomial relaxation technique (for details see [138]). However, the method in [138] can only obtain an approximation optimal solution up to a priori precision, and we do not know if it is efficient for computing the smallest M-eigenvalues of higher order paired symmetric tensors.

8.6 Notes

The strong ellipticity condition is a very important property for elasticity tensors, which have been frequently used in mechanics, including piezooptical tensor, the second order electrooptical effect, electrostriction and second order magnetostriction. Therefore, to verify whether or not the strong ellipticity condition holds for a given elastic material is essential in the theory of elasticity. Actually, a lot of researchers in the field of elasticity have paid much attention to this problem [59, 106, 112, 158, 159, 236, 244, 271, 276]. To be different with the previous methods, in this chapter, we mainly study the strong ellipticity of elasticity tensors by spectral theory of tensors.

Section 8.1: The definition of M-eigenvalues for elasticity tensors was first defined by Qi, Dai and Han in [226]. It is also proved that the fourth order three dimensional elasticity tensor satisfies strong ellipticity condition if and only if its smallest M-eigenvalue is positive. The main sufficient condition for M-positive definiteness is the S-positive definiteness. The S-positive definiteness of elasticity tensors has been studied by Lord Kelvin in [261, 262] 160 years ago. In the special case such that

$m = n = 2$, a direct method was proposed to compute the smallest M-eigenvalue. Details about resultant of a system for two variables can be obtained in [102].

Section 8.2: For the strong ellipticity condition of fourth order three dimensional elasticity tensors, an efficient sufficient and necessary condition was proposed via Z-eigenvalues of symmetric tensors by Han, Dai and Qi [114]. The notion of Z-eigenvalue of symmetric tensors was first defined by Qi in [221]. More results about Z-eigenvalues for symmetric tensors can be found in [228]. A special kinds of ellipticity tensor for the rhombic system has been studied in [59]. Here, we present an sufficient and necessary condition for the rhombic system via a symmetric copositive matrix. The concrete sufficient and necessary conditions for a 3×3 symmetric matrix to be copositive was given by Simpson and Spector [244].

Section 8.3: The content of this section was originally given by Ding, Qi and Yan [85] and Ding, Liu, Qi and Yan [79]. The related M-tensors were first studied by Ding, Qi and Wei [82] and Zhang, Qi and Zhou [305]. More results about M-tensor can be found in the book [228].

Section 8.4: The practical algorithm studied in this section was given by Wang, Qi and Zhang in [280]. The power method was first used to compute the largest H-eigenvalue by Ng, Qi and Zhou in 2009 [197].

Section 8.5: The study in this section was originally given by Huang and Qi in 2017 [138]. Some concrete models for higher order elasticity tensors from solid mechanics and elasticity mechanics can be found in Hiki [123], Jong et al. [155], and Thurston and Brugger [263].

8.7 Exercises

1 Suppose $\mathscr{A} = (a_{ijkl}) \in \mathbb{E}_{4,3}$ be given by

$$\begin{aligned} &a_{1111} = 1,\ a_{2222} = 1,\ a_{3333} = 1,\ a_{1122} = 2,\ a_{2233} = 2,\ a_{3311} = 2,\\ &a_{1212} = a_{1221} = a_{2112} = a_{2121} = -\tfrac{1}{2},\ a_{2323} = a_{2332} = a_{3223} = a_{3232} = -\tfrac{1}{2},\\ &a_{3131} = a_{3113} = a_{1331} = a_{1313} = -\tfrac{1}{2}. \end{aligned}$$

Please prove that the given tensor is M-positive definite.

2 Suppose that $\lambda = 0$ is an M-eigenvalue of the elasticity tensor $\mathscr{A}$ in last question. By the definition of M-eigenvalue for elasticity tensors, write the M-eigenvectors corresponding to the M-eigenvalue $\lambda = 0$.

3 Suppose $\mathscr{A} \in \mathbb{E}_{4,3}$ is an elasticity M-tensor. Prove that $\mathscr{A}$ is a nonsingular elasticity M-tensor if and only if the unfolded matrix $(\mathscr{A} \cdot \mathbf{y}^2)$ is a nonsingular M-matrix for each $\mathbf{y} \geq \mathbf{0}$.

4 Suppose $\mathscr{A} \in \mathbb{E}_{4,3}$ is an elasticity tensor such that

$$\begin{aligned} &a_{1111} = a_{1122} = a_{1133} = a_{2211} = a_{2222} = a_{2233} = a_{3311} = 1,\\ &a_{3322} = 3,\ a_{3333} = 3,\ a_{3323} = a_{3332} = -1, \quad \text{other } a_{ijkl} = 0. \end{aligned}$$

Prove that $\lambda = 1$ is the smallest M-eigenvalue for tensor $\mathscr{A}$. Furthermore, give the M-eigenvectors corresponding to $\lambda = 1$.

Chapter 9
Higher Order Tensors in Quantum Physics

In this chapter, we will apply tensor analysis to the quantum entanglement problem and the classicality problem of spin states in quantum physics.

The quantum entanglement problem is one of the most important problems in quantum physics and quantum information technology. The central question is: giving a quantum state, can one determine whether it is entangled or not? In the literature, several approaches to measure entanglement of a state are proposed, e.g., hyperdeterminant, tensor rank, etc. Among others, *geometric measure* is a newly proposed measure for quantum entanglement. In this chapter, we will study the geometric measure of a quantum state by using spectral theory of tensors. It will establish a bridge between the recently developed spectral theory of tensors and a central problem in quantum physics.

Mathematically, a general n-partite state of a composite quantum system can be regarded as an element in a Hilbert tensor product space. Generally speaking, a separable state can be viewed as a rank-one tensor in this tensor space, and the geometric measure is a measure of the distance between a given state and the set of separable states. This distance can be characterized by taking the maximization of a homogeneous form over the set of separable states, which is called *Hartree value*.

We will show how the spectral theory of non-negative tensors can be applied to the study of the geometric measure of entanglement for pure states. For symmetric pure multipartite qubit or qutrit states, an elimination method is given. For symmetric pure multipartite qudit states, a numerical algorithm with randomization is presented. We will also illustrate that a nonsymmetric pure state can be augmented to a symmetric one whose amplitudes can be encoded in a non-negative symmetric tensor, and then the geometric measure of entanglement can be calculated. Several examples, such as mGHZ states, W states, inverted W states, qudits, and nonsymmetric states, are used to demonstrate the power of the proposed methods. Given a pure state, one may find a change of basis (a unitary transformation) so that all the probability amplitudes of the pure state are non-negative under the new basis. Therefore, the methods proposed here can be applied to a very wide class of multipartite pure states.

L. Qi et al., *Tensor Eigenvalues and Their Applications*, Advances in Mechanics and Mathematics 39, https://doi.org/10.1007/978-981-10-8058-6_9

Furthermore, we will study the connection between the smallest Z-eigenvalue of a tensor and the entanglement of a pure or mixed state. This originates in the fact that the entanglement of a state is related to the positive semi-definiteness of a tensor, which in turn is linked to the sign of its smallest Z-eigenvalue.

We will define the *unitary eigenvalue* (U-eigenvalue) of a complex tensor, the *unitary symmetric eigenvalue* (US-eigenvalue) of a symmetric complex tensor, and the best complex rank-one approximation. An upper bound on the number of distinct US-eigenvalues of symmetric tensors will be given, and we will count all US-eigenpairs with nonzero eigenvalues of symmetric tensors. We will convert the geometric measure of the entanglement problem to an algebraic equation system problem. A numerical example shows that a symmetric real tensor may have a best complex rank-one approximation that is better than its best real rank-one approximation, which implies that the absolute-value of the largest Z-eigenvalue is not always the geometric measure of entanglement.

We will then consider mixed states. The geometric measure of quantum entanglement of a pure state, defined by its distance to the set of pure separable states, can be extended to multipartite mixed states. We will characterize the nearest disentangled mixed state to a given mixed state with respect to the geometric measure by means of a system of equations. The entanglement eigenvalue of a mixed state will be introduced. And we will show that, for a given mixed state, its nearest disentangled mixed state is associated with its entanglement eigenvalue.

A geometrical picture of quantum states often helps us get some insight on underlying physical properties. For arbitrary pure spin states, such a geometrical representation was developed by Ettore Majorana [192]: a spin-j state is visualized as $N = 2j$ points on the unit sphere S^2, called in this context the Bloch sphere. The advantage of such a picture is a direct interpretation of certain unitary operations: namely, if a quantum spin-j state is mapped to another one by a unitary operation that corresponds to a $(2j + 1)$-dimensional representation of a spatial rotation, its Majorana points are mapped to points obtained by that spatial rotation. Recently a tensor representation of an arbitrary mixed or pure spin-j state was proposed to generalizes this picture [103]. It consists of a real symmetric tensor of order $N = 2j$ and dimension 4. A spin-j state corresponds to a boson if j is a positive integer, and corresponds to a fermion if j is a positive half-integer. Thus, a boson corresponds to an even order four dimensional tensor, while a fermion corresponds to an odd order four dimensional tensor.

Recently, Qi, Zhang, Braun, Bohnet-Waldraff and Giraud [233] introduced regularly decomposable tensors and showed that a spin-j state is classical if and only if its representing tensor is a regularly decomposable tensor. An algorithm was proposed to determine if the representing tensor of a spin state is regularly decomposable or not.

9.1 Quantum Entanglement Problems

In 1935, Albert Einstein, Boris Podolsky and Nathan Rosen described a "spooky" phenomenon, which is known as EPR paradox. They demonstrated that in quantum mechanics the knowledge of one physical quantity precluded the knowledge of the other one even when they were described by non-commuting operators, and thus believed "the quantum mechanical description of physical reality given by wave functions is not complete [90]." The word "entanglement" was first given by Erwin Schrödinger when writing a letter to Einstein to discuss the correlations between two particles. This problem lay dormant for many years until 1964, when John Stewart Bell showed that if the locality assumption held, then the completion of quantum mechanics did not exist in the sense of EPR.

Bell's work provided us the chance of regarding entanglement as a resource for application. Quantum entanglement is an extremely active topic in physics community. Experimental entanglement of photons, neutrinos, electrons, and even small diamonds is widely studied and effectively demonstrated [15, 209]. Furthermore, quantum entanglement is applied in communication and computation. The big achievement of quantum entanglement application was made in 2016, when Chinese quantum satellite sent 'spooky' messages over 1,200 km — 12 times more than ever before — in a move that could enable massive breakthroughs in secure communications.

Next let us make the concept of quantum entanglement concrete. In quantum mechanics, it means pairs or groups of particles interacted mutually and the quantum state of each particle cannot be described independently of the others, even when the distance between the particles is large. This physical phenomenon is called quantum entanglement. In fact, when we say entanglement, we refer to a multipartite system, whose Hilbert space can be presented as a product of two or more tensor factors according to physical subsystems of this system. If a global state of a composite system can be expressed as a product of the states of subsystems, it is called a disentangled or separable system. Otherwise, we call this state entangled.

In this section, we mainly focus on the entanglement problem of pure states. Mathematically, a quantum state corresponds to a vector of norm 1 in a Hilbert space over the complex numbers, while a mixed state is a probabilistic mixture of pure states. To describe a state vector, i.e., a pure quantum state, we follows the custom in quantum mechanics by using bra–ket notation, which contains the angle brackets "$\langle$" and "$\rangle$", and vertical bar "$|$". Each bra $\langle\Psi|$ is the conjugate transpose of the so-called ket $|\Psi\rangle$, which are related to the same quantum state.

Definition 9.1 Suppose that $\mathscr{H}$ is a Hilbert tensor product space

$$\mathscr{H} = \mathscr{H}_1 \otimes \cdots \otimes \mathscr{H}_m,$$

and the dimension of $\mathscr{H}_k$ is d_k for $k \in [m]$. We denote $|\Psi\rangle$ as a general m-partite state of a composite quantum system. Then, $|\Psi\rangle$ may be regarded as an element of $\mathscr{H}$, i.e.,

$$|\Psi\rangle \in \mathscr{H} \qquad \text{and} \qquad \langle\Psi|\Psi\rangle = 1,$$

in which $\langle\cdot|\cdot\rangle$ means the inner product. If an m-partite state $|\phi\rangle \in \mathscr{H}$ could be represented as

$$|\phi\rangle = |\phi^{(1)}\rangle \otimes \cdots \otimes |\phi^{(m)}\rangle, \tag{9.1}$$

where $|\phi^{(k)}\rangle \in \mathscr{H}_k$ and $\||\phi^{(k)}\rangle\| = 1$ for $k \in [m]$, then $|\phi\rangle$ is called a **separable (i.e., Hartree) m-partite state**. The set of all separable m-partite states in $\mathscr{H}$ is denoted as $\mathrm{Sep}(\mathscr{H})$. A pure state of a composite quantum system is **entangled** if it is not separable.

For example, given two basis vectors $\{|0\rangle_A, |1\rangle_A\}$ of $\mathscr{H}_1$ and two basis vectors $\{|0\rangle_B, |1\rangle_B\}$ of $\mathscr{H}_2$, one of the famous Bell state

$$|\Psi\rangle = \frac{1}{\sqrt{2}}(|0\rangle_A \otimes |1\rangle_B - |1\rangle_A \otimes |0\rangle_B)$$

is an entangled state.

Measuring physical properties such as position, momentum, spin, and polarization, we find that entangled particles are appropriately correlated. For example, if the total spin of two entangled particles is zero, we put them in Beijing and Hongkong separately and measure the spin of each particle on the same axis. When the spin of the particle in Hongkong (Beijing) is found to be clockwise (counterclockwise), the spin of the particle in Beijing (Hongkong) will be discovered as counterclockwise (clockwise). Since the outcome of the measurement is random and no one transmit information in the experiment, it seems that when we operate on one particle, the other one "knows" the operation and "alter" its spin according to the result of its partner, even when they are separated by large distance. This is also the basic quantum entanglement theory that ensure truly secure communication in the Chinese quantum satellite.

9.2 Geometric Measure of Entanglement of Multipartite Pure States

Assume that $d_1 \leq \cdots \leq d_m$ without loss of generality in Definition 9.1. The degree of which a general m-partite state $|\Psi\rangle \in \mathscr{H}$ is entangled could be characterized by the distance to the set $\mathrm{Sep}(\mathscr{H})$ [243, 284]:

$$\mathrm{dist} = \||\Psi\rangle - |\phi_\Psi\rangle\| = \min\{\||\Psi\rangle - |\phi\rangle\| : |\phi\rangle \in \mathrm{Sep}(\mathscr{H})\}, \tag{9.2}$$

where $|\phi_\Psi\rangle \in \mathrm{Sep}(\mathscr{H})$ is the nearest separable m-partite state of $|\Psi\rangle$. The distance in (9.2) stands for a natural geometric measure of the entanglement content. The farther a state from the set of separable m-partite states is, the more entangled it is. In addition, we remark that since the objective of (9.2) is a continuous function and the feasible region is a compact set in a finite dimensional space, there always exists the nearest separable m-partite state $|\phi_\Psi\rangle$.

For a general m-partite state $|\Psi\rangle$ and an arbitrary separable m-partite state $|\phi\rangle = \otimes_{k\in[m]}|\phi^{(k)}\rangle$ in (9.1), we have

$$\||\Psi\rangle - |\phi\rangle\|^2 = 2 - \langle\Psi|\phi\rangle - \langle\phi|\Psi\rangle. \tag{9.3}$$

To characterise the minimization problem (9.2), we study its dual maximization problem

$$\begin{aligned} &\max\{\langle\Psi|\phi\rangle + \langle\phi|\Psi\rangle : |\phi\rangle \in \mathrm{Sep}(\mathscr{H})\} \\ &= \max\left\{\langle\Psi|(\otimes_{k\in[m]}|\phi^{(k)}\rangle) + (\otimes_{k\in[m]}\langle\phi^{(k)}|)|\Psi\rangle : \langle\phi^{(k)}|\phi^{(k)}\rangle = 1 \text{ for } k \in [m]\right\}. \end{aligned} \tag{9.4}$$

By introducing Lagrange multipliers λ_k for $k \in [m]$, we obtain

$$\langle\Psi|(\otimes_{j\neq k}|\phi^{(j)}\rangle) = \lambda_k\langle\phi^{(k)}| \quad \text{and} \quad (\otimes_{j\neq k}\langle\phi^{(j)}|)|\Psi\rangle = \lambda_k|\phi^{(k)}\rangle. \tag{9.5}$$

Multiplying them by $|\phi^{(k)}\rangle$ and $\langle\phi^{(k)}|$, we see that

$$\lambda := \lambda_k = \langle\Psi|\phi\rangle = \langle\phi|\Psi\rangle \tag{9.6}$$

is a real number in $[-1, 1]$.

Motivated by (9.3) and (9.6), we denote

$$G_\Psi := \max\{|\langle\Psi|\phi\rangle| : |\phi\rangle \in \mathrm{Sep}(\mathscr{H})\} \tag{9.7}$$

the maximal overlap of a fixed m-partite pure state $|\Psi\rangle \in \mathscr{H}$. We see that the maximal overlap in (9.7) is equal to the largest **entanglement eigenvalue** λ satisfying the following polynomial system

$$\begin{cases} \langle\Psi|(\otimes_{j\neq k}|\phi^{(j)}\rangle) = \lambda\langle\phi^{(k)}|, \\ (\otimes_{j\neq k}\langle\phi^{(j)}|)|\Psi\rangle = \lambda|\phi^{(k)}\rangle, \\ \||\phi^{(k)}\rangle\| = 1, \quad k \in [m]. \end{cases} \tag{9.8}$$

Next, we introduce the minimum Hartree value, and show that the geometry measure of the entanglement content of $|\Psi\rangle$ is bounded up by the Hartree value [222].

Theorem 9.1 *We call*

$$\sigma := \min\{\langle\Psi|\phi_\Psi\rangle : |\Psi\rangle \in \mathscr{H}, \langle\Psi|\Psi\rangle = 1\}$$

the minimum Hartree value of $\mathscr{H}$, where $|\phi_\Psi\rangle$ is the nearest separable m-partite state to $|\Phi\rangle$. Therefore, $\sigma > 0$, and $\langle\Psi|\phi_\Psi\rangle \geq \sigma$ for any $|\Psi\rangle \in \mathscr{H}$. Furthermore, the geometry measure of the entanglement content of $|\Psi\rangle$ satisfies

$$\text{dist} = \||\Psi\rangle - |\phi_\Psi\rangle\| \leq \sqrt{2-2\sigma}. \tag{9.9}$$

Proof For the function

$$g(|z\rangle) := \max\{|\langle z|\phi\rangle| : |\phi\rangle \in \text{Sep}(\mathscr{H})\}, \qquad \text{where } |z\rangle \in \mathscr{H},$$

we see that $g(|z\rangle) \geq 0$ and $g(|z\rangle) = 0$ if and only if $|z\rangle = 0$. Moreover, for $|z\rangle, |w\rangle \in \mathscr{H}$, we have $g(|z\rangle + |w\rangle) \leq g(|z\rangle) + g(|w\rangle)$. Hence, $g(\cdot)$ defines a norm in the finite dimensional space $\mathscr{H}$. Recalling another norm $h(|z\rangle) = \sqrt{\langle z|z\rangle}$ defined also in $\mathscr{H}$, and according to the norm equivalence theorem in the finite dimensional space, we get

$$\begin{aligned}\sigma &= \min\{\langle\Psi|\phi_\Psi\rangle : |\Psi\rangle \in \mathscr{H}, \langle\Psi|\Psi\rangle = 1\} \\ &\geq \min\{g(|\Psi\rangle) : |\Psi\rangle \in \mathscr{H}, h(|\Psi\rangle) = 1\} \\ &> 0.\end{aligned}$$

Hence $\langle\Psi|\phi_\Psi\rangle \geq \sigma > 0$ for any $|\Psi\rangle \in \mathscr{H}$.

Now, for the dual problem (9.4), we have

$$\max\{\langle\Psi|\phi\rangle + \langle\phi|\Psi\rangle : |\phi\rangle \in \text{Sep}(\mathscr{H})\} = \langle\Psi|\phi_\Psi\rangle + \langle\phi_\Psi|\Psi\rangle \geq 2\sigma.$$

Hence, the inequality (9.9) for the primal problem is valid. □

The following theorem reveals a lower bound for the minimum Hartree value [222].

Theorem 9.2 *A lower bound for the minimum Hartree value σ of $\mathscr{H}$ is*

$$\sigma \geq \frac{1}{\sqrt{d_1 \cdots d_{m-1}}}. \tag{9.10}$$

Proof In general, we suppose that $|e_{i_k}^{(k)}\rangle$ for $i_k \in [d_k]$ is an orthonormal basis of $\mathscr{H}_k$ for $k \in [m]$. Under these bases, we could write

$$|\Psi\rangle = \sum_{i_1,\ldots,i_m} \overline{a_{i_1\ldots i_m}} |e_{i_1}^{(1)}\rangle \cdots |e_{i_m}^{(m)}\rangle \quad \textit{and} \quad |\phi^{(k)}\rangle = \sum_{i_k \in [d_k]} u_{i_k}^{(k)} |e_{i_k}^{(k)}\rangle.$$

Let $\mathscr{A} = (a_{i_1 \ldots i_m})$ be a supermatrix and $u^{(k)} = (u^{(k)}_{i_k})$ for $k \in [m]$. We define

$$\rho(\mathscr{A}) := \max \left\{ \left| \sum_{i_1, \ldots, i_m} a_{i_1 \ldots i_m} u^{(1)}_{i_1} \cdots u^{(m)}_{i_m} \right| : \|u^{(k)}\| = 1 \text{ for } k \in [m] \right\}. \quad (9.11)$$

In fact, we say $\rho(\mathscr{A}) = \langle \Psi | \phi_\Psi \rangle = \max\{|\langle \Psi | \phi \rangle| : \phi = \otimes_{k \in [m]} |\phi^{(k)}\rangle \in \mathrm{Sep}(\mathscr{H})\}$.

Now, we consider slice matrices $A_{i_1 \ldots i_{m-2}}$ of which the (i, j)th element is $a_{i_1 \ldots i_{m-2} ij}$ for $i \in [d_{m-1}]$ and $j \in [d_m]$. By (9.11), we see

$$\rho(A_{i_1 \ldots i_{m-2}}) \leq \rho(\mathscr{A}), \qquad \forall\, i_1, \ldots, i_{m-2}.$$

We say that $\rho(A_{i_1 \ldots i_{m-2}})$ is the spectral norm of a matrix $A_{i_1 \ldots i_{m-2}}$ and

$$\|A_{i_1 \ldots i_{m-2}}\|_F \leq \sqrt{d_{m-1}}\, \rho(A_{i_1 \ldots i_{m-2}}) \qquad \forall\, i_1, \ldots, i_{m-2}.$$

Because of $\langle \Psi | \Psi \rangle = 1$, we have $\|\mathscr{A}\|_F = 1$. Combining these inequalities, we get

$$\begin{aligned} 1 &= \|\mathscr{A}\|_F^2 \\ &= \sum_{i_1, \ldots, i_{m-2}} \|A_{i_1 \ldots i_{m-2}}\|_F^2 \\ &\leq \sum_{i_1, \ldots, i_{m-2}} d_{m-1} \rho(A_{i_1 \ldots i_{m-2}})^2 \\ &\leq \sum_{i_1, \ldots, i_{m-2}} d_{m-1} \rho(\mathscr{A})^2 \\ &= d_1 \cdots d_{m-1} \rho(\mathscr{A})^2. \end{aligned}$$

Hence, for any $\Psi \in \mathscr{H}$ with $\langle \Psi | \Psi \rangle = 1$, we have

$$\langle \Psi | \phi_\Psi \rangle = \rho(\mathscr{A}) \geq \frac{1}{\sqrt{d_1 \cdots d_{m-1}}}.$$

Because $\sigma = \min\{\langle \Psi | \phi_\Psi \rangle : \Psi \in \mathscr{H}, \langle \Psi | \Psi \rangle = 1\}$, we obtain the lower bound (9.10). □

We note that when $m = 2$, the minimum Hartree value is $\sigma = \frac{1}{\sqrt{d_1}}$. The lower bound is attainable. Let $\mathscr{A}$ be a diagonal tensor with diagonal elements $a_{ii} = \frac{1}{\sqrt{d_1}}$ for $i \in [d_1]$. Then we see that $|\Psi\rangle$ is a pure state. By setting $\phi_\Psi = |e_1^{(1)}\rangle \otimes |e_1^{(2)}\rangle$, we obtain $\sigma = \langle \Psi | \phi_\Psi \rangle = \frac{1}{\sqrt{d_1}}$ straightforwardly.

Using these two theorems, we say that the geometric measure of the entanglement content of $|\Psi\rangle \in \mathscr{H}$ is bounded above

$$\text{dist} = \| |\Psi\rangle - |\phi_\Psi\rangle \| \le \sqrt{2 - 2\sigma} \le \sqrt{2 - \frac{2}{\sqrt{d_1 \cdots d_{m-1}}}}.$$

9.3 Z-Eigenvalues and Entanglement of Symmetric States

We consider a special Hilbert tensor product space $\mathscr{H}^{\otimes m}$ and the dimension of $\mathscr{H}$ is n. Let $|\Psi\rangle \in \mathscr{H}^{\otimes m}$ be an m-partite symmetric pure state with $\langle\Psi|\Psi\rangle = 1$. An m-partite symmetric pure separate state could be represented as $|\phi\rangle^{\otimes m}$, where $|\phi\rangle \in \mathscr{H}$ and $\langle\phi|\phi\rangle = 1$. By (9.8), the **quantum eigenvalue** problem could be rewritten as

$$\begin{cases} \langle\Psi|(|\phi\rangle^{\otimes m-1}) = \lambda\langle\phi|, \\ (\langle\phi|^{\otimes m-1})|\Psi\rangle = \lambda|\phi\rangle, \\ \langle\phi|\phi\rangle = 1. \end{cases} \tag{9.12}$$

By choosing an orthonormal basis of $\mathscr{H}$, the symmetric pure state $|\Psi\rangle$ is represented by an mth order n dimensional symmetric complex tensor $\mathscr{A}_\Psi$. The state $|\phi\rangle$ is represented by an n dimensional complex vector $\mathbf{z}$. The equality $\langle\phi|\phi\rangle = 1$ reduces to $\mathbf{z}^H\mathbf{z} = 1$, where $(\cdot)^H$ means the conjugate transpose of complex vectors. Then, the quantum eigenvalue problem (9.12) could be formulated as [309]

$$\begin{cases} \mathscr{A}_\Psi\mathbf{z}^{m-1} = \lambda\bar{\mathbf{z}}, \\ \overline{\mathscr{A}_\Psi}\bar{\mathbf{z}}^{m-1} = \lambda\mathbf{z}, \\ \mathbf{z}^H\mathbf{z} = 1. \end{cases} \tag{9.13}$$

Obviously, if there exist $\lambda \in \mathbf{C}$ and $\mathbf{z} \in \mathbf{C}$ satisfying (9.13), then λ is real. We call λ a quantum eigenvalue of the tensor $\mathscr{A}_\Psi$ and $\mathbf{z}$ the associated quantum eigenvector. The largest quantum eigenvalue of $\mathscr{A}_\Psi$ is named as the entanglement eigenvalue of $\mathscr{A}_\Psi$, which is associated with the nearest separable state of the state Ψ.

Since λ is real, the system (9.13) is equivalent to

$$\begin{cases} \mathscr{A}_\Psi\mathbf{z}^{m-1} = \lambda\bar{\mathbf{z}}, \\ \mathbf{z}^H\mathbf{z} = 1, \\ \lambda \in \mathbf{R}. \end{cases} \tag{9.14}$$

Next, we show that quantum eigenvalues appear in pairs [309].

Proposition 9.1 *If $\lambda \in \mathbf{R}$ is a quantum eigenvalue of a symmetric tensor $\mathscr{A}_\Psi$, then $-\lambda$ is a quantum eigenvalue of $\mathscr{A}_\Psi$, too.*

Proof Suppose that λ is a quantum eigenvalue of $\mathscr{A}_\Psi$ and $\mathbf{z}$ is the associated quantum eigenvector. By direct calculations, $-\lambda$ is also a quantum eigenvalue of $\mathscr{A}_\Psi$ with an associated quantum eigenvector $\mathbf{z}\exp(\frac{\pi}{m}\mathrm{i})$, where i is the imaginary unit. □

The above property means that the quantum spectral radius of $\mathscr{A}_\Psi$ is the largest quantum eigenvalue of $\mathscr{A}_\Psi$. For convenience, we denote $Q(\Psi)$ as the quantum spectral radius of $\mathscr{A}_\Psi$ and $Z(\Psi)$ as the largest quantum eigenvalue of $\mathscr{A}_\Psi$. Then we have the following theorem [132, 309].

Theorem 9.3 *If the symmetric tensor $\mathscr{A}_\Psi$ of an m-partite pure state $|\Psi\rangle$ is real, then*

$$Q(\Psi) \geq Z(\Psi).$$

In addition, the equality holds in the following five cases:

(1) $m = 2$;
(2) $\mathscr{A}_\Psi$ *is diagonal;*
(3) $\mathscr{A}_\Psi$ *is nonnegative;*
(4) $\mathscr{A}_\Psi$ *is nonpositive;*
(5) $\mathscr{A}_\Psi = \sum_{j\in[n]} \alpha_j (\mathbf{y}^{(j)})^m$, *where* α_j, $j \in [n]$, *are real numbers,* $\{\mathbf{y}^{(j)}\}_{j\in[n]}$ *is an orthonormal basis of* $\mathbf{R}^n$.

Proof According to definitions of quantum eigenvalues and Z-eigenvalues of a real tensor $\mathscr{A}_\Psi$, a Z-eigenvalue λ of $\mathscr{A}_\Psi$ is also a quantum eigenvalue. In addition, by Proposition 9.1, $|\lambda|$ is a quantum eigenvalue of $\mathscr{A}_\Psi$. Hence, we have $Q(\Psi) \geq Z(\Psi)$.

To prove the equivalence of $Q(\Psi)$ and $Z(\Psi)$, we only need to prove $Q(\Psi) \leq Z(\Psi)$.

(1) Let λ be an entanglement eigenvalue of $\mathscr{A}_\Psi$ and $\mathbf{z} = \mathbf{x} + i\mathbf{y}$ be the associated entanglement eigenvector, where $\mathbf{x}, \mathbf{y}$ are real vectors and one of them are nonzero at least. Hence, we have $\mathscr{A}_\Psi \mathbf{x} = \lambda \mathbf{x}$ and $\mathscr{A}_\Psi \mathbf{y} = -\lambda \mathbf{y}$. If $\mathbf{x}$ is nonzero, $(\lambda, \frac{\mathbf{x}}{\|\mathbf{x}\|})$ is a Z-eigenpair of $\mathscr{A}_\Psi$. If $\mathbf{y}$ is nonzero, $(-\lambda, \frac{\mathbf{y}}{\|\mathbf{y}\|})$ is a Z-eigenpair of $\mathscr{A}_\Psi$. Recalling Proposition 9.1, we claim that $Q(\Psi) \leq Z(\Psi)$.

(2) We consider a real diagonal tensor $\mathscr{A}_\Psi = \mathrm{diag}(a_1, \ldots, a_n)$. Denote $|a_i| = \max\{|a_k| : k \in [n]\}$. Obviously, a_i is a Z-eigenvalue of $\mathscr{A}_\Psi$ with its associated Z-eigenvector $\mathbf{e}_i$. This means that $Z(\Psi) \geq |a_i|$.

Suppose that $(\lambda, \mathbf{z})$ is an entanglement eigenpair of $\mathscr{A}_\Psi$. Since $\mathbf{z} \neq 0$, there exists a component $z_j \neq 0$. From $\mathbf{z}^H\mathbf{z} = 1$, we know $|z_j| \leq 1$. By $a_j z_j^{m-1} = \lambda \overline{z_j}$, we have

$$Q(\Psi) = \lambda \leq |a_j||z_j|^{m-2} \leq |a_i| \leq Z(\Psi).$$

(3) By the spectral theory of nonnegative tensors, the Z-spectral radius $Z(\Psi)$ is a Z-eigenvalue of the nonnegative tensor $\mathscr{A}_\Psi$ and

$$Z(\Psi) = \max_{\mathbf{x}\in\mathbf{R}^n_+\cap S^{n-1}} \mathscr{A}_\Psi \mathbf{x}^m = \max_{\mathbf{x}\in S^{n-1}} \mathscr{A}_\Psi \mathbf{x}^m = Q(\Psi),$$

where $S^{n-1} = \{\mathbf{x} \in \mathbf{R}^n : \mathbf{x}^\top \mathbf{x} = 1\}$.

(4) Since $-\mathscr{A}_\Psi$ is nonnegative, we have $Z(\Psi) = Z(-\Psi) = Q(-\Psi) = Q(\Psi)$ by the above assertion.

(5) Suppose that λ is an entanglement eigenvalue of $\mathscr{A}_\Psi$ with the associated eigenvector $\mathbf{z}$. Then, we have

$$\sum_{j\in[n]} \alpha_j \left(\mathbf{y}^{(j)\top}\mathbf{z}\right)^{m-1} \mathbf{y}^{(j)} = \lambda\bar{\mathbf{z}}. \tag{9.15}$$

Since $\{\mathbf{y}^{(j)}\}$ is an orthonormal basis of $\mathbf{R}^n$, there exists an index k such that $\mathbf{y}^{(k)\top}\mathbf{z} \neq 0$. By multiplying $\mathbf{y}^{(k)}$ on both sides of (9.15), we get

$$\begin{aligned}\lambda\mathbf{y}^{(k)\top}\bar{\mathbf{z}} &= \sum_{j\in[n]} \alpha_j \left(\mathbf{y}^{(j)\top}\mathbf{z}\right)^{m-1} \left(\mathbf{y}^{(k)\top}\mathbf{y}^{(j)}\right) \\ &= \alpha_k \left(\mathbf{y}^{(k)\top}\mathbf{z}\right)^{m-1}.\end{aligned}$$

Hence, we get

$$Q(\Psi) = \lambda \le |\alpha_k| \left|\mathbf{y}^{(k)\top}\mathbf{z}\right|^{m-2} \le |\alpha_k|.$$

Let $\mathbf{x} = \mathbf{y}^{(k)} \in \mathbf{R}^n$. Then, $\mathbf{x}^\top\mathbf{x} = 1$ and $\mathscr{A}_\Psi \mathbf{x}^{m-1} = \alpha_k \mathbf{x}$. Thus, α_k is a Z-eigenvalue of $\mathscr{A}_\Psi$ and hence $Z(\Phi) \ge |\alpha_k| \ge Q(\Psi)$. The proof is complete. $\square$

Generally speaking, the inequality $Q(\Psi) \neq Z(\Psi)$ holds. In [140], a counter example is given and verifies this conclusion.

If $Q(\Psi) = Z(\Psi)$ holds, the problem of finding the entanglement eigenvalue of an m-partite pure state Ψ could be transformed into the problem of finding the Z-spectral radius of $\mathscr{A}_\Psi$. Whereafter, we show some examples [132]. For $0 \le k \le m$, we define

$$|S(m,k)\rangle := \sqrt{\frac{k!(m-k)!}{m!}} \sum_{\tau\in\mathfrak{S}_m} \left|\tau(\underbrace{0\ldots0}_{k}\underbrace{1\ldots1}_{m-k})\right\rangle,$$

where $\mathfrak{S}_m$ is the symmetric group on m elements.

Example 1. The mGHZ state is of the form

$$|\text{mGHZ}\rangle = \frac{|S(m,0)\rangle + |S(m,m)\rangle}{\sqrt{2}}.$$

Under the basis $\{|0\rangle, |1\rangle\}$, we get $\mathscr{A}_{mGHZ} \in S_{m,2}$. By calculations, its Z-eigenpairs are

$$\left(\frac{1}{\sqrt{2}}, (1,0)^\top\right),$$
$$\left(\frac{1}{\sqrt{2}}, (0,1)^\top\right),$$
$$\left(\frac{1}{\sqrt{2}^{m-1}}, \left(\frac{1}{\sqrt{2}}, \frac{1}{\sqrt{2}}\right)^\top\right),$$

and five more when m is odd

$$\left(\frac{1}{\sqrt{2}}, (-1,0)^\top\right),$$
$$\left(\frac{1}{\sqrt{2}}, (0,-1)^\top\right),$$
$$\left(\frac{1}{\sqrt{2}^{m-1}}, \left(-\frac{1}{\sqrt{2}}, \frac{1}{\sqrt{2}}\right)^\top\right),$$
$$\left(\frac{1}{\sqrt{2}^{m-1}}, \left(\frac{1}{\sqrt{2}}, -\frac{1}{\sqrt{2}}\right)^\top\right),$$
$$\left(\frac{1}{\sqrt{2}^{m-1}}, \left(-\frac{1}{\sqrt{2}}, -\frac{1}{\sqrt{2}}\right)^\top\right).$$

Hence, we obtain $Q(mGHZ) = Z(mGHZ) = \frac{1}{\sqrt{2}}$. The corresponding nearest separable pure state is $|\phi\rangle^{\otimes m}$ where $|\phi\rangle = |0\rangle$ or $|\phi\rangle = |1\rangle$.

Example 2. For a three-partite qubit, the W state is of the form

$$|W\rangle = |S(3,2)\rangle = \frac{|001\rangle + |010\rangle + |100\rangle}{\sqrt{3}}.$$

Under the basis $\{|0\rangle, |1\rangle\}$, we get $\mathscr{A}_W \in S_{3,2}$. By calculations, its Z-eigenpairs are

$$(0, (0,1)^\top),$$
$$\left(\frac{2}{3}, \left(\sqrt{\frac{2}{3}}, \sqrt{\frac{1}{3}}\right)^\top\right),$$
$$\left(\frac{2}{3}, \left(-\sqrt{\frac{2}{3}}, \sqrt{\frac{1}{3}}\right)^\top\right),$$
$$\left(-\frac{2}{3}, \left(\sqrt{\frac{2}{3}}, -\sqrt{\frac{1}{3}}\right)^\top\right),$$
$$\left(-\frac{2}{3}, \left(-\sqrt{\frac{2}{3}}, -\sqrt{\frac{1}{3}}\right)^\top\right).$$

Hence, $Q(W) = Z(W) = \frac{2}{3}$. The corresponding nearest separable pure state is $|\phi\rangle^{\otimes 3}$ with $|\phi\rangle = \sqrt{\frac{2}{3}}|0\rangle + \sqrt{\frac{1}{3}}|1\rangle$.

Example 3. For a three-partite qubit, the inverted-W state is of the form

$$|\widetilde{W}\rangle = |S(3,1)\rangle = \frac{|110\rangle + |101\rangle + |011\rangle}{\sqrt{3}}.$$

Under the basis $\{|0\rangle, |1\rangle\}$, we get $\mathscr{A}_{\widetilde{W}} \in S_{3,2}$. By switching components of Z-eigenvectors, the Z eigenpairs for $\mathscr{A}_W$ become that for $\mathscr{A}_{\widetilde{W}}$. Hence, $Q(\widetilde{W}) = Z(\widetilde{W}) = \frac{2}{3}$ and the corresponding nearest separable pure state is $|\phi\rangle^{\otimes 3}$ with $|\phi\rangle = \sqrt{\frac{1}{3}}|0\rangle + \sqrt{\frac{2}{3}}|1\rangle$.

Example 4. For a three-partite qutrit, the general GHZ state is of the form

$$|\Psi\rangle = \alpha|111\rangle + \beta|222\rangle + \gamma|333\rangle, \qquad \alpha^2 + \beta^2 + \gamma^2 = 1.$$

Under the basis $\{|1\rangle, |2\rangle, |3\rangle\}$, we get $\mathscr{A}_\Psi \in S_{3,3}$ and it is nonnegative when $\alpha, \beta, \gamma \geq 0$. By calculations, its nonnegative Z-eigenpairs are

$$\begin{gathered}
(\alpha, (1,0,0)^\top), \\
(\beta, (0,1,0)^\top), \\
(\gamma, (0,0,1)^\top), \\
\left(\frac{\alpha\beta}{\sqrt{\alpha^2+\beta^2}}, \left(\frac{\beta}{\sqrt{\alpha^2+\beta^2}}, \frac{\alpha}{\sqrt{\alpha^2+\beta^2}}, 0\right)^\top\right), \\
\left(\frac{\alpha\gamma}{\sqrt{\alpha^2+\gamma^2}}, \left(\frac{\gamma}{\sqrt{\alpha^2+\gamma^2}}, 0, \frac{\alpha}{\sqrt{\alpha^2+\gamma^2}}\right)^\top\right), \\
\left(\frac{\beta\gamma}{\sqrt{\beta^2+\gamma^2}}, \left(0, \frac{\gamma}{\sqrt{\beta^2+\gamma^2}}, \frac{\beta}{\sqrt{\beta^2+\gamma^2}}\right)^\top\right), \\
\left(\frac{\alpha\beta\gamma}{\tau}, \left(\frac{\beta\gamma}{\tau}, \frac{\alpha\gamma}{\tau}, \frac{\alpha\beta}{\tau}\right)^\top\right),
\end{gathered}$$

where $\tau := \sqrt{\alpha^2\beta^2 + \beta^2\gamma^2 + \alpha^2\gamma^2}$.

Hence, $Q(W) = Z(W) = \max\{\alpha, \beta, \gamma\}$ when $\alpha, \beta, \gamma \geq 0$. The corresponding nearest separable pure state is $|\phi\rangle^{\otimes 3}$ with $|\phi\rangle = |1\rangle$ when $Q(W) = \alpha$, $|2\rangle$ when $Q(W) = \beta$, or $|3\rangle$ when $Q(W) = \gamma$.

Example 5. Given a multipartite qudit state $|\Phi\rangle$ as follows

$$|\Phi\rangle = \frac{1}{2}\sum_{i=1}^{4} \underbrace{|i\rangle \otimes \cdots \otimes |i\rangle}_{4}.$$

By the algorithm proposed in [66], we get $Q(\Phi) = Z(\Phi) = 0.5$ and the associated Z-eigenvectors are $(1,0,0,0)^\top$, $(0,1,0,0)^\top$, $(0,0,1,0)^\top$, and $(0,0,0,1)^\top$.

9.4 Geometric Measure and U-Eigenvalues of Tensors

Let $\mathscr{A} \in \mathbf{C}^{n_1\times\cdots\times n_m}$ be an mth order complex tensor and $\mathbf{x}^{(i)} \in \mathbf{C}^{n_i}$ be complex vectors for $i \in [m]$. Then $\otimes_{i\in[m]}\mathbf{x}^{(i)}$ forms an mth order complex tensor with rank-one. We define

$$\langle \mathscr{A}, \otimes_{i\in[m]}\mathbf{x}^{(i)}\rangle = \sum_{i_1,\ldots,i_m} \overline{\mathscr{A}_{i_1\ldots i_m}} x_{i_1}^{(1)} \cdots x_{i_m}^{(m)}.$$

Moreover, we define vectors $\langle \mathscr{A}, \otimes_{i\neq k}\mathbf{x}^{(i)}\rangle$ and $\langle \otimes_{i\neq k}\mathbf{x}^{(i)}, \mathscr{A}\rangle$, whose components are

$$\langle \mathscr{A}, \otimes_{i\neq k}\mathbf{x}^{(i)}\rangle_{i_k} := \sum_{i_1,\ldots,i_{k-1},i_{k+1},\ldots,i_m} \overline{\mathscr{A}_{i_1\ldots i_k\ldots i_m}} x_{i_1}^{(1)} \cdots x_{i_{k-1}}^{(k-1)} x_{i_{k+1}}^{(k+1)} \cdots x_{i_m}^{(m)},$$

$$\langle \otimes_{i\neq k}\mathbf{x}^{(i)}, \mathscr{A}\rangle_{i_k} := \sum_{i_1,\ldots,i_{k-1},i_{k+1},\ldots,i_m} \mathscr{A}_{i_1\ldots i_k\ldots i_m} \overline{x}_{i_1}^{(1)} \cdots \overline{x}_{i_{k-1}}^{(k-1)} \overline{x}_{i_{k+1}}^{(k+1)} \cdots \overline{x}_{i_m}^{(m)}.$$

A complex number λ is called a **unitary eigenvalue** of a tensor $\mathscr{A} \in \mathbf{C}^{n_1\times\cdots\times n_m}$ [199] if there exist complex vectors $\mathbf{x}^{(k)} \in \mathbf{C}^{n_k}$ satisfying the following system

$$\begin{cases} \langle \mathscr{A}, \otimes_{i\neq k}\mathbf{x}^{(i)}\rangle = \lambda \overline{\mathbf{x}}^{(k)}, \\ \langle \otimes_{i\neq k}\mathbf{x}^{(i)}, \mathscr{A}\rangle = \lambda \mathbf{x}^{(k)}, \quad \forall k \in [m]. \\ \|\mathbf{x}^{(k)}\| = 1, \end{cases} \tag{9.16}$$

In fact, the largest $|\lambda|$ is the entanglement eigenvalue of $\mathscr{A}$, and the corresponding rank-one tensor $\otimes_{i\in[m]}\mathbf{x}^{(i)}$ is the nearest separable state.

More specifically, for the symmetric case, we assume that $\mathscr{S} \in \mathbf{C}^{n\times\cdots\times n}$ is a symmetric tensor. For $\mathbf{x} \in \mathbf{C}^n$, we denote

$$\overline{\mathscr{S}}\mathbf{x}^m := \langle \mathscr{S}, \otimes_{i\in[m]}\mathbf{x}\rangle = \sum_{i_1,\ldots,i_m} \overline{\mathscr{S}_{i_1\ldots i_m}} x_{i_1} \cdots x_{i_m},$$

and a vector $\overline{\mathscr{S}}\mathbf{x}^{m-1}$ whose components are

$$(\overline{\mathscr{S}}\mathbf{x}^{m-1})_j := \sum_{i_2,\ldots,i_m} \overline{\mathscr{S}_{ji_2\ldots i_m}} x_{i_2} \cdots x_{i_m}.$$

A complex number λ is called a **unitary symmetric eigenvalue** of $\mathscr{S}$ [199], if there exists a vector $\mathbf{x} \in \mathbf{C}^n$ satisfying

$$\begin{cases} \overline{\mathscr{S}}\mathbf{x}^{m-1} = \lambda \overline{\mathbf{x}}, \\ \mathscr{S}\overline{\mathbf{x}}^{m-1} = \lambda \mathbf{x}, \\ \|\mathbf{x}\| = 1. \end{cases} \tag{9.17}$$

The complex vector $\mathbf{x}$ is named a **unitary symmetric eigenvector** of $\mathcal{S}$ associated with the unitary symmetric eigenvalue λ, and $(\lambda, \mathbf{x})$ is called a **unitary symmetric eigenpair**. The largest $|\lambda|$ is the entanglement eigenvalue of $\mathcal{S}$, and the corresponding rank-one tensor $\otimes_{i\in[m]}\mathbf{x}$ is the nearest symmetric separable state.

Theorem 9.4 *Suppose that tensors $\mathcal{S} \in \mathbf{C}^{n_1\times\cdots\times n_m}$. Then,*
(a) all unitary eigenvalues of $\mathcal{S}$ are real;
(b) if $\mathcal{S}$ is a complex symmetric tensor, its unitary symmetric eigenpair $(\lambda, \mathbf{x})$ could also be defined as

$$\begin{cases} \overline{\mathcal{S}}\mathbf{x}^{m-1} = \lambda\overline{\mathbf{x}}, \\ \|\mathbf{x}\| = 1, \\ \lambda \in \mathbf{R}. \end{cases} \tag{9.18}$$

or

$$\begin{cases} \mathcal{S}\overline{\mathbf{x}}^{m-1} = \lambda\mathbf{x}, \\ \|\mathbf{x}\| = 1, \\ \lambda \in \mathbf{R}. \end{cases} \tag{9.19}$$

Proof (a) Suppose that λ is a unitary eigenvalue of $\mathcal{A}$ with an associated rank-one tensor $\otimes_{i\in[m]}\mathbf{x}^{(i)}$. By multiplying $\mathbf{x}^{(k)}$ and $\overline{\mathbf{x}}^{(k)}$ to the first and second equations of (9.16), we have

$$\langle \mathcal{A}, \otimes_{i\in[m]}\mathbf{x}^{(i)}\rangle = \lambda = \langle \otimes_{i\in[m]}\mathbf{x}^{(i)}, \mathcal{A}\rangle.$$

From the definition of inner product of complex tensors, we have

$$\langle \mathcal{A}, \otimes_{i\in[m]}\mathbf{x}^{(i)}\rangle = \overline{\langle \otimes_{i\in[m]}\mathbf{x}^{(i)}, \mathcal{A}\rangle}.$$

Hence, the unitary eigenvalue λ is a real number.

(b) Since λ is real, the second equation in (9.17) is the conjugate of the first equation in (9.17). Hence, (9.18) and (9.19) are equivalent to (9.17). □

For a complex symmetric tensor $\mathcal{S}$, if $(\lambda, \mathbf{x})$ is its unitary symmetric eigenpair, then $(\lambda, \overline{\mathbf{x}})$ is its quantum eigenpair. Note that $\otimes_{i\in[m]}\mathbf{x}$, but not $\otimes_{i\in[m]}\overline{\mathbf{x}}$, is related to the nearest separable state. This is the main advantage of the unitary eigenpair of complex tensors.

Theorem 9.5 *Suppose that $\mathcal{S}$ is a complex symmetric tensor of order m.*
(a) If $m \geq 3$ is an odd integer and $\lambda \neq 0$, then the system (9.17) is equivalent to

$$\overline{\mathcal{S}}\mathbf{x}^{m-1} = \overline{\mathbf{x}}, \qquad \mathbf{x} \neq \mathbf{0}.$$

(b) If $m \geq 3$ is an even integer and $\lambda \neq 0$, then the system (9.17) is equivalent to

$$\overline{\mathcal{S}}\mathbf{x}^{m-1} = \pm\overline{\mathbf{x}}, \qquad \mathbf{x} \neq \mathbf{0}.$$

Proof (a) Suppose that $(\lambda, \mathbf{x})$ is a unitary symmetric eigenpair of $\mathscr{S}$. Then, λ is real. If $\lambda \neq 0$, we define $\mu := \lambda^{-1/(m-2)}$ which is also real because $m \geq 3$ is odd. By setting $\mathbf{y} = \mu\mathbf{x}$, (9.18) equals

$$\overline{\mathscr{S}\mathbf{y}^{m-1}} = \bar{\mathbf{y}}.$$

On the other hand, if there exists a nonzero complex vector $\mathbf{y}$ satisfying the above equation, the pair $(\lambda, \mathbf{x})$ with

$$\lambda = \pm\|\mathbf{y}\|^{-(m-2)} \quad \text{and} \quad \mathbf{x} = \frac{\mathbf{y}}{\|\mathbf{y}\|}.$$

is a unitary eigenpair of $\mathscr{S}$.

(b) It can be proved by a similar approach for part (a). □

Theorem 9.6 *Suppose that $\mathscr{S}$ is a symmetric complex tensor and λ is its unitary symmetric eigenvalue. Then*
(a) $-\lambda$ is also a unitary symmetric eigenvalue of $\mathscr{S}$;
(b) $Q(S) = \lambda_{\max}$, where $\lambda_{\max}$ is the largest unitary symmetric eigenvalue of $\mathscr{S}$.

Proof (a) By (9.18), if $(\lambda, \mathbf{x})$ is a unitary symmetric eigenpair of $\mathscr{S}$, $(-\lambda, \exp(\frac{\pi}{m}\mathrm{i})\mathbf{x})$ is also a unitary symmetric eigenpair of $\mathscr{S}$.

(b) The assertion is straightforward by (a). □

Next, we turn to the best symmetric complex rank-one approximation of symmetric tensors, i.e., the closest symmetric product state problems. Let $\mathscr{S}$ be a complex symmetric tensor. We try to find a unit vector $\mathbf{x}^*$ such that

$$\overline{\mathscr{S}}(\mathbf{x}^*)^m = \max\{|\overline{\mathscr{S}}\mathbf{x}^m| : \mathbf{x}^H\mathbf{x} = 1\}. \tag{9.20}$$

Then, we have the following theorem [199].

Theorem 9.7 *The best symmetric complex rank-one approximation problem* (9.20) *is equivalent to the following optimization problem*

$$\min\{\|\mathbf{x}\| : \overline{\mathscr{S}\mathbf{x}^{m-1}} = \bar{\mathbf{x}}, \mathbf{0} \neq \mathbf{x} \in \mathbf{C}^n\}.$$

Proof It is clear that $\overline{\mathscr{S}}(\mathbf{x}^*)^m = \lambda_{\max}$ is the largest unitary symmetric eigenvalue of $\mathscr{S}$. Then, this theorem follows from Theorems 9.5 and 9.4. □

9.5 Regularly Decomposable Tensors and Classical Spin States

In quantum theory and many related areas such as solid-state physics, molecular, atomic, high-energy physics, the concept of spin is widely used. With the recent advance of quantum information theory, the classicality of spin states becomes more and more important and has gained much attention. In literature, researchers proposed several different definitions of classicality of a quantum state, such as the one based upon the positivity of the Wigner function, or the absence of entanglement in multipartite systems. Analogous to the classicality of the harmonic oscillator states of the electromagnetic field, the classicality of a spin state is well-defined in terms of its density matrix which can be decomposed as a positive weighted sum of angular momentum coherent states. Recently, a compact and elegant representation of spin density matrices in terms of 4-dimensional tensors that share the most important properties of Bloch vectors was introduced. Especially, the classicality property of a spin state, either a boson or a fermion, can be fully characterized by the regular decomposability of the corresponding representing tensor [233]. In this regard, the identification of the classicality of a spin state can be fully accomplished by the verification of the regular decomposability of the corresponding tensor.

Suppose $\mathscr{A} \in S_{m,n+1}$ is a real symmetric tensor. We call $\mathscr{A}$ a completely decomposable tensor if there exist vectors $\mathbf{u}^{(1)}, \ldots, \mathbf{u}^{(r)} \in \mathbf{R}^{n+1}$ such that

$$\mathscr{A} = \sum_{i\in[r]} (\mathbf{u}^{(i)})^{\otimes m}, \tag{9.21}$$

where $(\mathbf{u}^{(i)})^{\otimes m} = \mathbf{u}^{(i)} \otimes \cdots \otimes \mathbf{u}^{(i)}$ is a rank-one tensor.

Definition 9.2 A vector $\mathbf{x} = (x_0, x_1, \ldots, x_n)^\top \in \mathbf{R}^{n+1}$ is called a **regular vector** if $x_0 \neq 0$ and $x_0^2 = x_1^2 + \cdots + x_n^2$.

Definition 9.3 Let $\mathscr{A} = (a_{i_1 i_2 \ldots i_m}) \in S_{m,n+1}$. We define its jth row tensor $\mathscr{A}_j = (a_{j,i_2 \ldots i_m})$ as a symmetric tensor in $S_{m-1,n+1}$ for $j \in [n]$.

Definition 9.4 (a) Let $m = 2\ell$ be an even integer and $\mathscr{A} \in S_{m,n+1}$. We say that $\mathscr{A} \in \mathbf{R}_{2\ell,n+1}$ is a **regularly decomposable** tensor of even order if $\mathscr{A}$ is completely decomposable as (9.21) and $\mathbf{u}^{(1)}, \ldots, \mathbf{u}^{(r)} \in \mathbf{R}^{n+1}$ therein are regular vectors.

(b) Let $m = 2\ell + 1$ be an odd integer and $\mathscr{A} \in S_{m,n+1}$. If its row tensor $\mathscr{A}_0$ is a regularly decomposable tensor with the regularly decomposition

$$\mathscr{A}_0 = \sum_{k\in[r]} (\mathbf{u}^{(k)})^{\otimes 2\ell},$$

where $\mathbf{u}^{(k)} = (u_0^{(k)}, \ldots, u_n^{(k)})^\top$ for $k \in [r]$ are regular vectors, and the other row tensors of $\mathscr{A}$ are induced by the following formula

$$\mathscr{A}_j = \sum_{k\in[r]} \frac{u_j^{(k)}}{u_0^{(k)}} (\mathbf{u}^{(k)})^{\otimes 2\ell},$$

for $j \in [n]$, we call $\mathscr{A}$ a regularly decomposable tensor of odd order.

The regular decomposability of a symmetric tensor was originally defined separately in even order case and in odd order case via special rank-one tensor decompositions. However, a unified characterization for both even and odd order tensors was further established.

Theorem 9.8 *A tensor* $\mathscr{A} = (a_{i_1 \ldots i_m}) \in S_{m,n+1}$ *is a regularly decomposable tensor if and only if there exist an integer* r, *some* $\alpha_k > 0$ *and* $\mathbf{v}^{(k)} = \left(1, v_1^{(k)}, \ldots, v_n^{(k)}\right)^\top$ *satisfying* $\sum_{i\in[n]} \left(v_i^{(k)}\right)^2 = 1$ *for* $k \in [r]$ *such that*

$$\mathscr{A} = \sum_{k\in[r]} \alpha_k \left(\mathbf{v}^{(k)}\right)^{\otimes m}. \tag{9.22}$$

Proof Suppose that $m = 2\ell$ is even, $\mathscr{A}$ is defined by (9.21), and $\mathbf{u}^{(1)}, \ldots, \mathbf{u}^{(r)}$ are regular vectors. By setting

$$\mathbf{v}^{(k)} = \frac{\mathbf{u}^{(k)}}{u_0^{(k)}}$$

and

$$\alpha_k = \left(u_0^{(k)}\right)^{2\ell} > 0$$

for $k \in [r]$, we obtain the expression (9.22). On the other hand, $\mathscr{A}$ can be written as (9.22). Since $\alpha_k^{\frac{1}{2\ell}} \mathbf{v}^{(k)}$ are regular vectors, $\mathscr{A}$ is a regularly decomposable tensor. Thus, this theorem is valid in the even order case.

The odd case can be proved similarly. □

If a tensor $\mathscr{A} = (a_{i_1 \ldots i_m}) \in S_{m,n+1}$ satisfies

$$a_{00i_3 \ldots i_m} = \sum_{i\in[n]} a_{iii_3 \ldots i_m},$$

then $\mathscr{A}$ is called a **regular symmetric tensor**. Then, we have the following critical theorem [233].

Theorem 9.9 *A spin-j state can be represented by a regular symmetric tensor* $\mathscr{A} \in S_{2j,4}$. *A spin-j state is classical if and only if its representing tensor* $\mathscr{A}$ *is a regularly decomposable tensor.*

By this theorem, the physical problem to determine whether a spin-j state is classical or not is converted to a mathematical problem of identifying the regular decomposability of its representing tensor.

We call R a normalized orthogonal matrix if

$$R = \begin{pmatrix} 1 & \mathbf{0}^\top \\ \mathbf{0} & Q \end{pmatrix},$$

where Q is an orthogonal matrix and $\mathbf{0} \in \mathbf{R}^n$ is a zero vector.

Theorem 9.10 *Suppose that $\mathscr{A}, \mathscr{B} \in S_{m,n+1}$, R is a normalized orthogonal matrix, and $\mathscr{B} = R^m \mathscr{A}$. Then, $\mathscr{B}$ is regularly decomposable if and only if $\mathscr{A}$ is regularly decomposable.*

Proof Suppose that $\mathscr{A}$ is regularly decomposable. From Theorem 9.8, there exist positive numbers α_k and regular vectors $\mathbf{v}^{(k)} = \left(1, \hat{\mathbf{v}}^{(k)}\right)^\top$ with $\hat{\mathbf{v}}^{(k)} \in \mathbf{R}^n$ being unit vectors for $k \in [r]$ satisfying (9.22). Then

$$\mathscr{B} = R^m \mathscr{A} = \sum_{k \in [r]} \alpha_k \left(R\mathbf{v}^{(k)}\right)^{\otimes m}.$$

Clearly, since R is a normalized orthogonal matrix, we have

$$R\mathbf{v}^{(k)} = \begin{pmatrix} 1 & \mathbf{0}^\top \\ \mathbf{0} & Q \end{pmatrix} \begin{pmatrix} 1 \\ \hat{\mathbf{v}}^{(k)} \end{pmatrix} = \begin{pmatrix} 1 \\ Q\hat{\mathbf{v}}^{(k)} \end{pmatrix}$$

and $Q\hat{\mathbf{v}}^{(k)}$ are also unit vectors for $k \in [r]$. Hence, $R\mathbf{v}^{(k)}$ are regular vectors for $k \in [r]$ and $\mathscr{B}$ is a regularly decomposable tensor.

The other part can be proved similarly by using the normalized orthogonal matrix R^{-1}. □

For notational convenience, we use $RD_{m,n+1}$ to denote the set of all regularly decomposable tensors in $S_{m,n+1}$.

Theorem 9.11 *$RD_{m,n+1}$ is a closed convex cone.*

Proof By Theorem 9.8, it is easy to see that $RD_{m,n+1}$ is a convex cone. Next, we prove its closeness.

Let $\{\mathscr{A}^{(p)}\}_{p=1,2,\ldots}$ be a convergent sequence of regularly decomposable tensors in $RD_{m,n+1}$, and

$$\lim_{p\to\infty} \mathscr{A}^{(p)} = \mathscr{A}.$$

According to Theorem 9.8, we assume that

$$\mathscr{A}^{(p)} = \sum_{k \in [r_p]} \alpha_{k,p} \left(\mathbf{v}^{(k,p)}\right)^{\otimes m},$$

where $\alpha_{k,p} \geq 0, \mathbf{v}^{(k,p)} = (1, \hat{\mathbf{v}}^{(k,p)})^\top$ with $\|\hat{\mathbf{v}}^{(k,p)}\| = 1$ for $k \in [r_p]$ and $p = 1, 2, \ldots$. By the Carathéodory theory, we have

$$r_p \leq \binom{m+n+2}{m} + 1.$$

Hence, without loss of generality, there is a common $r = r_p \le \binom{m+n+2}{m} + 1$ for all p. By taking a subsequence if necessary, there exist $\alpha_k \ge 0$, regular vectors $\mathbf{v}^{(k)} = (1, \hat{\mathbf{v}}^{(k)})^\top$ such that

$$\lim_{p\to\infty} \alpha_{k,p} = \alpha_k,$$

$$\lim_{p\to\infty} \hat{\mathbf{v}}^{(k,p)} = \hat{\mathbf{v}}^{(k)},$$

and

$$\mathscr{A} = \sum_{k\in[r]} \alpha_k \left(\mathbf{v}^{(k)}\right)^{\otimes m} \in RD_{m,n+1}.$$

Hence, $RD_{m,n+1}$ is closed. □

However, to verify the membership of a tensor $\mathscr{A} \in RD_{m,n+1}$ is generally NP-hard.

It is known that the classicality of a spin-j state is equivalent to the regular decomposibility of its representing tensor $\mathscr{A} \in S_{2j,4}$. Together with the equivalent characterization of regularly decomposable tensors in Theorem 9.8, we will discuss an E-truncated K-moment problem model for the identification of the desired classicality in this section.

Before transforming the characterization to an E-truncated K-moment problem, some necessary preliminaries on polynomials are briefly recalled. We use $\mathbf{R}[\mathbf{x}] := \mathbf{R}[x_1, \ldots, x_n]$ to denote the ring of all polynomials in $\mathbf{x} := (x_1, \ldots, x_n)^\top \in \mathbf{R}^n$ with real coefficients, and $\mathbf{R}[\mathbf{x}]_d$ the space of all polynomials in $\mathbf{R}[\mathbf{x}]$ whose degrees are at most d. Apparently, the dimension of $\mathbf{R}[\mathbf{x}]_d$ is $\binom{n+d}{d}$. An ideal of $\mathbf{R}[\mathbf{x}]$ is a subset J of $\mathbf{R}[\mathbf{x}]$ such that $J \cdot \mathbf{R}[\mathbf{x}] \subseteq J$ and $J + J \subseteq J$. Let $h := (h_1, \ldots, h_r)$ with each h_i in $\mathbf{R}[\mathbf{x}]$. The ideal generated by h, termed as $I(h)$, is exactly the set $\sum_{i=1}^{r} h_i \mathbf{R}[\mathbf{x}]$. The kth truncated ideal generated by h, denoted by $I_k(h)$, is defined as the set $\sum_{i=1}^{r} h_i \mathbf{R}[\mathbf{x}]_{k-\deg(h_i)}$, where $\deg(h_i)$ denotes the degree of the polynomial h_i. Obviously, $\cup_{k\in\mathbf{N}} I_k(h) = I(h)$, where $\mathbf{N}$ is the set of all nonnegative integers. A polynomial $p \in \mathbf{R}[\mathbf{x}]$ is called a sum of squares (SOS) if there exist $q_1, \ldots, q_m \in \mathbf{R}[\mathbf{x}]$ such that $p = \sum_{i=1}^{m} q_i^2$. We use $\Sigma[\mathbf{x}]$ to denote the set of all SOS polynomials and $\Sigma[\mathbf{x}]_k$ to denote the intersection of $\Sigma[\mathbf{x}]$ and $\mathbf{R}[\mathbf{x}]_k$. The set $Q_k(h) := \Sigma[\mathbf{x}]_{2k} + \sum_{i=1}^{r} h_i \Sigma[\mathbf{x}]_{2k-\deg(h_i)}$ is called the kth quadratic module generated by h and the union $Q(h) := \cup_{k\ge 0} Q_k(h)$ is called the quadratic module generated by h.

For any vector $\mathbf{x} = (x_1, \ldots, x_n)^\top$ and any $\alpha := (\alpha_1, \ldots, \alpha_n)$, denote $\mathbf{x}^\alpha := x_1^{\alpha_1} \cdots x_n^{\alpha_n}$ and $|\alpha| := \sum_{i=1}^{n} \alpha_i$. Denote $\mathbf{N}_d^n := \{\alpha \in \mathbf{N}^n : |\alpha| \le d\}$, e.g., $\mathbf{N}_2^2 = \{(0,0), (0,1), (1,0), (2,0), (1,1), (0,2)\}$. Let $\mathbf{R}^{\mathbf{N}_d^n}$ be the real sequences indexed by $\alpha \in \mathbf{N}_d^n$, i.e., $\mathbf{R}^{\mathbf{N}_d^n} := \{\mathbf{y} : \mathbf{y} = (y_\alpha)_{\alpha \in \mathbf{N}_d^n}\}$, e.g., $\mathbf{y} = (y_{00}, y_{01}, y_{10}, y_{20}, y_{11}, y_{02})^\top \in \mathbf{R}^{\mathbf{N}_2^2}$. Each $\mathbf{y} \in \mathbf{R}^{\mathbf{N}_d^n}$ is called a *truncated moment sequence (tms) of degree d*.

Let $E := \{\alpha \in \mathbf{N}^{n+1} : |\alpha| = m\}$. Then, each index $(i_1, \ldots, i_m)$ corresponds to a vector $\sum_{j \in [m]} \mathbf{1}^{(i_j)} \in E$, where each $\mathbf{1}^{(i_j)} \in \mathbf{R}^{n+1}$ is the i_jth unit vector. In this regard, each symmetric tensor corresponds to a unique identifying vector

$$\mathbf{a} = (a_\alpha)_{\alpha \in E} \in \mathbf{R}^E := \{\mathbf{a} : \mathbf{a} = (a_\alpha)_{\alpha \in E}, a_\alpha \in \mathbf{R}\}.$$

The vector $\mathbf{a}$ is called an E-truncated moment sequence (E-tms for short) of $\mathscr{A}$.

Let

$$K := \{\mathbf{x} \in \mathbf{R}^{n+1} : \mathbf{x}^\top \mathbf{x} - 2 = 0, x_1 - 1 = 0\}. \tag{9.23}$$

Obviously, K is a compact set in $\mathbf{R}^{n+1}$. By employing Theorem 9.8, a tensor $\mathscr{A} \in RD_{m,n+1}$ if and only if there exist $\alpha_k > 0$ and $\mathbf{v}^{(k)} = \left(1, v_1^{(k)}, \ldots, v_n^{(k)}\right)^\top \in K$ for $k = 1, \ldots, r$ such that

$$\mathscr{A} = \sum_{k=1}^{r} \alpha_k \left(\mathbf{v}^{(k)}\right)^{\otimes m}. \tag{9.24}$$

For any given $\mathbf{a} \in \mathbf{R}^E$, a nonnegative Borel measure μ supported in K is called a *K-representing measure* for $\mathbf{a}$ if

$$\mathbf{a}_\alpha = \int_K \mathbf{x}^\alpha d\mu, \quad \forall \alpha \in E.$$

Recall that a measure is called *finitely atomic* if its support is a finite set, and is called a *r-atomic* if its support consists of at most r distinct points (more details see [202]). Therefore,

$$\mathscr{A} \in RD_{m,n+1} \Longleftrightarrow \tag{9.25}$$
$$\text{the identifying vector } \mathbf{a} \text{ of } \mathscr{A} \text{ admits a } K\text{-representing measure,}$$

where K is defined in (9.23).

Denote $\mathbf{R}[\mathbf{x}]_E := span\{\mathbf{x}^\alpha : \alpha \in E\}$. Obviously, $\mathbf{R}[\mathbf{x}]_E$ is a linear subspace of $\mathbf{R}[\mathbf{x}]_m$. For a polynomial $p \in \mathbf{R}[\mathbf{x}]_E$, $p|_K \geq 0 (> 0)$ denotes that $p(\mathbf{x}) \geq 0 (> 0)$ for any $\mathbf{x} \in K$. Recall from [96] that $\mathbf{R}[\mathbf{x}]_E$ is *K-full* if there exists a polynomial $p \in \mathbf{R}[\mathbf{x}]_E$ such that $p|_K > 0$. By choosing $p(x) := x_1^m > 0$ in our case, it is easily seen that $\mathbf{R}[\mathbf{x}]_E$ is K-full.

Given a tms $\mathbf{y}$, we define the Riesz functional $\mathbf{L}_\mathbf{y}$ acting on $\mathbf{R}[\mathbf{x}]_d$ in the manner that

$$\mathbf{L}_\mathbf{y}(p(\mathbf{x})) = \mathbf{L}_\mathbf{y}(\sum_{\alpha \in \mathcal{N}_d^n} p_\alpha \mathbf{x}^\alpha) := \sum_{\alpha \in \mathcal{N}_d^n} p_\alpha y_\alpha,$$

where p_α is the coefficient of $\mathbf{x}^\alpha$ in p. For example, when $n = 2$, $d = 2$, $\mathbf{y} = (1, 2, 3, 4, 5, 6)^\top$, and $p(\mathbf{x}) = 1 + x_1 - x_2^2$, we have

$$\mathbf{L}_{\mathbf{y}}(p) = 1 \times 1 + 1 \times 3 + (-1) \times 6 = -2.$$

For any $q \in \mathbf{R}[\mathbf{x}]$ with $\deg(q) \le 2k$, the kth localizing matrix of q, generated by a tms $\mathbf{y} \in \mathbf{R}^{\mathcal{N}^n_{2k}}$, is the symmetric matrix $L_q^{(k)}(\mathbf{y})$ such that

$$\mathbf{L}_{\mathbf{y}}(qpp') = vec(p)^\top \left(L_q^{(k)}(\mathbf{y})\right) vec(p'),$$

for all $p, p' \in \mathbf{R}[\mathbf{x}]$ with $\deg(p), \deg(p') \le k - \lceil \deg q/2 \rceil$. Here $vec(p)$ denotes the coefficient vector of the polynomial p and $\lceil t \rceil$ is the ceil function taking the smallest integer no less than t. For example, take $n = 2$, $k = 2$ and $q = 1 + x_1 - x_2^2$, then

$$L_q^{(k)}(\mathbf{y}) = \begin{pmatrix} y_{00} + y_{10} - y_{02} & y_{10} + y_{20} - y_{12} & y_{01} + y_{11} - y_{03} \\ y_{10} + y_{20} - y_{12} & y_{20} + y_{30} - y_{22} & y_{11} + y_{21} - y_{13} \\ y_{01} + y_{11} - y_{03} & y_{11} + y_{21} - y_{13} & y_{02} + y_{12} - y_{04} \end{pmatrix}.$$

When $q = 1$, $L_1^{(k)}(\mathbf{y})$ is called the *kth moment matrix* generated by $\mathbf{y}$ and is simply denoted as $M_k(\mathbf{y})$. Further, when $n = 2$, we have

$$M_1(\mathbf{y}) = L_1^{(1)}(\mathbf{y}) = \begin{pmatrix} y_{00} & y_{10} & y_{01} \\ y_{10} & y_{20} & y_{11} \\ y_{01} & y_{11} & y_{02} \end{pmatrix}$$

and

$$M_2(\mathbf{y}) = L_1^{(2)}(\mathbf{y}) = \begin{pmatrix} y_{00} & y_{10} & y_{01} & y_{20} & y_{11} & y_{02} \\ y_{10} & y_{20} & y_{11} & y_{30} & y_{21} & y_{12} \\ y_{01} & y_{11} & y_{02} & y_{21} & y_{12} & y_{03} \\ y_{20} & y_{30} & y_{21} & y_{40} & y_{31} & y_{22} \\ y_{11} & y_{21} & y_{12} & y_{31} & y_{22} & y_{13} \\ y_{02} & y_{12} & y_{03} & y_{22} & y_{13} & y_{04} \end{pmatrix}.$$

Note that the set K defined in (9.23) can be rewritten as

$$K = \left\{\mathbf{x} \in \mathbf{R}^{n+1} : h_1(\mathbf{x}) = 0, h_2(\mathbf{x}) = 0\right\}$$

with $h_1(\mathbf{x}) := x_1 - 1$ and $h_2(\mathbf{x}) := \mathbf{x}^\top\mathbf{x} - 2$. As shown in [202], a necessary condition for $\mathbf{y} \in \mathbf{R}^{\mathcal{N}^n_{2k}}$ to admit a K-representing measure is

$$L_{h_1}^{(k)}(\mathbf{y}) = 0, \text{ and } L_{h_2}^{(k)}(\mathbf{y}) = 0. \tag{9.26}$$

Additionally, if the following rank condition

$$rank(M_{k-1}(\mathbf{y})) = rank(M_k(\mathbf{y})) \tag{9.27}$$

holds at $\mathbf{y}$, we say that $\mathbf{y}$ is *flat* and in this case, $\mathbf{y}$ admits a unique K-representing measure. Recall from [202] that a tms $\mathbf{z} \in \mathbf{R}^{\mathbf{N}_d^{n+1}}$ is said to be an extension of $\mathbf{y} \in \mathbf{R}^{\mathbf{N}_m^{n+1}}$ if $m \geq d$ and $y_\alpha = z_\alpha$ for all $\alpha \in \mathbf{N}_d^{n+1}$. If $\mathbf{z}$ is flat and extends $\mathbf{y}$, then we say $\mathbf{z}$ is a *flat extension* of $\mathbf{y}$. An E-tms $\mathbf{a} \in \mathbf{R}^E$ admits a K-measure if and only if it has a flat extension $\mathbf{z} \in \mathbf{R}^{\mathbf{N}_{2k}^{n+1}}$ for some k. Therefore, checking whether a symmetric tensor $\mathscr{A}$ is in $RD_{m,n+1}$ is equivalent to checking whether its identifying vector $\mathbf{a} \in \mathbf{R}^E$ has a flat extension or not. Thus, we have the following theorem [21].

Theorem 9.12 (Bohnet-Waldraff, Braun, Giraud 2017) *A spin-j state is classical if and only if the identifying vector of its representing tensor has a flat extension.*

Now, the identification of the classicality of a spin-j state is transformed to the existence problem of a flat extension for the identifying vector of its corresponding tensor representation. Let $t > m$ be any even integer and choose a polynomial $p \in \mathbf{R}[\mathbf{x}]_t$ with $p(\mathbf{x}) = \sum_{\alpha \in \mathbf{N}_t^{n+1}} p_\alpha \mathbf{x}^\alpha$. Consider the following optimization problem

$$\min_{\mathbf{z}} \left\{ \sum_{\alpha \in \mathbf{N}_t^{n+1}} p_\alpha z_\alpha : \mathbf{z}|_E = \mathbf{a}, \mathbf{z} \in \mathbf{R}^{\mathbf{N}_t^{n+1}}, \mathbf{z} \text{ admits a } K\text{-representing measure} \right\}, \tag{9.28}$$

where $\mathbf{a} \in \mathbf{R}^{\mathbf{N}_m^{n+1}}$ is given. Since K is compact and $\mathbf{R}[\mathbf{x}]_E$ is K-full, the feasible set of problem (9.28) is compact and convex. By convex programming theory, the solution set of (9.28) is nonempty. To make the problem tractable, we usually choose an SOS polynomial p in $n+1$ variables with degree t, and relax the feasible region to

$$\mathscr{F}_k(h_1, h_2) := \left\{ \mathbf{z} \in \mathbf{R}^{\mathbf{N}_{2k}^{n+1}} : \mathbf{z}|_E = \mathbf{a}, L_{h_1}^{(k)}(\mathbf{z}) = 0, L_{h_2}^{(k)}(\mathbf{z}) = 0 \right\} \tag{9.29}$$

with $k \geq d/2$ being an integer. The kth ($k = d/2, d/2+1, \ldots$) order relaxation of (9.28) is

$$\min_{\mathbf{z}} \left\{ \sum_{\alpha \in \mathbf{N}_{2k}^{n+1}} p_\alpha z_\alpha : \mathbf{z} \in \mathscr{F}_k(h_1, h_2) \right\}. \tag{9.30}$$

The corresponding semidefinite relaxation algorithm is established in Algorithm 3 based upon solving the hierarchy of (9.30) for $k = d/2, d/2+1, \ldots$.

Additional remarks are needed for the above algorithm. In Step 0, we choose $p(\mathbf{x}) = [\mathbf{x}]_{d/2}^\top G^\top G[\mathbf{x}]_{d/2}$, where G is a random square matrix obeying Gaussian distribution. In Step 2, we use singular value decomposition to numerically compute

---

Algorithm 3 A relaxation algorithm to identify the classicality of a spin-j state

Input: The tensor representation $\mathscr{A}$ of a spin-j state;

Output: The certification of non-classicality or the coefficients α_k's and the vectors $\mathbf{v}_k$'s to generate a decomposition of $\mathscr{A}$ as described in (9.22);

0 Compute the identifying vector $\mathbf{a}$ from $\mathscr{A}$. Choose a generic SOS polynomial p and set $k = d/2$.

1 Solve (9.30). If (9.30) is infeasible, then output a certificate that the spin-j state is not classical and stop. Otherwise, compute a solution $\mathbf{z}^*(k)$. Let $l = 1$.

2 compute $r_1 := rank(M_{k-1}(\mathbf{z}^*(k)|_{2l})$ and $r_2 := rank(M_k(\mathbf{z}^*(k)|_{2l})$. If $r_1 \neq r_2$, then go to Step 4, otherwise go to Step 3.

3 Compute the finite atomic measure μ admitted by $\mathbf{z}^*(k)|_{2l}$ as follows:

$$\mu = \sum_{k\in[r]} \alpha_k \delta(\mathbf{v}^{(k)}),$$

where $\alpha_k > 0$, $\mathbf{v}^{(k)} \in K$, $r = rank(M_l(\mathbf{z}^*(k)|_{2l})$, and $\delta(\mathbf{v}^{(k)})$ is the Dirac measure supported on point $\mathbf{v}^{(k)}$ $(k = 1, \ldots, r)$. Stop.

4 If $l < k$, set $l = l + 1$ and go to Step 2; Otherwise, set $k = k + 1$ and go to Step 1.

Return: A certificate or α_k, $\mathbf{v}^{(k)}$ for $k \in [r]$.

---

the desired rank values by counting the number of the singular values greater than 10^{-6}. In Step 3, the method was proposed in [121] to get the measure μ. By invoking that $\mathbf{R}[\mathbf{x}]_E$ is K-full, the following convergence results follow readily from [202].

Theorem 9.13 *For Algorithm 3, we have*

(a) If (9.30) is infeasible for some k, then the spin-j state is not classical;
(b) If the spin-j state is not classical, then (9.30) is infeasible for all k big enough;
(c) If the spin-j state is classical, then for almost all generated p, we can asymptotically get a flat extension of **a** *by solving the hierarchy of (9.30).*

The first experiment is for the following randomly generated cases with $n = 3$ and (m, r) in different positive integer values.

(a) Randomly generate $\mathbf{x}^{(k)} = randn(1, n)$ and set $\mathbf{v}^{(k)} = \left(1, \frac{\mathbf{x}^{(k)\top}}{\|\mathbf{x}^{(k)}\|}\right)^\top$ for $k = 1, \ldots, r$ to get the corresponding representation tensor $\mathscr{A}$. (Obviously, all such generated tensors are regularly decomposable tensors.) Results are listed in Table 9.1.

(b) Randomly generate symmetric tensors $\mathscr{A} \in S_{m,n}$ with $a_{11i_3\ldots i_m} = \sum_{i=2}^{n+1} a_{iii_3\ldots i_m} + 1$ for some randomly chosen index $(i_3, \ldots, i_m)$. (It is known from Theorem 9.8 that none of such $\mathscr{A}$'s are regularly decomposable.) Results are listed in Table 9.2.

The second experiment is testing on the data of entangled states and classical states. Each class contains 100 spin-j states for different j ranging from 1 to 3. As we know, tensors corresponding to the classical states are regularly decomposable, while

Table 9.1 Identification for Case (i) with different m's and r's

(m, r)	Instance no.	Success ratio (%)	Average time (s)	Average error
(4, 3)	1000	100	0.998895	4.633782e-08
(4, 6)	1000	100	1.068820	9.009269e-06
(4, 10)	1000	100	3.461200	1.314734e-02
(5, 3)	1000	100	1.064470	5.598596e-07
(5, 6)	1000	100	1.148770	5.413121e-06
(5, 10)	1000	100	4.116700	4.841370e-04
(8, 3)	1000	100	10.207400	4.155197e-07
(8, 6)	1000	100	10.580800	6.702206e-07
(8, 10)	1000	100	12.043500	7.853554e-06
(10, 3)	1000	100	43.965400	1.071692e-07
(10, 6)	1000	100	47.134200	1.401306e-05
(10, 10)	1000	100	47.501800	2.199560e-05
(12, 15)	1000	100	176.678100	9.420467e-06

Table 9.2 Identification for Case (ii) with different m's

(m, r)	Instance no.	Success ratio (%)	Average time (s)
4	1000	100	0.715840
5	1000	100	1.029300
8	1000	100	6.573300
10	1000	100	29.981300
12	1000	100	118.626000

tensors corresponding to entangled states should not be regularly decomposable. For example, for a classical spin-5/2 state with density matrix

$$\rho = \begin{pmatrix} 0.1621 + 0.0000i & 0.0581 - 0.0020i & -0.0074 - 0.0050i & -0.0073 + 0.0254i & -0.0010 + 0.0109i & -0.0016 + 0.0002i \\ 0.0581 + 0.0020i & 0.1679 + 0.0000i & 0.0734 - 0.0099i & -0.0081 + 0.0112i & -0.0145 + 0.0188i & 0.0021 + 0.0052i \\ -0.0074 + 0.0050i & 0.0734 + 0.0099i & 0.1323 + 0.0000i & 0.0456 + 0.0126i & -0.0009 + 0.0168i & -0.0146 + 0.0071i \\ -0.0073 - 0.0254i & -0.0081 - 0.0112i & 0.0456 - 0.0126i & 0.1084 + 0.0000i & 0.0010 + 0.0466i & 0.0112 + 0.0031i \\ -0.0010 - 0.0109i & -0.0145 - 0.0188i & -0.0009 - 0.0168i & 0.0010 - 0.0466i & 0.1572 + 0.0000i & -0.0488 + 0.0865i \\ -0.0016 - 0.0002i & 0.0021 - 0.0052i & -0.0146 - 0.0071i & 0.0112 - 0.0031i & -0.0488 - 0.0865i & 0.2720 - 0.0000i \end{pmatrix},$$

its corresponding tensor $\mathscr{A} \in S_{5,4} = \left(a_{i_1 \ldots i_5}\right)$ has the following representing nonzero entries

Table 9.3 The α_k's and $\mathbf{v}^{(k)}$'s of the regular decomposition of $\mathscr{A}$

α_k's	$(\mathbf{v}^{(k)})^\top$'s
0.1153	(1.0000 −0.6399 −0.1329 −0.7568)
0.0744	(1.0000 −0.2910 −0.3733 −0.8808)
0.0073	(1.0000 −0.5795 −0.8084 0.1027)
0.0352	(1.0000 −0.7261 0.1853 −0.6619)
0.0950	(1.0000 0.0344 −0.6409 −0.7667)
0.0424	(1.0000 0.1770 −0.9618 −0.2088)
0.0714	(1.0000 0.1108 0.0316 −0.9933)
0.0646	(1.0000 0.7443 −0.3981 −0.5361)
0.0751	(1.0000 −0.2113 −0.4739 0.8549)
0.1314	(1.0000 0.7745 −0.5572 0.2995)
0.0307	(1.0000 −0.0240 0.3697 0.9289)
0.1271	(1.0000 0.1990 0.8833 0.4246)
0.0800	(1.0000 0.7009 −0.0452 0.7118)
0.0504	(1.0000 0.7741 0.4804 0.4123)

$a_{11111} = 1$, $a_{11112} = 0.1472$, $a_{11122} = 0.2693$, $a_{11222} = 0.0923$, $a_{12222} = 0.1354$,
$a_{22222} = 0.0588$, $a_{11113} = -0.1322$, $a_{11123} = -0.0226$, $a_{11223} = -0.0511$, $a_{12223} = -0.0303$,
$a_{22223} = -0.0280$, $a_{11133} = 0.2797$, $a_{11233} = 0.0647$, $a_{12233} = 0.0471$, $a_{22233} = 0.0273$,
$a_{11333} = -0.0106$, $a_{12333} = -0.0014$, $a_{13333} = 0.1556$, $a_{22333} = -0.0123$, $a_{23333} = 0.0324$,
$a_{33333} = 0.0121$, $a_{11114} = -0.0987$, $a_{11124} = 0.1366$, $a_{11224} = -0.0047$, $a_{12224} = 0.0661$,
$a_{22224} = 0.0034$, $a_{11134} = 0.1111$, $a_{11234} = 0.0067$, $a_{12234} = 0.0044$, $a_{22234} = -0.0019$,
$a_{11334} = 0.0229$, $a_{12334} = 0.0159$, $a_{13334} = 0.0596$, $a_{22334} = 0.0076$, $a_{23334} = 0.0082$,
$a_{33334} = 0.0198$, $a_{11144} = 0.4510$, $a_{11244} = -0.0099$, $a_{12244} = 0.0868$, $a_{22244} = 0.0062$,
$a_{11344} = -0.0705$, $a_{12344} = 0.0091$, $a_{13344} = 0.0770$, $a_{22344} = -0.0108$, $a_{23344} = 0.0051$,
$a_{33344} = -0.0104$, $a_{11444} = -0.1169$, $a_{12444} = 0.0546$, $a_{13444} = 0.0471$, $a_{14444} = 0.2872$,
$a_{22444} = -0.0157$, $a_{23444} = 0.0004$, $a_{24444} = -0.0212$, $a_{33444} = -0.0044$, $a_{34444} = -0.0493$,
$a_{44444} = -0.0968$.

Using Algorithm 3, we can identify its classicality and obtain an approximate decomposition $\mathscr{B} := \sum_{k=1}^{14} \alpha_k (\mathbf{v}^{(k)})^{\otimes 5}$ with the relative error $\eta := \frac{\|\mathscr{A} - \mathscr{B}\|_F}{\|\mathscr{A}\|_F} = 5.495371\text{e-}05$.
The α_k's and $\mathbf{v}^{(k)}$'s are listed in Table 9.3.
And for an entangled spin-2 state with density matrix

$$\rho = \begin{pmatrix} 0.1852 + 0.0000i & -0.0737 - 0.0362i & -0.0482 + 0.0792i & 0.0579 + 0.0692i & 0.0812 - 0.0258i \\ -0.0737 + 0.0362i & 0.0997 + 0.0000i & -0.0231 - 0.0642i & -0.0860 + 0.0117i & -0.0572 - 0.0316i \\ -0.0482 - 0.0792i & -0.0231 + 0.0642i & 0.2556 + 0.0000i & 0.0465 - 0.1158i & 0.0445 - 0.0175i \\ 0.0579 - 0.0692i & -0.0860 - 0.0117i & 0.0465 + 0.1158i & 0.1144 + 0.0000i & 0.0274 + 0.0255i \\ 0.0812 + 0.0258i & -0.0572 + 0.0316i & 0.0445 + 0.0175i & 0.0274 - 0.0255i & 0.3452 + 0.0000i \end{pmatrix},$$

its corresponding tensor $\mathscr{A} \in S_{4,4} = \left(a_{i_1 \ldots i_4}\right)$ has the following representing nonzero entries

$$
\begin{array}{llll}
a_{1111} = 1, & a_{1112} = 0.0480, & a_{1122} = 0.1660, & a_{1222} = -0.0292, \\
a_{2222} = 0.0536, & a_{1113} = 0.0107, & a_{1123} = 0.0324, & a_{1223} = -0.0387, \\
a_{2223} = 0.0169, & a_{1133} = 0.3337, & a_{1233} = 0.0373, & a_{1333} = 0.0540, \\
a_{2233} = 0.0522, & a_{2333} = 0.0206, & a_{3333} = 0.2100, & a_{1114} = 0.0580, \\
a_{1124} = 0.0231, & a_{1224} = -0.0026, & a_{2224} = 0.0038, & a_{1134} = -0.0604, \\
a_{1234} = -0.0032, & a_{1334} = 0.0068, & a_{2234} = -0.0086, & a_{2334} = 0.0031, \\
a_{3334} = -0.0291, & a_{1144} = 0.5003, & a_{1244} = 0.0399, & a_{1344} = -0.0046, \\
a_{1444} = 0.0538, & a_{2244} = 0.0602, & a_{2344} = -0.0051, & a_{2444} = 0.0162, \\
a_{3344} = 0.0714, & a_{3444} = -0.0227, & a_{4444} = 0.3686. &
\end{array}
$$

Implement Algorithm 3, a certificate of non-classicality is given for such an entangled state.

9.6 Notes

Recently, Chinese quantum satellite, named Micius, has sent "spooky action" messages over 1,200 km. This news motivates and encourages us to study in quantum physics and related areas.

Section 9.1 The physical phenomenon of quantum entanglement studied in [20, 89, 209] is different from classical physics.

Section 9.2 The conception of geometric measure of entanglement was proposed by Shimony [243]. Qi [222] gave bounds for the minimum Hartree value of entanglement states. The relationship between the minimum Hartree value and the geometric measure of entanglement was also addressed. Several numerical algorithms have been developed for determining whether a quantum state is separable, which is fundamental in quantum science [37, 115, 279].

Section 9.3 Hu, Qi and Zhang [132] applied spectral theory of nonnegative tensors for the geometric measure of entanglement of multipartite pure states. Zhang and Qi [309] considered the cases of real symmetric tensors. From the viewpoint of geometric measure of quantum entanglement, this problem is related to the entanglement eigenvalue [131] and the best rank-one complex approximation of tensors [199].

Section 9.4 Ni, Qi and Bai [199] introduced unitary eigenvalues and unitary symmetric eigenvalues. They also addressed the relationship between the unitary symmetric eigenvalue and the quantum spectral radius of a representing tensor of quantum states. Several numerical algorithms for computing unitary eigenvalues of complex tensors were proposed in [38, 133, 198].

Section 9.5 Qi, Zhang, Braun, Bohnet-Waldraff and Giraud [233] studied the classicality of spin states from the viewpoint of spectral tensor theory. They introduced a new class of tensors named regularly decomposable tensors. A spin-j state is classical if and only if its representing tensor is a regularly decomposable tensor. Braun, Bohnet-Waldraff and Giraud [21] proposed a polynomial optimization algorithm for identifying whether a tensor is regularly decomposable or not.

9.7 Exercises

1 Suppose that the symmetric tensor $\mathscr{A}_\Psi$ of an m-partite pure state $|\Psi\rangle$ is real. The even number $m \geq 4$, an element of $\mathscr{A}_\Psi$ is nonzero if and only if (a) one half of its indices are equal and the other half are also constant, (b) the absolute value of its diagonal elements are greater than or equal to $\binom{m-1}{\frac{m}{2}}$ times of the absolute value of any off-diagonal element whose indices overlap with that diagonal element. Prove that $Q(\Psi) = Z(\Psi)$, where $Q(\Psi)$ is the quantum spectral radius of $\mathscr{A}_\Psi$ and $Z(\Psi)$ is the largest quantum eigenvalue of $\mathscr{A}_\Psi$.

[Hint: This is from [309]].

2 Prove part (b) of Theorem 9.5.

[Hint: This is from [199]].

References

1. Abo, H., Seigal, A., Sturmfels, B.: Eigenconfigurations of tensors. Algebr. Geom. Methods Discret. Math. **685**, 1–25 (2017)
2. Afzali, M., Sardouie, S.H., Fatemizadeh, E., Soltanian-Zadeh, H.: Canonical polyadic decomposition for principal diffusion direction extraction in diffusion weighted imaging. In: 2017 Iranian Conference on Electrical Engineering (ICEE), pp. 122–127 (2017)
3. Agangic, M., Cottle, R.: A note on Q-matrices. Math. Program. **16**, 374–377 (1979)
4. Aubin, J., Frankowska, H.: Set-Valued Analysis. Springer, Berlin (2009)
5. Bai, S., Lu, L.: A bound on the spectral radius of hypergraphs with e edges (2017). arXiv:1705.01593
6. Bai, X., Huang, Z., Wang, Y.: Global uniqueness and solvability for tensor complementarity problems. J. Optim. Theory Appl. **170**, 72–84 (2016)
7. Banerjee, A., Char, A., Mondal, B.: Spectra of general hypergraphs. Linear Algebra Appl. **518**, 14–30 (2017)
8. Barmpoutis, A., Jian, B., Vemuri, B., Shepherd, T.: Symmetric positive 4th order tensors and their estimation from diffusion weighted MRI. In: Information Processing in Medical Imaging. Springer, Berlin, pp. 308–319 (2007)
9. Basser, P.J., Mattiello, J., LeBihan, D.: MR diffusion tensor spectroscopy and imaging. Biophys. J. **66**(1), 259–267 (1994)
10. Basser, P.J., Mattiello, J., LeBihan, D.: Estimation of the effective self-diffusion tensor from the NMR spin echo. J. Magn. Reson. Ser. B **103**(3), 247–254 (1994)
11. Batselier, K., Chen, Z., Wong, N.: Tensor network alternating linear scheme for MIMO volterra system identification. Automatica **84**, 26–35 (2017)
12. Batselier, K., Wong, N.: Symmetric tensor decomposition by an iterative eigendecomposition algorithm. J. Comput. Appl. Math. **308**, 69–82 (2016)
13. Behera, R., Mishra, D.: Further results on generalized inverses of tensors via the Einstein product. Linear Multilinear Algebra **65**(8), 1662–1682 (2017)
14. Behrens, T.E., Berg, H.J., Jbabdi, S., Rushworth, M.F.S., Woolrich, M.W.: Probabilistic diffusion tractography with multiple fibre orientations: What can we gain? Neuroimage **34**(1), 144–155 (2007)
15. Bengtsson, I., Zyczkowski, K.: Geometry of Quantum States: an Introduction to Quantum Entanglement. Cambridge University Press, Cambridge (2007)
16. Benson, A.R., Gleich, D.F., Lim, L.H.: The spacey random walk: a stochastic process for higher-order data. SIAM Rev. **59**(2), 321–345 (2017)
17. Berman, A., Plemmons, R.J.: Nonnegative Matrices in the Mathematical Sciences. Classics in Applied Mathematics, Revised reprint of the 1979 original, vol. 9. Society for Industrial and Applied Mathematics (SIAM), Philadelphia (1994)

L. Qi et al., *Tensor Eigenvalues and Their Applications*, Advances in Mechanics and Mathematics 39, https://doi.org/10.1007/978-981-10-8058-6

18. Bloy, L., Verma, R.: On computing the underlying fiber directions from the diffusion orientation distribution function. In: Metaxas, D., Axel, L., Fichtinger, G., Székeley, G. (eds.) Medical Image Computing and Computer-Assisted Intervention, pp. 1–8. Springer, Berlin (2008)
19. Bochnak, J., Coste, M., Roy, M.: Real Algebraic Geometry. Springer, Berlin (1998)
20. Bohnet-Waldraff, F., Braun, D., Giraud, O.: Tensor eigenvalues and entanglements of symmetric states. Phys. Rev. A **94**(4), 042324 (2016)
21. Bohnet-Waldraff, F., Braun, D., Giraud, O.: Entanglement and the truncated moment problem. Phys. Rev. A **96**(3), 032312 (2017)
22. Bolte, J., Daniilidis, A., Lewis, A.: The Łojasiewicz inequality for nonsmooth subanalytic functions with applications to subgradient dynamical systems. SIAM J. Optim. **17**, 1205–1223 (2007)
23. Boyer, R., De Lathauwer, L., Abed-Meraim, K.: Higher order tensor-based method for delayed exponential fitting. IEEE Trans. Signal Process. **55**, 2795–2809 (2007)
24. Brand, H.R., Pleiner, H., Cladis, P.E.: Flow properties of the optically isotropic tetrahedratic phase. Eur. Phys. J. E **7**(2), 163–166 (2002)
25. Brazell, M., Li, N., Navasca, C., Tamon, C.: Solving multilinear systems via tensor inversion. SIAM J. Matrix Anal. Appl. **34**, 542–570 (2013)
26. Breiding, P.: The expected number of eigenvalues of a real gaussian tensor. SIAM J. Appl. Algebra Geom. **1**(1), 254–271 (2017)
27. Bu, C., Jin, X., Li, H., Deng, C.: Brauer-type eigenvalue inclusion sets and the spectral radius of tensors. Linear Algebra Appl. **512**, 234–248 (2017)
28. Buckingham, A.D.: Angular correlation in liquids. Discuss. Faraday Soc. **43**, 205–211 (1967)
29. Cartwright, D., Sturmfels, B.: The number of eigenvalues of a tensor. Linear Algebra Appl. **438**, 945–962 (2013)
30. Chang, J., Chen, Y., Qi, L.: Computing eigenvalues of large scale sparse tensors arising from a hypergraph. SIAM J. Sci. Comput. **38**, A3618–A3643 (2016)
31. Chang, J., Ding, W., Qi, L., Yan, H.: Computing the p-spectral radii of uniform hypergraphs with applications. J. Sci. Comput. **75**, 1–25 (2018)
32. Chang, K.C., Pearson, K., Zhang, T.: Perron-Frobenius theorem for nonnegative tensors. Commun. Math. Sci. **6**, 507–520 (2008)
33. Chang, K.C., Pearson, K., Zhang, T.: On eigenvalue problems of real symmetric tensors. J. Math. Anal. Appl. **350**, 416–422 (2009)
34. Chang, K.C., Qi, L., Zhang, T.: A survey on the spectral theory of nonnegative tensors. Numer. Linear Algebra Appl. **20**, 891–912 (2013)
35. Che, M., Cichocki, A., Wei, Y.: Neural networks for computing best rank-one approximations of tensors and its applications. Neurocomputing **267**, 114–133 (2017)
36. Che, M., Li, G., Qi, L., Wei, Y.: Pseudo-spectra theory of tensors and tensor polynomial eigenvalue problems. Linear Algebra Appl. **533**, 536–572 (2017)
37. Che, M., Qi, L., Wei, Y.: Positive definite tensors to nonlinear complementarity problems. J. Optim. Theory Appl. **168**, 475–487 (2016)
38. Che, M., Qi, L., Wei, Y.: Iterative algorithms for computing US-and U-eigenpairs of complex tensors. J. Comput. Appl. Math. **317**, 547–564 (2017)
39. Che, M., Wei, Y.: An inequality for the Perron pair of an irreducible and symmetric nonnegative tensor with application. J. Oper. Res. Soc. China **5**(1), 65–82 (2017)
40. Chen, B., Chen, X., Kanzow, C.: A penalized Fischer-Burmeister NCP-function. Math. Program. **88**, 211–216 (2000)
41. Chen, B., He, S., Li, Z., Zhang, S.: On new classes of nonnegative symmetric tensors. SIAM J. Optim. **27**(1), 292–318 (2017)
42. Chen, D., Chen, Z., Zhang, X.D.: Spectral radius of uniform hypergraphs and degree sequences. Front. Math. China **12**(6), 1279–1288 (2017)
43. Chen, H., Chen, Y., Li, G., Qi, L.: A semi-definite program approach for computing the maximum eigenvalue of a class of structured tensors and its applications in hypergraphs and copositivity test. Numer. Linear Algebra Appl. **25**(1), e2125 (2018)

44. Chen, H., Huang, Z., Qi, L.: Copositivity detection of tensors: theory and algorithm. J. Optim. Theory Appl. **174**, 746–761 (2017)
45. Chen, H., Huang, Z., Qi, L.: Copositive tensor detection and its applications in physics and hypergraphs. Comput. Optim. Appl. **69**(1), 133–158 (2018)
46. Chen, L., Han, L., Yin, H., Zhou, L.: A homotopy method for computing the largest eigenvalue of an irreducible nonnegative tensor (2017). arXiv:1701.07534
47. Chen, L., Han, L., Zhou, L.: Computing tensor eigenvalues via homotopy continuation. SIAM J. Matrix Anal. Appl. **37**, 290–319 (2016)
48. Chen, T.: Unmixing the mixed volume computation (2017). arXiv:1703.01684
49. Chen, Y., Dai, Y., Han, D.: Fiber orientation distribution estimation using a Peaceman-Rachford splitting method. SIAM J. Imaging Sci. **9**, 573–604 (2016)
50. Chen, Y., Dai, Y., Han, D., Sun, W.: Positive semidefinite generalized diffusion tensor imaging via quadratic semidefinite programming. SIAM J. Imaging Sci. **6**, 1531–1552 (2013)
51. Chen, Y., Jákli, A., Qi, L.: Spectral analysis of piezoelectric-type tensors (2017). arXiv:1703.07937
52. Chen, Y., Qi, L., Virga, E.G.: Octupolar tensors for liquid. J. Phys. A: Math. Theor. **51**, 025206 (2018)
53. Chen, Y., Qi, L., Wang, Q.: Computing extreme eigenvalues of large scale Hankel tensors. J. Sci. Comput. **68**, 716–738 (2016)
54. Chen, Y., Qi, L., Wang, Q.: Positive semi-definiteness and sum-of-squares property of fourth order four dimensional Hankel tensors. J. Comput. Appl. Math. **302**, 356–368 (2016)
55. Chen, Y., Qi, L., Zhang, X.: The Fiedler vector of a Laplacian tensor for hypergraph partitioning. SIAM J. Sci. Comput. **39**, A2508–A2537 (2017)
56. Chen, Z., Qi, L.: A semismooth Newton method for tensor eigenvalue complementarity problem. Comput. Optim. Appl. **65**, 109–126 (2016)
57. Chen, Z., Yang, Q., Ye, Y.: Generalized eigenvalue complementarity problem for tensors. Pac. J, Optim (2016)
58. Cheung, M.M., Hui, E.S., Chan, K.C., Helpern, J.A., Qi, L., Wu, E.X.: Does diffusion kurtosis imaging lead to better neural tissue characterization? A rodent brain maturation study. Neuroimage **45**, 386–392 (2009)
59. Chiriţă, S., Danescu, A., Ciarletta, M.: On the strong ellipticity of the anisotropic linearly elastic materials. J. Elast. **87**, 1–27 (2007)
60. Clarke, F.: Nonsmooth analysis and optimization. In: Proceedings of the International Congress of Mathematicians, pp. 847–853 (1983)
61. Comon, P., Golub, G., Lim, L., Mourrain, B.: Symmetric tensors and symmetric tensor rank. SIAM J. Matrix Anal. Appl. **30**, 1254–1279 (2008)
62. Cottle, R.W.: Nonlinear programs with positively bounded Jacobians. SIAM J. Appl. Math. **14**, 147–158 (1966)
63. Cottle, R.W., Pang, J.S., Stone, R.E.: The Linear Complementarity Problem. Academic Press, Boston (1992)
64. Cox, D., Little, J., O'Shea, D.: Using Algebraic Geometry, 2nd edn. Springer, Berlin (2004)
65. Cox, D., Little, J., O'Shea, D.: Ideals, Varieties, and Algorithms: An Introduction to Computational Algebraic Geometry and Commutative Algebra. Springer, Berlin (2007)
66. Cui, C., Dai, Y., Nie, J.: All real eigenvalues of symmetric tensors. SIAM J. Matrix Anal. Appl. **35**, 1582–1601 (2014)
67. Cui, C., Li, Q., Qi, L., Yan, H.: A quadratic penalty method for hypergraph matching. J. Global Optim. **70**(1), 237–259 (2018)
68. Cui, J., Peng, G., Lu, Q., Huang, Z.: New iterative criteria for strong H-tensors and an application. J. Inequal. Appl. **2017**(1), 49 (2017)
69. Culp, J., Pearson, K., Zhang, T.: On the uniqueness of the Z_1-eigenvector of transition probability tensors. Linear Multilinear Algebra **65**(5), 891–896 (2017)
70. Curie, J., Curie, P.: Développement, par pression, de l'électricité polaire dans les cristaux hémièdres à faces inclinées. Comptes rendus (in French) **91**, 294–295 (1880)

71. Curie, J., Curie, P.: Contractions et dilatations produites par des tensions électriques dans les cristaux hémièdres à faces inclinées. Comptes rendus (in French) **93**, 1137–1140 (1881)
72. Curto, R., Fialkow, L.: Truncated K-moment problems in several variables. J. Oper. Theory **54**(1), 189–226 (2005)
73. Da Costa, A., Seeger, A.: Cone-constrained eigenvalue problems: theory and algorithms. Comput. Optim. Appl. **45**(1), 25–57 (2010)
74. de Jong, M., Chen, W., Geerlings, H., Asta, M., Persson, K.A.: A database to enable discovery and design of piezoelectric materials. Sci. Data **2**, 150053 (2015)
75. De Lathauwer, L., De Moor, B., Vandewalle, J.: On the best rank-1 and rank-$(R_1, R_2, \ldots, R_N)$ approximation of higher-order tensors. SIAM J. Matrix Anal. Appl. **21**, 1324–1342 (2000)
76. De Luca, T., Facchinei, F., Kanzow, C.: A semismooth equation approach to the solution of nonlinear complementarity problems. Math. program. **75**(3), 407–439 (1996)
77. Descoteaux, M., Angelino, E., Fitzgibbons, S., Deriche, R.: Apparent diffusion coefficients from high angular resolution diffusion imaging: estimation and applications. Magn. Reson. Med. **56**(2), 395–410 (2006)
78. Descoteaux, M., Angelino, E., Fitzgibbons, S., Deriche, R.: Regularized, fast, and robust analytical Q-ball imaging. Magn. Reson. Med. **58**(3), 497–510 (2007)
79. Ding, W., Liu, J., Qi, L., Yan, H.: Elasticity m-tensors and the strong ellipticity condition (2017). arXiv:1705.09911
80. Ding, W., Liu, J., Qi, L., Yan, H.: Bisymmetric M-tensors (2017). Preprint
81. Ding, W., Luo, Z., Qi, L.: P-tensors, P_0-tensors and tensor complementarity problem (2015). arXiv:1507.06731
82. Ding, W., Qi, L., Wei, Y.: M-tensors and nonsingular M-tensors. Linear Algebra Appl. **439**, 3264–3278 (2013)
83. Ding, W., Qi, L., Wei, Y.: Fast Hankel tensor-vector product and its application to exponential data fitting. Numer. Linear Algebra Appl. **22**, 814–832 (2015)
84. Ding, W., Qi, L., Wei, Y.: Inheritance properties and sum-of-squares decomposition of Hankel tensors: theory and algorithms. BIT **57**, 169–190 (2017)
85. Ding, W., Qi, L., Yan, H.: On some sufficient conditions for strong ellipticity (2017). arXiv:1705.05081
86. Ding, W., Wei, Y.: Generalized tensor eigenvalue problems. SIAM J. Matrix Anal. Appl. **36**, 1073–1099 (2015)
87. Ding, W., Wei, Y.: Solving multilinear systems with M-tensors. J. Sci. Comput. **68**, 689–715 (2016)
88. Du, S., Zhang, L., Chen, C., Qi, L.: Tensor absolute value equations. Sci. China Math. https://doi.org/10.1007/s11425-017-9238-6
89. Einstein, A.: The foundation of the general theory of relativity. In: Kox, A.J., Klein, M.J., Schulmann, R. (eds.) The Collected Papers of Albert Einstein 6, pp. 146–200. Princeton University Press, Princeton (2007)
90. Einstein, A., Podolsky, B., Rosen, N.: Can quantum-mechanical description of physical reality be considered complete? Phys. Rev. **47**(10), 777–780 (1935)
91. Facchinei, F., Pang, J.S.: Finite Dimensional Variational Inequalities and Complementarity Problems. Springer, New York (2003)
92. Fan, H., Kuang, G., Qiao, L.: Fast tensor principal component analysis via proximal alternating direction method with vectorized technique. Appl. Math. **8**, 77–86 (2017)
93. Fan, J., Nie, J., Zhou, A.: Tensor eigenvalue complementarity problems. Math. Program. (2017). https://doi.org/10.1007/s10107-017-1167-y
94. Fan, Y.Z., Huang, T., Bao, Y.H., Zhuan-Sun, C.L., Li, Y.P.: The spectral symmetry of weakly irreducible nonnegative tensors and connected hypergraphs (2017). arXiv:1704.08799
95. Ferreira, O., Németh, S.: On the spherical convexity of quadratic functions (2017). arXiv:1704.07665
96. Fialkow, L., Nie, J.: The truncated moment problem via homogenization and flat extensions. J. Funct. Anal. **263**(6), 1682–1700 (2012)

97. Fiedler, M., Pták, V.: On matrices with nonpositive off-diagonal elements and positive principal minors. Czechoslovak Math. J. **12**, 163–172 (1962)
98. Friedland, S., Gaubert, S., Han, L.: Perron-Frobenius theorem for nonnegative multilinear forms and extensions. Linear Algebra Appl. **438**, 738–749 (2013)
99. Gaeta, G., Virga, E.G.: Octupolar order in three dimensions. Eur. Phys. J. E **39**, 113 (2016)
100. Gao, D.Y.: On unified modeling, canonical duality-triality theory, challenges and breakthrough in optimization (2016). arXiv:1605.05534
101. Gautier, A., Tudisco, F., Hein, M.: The Perron-Frobenius theorem for multihomogeneous maps (2017). arXiv:1702.03230
102. Gelfand, I.M., Kapranov, M.M., Zelevinsky, A.V.: Discriminants. Resultants and Multidimensional Determinants. Birkhäuser, Boston (1994)
103. Giraud, O., Braun, D., Baguette, D., Bastin, T., Martin, J.: Tensor representation of spin states. Phys. Rev. Lett. **114**(8), 080401 (2015)
104. Gnang, E.K., Filmus, Y.: On the spectra of hypermatrix direct sum and Kronecker products constructions. Linear Algebra Appl. **519**, 268–277 (2017)
105. Golub, G.H., Van Loan, C.F.: Matrix Computations, 4th edn. Johns Hopkins University Press, Baltimore (2013)
106. Gourgiotis, P.A., Bigoni, D.: Stress channelling in extreme couple-stress materials part I: strong ellipticity, wave propagation, ellipticity, and discontinuity relations. J. Mech. Phys. Solids **88**, 150–168 (2016)
107. Gowda, M.S.: Polynomial complementarity problems. Pac. J. Optim. **13**(2), 227–241 (2017)
108. Gowda, M.S., Luo, Z., Qi, L., Xiu, N.: Z-tensors and complementarity problems (2015). arXiv:1510.07933
109. Guha, A., Pydi, M.S., Paria, B., Dukkipati, A.: Analytic connectivity in general hypergraphs (2017). arXiv:1701.04548
110. Guo, C.H., Lin, W.W., Liu, C.S.: A modified Newton iteration for finding nonnegative Z-eigenpairs of a nonnegative tensor (2017). arXiv:1705.07487
111. Guo, P.C.: A modified Newton method for multilinear pagerank (2017). arXiv:1701.05673
112. Gurtin, M.E.: The linear theory of elasticity. Linear Theories of Elasticity and Thermoelasticity, pp. 1–295. Springer, Berlin (1973)
113. Hakula, H., Ilmonen, P., Kaarnioja, V.: Computation of extremal eigenvalues of high-dimensional lattice-theoretic tensors via tensor-train decompositions (2017). arXiv:1705.05163
114. Han, D., Dai, H.H., Qi, L.: Conditions for strong ellipticity of anisotropic elastic materials. J. Elast. **97**, 1–13 (2009)
115. Han, D., Qi, L.: A successive approximation method for quantum separability. Front. Math. China **8**, 1275–1293 (2013)
116. Han, D., Qi, L., Wu, E.X.: Extreme diffusion values for non-Gaussian diffusions. Optim. Methods Softw. **23**, 703–716 (2008)
117. Han, L.: A homotopy method for solving multilinear systems with M-tensors. Appl. Math. Lett. **69**, 49–54 (2017)
118. Harker, P.T., Pang, J.S.: Finite-dimensional variational inequality and nonlinear complementarity problems: a survey of theory, algorithms and applications. Math. Program. **48**, 161–220 (1990)
119. Haussühl, S.: Physical Properties of Crystals: An Introduction. Wiley-VCH Verlag, Weinheim (2007)
120. Helton, J., Nie, J.: A semidefinite approach for truncated K-moment problems. Found. Comput. Math. **12**(6), 851–881 (2012)
121. Henrion, D., Lasserre, J.: Detecting global optimality and extracting solutions in GloptiPoly. Positive Polynomials in Control, pp. 293–310. Springer, Berlin (2005)
122. Henrion, D., Lasserre, J., Lofberg, J.: GloptiPoly 3: moments, optimization and semidefinite programming. Optim. Methods Softw. **24**(4–5), 761–779 (2009)
123. Hiki, Y.: High order elastic constants of solids. Annu. Rev. Mater. Sci. **11**, 51–73 (1981)

124. Hilbert, D.: Über die Darstellung definiter Formen als Summe von Formenquadraten. Math. Ann. **32**, 342–350 (1888)
125. Hiriart-Urruty, J.B., Lemaréchal, C.: Convex Analysis and Minimization Algorithms. Springer, Berlin (1993)
126. Hou, J., Ling, C., He, H.: A class of second-order cone eigenvalue complementarity problems for higher-order tensors. J. Oper. Res. Soc. China **5**(1), 45–64 (2017)
127. Hu, S., Huang, Z., Ling, C., Qi, L.: On determinants and eigenvalue theory of tensors. J. Symb. Comput. **50**, 508–531 (2013)
128. Hu, S., Huang, Z., Ni, H., Qi, L.: Positive definiteness of diffusion kurtosis imaging. Inverse Probl. Imaging **6**, 57–75 (2012)
129. Hu, S., Huang, Z., Qi, L.: Strictly nonnegative tensors and nonnegative tensor partition. Sci. China Math. **57**, 181–195 (2014)
130. Hu, S., Qi, L.: A necessary and sufficient condition for existence of a positive Perron vector. SIAM J. Matrix Anal. Appl. **37**(4), 1747–1770 (2016)
131. Hu, S., Qi, L., Song, Y., Zhang, G.: Geometric measure of entanglement of multipartite mixed states. Int. J. Softw. Inform. **8**, 317–326 (2014)
132. Hu, S., Qi, L., Zhang, G.: Computing the geometric measure of entanglement of multipartite pure states by means of non-negative tensors. Phys. Rev. A **93**, 012304 (2016)
133. Hua, B., Ni, G.Y., Zhang, M.S.: Computing geometric measure of entanglement for symmetric pure states via the Jacobian SDP relaxation technique. J. Oper. Res. Soc. China **5**(1), 111–121 (2017)
134. Huang, K., Xie, M., Lu, M.: Tensor Analysis (in Chinese), 2nd edn. Tsinghua University Press, Beijing (2003)
135. Huang, Z.: Locating a maximally complementary solution of the monotone NCP by using non-interior-point smoothing algorithms. Math. Methods Oper. Res. **61**(1), 41–55 (2005)
136. Huang, Z., Han, J., Xu, D., Zhang, L.: The non-interior continuation methods for solving the P_0 function nonlinear complementarity problem. Sci. China Ser. A **44**(9), 1107–1114 (2001)
137. Huang, Z., Qi, L.: Formulating an n-person noncooperative game as a tensor complementarity problem. Comput. Optim. Appl. **66**, 557–576 (2017)
138. Huang, Z., Qi, L.: Positive definiteness of paired symmetric tensors and elasticity tensors. J. Comput. Appl. Math. **338**, 22–43 (2018)
139. Huang, Z., Suo, Y., Wang, J.: On Q-tensors. To appear in Pac. J. Optim. (2018)
140. Hübener, R., Kleinmann, M., Wei, T.C., González-Guillén, C., Gühne, O.: Geometric measure of entanglement for symmetric states. Phys. Rev. A **80**(3), 032324 (2009)
141. Hui, E.S., Cheung, M.M., Qi, L., Wu, E.X.: Towards better MR characterization of neural tissues using directional diffusion kurtosis analysis. Neuroimage **42**, 122–134 (2008)
142. Hui, E.S., Cheung, M.M., Qi, L., Wu, E.X.: Advanced MR diffusion characterization of neural tissue using directional diffusion kurtosis analysis. Conf. Proc. IEEE Eng. Med. Biol. Soc. **2008**, 3941–3944 (2008)
143. Isac, G.: Complementarity Problems. Springer, Berlin (1992)
144. Isac, G.: Exceptional families of elements, feasibility and complementarity. J. Optim. Theory Appl. **104**(3), 577–588 (2000)
145. Isac, G., Bulavski, V., Kalashnikov, V.: Exceptional families, topological degree and complementarity problems. J Global Optim. **10**, 207–225 (1997)
146. Isac, G., Obuchowska, W.: Functions without EFE and complementarity problems. J. Optim. Theory Appl. **99**(1), 147–163 (1998)
147. Jaffe, A., Weiss, R., Nadler, B.: Newton correction methods for computing real eigenpairs of symmetric tensors. To appear in SIAM J. Matrix Anal. Appl. (2018)
148. Jensen, J.H., Helpern, J.A., Ramani, A., Lu, H., Kaczynski, K.: Diffusional kurtosis imaging: the quantification of non-Gaussian water diffusion by means of magnetic resonance imaging. Magn. Reson. Med. **53**, 1432–1440 (2005)
149. Jia, J.J., Yang, Q.Z.: Upper bounds for the spectral radii of nonnegative tensors. J. Oper. Res. Soc. China **5**(1), 83–98 (2017)

150. Jiang, B., Li, Z., Zhang, S.: On cones of nonnegative quartic forms. Found. Comput. Math. **17**(1), 161–197 (2017)
151. Jiang, B., Yang, F., Zhang, S.: Tensor and its Tucker core: the invariance relationships. Numer. Linear Algebra Appl. **24**(3), e2086 (2017)
152. Jin, H.: The location of H-eigenvalues of real even order symmetry tensors. Linear Multilinear Algebra **65**(3), 623–634 (2017)
153. Jin, H., Benítez, J.: Some generalizations and probability versions of Samuelsons inequality. Math. Inequal. Appl. **20**(1), 1–12 (2017)
154. Jin, H., Rajesh Kannan, M., Bai, M.: Lower and upper bounds for H-eigenvalues of even order real symmetric tensors. Linear Multilinear Algebra **65**(7), 1402–1416 (2017)
155. Jong, M.D., Chen, W., Angsten, T., Jain, A., Notestine, R., Gamst, A., Sluiter, M., Ande, C.K., van der Zwaag, S., Plata, J.J., Toher, C., Curtarolo, S., Ceder, G., Persson, K.A., Asta, M.: Charting the complete elastic properties of inorganic crystalline compounds. Sci. Data **2**, 150009 (2015)
156. Kaarnioja, V.: On the structure of join tensors with applications to tensor eigenvalue problems (2017). arXiv:1705.06313
157. Kannike, K.: Vacuum stability of a general scalar potential of a few fields. Eur. Phys. J. C **76**, 324 (2016)
158. Knowles, J.K., Sternberg, E.: On the ellipticity of the equations of nonlinear elastostatics for a special material. J. Elast. **5**, 341–361 (1975)
159. Knowles, J.K., Sternberg, E.: On the failure of ellipticity of the equations for finite elastostatic plane strain. Arch. Rational Mech. Anal. **63**, 321–336 (1976)
160. Kozhasov, K.: On fully real eigenconfigurations of tensors (2017). arXiv:1707.04005
161. Kuo, Y.C., Lin, W.W., Liu, C.S.: Continuation methods for computing Z-/H-eigenpairs of nonnegative tensors. J. Comput. Appl. Math. (2018). https://doi.org/10.1016/j.cam.2018.02.027
162. Lasserre, J.B.: Global optimization with polynomials and the problem of moments. SIAM J. Optim. **11**(3), 796–817 (2001)
163. Lasserre, J.B.: Moments, Positive Polynomials and Their Applications. Imperial College Press, London (2009)
164. Le Bihan, D., Mangin, J.F., Poupon, C., Clark, C.A., Pappata, S., Molko, N., Chabriat, H.: Diffusion tensor imaging: concepts and applications. J. Magn. Reson. Imaging **13**(4), 534–546 (2001)
165. Lei, X.: ${}^{\alpha}_{\beta}\Psi_S$-inclusion sets for eigenvalues of a tensor. Linear Multilinear Algebra **66**(5), 942-960 (2018)
166. Li, C., Li, Y.: C-eigenvalues intervals for piezoelectric-type tensors (2017). arXiv:1704.02153
167. Li, C., Wang, F., Zhao, J., Zhu, Y., Li, Y.: Criterions for the positive definiteness of real supersymmetric tensors. J. Comput. Appl. Math. **255**, 1–14 (2014)
168. Li, D., Xie, L., Xu, R.: Splitting methods for tensor equations. Numer. Linear Algebra Appl. **24**(5), e2102 (2017)
169. Li, T.: Homotopy methods. In: Engquist, B. (ed.) Encyclopedia of Applied and Computational Mathematics, pp. 653–656. Springer, Berlin (2015)
170. Li, X., Ng, M.: Solving sparse non-negative tensor equations: algorithms and applications. Front. Math. China **10**, 649–680 (2015)
171. Li, W., Cooper, J., Chang, A.: Analytic connectivity of k-uniform hypergraphs. Linear Multilinear Algebra **65**(6), 1247–1259 (2017)
172. Li, W., Liu, W.H., Vong, S.W.: On the bound of the eigenvalue in module for a positive tensor. J. Oper. Res. Soc. China **5**(1), 123–129 (2017)
173. Li, Y., Liu, Q., Qi, L.: Programmable criteria for strong H-tensors. Numer. Algorithms **74**(1), 199–221 (2017)
174. Li, Z., Dai, Y.: Alternating projection method for tensor equation (2017). Preprint
175. Li, Z., Zhang, F., Zhang, X.: On the number of vertices of the stochastic tensor polytope. Linear Multilinear Algebra **65**(10), 2064–2075 (2017)

176. Liang, L., Zheng, B., Tian, Y.: Algebraic Lyapunov and Stein stability results for tensors. Linear Multilinear Algebra **66**(4), 731–741 (2017)
177. Lim, L.H.: Singular values and eigenvalues of tensors: a variational approach. In: Proceedings of the 1st IEEE International Workshop on Computational Advances in Multi-Sensor Adaptive Processing (CAMSAP), pp. 129–132 (2005)
178. Lin, C.P., Tseng, W.Y.I., Cheng, H.C., Chen, J.H.: Validation of diffusion tensor magnetic resonance axonal fiber imaging with registered manganese-enhanced optic tracts. NeuroImage **14**(5), 1035–1047 (2001)
179. Ling, C., He, H., Qi, L.: On the cone eigenvalue complementarity problem for higher-order tensors. Comput. Optim. Appl. **63**, 143–168 (2016)
180. Ling, C., He, H., Qi, L.: Higher-degree eigenvalue complementarity problems for tensors. Comput. Optim. Appl. **64**, 149–176 (2016)
181. Ling, C., He, H., Qi, L.: Improved approximation results on standard quartic polynomial optimization. Optim. Lett. **11**(8), 1767–1782 (2017)
182. Lippmann, G.: Principe de la conservation de l'électricité. Annales de chimie et de physique **24**, 145–178 (1881)
183. Liu, D., Li, W., Ng, M., Vong, S.: The uniqueness and perturbation bounds for sparse nonnegative tensor equations (2017). Preprint
184. Liu, K., Nissinen, J., Slager, R.-J., Wu, K., Zaanen, J.: Generalized liquid crystals: Giant fluctuations and the vestigial chiral order of I, O, and T matter. Phys. Rev. X **6**, 041025 (2016)
185. Lu, H., Jensen, J.H., Ramani, A., Helpern, J.A.: Three-dimensional characterization of non-Gaussian water diffusion in humans using diffusion kurtosis imaging. NMR Biomed. **19**, 236–247 (2006)
186. Lubensky, T.C., Radzihovsky, L.: Theory of bent-core liquid-crystal phases and phase transitions. Phys. Rev. E **66**, 031704 (2002)
187. Luo, Z., Qi, L.: Completely positive tensors: properties, easily checkable subclasses, and tractable relaxations. SIAM J. Matrix Anal. Appl. **7**(4), 1675–1698 (2016)
188. Luo, Z., Qi, L.: Positive semidefinite tensors (in Chinese). Sci. Sin. Math. **46**, 639–654 (2016)
189. Luo, Z., Qi, L., Xiu, N.: The sparsest solutions to Z-tensor complementarity problems. Optim. Lett. **11**, 471–482 (2017)
190. Luo, Z., Qi, L., Ye, Y.: Linear operators and positive semidefiniteness of symmetric tensor spaces. Sci. China Math. **58**, 197–212 (2015)
191. Maccioni, M.: Tensor rank and eigenvectors. Ph.D. dissertation, in progress, University of Flo (2017)
192. Majorana, E.: Atomi orientati in campo magnetico variabile. Il Nuovo Cimento (1924–1942) **9**(2), 43–50 (1932)
193. Mangsarian, O., Fromovitz, S.: The Fritz John necessary optimality conditions in the presence of equality and inequality constraints. J. Math. Anal. Appl. **17**, 37–47 (1967)
194. Mei, W., Song, Y.: Infinite and finite dimensional generalized Hilbert tensors. Linear Algebra Appl. **532**, 8–24 (2017)
195. Mifflin, R.: Semismooth and semiconvex functions in constrained optimization. SIAM J. Control Optim. **15**(6), 959–972 (1997)
196. Moré, J.J.: Classes of functions and feasibility conditions in nonlinear complementarity problems. Math. Program. **6**, 327–338 (1974)
197. Ng, M., Qi, L., Zhou, G.: Finding the largest eigenvalue of a nonnegative tensor. SIAM J. Matrix Anal. Appl. **31**, 1090–1099 (2009)
198. Ni, G., Bai, M.: Spherical optimization with complex variables for computing US-eigenpairs. Comput. Optim. Appl. **65**(3), 799–820 (2016)
199. Ni, G., Qi, L., Bai, M.: Geometric measure of entanglement and U-eigenvalues of tensors. SIAM J. Matrix Anal. Appl. **35**, 73–87 (2014)
200. Nie, J.: Sum of squares methods for minimizing polynomial forms over spheres and hypersurfaces. Front. Math. China **7**, 321–346 (2012)
201. Nie, J.: Polynomial optimization with real varieties. SIAM J. Optim. **23**(3), 1634–1646 (2013)

202. Nie, J.: The $\mathscr{A}$-truncated K-moment problem. Found. Comput. Math. **14**(6), 1243–1276 (2014)
203. Nie, J.: Optimality conditions and finite convergence of Lasserre's hierarchy. Math. Program. **146**(1–2), 97–121 (2014)
204. Nie, J.: The hierarchy of local minimums in polynomial optimization. Math. Program. Ser. B **151**, 555–583 (2015)
205. Nie, J.: Linear optimization with cones of moments and nonnegative polynomials. Math. Program. **153**(1), 247–274 (2015)
206. Nie, J., Ranestad, K.: Algebraic degree of polynomial optimization. SIAM J. Optim. **20**(1), 485–502 (2009)
207. Nie, J., Ye, K.: Hankel tensor decompositions and ranks (2017). arXiv:1706.03631
208. Nie, J., Zhang, X.: Real eigenvalues of nonsymmetric tensors. Comput. Optim. Appl. (2017). https://doi.org/10.1007/s10589-017-9973-y
209. Nielsen, M.A., Chuang, I.L.: Quantum Computation and Quantum Information. Cambridge Series on Information and the Natural Sciences. Cambridge University Press, Cambridge (2004)
210. Nikiforov, V.: Combinatorial methods for the spectral p-norm of hypermatrices. Linear Algebra Appl. **529**, 324–354 (2017)
211. Nocedal, J., Wright, S.J.: Numerical Optimization. Springer, Berlin (1999)
212. Nye, J.F.: Physical Properties of Crystals: Their Representation by Tensors and Matrices, 2nd edn. Clarendon Press, Oxford (1985)
213. Ottaviani, G., Tocino, A.: Best rank k approximation for binary forms. Collect. Math. **69**(1), 163–171 (2018)
214. Ouyang, C., Qi, L., Yuan, X.: The first few unicyclic and bicyclic hypergraphs with largest spectral radii. Linear Algebra Appl. **527**, 141–162 (2017)
215. Özarslan, E., Mareci, T.H.: Generalized diffusion tensor imaging and analytical relationships between diffusion tensor imaging and high angular resolution diffusion imaging. Magn. Reson. Med. **50**, 955–965 (2003)
216. Pang, J.: Newton's method for B-differentiable equations. Math. Oper. Res. **15**(2), 311–341 (1990)
217. Papy, J.M., De Lathauwer, L., Van Huffel, S.: Exponential data fitting using multilinear algebra: the single-channel and multi-channel case. Numer. Linear Algebra Appl. **12**, 809–826 (2005)
218. Papy, J.M., De Lathauwer, L., Van Huffel, S.: Exponential data fitting using multilinear algebra: the decimative case. J. Chemom. **23**, 341–351 (2009)
219. Phan, A.H., Yamagishi, M., Cichocki, A.: An augmented Lagrangian algorithm for decomposition of symmetric tensors of order-4. In: 2017 IEEE International Conference on Acoustics, Speech and Signal Processing (ICASSP), pp. 2547–2551 (2017)
220. Qi, L.: Convergence analysis of some algorithms for solving nonsmooth equations. Math. Oper. Res. **18**(1), 227–244 (1993)
221. Qi, L.: Eigenvalues of a real supersymmetric tensor. J. Symb. Comput. **40**, 1302–1324 (2005)
222. Qi, L.: The minimum Hartree value for the quantum entanglement problem (2012). arXiv:1202.2983
223. Qi, L.: Symmetric nonnegative tensors and copositive tensors. Linear Algebra Appl. **439**, 228–238 (2013)
224. Qi, L.: Hankel tensors: associated Hankel matrices and Vandermonde decomposition. Commun. Math. Sci. **13**, 113–125 (2015)
225. Qi, L.: Transposes, L-inverses and spectral analysis of third order tensors (2017). arXiv: 1701.06761
226. Qi, L., Dai, H.H., Han, D.: Conditions for strong ellipticity and M-eigenvalues. Front. Math. China **4**, 349–364 (2009)
227. Qi, L., Han, D., Wu, E.X.: Principal invariants and inherent parameters of diffusion kurtosis tensors. J. Math. Anal. Appl. **349**, 165–180 (2009)

228. Qi, L., Luo, Z.: Tensor Analysis: Spectral Theory and Special Tensors. SIAM, Philadelpia (2017)
229. Qi, L., Wang, Y., Wu, E.X.: D-eigenvalues of diffusion kurtosis tensors. J. Comput. Appl. Math. **221**, 150–157 (2008)
230. Qi, L., Xu, C., Xu, Y.: Nonnegative tensor factorization, completely positive tensors, and a hierarchical elimination algorithm. SIAM J. Matrix Anal. Appl. **35**(4), 1227–1241 (2014)
231. Qi, L., Yu, G., Wu, E.X.: Higher order positive semi-definite diffusion tensor imaging. SIAM J. Imaging Sci. **3**, 416–433 (2010)
232. Qi, L., Yu, G., Xu, Y.: Nonnegative diffusion orientation distribution function. J. Math. Imaging Vis. **45**, 103–113 (2013)
233. Qi, L., Zhang, G., Braun, D., Bohnet-Waldraff, F., Giraud, O.: Regularly decomposable tensors and classical spin states. Commun. Math. Sci. **15**, 1651–1665 (2017)
234. Rheinboldt, W.C.: Methods for Solving Systems of Nonlinear Equations. CBMS-NSF Regional Conference Series in Applied Mathematics, vol. 70, 2nd edn. Society for Industrial and Applied Mathematics (SIAM), Philadelphia (1998)
235. Rockafellar, R.T.: Convex Analysis. Princeton University Press, Princeton (1970)
236. Rosakis, P.: Ellipticity and deformations with discontinuous gradients in finite elastostatics. Arch. Rational Mech. Anal. **109**, 1–37 (1990)
237. Sang, C., Zhao, J.: A new eigenvalue inclusion set for tensors with its applications. Cogent Math. **4**(1), 1320831 (2017)
238. Schnabel, R.B., Frank, P.D.: Tensor methods for nonlinear equations. SIAM J. Numer. Anal. **21**(5), 815–843 (1984)
239. Schultz, T., Fuster, A., Ghosh, A., Deriche, R., Florack, L., Lim, L.: Higher-order tensors in diffusion imaging. In: Westin, C.F., Vilanova, A., Burgeth, B. (eds.) Visualization and Processing of Tensors and Higher Order Descriptors for Multi-valued Data, pp. 129–161. Springer, Berlin (2014)
240. Shao, J.: A general product of tensors with applications. Linear Algebra Appl. **439**, 2350–2366 (2013)
241. Shao, J., Yuan, X.: Some properties of the Laplace and normalized Laplace spectra of uniform hypergraphs. Linear Algebra Appl. **531**, 98–117 (2017)
242. Shekhawat, H.S., Weiland, S.: A locally convergent Jacobi iteration for the tensor singular value problem. Multidimens. Syst. Signal Process. (2017). https://doi.org/10.1007/s11045-017-0485-9
243. Shimony, A.: Degree of entanglement. Ann. N. Y. Acad. Sci. **755**, 675–679 (1995)
244. Simpson, H.C., Spector, S.J.: On copositive matrices and strong ellipticity for isotropic elastic materials. Arch. Rational Mech. Anal. **84**, 55–68 (1983)
245. Song, Y., Mei, W.: B tensors and tensor complementarity problems (2017). arXiv:1707.01173
246. Song, Y., Qi, L.: Properties of tensor complementarity problem and some classes of structured tensors. Ann. Appl. Math. **3**, 308–323 (2017)
247. Song, Y., Qi, L.: Property of some classes of structured tensors. J. Optim. Theory Appl. **165**, 854–873 (2015)
248. Song, Y., Qi, L.: Necessary and sufficient conditions for copositive tensors. Linear Multilinear Algebra **63**(1), 120–131 (2015)
249. Song, Y., Qi, L.: Eigenvalue analysis of constrained minimization problem for homogeneous polynomials. J. Global Optim. **64**, 563–575 (2016)
250. Song, Y., Qi, L.: Tensor complementarity problem and semi-positive tensors. J. Optim. Theory Appl. **169**, 1069–1078 (2016)
251. Song, Y., Qi, L.: Strictly semi-positive tensors and the boundedness of tensor complementarity problems. Optim. Lett. **11**(7), 1407–1426 (2017)
252. Song, Y., Qi, L.: Infinite dimensional Hilbert tensors on spaces of analytic functions. Commun. Math. Sci. **15**, 1897–1912 (2017)
253. Song, Y., Yu, G.: Properties of solution set of tensor complementarity problem. J. Optim. Theory Appl. **170**, 85–96 (2016)

254. Stejskal, E.O., Tanner, J.E.: Spin diffusion measurements: spin echoes in the presence of a time-dependent field gradient. J. Chem. Phys. **42**(1), 288–292 (1965)
255. Sturm, J.: Using SeDuMi 1.02, a MATLAB toolbox for optimization over symmetric cones. Optim. Methods Softw. **11**(1–4), 625–653 (1999)
256. Sturmfels, B.: Tensors and their eigenvectors. Not. AMS **63**(6), 604–606 (2016)
257. Sun, D., Qi, L.: On NCP-functions. Comput. Optim. Appl. **13**(1–3), 201–220 (1999)
258. Sun, L., Zheng, B., Bu, C., Wei, Y.: Moore-Penrose inverse of tensors via Einstein product. Linear Multilinear Algebra **64**, 686–698 (2016)
259. Sun, L., Zheng, B., Bu, C., Wei, Y.: Generalized inverses of tensors via a general product of tensors. To appear in Front. Math. China (2018)
260. Sun, L., Zheng, B., Zhou, J., Yan, H.: Some inequalities for the Hadamard product of tensors. Linear Multilinear Algebra (2017). https://doi.org/10.1080/03081087.2017.1346060
261. Thomson, W. (Lord Kelvin): Elements of a mathematical theory of elasticity. Philos. Trans. R. Soc. Lond. **146**, 481–498 (1856)
262. Thomson, W. (Lord Kelvin): Elasticity. Encyclopedia Briannica, vol. 7, 9th edn, pp. 796–825. Adam and Charles Black, London (1878)
263. Thurston, R.N., Brugger, K.: Third-order elastic constants and the velocity of small amplitude elastic waves in homogeneously stressed media. Phys. Rev. **133**, A1604 (1964); Erratum: Phys. Rev. **135**, AB3 (1964)
264. Toh, K.C., Todd, M.J., Tütüncü, R.H.: SDPT3-a MATLAB software package for semidefinite programming, version 1.3. Optim. Methods Softw. **11**(1–4), 545–581 (1999)
265. Tournier, J.D., Calamante, F., Gadian, D.G., Connelly, A.: Direct estimation of the fiber orientation density function from diffusion-weighted MRI data using spherical deconvolution. NeuroImage **23**(3), 1176–1185 (2004)
266. Tuch, D.S.: Q-ball imaging. Magn. Reson. Med. **52**, 1358–1372 (2004)
267. Tuch, D.S., Reese, T.G., Wiegell, M.R., Makris, N.G., Belliveau, J.W., Wedeen, V.J.: High angular resolution diffusion imaging reveals intravoxel white matter fiber heterogeneity. Magn. Reson. Med. **48**, 454–459 (2002)
268. Van Huffel, S., Chen, H., Decanniere, C., Van Hecke, P.: Algorithm for time-domain NMR data fitting based on total least squares. J. Magn. Reson. Ser. A **110**, 228–237 (1994)
269. Varga, R.: Matrix Iterative Analysis, 2nd edn. Springer, Berlin (2000)
270. Virga, E.G.: Octupolar order in two dimensions. Eur. Phys. J. E **38**, 63 (2015)
271. Walton, J.R., Wilber, J.P.: Sufficient conditions for strong ellipticity for a class of anisotropic materials. Int. J. Nonlinear Mech. **38**, 441–455 (2003)
272. Wang, F., Sun, D., Zhao, J., Li, C.: New practical criteria for H-tensors and its application. Linear Multilinear Algebra **65**(2), 269–283 (2017)
273. Wang, G., Zhou, G., Caccetta, L.: Z-eigenvalue inclusion theorems for tensors. Discret. Contin. Dyn. Syst. Ser. B **22**(1), 187–198 (2017)
274. Wang, M., Duc, K.D., Fischer, J., Song, Y.S.: Operator norm inequalities between tensor unfoldings on the partition lattice. Linear Algebra Appl. **520**, 44–66 (2017)
275. Wang, Q., Li, G., Qi, L., Xu, Y.: New classes of positive semi-definite Hankel tensors. Minimax Theory Appl. **2**(2), 231–248 (2017)
276. Wang, Y., Aron, M.: A reformulation of the strong ellipticity conditions for unconstrained hyperelastic media. J. Elast. **44**, 89–96 (1996)
277. Wang, Y., Huang, Z., Bai, X.: Exceptionally regular tensors and tensor complementarity problems. Optim. Methods Softw. **31**, 815–828 (2016)
278. Wang, Y., Huang, Z., Qi, L.: Global uniqueness and solvability of tensor variational inequalities. J. Optim. Theory Appl. (2018). https://doi.org/10.1007/s10957-018-1233-5
279. Wang, Y., Qi, L., Luo, S., Xu, Y.: An alternative steepest direction method for the optimization in evaluating the geometric discord. Pac. J. Optim. **10**, 137–150 (2014)
280. Wang, Y., Qi, L., Zhang, X.: A practical method for computing the largest M-eigenvalue of a fourth-order partially symmetric tensor. Numer. Linear Algebra Appl. **16**, 589–601 (2009)
281. Wang, Y., Wang, G.: Two S-type Z-eigenvalue inclusion sets for tensors. J. Inequal. Appl. **2017**(1), 152 (2017)

282. Wang, Y.J., Zhou, G.L.: A hybrid second-order method for homogenous polynomial optimization over unit sphere. J. Oper. Res. Soc. China **5**(1), 99–109 (2017)
283. Wang, Y., Zhou, G., Caccetta, L.: Nonsingular H-tensors and their criteria. J. Ind. Manag. Optim. **12**, 1173–1186 (2016)
284. Wei, T.C., Goldbart, P.M.: Geometric measure of entanglement and applications to bipartite and multipartite quantum states. Phys. Rev. A **68**(4), 042307 (2003)
285. Wei, Y., Ding, W.: Theory and Computation of Tensors: Multi-dimensional Arrays. Academic Press, Amsterdam (2016)
286. Weldeselassie, Y.T., Barmpoutis, A., Atkins, M.S.: Symmetric positive semi-definite cartesian tensor fiber orientation distributions (CT-FOD). Med. Image Anal. **16**(6), 1121–1129 (2012)
287. Wiant, D., Neupane, K., Sharma, S., Gleeson, J.T., Sprunt, S., Jákli, A., Pradhan, N., Iannacchione, G.: Observation of a possible tetrahedratic phase in a bent-core liquid crystal. Phys. Rev. E **77**, 061701 (2008)
288. Wu, E.D., Cheung, M.M.: MR diffusion kurtosis imaging for neural tissue characterization. NMR Biomed. **23**, 836–848 (2010)
289. Xiang, H., Qi, L., Wei, Y.: A note on the M-eigenvalues of the elasticity tensor and strong ellipticity (2017). arXiv:1708.04876
290. Xiao, P., Wang, L., Lu, Y.: The maximum spectral radii of uniform supertrees with given degree sequences. Linear Multilinear Algebra **523**, 33–45 (2017)
291. Xie, Z.J., Jin, X.Q., Wei, Y.M.: Tensor methods for solving symmetric M-tensor systems. J. Sci. Comput. **74**(1), 412–425 (2018)
292. Yang, Q., Yang, Y.: Further results for Perron-Frobenius theorem for nonnegative tensors II. SIAM J. Matrix Anal. Appl. **32**, 1236–1250 (2011)
293. Yang, S.D.: Description of second-order three-dimensional magnetic neutral points. Phys. Plasmas **24**(1), 012903 (2017)
294. Yang, W.W., Ni, Q.: A cubically convergent method for solving the largest eigenvalue of a nonnegative irreducible tensor. Numer. Algorithms (2017). https://doi.org/10.1007/s11075-017-0358-1
295. Yang, Y., Feng, Y., Huang, X., Suykens, J.A.K.: Rank-1 tensor properties with applications to a class of tensor optimization problems. SIAM J. Optim. **26**, 171–196 (2016)
296. Yang, Y., Yang, Q.: Further results for Perron-Frobenius theorem for nonnegative tensors. SIAM J. Matrix Anal. Appl. **31**, 2517–2530 (2010)
297. You, L., Chen, Y., Yuan, P.: Some results of strongly primitive tensors (2017). arXiv:1705.04554
298. Yu, G., Song, Y., Xu, Y., Yu, Z.: Spectral projected gradient methods for generalized tensor eigenvalue complementarity problem (2016). arXiv:1601.01738
299. Yu, G., Yu, Z., Xu, Y., Song, Y., Zhou, Y.: An adaptive gradient method for computing generalized tensor eigenpairs. Comput. Optim. Appl. **65**(3), 781–797 (2016)
300. Yuan, P., You, L.: Some remarks on P, P_0, B and B_0 tensors. Linear Algebra Appl. **459**, 511–521 (2014)
301. Yue, J.J., Zhang, L.P., Lu, M., Qi, L.Q.: The adjacency and signless Laplacian spectra of cored hypergraphs and power hypergraphs. J. Oper. Res. Soc. China **5**(1), 27–43 (2017)
302. Zhang, F., Zhou, B., Peng, L.: Detecting local illumination using skewness of oriented gradients from a single image. Appl. Mech. Mater. **58**, 2381–2386 (2011)
303. Zhang, F., Zhou, B., Peng, L.: Dynamic texture analysis using eigenvectors of gradient skewness tensors. In: 2012 International Conference on Computer Science and Service System (CSSS). IEEE (2012)
304. Zhang, F., Zhou, B., Peng, L.: Gradient skewness tensors and local illumination detection for images. J. Comput. Appl. Math. **237**, 663–671 (2013)
305. Zhang, L., Qi, L., Zhou, G.: M-tensors and some applications. SIAM J. Matrix Anal. Appl. **35**, 437–452 (2014)
306. Zhang, W., Kang, L., Shan, E., Bai, Y.: The spectra of uniform hypertrees. Linear Algebra Appl. **533**, 84–94 (2017)

307. Zhang, W., Liu, L., Kang, L., Bai, Y.: Some properties of the spectral radius for general hypergraphs. Linear Algebra Appl. **513**, 103–119 (2017)
308. Zhang, X., Ling, C., Qi, L., Wu, E.X.: The measure of diffusion skewness and kurtosis in magnetic resonance imaging. Pac. J. Optim. **6**, 391–404 (2010)
309. Zhang, X., Qi, L.: The quantum eigenvalue problem and Z-eigenvalues of tensors (2012). arXiv:1205.1342
310. Zhang, X., Zhou, G., Caccetta, L., Alqahtani, M.: Approximation algorithms for nonnegative polynomial optimization problems over unit spheres. Front. Math. China **12**(6), 1409–1426 (2017)
311. Zhao, J.: A new Z-eigenvalue inclusion theorem for tensors (2017). arXiv:1705.05187
312. Zhao, J.: A tighter Z-eigenvalue localization set for tensors and its applications (2017). arXiv:1704.03707
313. Zhao, J., Li, C.: Singular value inclusion sets for rectangular tensors. Linear Multilinear Algebra (2017). https://doi.org/10.1080/03081087.2017.1351518
314. Zhao, J., Sang, C.: A new S-type upper bound for the largest singular value of nonnegative rectangular tensors. J. Inequal. Appl. **2017**(1), 105 (2017)
315. Zhao, N., Yang, Q., Liu, Y.: Computing the generalized eigenvalues of weakly symmetric tensors. Comput. Optim. Appl. **66**(2), 285–307 (2017)
316. Zhao, R., Zheng, B., Liang, M.: On the estimates of the Z-eigenpair for an irreducible nonnegative tensor. J. Math. Anal. Appl. **450**(2), 1157–1179 (2017)
317. Zhou, J., Sun, L., Wei, Y., Bu, C.: Some characterizations of M-tensors via digraphs. Linear Algebra Appl. **495**, 190–198 (2016)
318. Zou, W., He, Q., Huang, M., Zheng, Q.: Eshelby's problem of non-elliptical inclusions. J. Mech. Phys. Solids **58**, 346–372 (2010)
319. Zubov, L.M., Rudev, A.N.: On necessary and sufficient conditions of strong ellipticity of equilibrium equations for certain classes of anisotropic linearly elastic materials. ZAMM-J. Appl. Math. Mech. **96**(9), 1096–1102 (2016)

Index

L. Qi et al., *Tensor Eigenvalues and Their Applications*, Advances in Mechanics and Mathematics 39, https://doi.org/10.1007/978-981-10-8058-6

The manufacturer's authorised representative in the EU is Springer Nature Customer Service Centre GmbH, Europaplatz 3, 69115 Heidelberg, Germany. If you have any concerns regarding our products, please contact ProductSafety@springernature.com

Printed and bound by CPI Group (UK) Ltd, Croydon, CR0 4YY
30/09/2026
02225446-0011